BIOTECHNOLOGY AND BIOINFORMATICS

Advances and Applications for
Bioenergy, Bioremediation,
and Biopharmaceutical Research

BIOTECHNOLOGY AND BIOINFORMATICS

Advances and Applications for Bioenergy, Bioremediation, and Biopharmaceutical Research

Edited by

Devarajan Thangadurai, PhD, and Jeyabalan Sangeetha, PhD

Apple Academic Press

TORONTO   NEW JERSEY

Apple Academic Press Inc. | Apple Academic Press Inc.
3333 Mistwell Crescent | 9 Spinnaker Way
Oakville, ON L6L 0A2 | Waretown, NJ 08758
Canada | USA

©2015 by Apple Academic Press, Inc.

First issued in paperback 2021

Exclusive worldwide distribution by CRC Press, a member of Taylor & Francis Group
No claim to original U.S. Government works

ISBN 13: 978-1-77463-326-7 (pbk)
ISBN 13: 978-1-77188-001-5 (hbk)

Library of Congress Control Number: 2014938133

Library and Archives Canada Cataloguing in Publication

Biotechnology and bioinformatics: advances and applications for bioenergy, bioremediation, and biopharmaceutical research/edited by Devarajan Thangadurai, PhD, Jeyabalan Sangeetha, PhD.

Includes bibliographical references and index.
ISBN 978-1-77188-001-5 (bound)

1. Biotechnology. 2. Bioinformatics. I. Thangadurai, D., editor II. Sangeetha, Jeyabalan, editor

TP248.2.B56 2014   660.6   C2014-902525-4

Apple Academic Press also publishes its books in a variety of electronic formats. Some content that appears in print may not be available in electronic format. For information about Apple Academic Press products, visit our website at **www.appleacademicpress.com** and the CRC Press website at **www.crcpress.com**

ABOUT THE EDITORS

Devarajan Thangadurai, PhD

Devarajan Thangadurai is a Senior Assistant Professor at Karnatak University in South India; President, Society for Applied Biotechnology; Secretary General, Association for the Advancement of Biodiversity Science; and, Journal Editor-in-Chief of several journals: (1) *Biotechnology, Bioinformatics and Bioengineering*, (2) *Acta Biologica Indica*, (3) *Biodiversity Research International*, and (4) *Asian Journal of Microbiology*. He received his PhD in Botany from Sri Krishnadevaraya University in South India (2003). During 2002–2004, he worked as CSIR Senior Research Fellow with funding from the Ministry of Science and Technology, Government of India. He served as Postdoctoral Fellow at the University of Madeira (Portugal), University of Delhi (India), and ICAR National Research Centre for Banana (India) from 2004 to 2006. He is the recipient of the Best Young Scientist Award with Gold Medal from Acharya Nagarjuna University (2003) and the VGST-SMYSR Young Scientist Award of Government of Karnataka, Republic of India (2011). He has edited/authored 15 books including *Genetic Resources and Biotechnology* (3 vols.), *Genes, Genomes and Genomics* (2 vols.) and *Mycorrhizal Biotechnology* with publishers of national and international reputation.

Jeyabalan Sangeetha, PhD

Jeyabalan Sangeetha earned a BSc in Microbiology (2001) and a PhD in Environmental Sciences (2010) from Bharathidasan University, Tiruchirappalli, Tamil Nadu, India. She holds also an MSc in Environmental Sciences (2003) from Bharathiar University, Coimbatore, Tamil Nadu, India. Between 2004 and 2008, she was the recipient of the Tamil Nadu Government Scholarship and the Rajiv Gandhi National Fellowship of University Grants Commission, Government of India for her doctoral studies. She has published approximately 20 manuscripts detailing the effect of pollutants on the environment and has organized conferences, seminars, workshops, and lectures. Her main research interests are in the areas of environmental microbiology and environmental biotechnology, with particular emphasis on solid waste management, environmental impact assessment, and microbial degradation of hydrocarbons. Her scientific and community leadership roles have included serving as an editor of the journal *Biodiversity Research International*, and, Secretary, Society for Applied Biotechnology. Presently, she is the UGC Kothari Postdoctoral Fellow at Karnatak University in South India.

CONTENTS

LIST OF CONTRIBUTORS

Mohd Azmuddin Abdullah
Department of Chemical Engineering, Universiti Teknologi Petronas, Bandar Seri Iskandar, 31750, Tronoh, Perak, Malaysia

Olufemi Adeluyi
Department of Computer Engineering, Chosun University, Gwangju, Republic of Korea

Rownock Afruza
Department of Biochemistry and Molecular Biology, University of Dhaka, Dhaka, 1000, Bangladesh

Ashfaq Ahmad
Department of Chemical Engineering, Universiti Teknologi Petronas, Bandar Seri Iskandar, 31750, Tronoh, Perak, Malaysia

Abdullah Al-Ashraf Amirul
School of Biological Sciences, Universiti Sains Malaysia, Penang, Malaysia

Neil Andrew D. Bascos
National Institute of Molecular Biology and Biotechnology, University of the Philippines - Diliman, Quezon City, 1101, Philippines

Kesaven Bhubalan
Department of Marine Science, Faculty of Maritime Studies and Marine Science, Universiti Malaysia Terengganu, Malaysia

Suchitra Baburao Borgave
Microbial Sciences Division, MACS' Agharkar Research Institute, G.G.Agarkar Road, Pune, Maharashtra, 411004, India

Rina María González-Cervantes
Biological Systems Department, Universidad Autónoma Metropolitana (Unidad Xochimilco), Calzada del Hueso 1100, Col. Villa Quietud, Delegación Coyoacán, C.P. 04960, D.F. México

Sajib Chakrobarty
Department of Biochemistry and Molecular Biology, University of Dhaka, Dhaka, 1000, Bangladesh

Rajib Chakrovorty
Victoria Research Laboratory, Department of Electrical and Electronic Engineering, National ICT Australia (NICTA), Victoria 3010, Australia

Paraskevi A. Farazi
Mediterranean Center for Cancer Research, Department of Life and Health Sciences, University of Nicosia, 46 Makedonitissas Avenue, P.O. Box 24005, Nicosia 1700, Cyprus

A.K.M. Mahbub Hasan
Department of Biochemistry and Molecular Biology, University of Dhaka, Dhaka, 1000, Bangladesh

Nobuyuki Hayashi
Department of Biochemistry and Applied Biosciences, United Graduate School of Agricultural Sciences, Kagoshima University, 1-21-24, Korimoto, Kagoshima city, Kagoshima 890-8580, Japan

Zakir Hossain
Industrial Microbiology Laboratory, Institute of Food Science & Technology, Bangladesh Council of Scientific and Industrial Research (BCSIR), Dhaka 1205, Bangladesh

Essam Mohammed Janahi
Department of Biology, College of Science, University of Bahrain, P.O. Box 32038, Sakhir, Kingdom of Bahrain

Lahiru Niroshan Jayakody
Department of Biochemistry and Applied Biosciences, United Graduate School of Agricultural Sciences, Kagoshima University, 1-21-24, Korimoto, Kagoshima city, Kagoshima 890-8580, Japan

Amaraja Abhay Joshi
Microbial Culture Collection Center, National Centre for Cell Science, University of Pune Campus, Ganeshkhind, Pune, Maharashtra, 411007, India

Pradnya Pralhad Kanekar
Microbial Sciences Division, MACS' Agharkar Research Institute, G. G. Agarkar Road, Pune 411004, Maharashtra, India

Thirumulu Ponnuraj Kannan
School of Dental Sciences, Universiti Sains Malaysia, 16150 Kubang Kerian, Kelantan, Malaysia

Anita Satish Kelkar
Microbial Sciences Division, MACS' Agharkar Research Institute, G.G.Agarkar Road, Pune, Maharashtra, 411004, India

Hiroshi Kitagaki
Department of Biochemistry and Applied Biosciences, United Graduate School of Agricultural Sciences, Kagoshima University, 1-21-24, Korimoto, Kagoshima city, Kagoshima 890-8580, Japan

Snehal Omkar Kulkarni
Microbial Sciences Division, MACS' Agharkar Research Institute, G.G.Agarkar Road, Pune, Maharashtra, 411004, India

Jeong-A Lee
Department of Computer Engineering, Chosun University, Gwangju, Republic of Korea

Marcos López-Pérez
Environmental Science Department, Universidad Autónoma Metropolitana (Unidad Lerma), Av. de las Garzas 10, Col. El Panteón, Lerma de Villada, Municipio de Lerma, C.P. 52005, Estado de México

Sathiya Maran
Human Genome Centre, School of Medical Sciences, Universiti Sains Malaysia, 16150 Kubang Kerian, Kelantan, Malaysia

Ana M. Martins
Department of Applied Biology, College of Sciences, University of Sharjah, P.O. Box 27272, Sharjah, United Arab Emirates

A. H. M. Nurun Nabi
Department of Biochemistry and Molecular Biology, University of Dhaka, Dhaka, 1000, Bangladesh

Smita Shrikant Nilegaonkar
Microbial Sciences Division, MACS' Agharkar Research Institute, G. G. Agarkar Road, Pune, Maharashtra, 411004, India

Gul-e-Saba
Department of Chemical Engineering, Universiti Teknologi Petronas, Bandar Seri Iskandar, 31750, Tronoh, Perak, Malaysia

Seema Shreepad Sarnaik
Microbial Sciences Division, MACS' Agharkar Research Institute, G. G. Agarkar Road, Pune 411004, Maharashtra, India

Teguh Haryo Sasongko
Human Genome Centre, School of Medical Sciences, Universiti Sains Malaysia, 16150 Kubang Kerian, Kelantan, Malaysia

Hamdy El-Sayed
Department of Chemical Engineering, Universiti Teknologi Petronas, Bandar Seri Iskandar, 31750, Tronoh, Perak, Malaysia

Syed Muhammad Usman Shah
Department of Chemical Engineering, Universiti Teknologi Petronas, Bandar Seri Iskandar, 31750, Tronoh, Perak, Malaysia

Fumiaki Suzuki
Faculty of Applied Biological Sciences, Gifu University, Gifu, Japan

Rebecca Sandeep Thombre
Microbial Sciences Division, MACS' Agharkar Research Institute, G. G. Agarkar Road, Pune, Maharashtra, 411004, India

Sevakumaran Vigneswari
Malaysian Institute of Pharmaceuticals and Nutraceuticals, MOSTI, Putrajaya, Malaysia

LIST OF ABBREVIATIONS

A	Adenosine
AAc	Acrylic Acid
ABE	Acetone-Butanol-Ethanol
ADME	Absorption, Distribution, Metabolism and Excretion
AI	Artificial Intelligence
AIDS	Acquired Immune Deficiency Syndrome
AOD	Alcohol Oxidase
AOX	Alcohol Oxidase Activity
ARS	Autonomous Replication Sequence
AS	Aggregation Substance
ASO	Allele-Specific Oligonucleotide
ATP	Adenosine Triphosphate
ATPase	Adenosine Triphosphatase
BCI	Brain Computer Interfaces
BIOML	BIOpolymer Markup Language
BLAST	Basic Local Alignment Search Tool
BMA	Butylmethacrylate
BMU	Biosignal Monitoring Unit
BoNT	Botulinum Neurotoxin
BOSC	Bioinformatics Open Source Conference
BS	Binding Substance
BSML	Bioinformatic Sequence Markup Language
CAH	Chronic Active Hepatitis
CaMV	Cauliflower Mosaic Virus
CASP	Critical Assessment of Structural Prediction
CC	Combinatorial Chemistry
CDD	Conserved Domain Database
CDs	Cyclodextrins
CE	Capillary Electrophoresis
C-MFA	C-based metabolic flux analysis
CGD	Chronic Granulomatous Disease
CGTase	Cyclodextrin Glycosyl Transferase
CGH	Comparative Genomic Hybridization

ChIP	Chromatin Immunoprecipitation
CHO	Chinese Hamster Ovary
CLogP	Calculated Log P
CMP	Calcium Metaphosphate
CNA	Copy Number Alterations
CP	Coupling Protein
CPH	Chronic Persistent Hepatitis
CPT	Camptothecin
CPU	Central Processing Unit
CT	Computer-Assisted Tomography
CTL	Cytotoxic T Lymphocytes
CUEs	Commercially Useful Enzymes
DCP	Dichlorophenol
DDBJ	DNA Data Bank of Japan
DDS	Drug Delivery System
DDT	Dichlorodiphenyl Trichloroethane
DEAM	Diethylacrylamide
DHA	Docosahexaenoic Acid
DHAS	Dihydroxyacetone Synthase
DHBV	Duck Hepatitis B virus
DMAc	Dimethylacetamide
DMF	Dimethylformamide
DMSO	Dimethyl sulfoxide
DNA	Deoxyribonucleic Acid
Dox	Doxorubicin
DR	Direct Repeats
DSC	Differential Scanning Calorimeter
DSP	Digital Signal Processing
DT	Disodium Terephthalate
DTD	Document Type Definition
Dtr	DNA Transfer and Replication
EBIS	Electrical Bioimpedance Spectroscopy
EGFR	Epidermal Growth Factor Receptors
EHEC	Entero Haemorrhagic *Escherichia coli*
ELISA	Enzyme-Linked Immunosorbent Assays
ELP	Elongator Holozyme Complex
EM	Expectation-Maximization
EMBL	European Molecular Biology Laboratory
EMG	Electromyogram

EPA	Eicosapentaenoic Acid
EPFL	École Polytechnique Fédérale de Lausanne
EPR	Enhanced Permeability and Retention
ER	Endoplasmatic Reticulum
ESBL	Extended-Spectrum Beta-Lactamase
ESC	Embryonic Stem Cell
ESI	Electrospray Ionization
EST	Expressed Sequence Tag
e-TAGs	e-Textile Attached Gadgets
ETU	Ethylene Thiourea
FAS	Fatty Acid Synthase
FASTA	Fast-all Sequence Alignment Algorithm
FBA	Flux Balance Analysis
Fd	Ferredoxin
FDA	Food and Drug Administration
FEV1	Forced Expiratory Volume in One Second
FFPE	Formalin-Fixed Paraffin Embedded
FLD	Formaldehyde Dehydrogenase
FMN	Flavinmonooxygenases
FPGAs	Field Programmable Gate Arrays
FVC	Forced Vital Capacity
G	Guanosine
G6P	Glucose-6-Phosphate
GAME	Genome Annotation Markup Elements
GC	Gas Chromatography
GDB	Genome Database
GEO	Gene Expression Omnibus
GI	Gastrointestinal
GMMs	Gaussian Mixture Models
GPCRs	G-Protein Coupled Receptors
GRAS	Generally Recognized As Safe
GRE	Glucocorticoid-Responsive Element
GSHV	Ground Squirrel Hepatitis Virus
GSMM	Genome-Scale Mathematical Model of the Metabolism
GSMMs	Genome-Scale Metabolic Models
GSR	Galvanic Skin Response
Has	Hydroxyacids
HB	Hydroxybutyrate
HBIG	Hepatitis B Immunoglobulin

HBV	Hepadnaviruses
HCC	Hepatocellular Carcinoma
HCH	Hexa Chlorocyclo Hexane
HDI	Hexamethylene Di Isocyanate
HER	Human Growth Factor Receptor
HGT	Horizontal Gene Transfer
HHBV	Heron Hepatitis B Virus
HIV	Human Immunodeficiency Virus
HMF	Hydroxy Methyl Furfural
HMM	Hidden Markov Model
HOPE	Hepes-glutamic acid buffer-mediated Organic solvent Protection Effect
HPLC	High Performance Liquid Chromatography
HRC	Heat Release Capacity
HRMA	High Resolution Melting Analysis
HSV	Herpes Simplex Virus
HTS	High Throughput Screening
HV	Hydroxyvalerate
IB	Inclusion Bodies
IFN	Interferon
IGF	Insulin Growth Factor
IHC	Immunohistochemistry
IMS	Imaging Mass Spectrometry
INPNC	Ice Nucleation Protein
IPA	Ingenuity Pathway Analysis
IPKB	Ingenuity Pathways Knowledge Base
IPP	Isopentenyl Pyrophosphate
IS	Insertion Sequences
IT	Information Technology
Kb	Kilobase
kDa	kilodalton
KEGG	Kyoto Encyclopedia of Genes and Genomes
LACS	Long Chain Acyl-CoA Synthases
LCM	Laser Capture Microdissection
LCST	Lower Critical Solution Temperature
LD	Linkage Disequilibrium
LDH	Layered Double Hydroxide
LGT	Lateral Gene Transfer

LNA	Locked Nucleic Acid
mAb	Monoclonal Antibodies
MCL	Medium-Chain-Length
MCMC	Markov Chain Monte Carlo
MCP	Monocrotophos
MDa	Megadalton
MeSH	Medical Subject Headings
MFA	Metabolic Flux Analysis
MGD	Mouse Genome Database
MGEs	Mobile Genetic Elements
MLPA	Multiple Ligation-dependent Probe Amplification
Mpf	Mating pair formation
MRI	Magnetic Resonance Imaging
MS	Mass Spectrometry
Mut	Methanol Utilization
MVA	Mevalonate
MVE	Methyl Vinyl Ether
MWT	Molecular Weight
NCBI	National Center for Biotechnology Information
NGS	Next Generation Sequencing
NMR	Nuclear Magnetic Resonance
NPS	Nitrite Pickling Salt
NRE	Negative Regulatory Element
NVCl	N-vinylcaprolactam
O|B|F	Open Bioinformatics Foundation
OAA	Oxaloacetate
OASIS	Open architecture for Accessible Services Integration and Standardization
OLA	Oligonucleotide Ligation Assay
OMIM	Online Mendelian Inheritance in Man
OP	Organophosphorus
Opd	Organophosphate Degrading
OPH	Organophosphorus Hydrolase
oriT	Origin of Transfer
OPP	Oxidative Pentose Phosphate Pathway
PAHs	Polycyclic Aromatic Hydrocarbons
PAMAM	Polyamidoamine
PBMCs	Peripheral Blood Mononuclear Cells
PBRs	Photobioreactors

PCA	Principal Component Analysis
PCB	Polychlorinated Biphenyls
PCR	Polymerase Chain Reaction
PDAs	Personal Digital Assistants
PDB	Protein Data Bank
PDTD	Potential Drug Target Database
PEF	Peak Expiratory Flow
PEG	Polyethylene Glycol
PEO	Poly-Ethylene Oxide
PEP	Phosphoenolpyruvate
PET	Positron Emission Tomography
PETIM	Poly-propyl ether imine
PFGE	Pulse Field Gel Electrophoresis
PgRNA	Pregenomic RNA
PHA	Poly Hydroxyalkanoate
PHB	Poly Hydroxybutirate
PHM	Personalized Health Monitoring
PHP	Hypertext Preprocessor
PIR	Protein Information Resource
PLA	Poly-Lactic Acid
PLGA	Polylactic Co-Glycolic Acid
PNIPAM	Poly-N-isopropyl acrylamide
PNPs	Polymeric Nanoparticles
PPG	Photoplethysmogram
PRE	Posttranscriptional Regulatory Element
PRIMA	*p53*-dependent Reactivation and Induction of Massive Apoptosis
PS	Photosystem
PUFAs	Polyunsaturated Fatty Acids
QF	Quantitative-Fluorescent
qPCR	Quantitative Real-Time Polymerase Chain Reaction
QSAR	Quantitative Structure-Activity Relationship
RC	Rolling Circle
RCR	Rolling-Circle Replication
RFLP	Restriction Fragment Length Polymorphism
RIA	Radioimmunoassay
RITA	Reactivation of *p53* and Induction of Tumour Cell Apoptosis
R-PC	R-phycocyanin

RPPA	Reverse Phase Protein Array
RT-PCR	Reverse Transcription Polymerase Chain Reaction
RuBP	Ribulose- Biphosphate
SCL	Short-Chain-Length
SE	Surface Exclusion
SGF	Short Glass Fibers
SMART	Simple Modular Architecture Research Tool
SNP	Single Nucleotide Polymorphism
SOG	Sudan Orange G
SRP	Signal Recognition Particle
TBM	Template Based Modeling
TDD	Therapeutic Drug Database
TEM	Transmission Electron Microscope
TF	Transcription Factors
TfR	Transferrin Receptors
THF	Tetrahydrofuran
Tn	Transposons
TNF	Tumour Necrosis Factor
TSHV	Tree Squirrel Hepatitis Virus
UCST	Upper Critical Solution Temperature
UNC	United Nation Convention
UNIVERSAAL	Universal open platform and Reference Specification for Ambient Assisted Living
UTI	Urinary Tract Infections
UTR	Untranslated Regions
VRE	Vancomycin-Resistant Enterococci
Vir	Virulence
WHO	World Health Organization
WHV	Woodchuck Hepatitis Virus

PREFACE

Biology is the study of living things, which helps to acquaint us about ourselves and the subsistence atmosphere around us with some scientific proofs. To do this, we need the unification of the various branches of biology. To discover, to identify problems, and to make complex things simple in biology, we are using various technologies that have developed through research. Among all the branches of biology, biotechnology plays a major role in research, which is aligned with all the branches of biology to develop new technologies. To make any modifications in existing biological systems, we use various methods of biotechnology, which helps in making impossible things possible, such as cloning, plant tissue culture, transplantation experiments, stem cell therapy, bioremediation, r-DNA technology in development of vaccines, drug discovery, etc.

Another branch of biology that plays a key and equal role along with biotechnology is bioinformatics. Bioinformatics is the field of science in which biology, computer science, and information technology merge to form a single discipline. Bioinformatics provides biological information of the organisms to make existing data available and to check the expected probability of results in advance, to show evolutionary relationships between organisms, to lead the identification, structure and sequence of particular proteins, to provide algorithm development and data mining, etc. These are all known techniques in use to improve technology and establish current trends as well as, at the same time, to see future perspectives.

In this context, we are introducing this book, *Biotechnology and Bioinformatics: Advances and Applications for Bioenergy, Bioremediation, and Biopharmaceutical Research* with sixteen interesting chapters written by eminent scientists and researchers in the current movement of science and technology. In the first chapter, Gul-e-Saba and Mohd Azmuddin Abdullah explain about the targeted-delivery of polymorphic nano-drugs to cancer cells. Rownock Afruza, Fumiaki Suzuki and A. H. M. Nurun Nabi elucidate pharmacogenomics in personalized medicine and the role of gene polymorphisms in the Chapter 2. In the third chapter a detailed discussion about natural history and molecular biology of hepatitis B virus has been given by Essam Mohammed Janahi. In the fourth chapter Olufemi Adeluyi and Jeong A. Lee interpret the reconfigurable parallel platforms for dependable personalized health monitoring. In the next chapter, Paraskevi A. Farazi examines high-throughput evaluation of gene expression from formalin-fixed paraffin embedded tissues. In Chapter 6, Pradnya Pralhad Kanekar and Seema Shreepad Sarnaik elaborate on the microbial detoxifying enzymes involved in bioremediation of organic chemopollutants. In the seventh chapter, Zakir Hossain expounds on understanding the role of bacterial plasmid to tackle widespread antimicrobial resistance. Marcos Lopez Perez and Rina María González-Cervantes discuss the production and biotechnological applications of recombinant proteins by methylotrophic yeast in Chapter 8. Identification of a novel fermentation inhibitor of bioethanol production, glycolaldehyde, and engineering of a resistant yeast strain toward it was interpreted by Hiroshi Kitagaki in Chapter 9. The tenth chapter, by Pradnya Pralhad Kanekar and her collaborators, consider the biotechnological potential of alkaliphillic microorganisms. Kesaven Bhubalan examines recent developments in polyhydroxyalkanoate biomaterials containing 4-hydroxy butyrate monomer in Chapter 11. In Chapter 12, Mohd Azmuddin Abdullah expounds on algal biotechnology for bioenergy, environmental remediation, and high value biochemical. Neil Andrew D. Bascos gives a detailed account on modeling molecular interactions in Chapter 13. In the fourteenth chapter, A. H. M. Nurun Nabi and her colleagues interpret the application of biotechnology and bioinformatics in drug discovery. In the next chapter Ana M. Martins elucidates the importance of bioinformatics in microbial biotechnology. In the last chapter, T. P. Kannan examines the role of bioinformatics and biotechnology in human genetic research.

The mission of editing and publishing this book has received the immeasurable support of leading experts in their respective fields. We are deeply indebted to them for greatness and accomplishments toward their succinct contribution of this work. We also wish to thank editorial staff, Sandy Jones Sickels, Vice President, and Ashish Kumar, Publisher and President, at Apple Academic Press, Inc., for making every effort to publish the book. Special thanks are due to the editorial staff for typesetting the entire manuscript and for the quality production of this book. We express thanks to our families for their understanding and collaboration during the preparation of this book.

We welcome constructive suggestions from readers that may help to improve the next edition.

— Devarajan Thangadurai, PhD, and Jeyabalan Sangeetha, PhD

CHAPTER 1

POLYMERIC NANOPARTICLE MEDIATED TARGETED DRUG DELIVERY TO CANCER CELLS

GUL-E-SABA and MOHD AZMUDDIN ABDULLAH

CONTENTS

1.1   INTRODUCTION

Nanotechnology deals with synthesis, characterization, assembly and controlled shape and size at nanoscale, one-billionth of a meter (10^{-9} m) and has wide range of applications. Enhanced solubility, effective permeability along with target ability of nanoparticles are unique properties in therapeutics delivery of drug to cancer cells. Nanomedicine has potential to move the health care forward. Drug market for nanotechnology is estimated at \$200 billion by 2015. Nanomedicines for deadly diseases such as cancer, acquired immune deficiency syndrome (AIDS), diabetes, and various other diseases are in different clinical trial phases (Kumari et al., 2009). Great advancement in microelectronics and materials science has resulted in an increase in a number of devices for diagnostics, biosensors and imaging technology, which assist rapid progress in the field of nano-carriers for drug delivery. Non-invasive imaging methods including X-ray-based computer-assisted tomography (CT), single-photon emission tomography, magnetic resonance imaging (MRI) and positron emission tomography (PET) are important tools for the detection of human cancer (Ferrari, 2005).

1.2   NANOTECHNOLOGY IN DRUG DELIVERY

A total of 1.6 million new cancer cases and more than 500,000 deaths from cancer are estimated in the United States in 2012. During the period 2004–2008, overall cancer incident rates decline slightly in men (by 0.6% per year) and are stable in women, while cancer death rates decrease by 1.8% per year in men and by 1.6% per year in women. Over the period 1999–2008, cancer death rates have declined by more than 1% per year in men and women of every racial/ethnic group with the exception of American Indians/Alaskan Natives, among whom rates have remained stable (Siegel et al., 2012). This shows that with regular screening and improved patient prognosis, survival rate can be improved. High equipment cost for various imaging technologies (Weissleder, 2002; Kingsley et al., 2006) and advancement in nano-medicines lead to increased use of chemotherapy. Current chemotherapy method, however, has been hampered by a number of limitations. Poor solubility of major number of anticancer drugs is the main cause of failure in drug delivery to cancer cells. On the other hand, the use of pharmaceutical solvents to increase solubility may have disastrous effects (Feng and Chien, 2003; Torchilin, 2004).

Effective anticancer drug should have important characteristics such as: (i) highly specific action of drug, i.e., more toxic towards cancer cells and least toxic to the rest of the body cells; (ii) the drug molecule should have affinity towards lipid bilayer cell membrane that would be the first target of drug to pass-through; (iii) a need for nano-drug delivery system to encapsulate the therapeutic molecules to reach the targeted site unhindered, as many potent therapeutics are destroyed by defense system of the body on the way to the target site. Inadequate amount of drug reaching the target site, or the need of high doses for greater potential of cancer cell death (Gringauz, 1997), subsequently increase the rate of patient mortality (Gringauz, 1997; Torchilin, 2004). The entrapment of anticancer agents in particles of nano- and micro-size range can affect controlled drug release at the desired site. The presence of targeting molecules at the surface of nanocarriers increases the targeting ability, resulting in higher accumulation at the target site (Orive et al., 2003; Kwon et al., 2006). However, severe side effects associated with conventional chemotherapy remains the major cause of patient discomfort (Brannon Peppas and Blanchette, 2004).

The major advantages of nano-carried mediated cancer chemotherapy are that the nano shape and size delivers enhanced cell entry with adequate accumulation at the tumor site. There will be increased solubility of hydrophobic anticancer agents and reduction of systemic and organ toxicity and renal elimination because of the protective nanoencapsulation. The greatest impact of nanotechnologies in cancer therapy lies in the field of controlled targeting of specific cancer pathway. Several signaling pathways have been the targets in cancer therapeutics (Guo et al., 2012). Therapeutic index of commercially available anticancer drugs can be improved via controlled and targeted delivery applications while the previously failed anticancer drugs in clinical trials can be reexamined with advanced nano-technology.

1.3 NANO-DELIVERY SYSTEMS

Pharmaceutical nanocarriers including nano-sphere, micelle, liposome, and dendrimer (Fig. 1.1) have been the focus of attention in drug delivery system (DDS). The encapsulating agent should not be immunogenic in nature or at least not showing any adverse immune response. Encapsulating the molecules could prevent reactive behavior towards floating proteins and other circulating molecules. The properties can be classified into two: the natural properties such as solubility, stability in vivo, and bio-distribution; and the additional properties such as long circulation in the blood, passive or active targeting to the pathological sites, and the responsiveness to local change in environmental conditions.

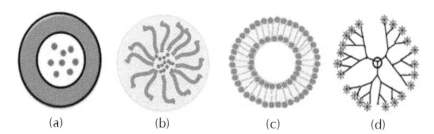

| (a) | (b) | (c) | (d) |

FIGURE 1.1 Pharmaceutical nanocarriers including nano-sphere (a), micelle (b), liposome (c) and dendrimer (d).

1.3.1 NANOPARTICLES AND MICELLES

Synthetic polymers have received increasing attention in drug delivery as carrier molecules. Polymeric nanoparticles encapsulated drug usually show improved pharmacokinetics with longer circulation time and the targeted ability to reach the tumor tissue in adequate amount as compared to small-drug molecule. This is termed polymer therapeutics or nano-medicines. Polymer therapeutics can be divided into five subclasses: polymeric drugs; polymer-drug conjugates; polymer-protein conjugates; polymeric micelles; and polyplexes (complexes of polymers and poly nucleic acids) and also polymer coatings for delivery of nucleic acid therapeutics (Duncan, 2003; Twaites *et al.*, 2005; Laga *et al.*, 2012; Varkouhi *et al.*, 2012).

Nanoparticles are defined as solid, colloidal particles with sizes in the range of 10–1000 nm (Couvreur, 1988; Kreuter, 1994). Polymeric nanoparticles (PNPs) include different types of nano-spheres or nano-capsules. Various monomers may first form polymers, which are then converted into polymeric nanoparticles; or the polymers themselves are converted into PNPs. Nano-spheres are typically spherical in shape, though some have also been reported with nonspherical shapes (Vauthier and Couvreur, 2000). In nano-capsules, the entrapped substances are restricted to a cavity consisting of water or oil surrounded by solid polymeric material shell (Couvreur *et al.*, 1995). Polymeric micelles, which are also of great interest in targeted drug delivery, are formed from block polymer or copolymers, which assemble in aqueous solution in such a way that they form outer hydrophilic layer and inner hydrophobic core. Hydrophobic anticancer agents are captured inside the core of the micelles, or they can also be conjugated with different ligands and antibodies to attain site-specific targeting. Various polymers such as polylactic coglycolic acid (PLGA), liposomes, polyethylene glycol (PEG), hyaluronic acid, dendrimers and other NPs are under intensive research in cancer therapy and imaging applications. Schematic representation of some PNPs is shown in Fig. 1.2 and the uni-targeted and multitargeted polymeric micelles are shown in Fig. 1.3.

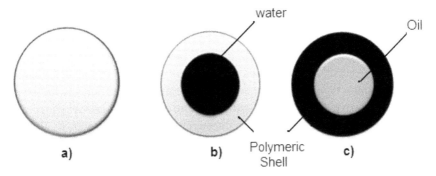

FIGURE 1.2   Schematic classification of nano-sphere (a), nano-capsule (water) (b) and nano-capsule (oil) (c).

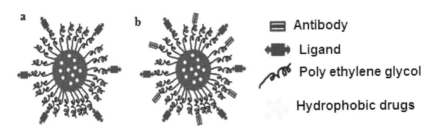

FIGURE 1.3   Schematic representation: (a) Uni-targeted polymeric micelle, where various ligands for overexpressed receptors could be tagged. This ligand-receptor targeting is effective for selective uptake of nanoparticle encapsulated drug molecules; (b) Multi-targeted polymeric micelle, along with some antibodies could be grafted on the surface of nanoparticles. This double tagged molecule could more effectively bind to cancer cells overexpressing some specific receptors, for example, CD44 and folate receptor.

1.3.2   LIPOSOME

Liposomes are phospholipid spheres of ~100 nm in diameter, bi-layered in structure and are excellent carriers of a variety of drugs (Torchilin, 2005). They could easily carry bulk amount of biologically active, hydrophilic, hydrophobic and neutral molecules. Hydrophilic drugs can be readily entrapped within the aqueous interior of the vesicles, while hydrophobic and neutral molecules are entrapped within the hydrophobic bilayers of the vesicles. Some commercially available liposome-based drugs are Caelyx™ and DaunoxomeZ™ containing doxorubicin and daunorubicin, respectively. Liposomes are extensively used in gene therapy and targeted drug delivery in cancer. With current liposome technology, it becomes easy to transfect a cell for the purpose of gene therapy, but the therapeutic gene may be degraded if it is not able to be transported out of the endosome, which is usually not large enough to ferry large molecules such as proteins. Several methods have been used to enhance the efficacy of gene therapy, such as the addition of synthetic pH-sensitive histidylated oligolysine to a drug-liposome complex to assist liposomal release from the endosome. Transfection efficiency has been shown to improve in the prostate and pancreatic cancer cell lines with increase in expression of the transgene without increasing toxicity (Yu *et al.*, 2004). With increased positive charge bearing, liposomes perform dual functions: (i) delivering genetic constructs to the tumor vasculature; (ii) antivascular effect with small molecule cytotoxic agents. A study shows that paclitaxl encapsulated in cationic liposomes diminishes tumor angiogenesis and inhibits orthotopic melanoma growth in mice (Kunstfeld *et al.*, 2003). PEG is used to increase circulation time of the positively charged liposomes. Inclusion of PEGylated lipids in the vesicles has the added advantage of reducing aggregate formation (Meyer *et al.*, 1998). Targeted codelivery of drug and gene has also been reported to enhance bioavailability and reduce toxicity (Wang *et al.*, 2010). However, liposomes have in-built problems, such as low encapsulation efficiency along with rapid leakage of water-soluble drug and poor storage stability that impede liposome-based drug delivery.

1.3.3 DENDRIMER

Dendrimer belongs to a group of polymers, versatile globular macromolecule with well-defined physical, chemical and biological characteristics. It possesses a highly branched 3D structure consisting of molecular chains branching out from a common center without any entanglement between molecules. Dendrimers have often been referred to as the "Polymers of the twenty-first century." The word "dendrimer" originated from the Greek word "Dendron," for 'tree,' and "meros," for 'part.' Due to multifunctional characteristics, dendrimers have stimulated wide interest in various field of chemistry and biology especially in the applications such as gene therapy, chemotherapy and drug delivery. Dendrimers contain internal cavities for any type of guest molecule (bio-particle). Drug-guest molecule can either be attached to the periphery of dendrimer or trapped between the molecular chains (Fig. 1.4). The most important aspect is the attached drug macromolecule in the interior to enhance its properties. The guest molecules to be entrapped depend on its shape and the architecture of the box and its cavities. The small molecules, such as rose bengal or *p*-nitrobenzoic acid, are trapped inside the 'dendritic box' such as poly(propylene imine) dendrimer with 64 branches on the periphery. A shell is formed on the surface of the dendrimer by reacting the terminal amines with an amino acid (L-phenylalanine) and the guest molecules are stably encapsulated inside the box. Hydrolysing the outer shell could liberate the guest molecules, which could be done by factors such as pH or specific microenvironment (Gul-e-Saba *et al.*, 2010). The presence of many chain-ends is responsible for high solubility, miscibility and high reactivity (Frechet, 1994).

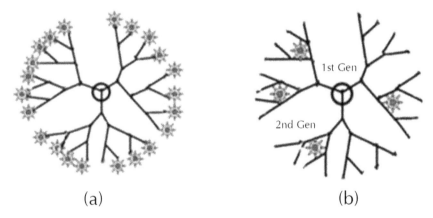

(a)   (b)

FIGURE 1.4   Dendrimers with drug attached to the periphery (a) and drug/guest molecule entangled inside (b) (Gul-e-Saba *et al.*, 2010).

There are many dendritic substances, which have strong therapeutic activities. However, due to their lack of solubility in desired solvents, they have not been used for therapeutic purposes. In a solubility test with tetrahydrofuran (THF), the solubility of dendritic polyester is remarkably higher than that of analogous linear polyester. A marked difference is also observed in chemical reactivity, where dendritic polyester is debenzylated by catalytic hydrogenolysis whereas linear polyester is unreactive. Similarly, lower generation dendrimers, which are large enough to be spherical but do not form tightly packed surface, have enormous surface areas in relation to volume (up to 1000 m^2/g) (Alper, 1991). Water-soluble dendrimers are capable of binding and solubilizing small acidic hydrophobic molecules with antifungal or antibacterial properties. Dendrimers having hydrophobic core and hydrophilic surface layer are termed unimolecular micelles. Unlike traditional micelles, dendrimers do not have critical micelle concentration, which offers the opportunity to solubilize poorly soluble drugs by encapsulating them within the dendritic structure at all concentrations of dendrimers. Drug-dendrimer conjugate formed as hydrophilic-hydrophobic coreshell dendrimer with polyamidoamine (PAMAM) interior and long alkane chain exterior has been shown to bind 5-fluorouracil, a water-soluble antitumor drug (Gul-e-Saba et al., 2010). The conjugates release free 5-fluorouracil on incubation in phosphate-buffered saline. Poly propyl ether imine (PETIM) dendrimer, is biocompatible and less cytotoxic in comparison to commercial PAMAM dendrimer (Jain et al., 2010). Specific delivery of methotrexate by dendrimer conjugated antiHER2 (Human Growth Factor Receptor-2) monoclonal antibodies (mAb) have been reported (Shukla et al., 2008).

1.4   PREPARATION OF POLYMERIC NANOPARTICLES

PNPs are prepared by several methods including polymerization of monomers. Dispersing preformed polymers method is used to prepare biodegradable PNPs such as poly lactic acid (PLA), PEG, PLGA and poly(e-caprolactone) (Vauthier et al., 1991; Allemann, 1992; Couvreur et al., 1995; Alonso, 1996). Several methods are involved in the preparation of PNPs.

1.4.1   SOLVENT EVAPORATION METHOD

Solvent evaporation is the first method developed to prepare PNPs from a preformed polymer. Organic solvents such as dichloromethane, chloroform or ethyl acetate, are used to dissolve the polymer. The formation of emulsion is then followed by evaporation. Hydrophobic drug is dissolved or dispersed into the preformed polymer solution, emulsified into aqueous solution to make oil-water (O-W) emulsion by using gelatin, poly(vinyl alcohol), polysorbate-80, or poloxamer-188 as surfactant-emulsifying agent. After the formation of a stable emulsion, the organic solvent is evaporated by increasing the temperature or under pressure. In the case of hydrophilic drugs, W-O-W method is used (Zambaux et al., 1998).

1.4.2 SPONTANEOUS EMULSIFICATION (SOLVENT DIFFUSION METHOD)

This is a modified version of the solvent evaporation method (Niwa *et al.*, 1993; Wehrle *et al.*, 1995; Murakami *et al.*, 1996). Both water-soluble solvent (acetone or methanol) along with water insoluble organic solvent (dichloromethane or chloroform) are used for oil phase. Because of spontaneous diffusion mechanism of water-soluble solvent, an interfacial turbulence is generated between the two phases, resulting in the formation of smaller particles. The reduced size of particle can also be achieved by increasing the concentration of water-soluble solvent.

1.4.3 SALTING OUT (EMULSIFICATION DIFFUSION METHOD)

These methods require the use of organic solvents, which are potentially hazardous to the environment as well as to the physiological system (Birnbaum *et al.*, 2000). Various modifications of emulsion have been developed that involve salting-out process, using water miscible solvent (acetone) instead of surfactants and chlorinated solvents. Two methods of preparing PNPs are (i) Salting-out method and (ii) emulsification – solvent diffusion method (Allemann *et al.*, 1992; Allemann *et al.*, 1993; Leroux *et al.*, 1995; Quintanar, 1998). Emulsification in the aqueous phase is achieved by dissolving high concentration of salt or sucrose, chosen for its strong salting-out effect in the aqueous phase, more like an Ouzo-effect without employing any high-shear force (Ganachaud and Katz, 2005). Various electrolytes such as magnesium chloride, calcium chloride and magnesium acetate are commonly used (Allemann *et al.*, 1992; De-Jaeghere *et al.*, 1999; Konan *et al.*, 2002; Perugini *et al.*, 2002; Su *et al.*, 2002; Zweers *et al.*, 2003; Nguyen *et al.*, 2003; Galindo-Rodriguez *et al.*, 2004; Zweers *et al.*, 2004; Galindo-Rodriguez, 2005; Zweers, 2006).

1.4.4 DIALYSIS

Dialysis offers a simple and effective method for the preparation of small, narrowly distributed PNPs (Fessi *et al.*, 1989; Jeong *et al.*, 2001). The mechanism of formation is assumed to be similar to that of nano-precipitation, first introduced by Fessi *et al.* (1989). The solvent used in the preparation of polymer solution affects the morphology and particle size distribution of the NPs. Dimethylformamide (DMF) has been used in the formation of poly(benzyl-glutamate)-*b*-poly(ethylene oxide) NPs (Oh *et al.*, 1999), and poly(lactide)-β-poly(ethylene oxide) NPs (Lee *et al.*, 2004). Other solvents such as Dimethyl sulfoxide (DMSO), DMF, Dimethylacetamide (DMAc), N-Methyl-2-pyrrolidone (NMPy) are used in the preparation of poly(glutamic acid) NPs (Akagi *et al.*, 2005). By using DMSO, the morphologies of the NPs are spherical with diameters ranging from about 100–200 nm. In the case of poly(l-lactic acid)-*b*-PEG, synthesized by using DMSO, particles having spherical shape ranging from 90–330 nm diameter can be produced (Na *et al.*, 2006). PNPs of various sizes are being produced with NMPy as solvent, suggesting the importance of solvent selection for intended sizes of NPs. A porous membrane of 12 kDa (or less) MW cut-off (MWCO) is usually used to separate the donor phase containing the drug nanoparticulate system, from the receiving phase where sink conditions for the drug with respect to the donor phase are maintained. The receiving phase is generally analyzed for the drug and the drug appearance in this phase is generally taken as the rate of drug release from NPs. Similar release pattern characterized by short-lasting burst release, followed by a longer-lasting sustained release has been reported (Essa *et al.*, 2011; Hao *et al.*, 2011; Jain *et al.*, 2011; Zambito *et al.*, 2012).

1.4.5 SUPERCRITICAL FLUID TECHNOLOGY

There is a need to use environmentally friendly solvents such as supercritical fluids to produce PNPs with high purity yield, without trace of organic solvents (York, 1999; Kawashima, 2001). Various studies have been done on particle formation with supercritical fluids and dense gases (Reverchon, 1999; Jung and Perrut, 2001; Shariati and Peters, 2003; Vemavarapu, 2005; Mishima, 2008). Examples of supercritical fluids are carbon monoxide, ammonia, n-pentane, and water (Hutchenson, 2002). The two most commonly used methods are as follows: (i) Rapid expansion of supercritical solution in air method, where the solute is first solubilized in supercritical fluid and then expanded via nozzle. The main obstacle in using supercritical fluids for the production of PNPs is the poor solubility or complete nonsolubility of polymers in the supercritical fluids. Due to this, only few studies have been carried out on the production of PNPs using this method such as the production of poly perfluoropolyether diamide droplets (Chernyak *et al.*, 2001), and poly heptadecafluorodecyl acrylate using concentrations of 0.5–5 wt.% in CO_2 (Blasig *et al.*, 2002); (ii) Rapid expansion of supercritical solution in a liquid solvent method. This is a modification to the first method where the expansion of the supercritical solution is in a liquid solvent instead of ambient air (Sun *et al.*, 2002).

1.5 TRANSPORT OF DRUG AND NANOPARTICLES TO CANCER CELLS

1.5.1 PASSIVE TARGETING

Passive transport of drug molecule is achieved by Enhanced Permeability and Retention (EPR) effect, first described by Matsumura and Maeda (1986), where the NPs are postulated to easily penetrate leaky vessels of tumor tissue and spare normal tissue. Anticancer drugs usually cause damages in healthy tissues that lead to systemic toxicity. Due to nontargeted action and degradation of drug by enzymes and other biological effects, drug concentration is not sufficient to kill cancer cells after reaching the cancer tissue. The EPR effect occurs as a result of inherited abnormalities in tumor vasculature that favors NPs to extravasate and finally accumulate inside the tissue as shown in Fig. 1.5. Lacking or ineffective lymphatic vessels in tumors lead to inefficient drainage of tumor tissue, resulting in EPR effect. This effect becomes important in the design of new and effective anticancer DDS.

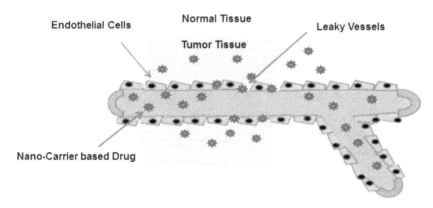

FIGURE 1.5   Transport of nanoparticles via leaky vessels.

1.5.2 ACTIVE TARGETING

Due to the advantages of nano-size, NPs can reach tumors by passive mean along with active targeting. Active targeting of NPs (Fig. 1.6) occurs by binding of ligands at the NPs surface to the receptor present on endothelial cells (Danhier *et al.*, 2010). Ligand binds to specific receptors that are over expressed by tumor cells or tumor vasculature but not expressed by normal cells. Examples are hyaluronic acid as major ligand of CD44 receptor of cancer cells (Gul-e-Saba *et al.*, 2010), epidermal growth factor receptors (EGFR) and transferrin receptors (TfR) (Broome *et al.*, 2012). The tumor endothelial cells can be recognized by NPs-tagged ligands that specifically bind to cellular receptors. Nano-carriers tagged with ligands kill angiogenic blood vessels, leading to tumor cell death (Kirpotin *et al.*, 2006). Folkman (1971) suggests that the destruction of tumoral endothelium is important in the death mechanism of the tumor cells owing to lack of oxygen and nutrients. Consequently, the receptor-mediated internalization confers major advantage to specific targeting of NPs. Carrier-mediated cellular uptake and coexistence of passive and carrier-mediated processes in drug transport are well-reported by Dobson and Kell (2008) and Sugano *et al.* (2010).

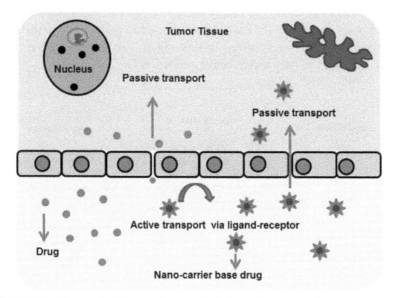

FIGURE 1.6   Transport of drug and nanoparticles by active and passive transport.

1.6   CANCER PATHWAYS

Caspase-8 is a signaling molecule in apoptotic pathway, but evidence also suggests the role of caspase-8 in several nonapoptotic functions. Caspase-8 plays a role in maintaining homeostasis of peripheral T-cells via the regulation of IL-2 production in order to control T-cell proliferation. In addition, caspase-8 is involved in the control of differentiation and proliferation of B cells, NK cells and hematopoietic progenitors (Algeciras-Schimnich *et al.*, 2002; Launay *et al.*, 2005). Rodríguez-Berriguete *et al.* (2012) have reported that caspase expression in prostate malignant cells is reduced in a substantial number of patients and that such an alteration occurs in the premalignant stage. Loss of caspase expression could constitute a useful marker for cancer diagnosis. Therapeutic approaches aimed to recover or enhance caspase expression may be effective against cancer. Several strategies have been developed to up-regulate caspase-8 expression in order to restore its function in human cancers. Studies show that the basel activity of caspase-8 promoter is controlled through SP1 and ETS-like transcription factors (Fulda, 2009). Given the frequent inactivation of caspase-8 by hypermethylation, one approach is in the use of demethylating agents. Exposure to demethylating agent, 5-aza-2 deoxycytidine (5-AZA), has resulted in demethylation of the regulatory sequence of caspase-8, an increase in caspase-8 promoter activity and in re-expression of caspase-8 in a number of tumors with epigenetically silenced caspase-8, including neuroblastoma, medulloblastoma, Ewing sarcoma and lung carcinoma (Fulda, 2009).

Strategies aimed to induce apoptosis or to prevent the destruction of normal cells by cytotoxic therapies may involve activation as well as inactivation of tumor suppressor protein, *p53*. Determining *p53* status is of great use both in cancer diagnosis and in the choice of drug therapy. *p53* has a key role in the signaling of DNA damage and in protecting genome integrity and most frequently mutated gene in human cancers (Hussain and Harris, 2000; Benard *et al.*, 2003). In certain cell types, activation of *p53* by therapeutic agents may induce cell cycle arrest (and DNA repair) rather than apoptosis, thus resulting in a form of protection of cancer cells against the effects of therapy. Thus, activation of *p53* may be seen either as a chemo-sensitizer or as a chemo-protective mechanism, depending on the cell type (Bouchet *et al.*, 2006). *p53* inactivation is obtained through direct mutation or deletion of the *p53* gene, and by disrupting the pathways that regulate *p53* protein activity (Deyoung and Ellisen, 2007; Machado-Silva *et al.*, 2010). Tumours containing mutant *p53* are more resistant to ionizing radiation than the ones with WT*p53*. New anticancer strategies are currently designed either to restore inactive/suppressed WT*p53* or "reactivate" mutant *p53*, that is, to reverse the *p53* mutant phenotype into WT*p53* while ensuring the approaches specifically affect tumor cells, without affecting normal cells. Some novel compounds targeting the *p53* pathway include small synthetic peptides and small molecules such as Ellipticine, PRIMA-1 (*p53*-dependent reactivation and induction of massive apoptosis), RITA (reactivation of *p53* and induction of tumor cell apoptosis), CP-31398, WR1065 and Nutlins (Bykov *et al.*, 2003; Bassett *et al.*, 2008; Machado-Silva *et al.*, 2010).

1.7   RESPONSIVE POLYMERIC DELIVERY SYSTEMS

Stimuli-responsive polymers respond to biological systems when there is a change in external stimulus (e.g., pH or temperature) in morphological and physiological manner such as change in conformation, release of bioactive molecule (e.g., anticancer drug), alteration in solubility, modification of hydrophilic/hydrophobic balance, or swelling/collapsing. More than one change may occur at the same time. Typical stimuli are temperature (Tanaka, 1978; Hirokawa and Tanaka, 1984; Amiya *et al.*, 1987; Chen and Hoffman, 1995), pH (Tanaka *et al.*, 1980; Osada *et al.*, 1992), electric field (Tanaka *et al.*, 1980), magnetic field (Szabo *et al.*, 1998), and light (Suzuki and Tanaka, 1990; Irie, 1993). The most important stimuli are pH, temperature, ionic strength, light and redox potential (Hoffman, 1991; Alarcon and Pennadam, 2005).

1.7.1   PH-RESPONSIVE POLYMER

pH plays an important role in targeted and controlled-release of drug. Certain cancers or wound tissue exhibit an acidic pH different from pH of circulatory area. Chronic wounds have been reported to have pH values between 5.4 and 7.4 (Dissemond *et al.*, 2001) and cancer tissue is also reported to be extracellularly acidic (Vaupel *et al.*, 1989; Rofstad *et al.*, 2006). Different cellular compartments have their own pH range (Grabe and Oster, 2001; Watson *et al.*, 2005). The pH of gastrointestinal (GI) tract is acidic (Florence and Attwood, 1998) and there is a pH shift to basic pH in the intestine (between pH 5–8) as shown in Fig. 1.7. PNPs or polymers are usually taken up by cells via receptor-mediated endocytosis as shown in Fig. 1.8. There is a drop in pH from 6.2 to 5.0 from early endosome to the late endosomes (lysosomes) in order to release the drug molecules to the cytosol (Duncan, 2003).

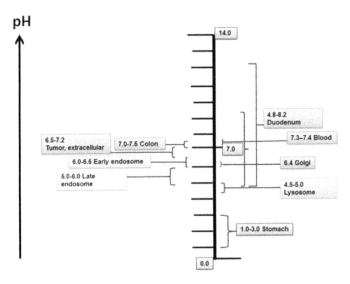

FIGURE 1.7   Schematic representation of pH scale showing various pH in different cell compartment.

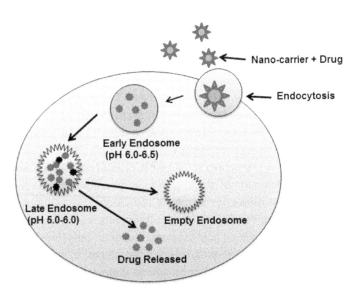

FIGURE 1.8   Schematic representation of rapid endo-lysosomal uptake of NPs by cancer cell. NPsenter the cell via receptor mediated endocytosis by firstly getting localized in the early endolysosomal compartment, then converted into late endosome, and finally release of drug triggered by a drop in pH.

Poly N-isopropyl acrylamide (PNIPAM) copolymers have been used for various drug deliveries, such as P-(NIPAM-*co*-BMAco-AAc) synthesized for the intestinal delivery of human calcitonin (Serres *et al.*, 1996; Ramkisson-Ganorkar *et al.*, 1999) and delivery of insulin (Yong-Hee *et al.*, 1994). The hydrophobic butylmethacrylate (BMA) and the acrylic acid (AAc) is stable or remain intact together at low pH of stomach. However, high pH favors the release mechanism (disintegration of beads) and finally release of drug. This is an example of PINAM copolymer pH response release.

1.7.2   THERMO-RESPONSIVE POLYMER

Temperature responsive polymer shows a response at certain range of temperature. Polymers, which become insoluble upon further heating, have a lower critical solution temperature (LCST), while the others, which become soluble upon heating, have an upper critical solution temperature (UCST). LCST and UCST systems are not limited to only an aqueous solvent environment, but also for other aqueous systems, which may be of relevance for various biomedical applications. Poly(acrylic acid-*co*-acrylamide) is a network of poly(acrylic acid) and poly(acrylamide) with transition temperature at 25°C (Aoki *et al.*, 1994). The UCST behavior is caused by the cooperative effects coming from hydrogen bonding between acrylic acid and acrylamide copolymer. Poly(methyl-vinyl-ether) has a transition temperature of 37°C, which makes it very promising for biomedical application. The polymer exhibits a typical type III demixing behavior (discontinuous swelling behavior), which is in contrast to the thermal behavior of PNIPAM (Moerkerke *et al.*, 1998). The elastin-like polypeptides have been comprehensively reviewed by Rodriguez-Cabello *et al.* (2006). Double responsive doxorubicin-polypeptide conjugate has been used for cancer therapy (Dreher *et al.*, 2003; Furgeson *et al.*, 2006) based on passive targeting by the EPR effect (Matsumura and Maeda, 1986). The LCST behavior of these polymers are biphasic: (i) the slightly higher temperature of tumor favors phase transition, the conjugate becomes insoluble after reaching the targeted tumor temperature site; (ii) the release of drug doxorubicin (Dox) via an acid-responsive linker. Poly(N-vinyl caprolactam) has not been studied as extensively but has shown interesting properties for medical and biotechnological applications especially solubility in water and organic solvents, biocompatibility, high absorption ability and a transition temperature of 33°C (Makhaeva *et al.*, 1998). Poly(N-ethyl oxazoline) has a transition temperature around 62°C, which is too high for any drug delivery application. However, double thermo-responsive system by graft polymerization of 2-ethyl-2-oxazoline (EtOx) onto a modified PNIPAM backbone (Rueda *et al.*, 2005) has great potential in drug delivery, because they tend to aggregate micelles above the LCST. Most common example of LCST polymers are PNIPAM (Schild, 1992; Shibayama, 1996), N, N-diethylacrylamide (DEAM) (Idziak *et al.*, 1999) methylvinylether (MVE) (Horne *et al.*, 1971; Mikheeva *et al.*, 1997) and N-vinylcaprolactam (NVCl) (Makhaeva *et al.*, 1998). The thermo-responsive monomer like PNIPAM can be combined with one of pH-responsive monomer that produces double-responsive copolymers (Dong and Hoffman, 1991).

1.8   POLY-LACTIC CO-GLYCOLIC ACID AND HYALURONIC ACID-BASED DRUG DELIVERY

Modifying specific polymers to the surface of nanocarriers through physical adsorption or chemical grafting is the most effective way to increase the circulation time *in vivo*, such as PEG or poly ethylene oxide (PEO). PLGA NP has been reported to be a promising carrier for anticancer drug administration while avoiding its side effects (Moreno *et al.*, 2010) such as PLGA-*b*-PEG-NPs for cisplatin and PLGA-PEG-HA for Curcumin (Yadav *et al.*, 2010; Dhar *et al.*, 2011). Grafting of various polymers such as PEG to the surface of nanoparticle containing PLGA will reduce the interaction between the NPs and digestive enzymes, increase uptake of the encapsulated drug in the bloodstream and lymphatic tissue, increase its half-life in the blood circulation (Dinarvand, 2011) and serve as efficient tools to ferry large doses of drug to tumor sites, with reduced access to nontumor tissues.

The results may vary though depending on the drugs and cell/tissue types. PLGA-PEG NPs have been developed for the selective delivery of cisplatin to tumor cells and results suggest significant reduction in drug toxicity by passive targeting to cancer cells (Avgoustakis *et al.*, 2002). PLGA-PEG as biodegradable nanoparticle exhibits enhanced cellular uptake and increase curcumin bioavailability and solubility as compared to free curcumin (Anand, 2009), enhance therapeutic efficacy of curcumin (Grabovac and Bernkop-Schnurch, 2007; Mukerjee *et al.*, 2009; Yallapu *et al.*, 2010; Tsai *et al.*, 2011), with 10-fold increase in the concentration of curcumin in blood, lungs and brain as compared to curcumin dissolved in PEG-400 formulation (Shahani *et al.*, 2010). Paclitaxl incorporated in PLGA NPs is a promising system to remove hypoxic tumor cells (Jin *et al.*, 2009), and in postsurgical chemotherapy against glioblastoma multiforme (Ong *et al.*, 2009). PEG-HA NPs loaded with camptothecin (CPT) and doxorubicin (DOX) show dose-dependent cytotoxicity to cancer cells (Choi *et al.*, 2011). CPT potentially inhibits DNA replication enzyme topoisomerase I that is overexpressed in advanced human colon, ovarian and oesophageal carcinomas and various other tumors (Giovanella *et al.*, 1989). PLGA microspheres maintain intratumor concentrations of CPT with remarkable reduction in tumor volume (Mallery *et al.*, 2001). DOX, an anthracycline antibiotic used as anticancer agent shows high antitumor activity, but its therapeutic effects may be limited due to its dose dependent cardio-toxicity and myelo-suppression (Misra and Sahoo, 2010). PLGA NPs containing DOX have been successfully formulated, characterized and evaluated *in vitro* where pH-dependent release of drug occurs as a result of degradation of the polymer and decreasing ionic interaction between the drug and the polymer at acidic pH. However, increased nuclear localization of DOX when delivered in the form of NPs does not result in increased therapeutic efficacy *in vitro* (Yoo *et al.*, 1999; Betancourt *et al.*, 2007). A study on PLGA-NPs formulated to encapsulate dexamethasone, a glucocorticoid, which acts as a chemotherapeutic agent having both antiproliferative and antiinflammatory effects, suggests that the drug binds to the cytoplasmic receptors. The dexamethasone-receptor complex is transported to the nucleus to regulate gene expression involved in cell proliferation (Gomez-Gaete *et al.*, 2007).

HA ($C_{14}H_{21}NO_{11}$), a disaccharide biopolymer (Fig. 1.9), has been studied to enhance reduction in the limiting properties of anticancer agents, and as a potential delivery vehicle for chemotherapeutics. HA may enhance activity of anticancer therapeutics with effective targeted delivery to cancer cells possessing CD44 and RHAMM receptors. Bio-conjugates of low molecular weight HA with cytotoxic agents could improve solubility of the cytotoxic agent and facilitate its intravenous administration.

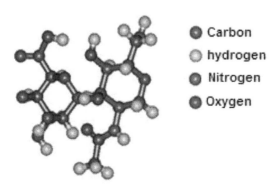

Carbon
hydrogen
Nitrogen
Oxygen

FIGURE 1.9   Molecular 3D structure of HA (Gul-e-Saba *et al.*, 2012b).

1.8.1 HA-CISPLATIN (PT)

Cisplatin, a platinum complex, $Pt(NH3)_2Cl_2$, is one of the most potent anti-cancer drug (Rosenberg, 1985). It consists of two isomers of square planar, which is the cis-isomer namely cis-diaminedichloroplatinum (II) or cis-platin, and transisomer or transplatin. The different geometries have resulted in different binding modes, resulting in only cisplatin having antitumor activity, even though transplatin produces the same complex with DNA. In our *in vitro* study (Fig. 1.10), cisplatin shows IC_{50} in colorectal cancer cell line (HT-29) at 8 ± 0.038 μg/mL, which is close to that reported at 20 μM = 6 μg/mL (Serova *et al.*, 2006); in breast cancer (MCF-7), $IC_{50} = 5 \pm 0.034$ μg/mL is similar to that reported by (Dhara and Lipparda, 2009); and in lung cancer (A549), $IC_{50} = 2 \pm 0.045$ μg/mL corresponds well to the previously reported values (Cafaggi *et al.*, 2007; St-Germain *et al.*, 2010). High IC_{50} values of cisplatin could be a result of cisplatin resistance to specific cell lines. Cisplatin resistance is multi-factorial, which can be attributed to (i) decreased drug accumulation at tumor tissue (Kelland *et al.*, 1992); (ii) better ability to repair (Parker *et al.*, 1991; Zhen *et al.*, 1992) and overcome/tolerate DNA damage (Johnson *et al.*, 1997); and (iii) increased drug detoxification (Godwin *et al.*, 1992). These eventually lead to various serious associated side effects including renal toxicity, nausea and vomiting in clinical trials.

Cisplatin reacts with hyaluronan salt to produce HA-cisplatin (HA-Pt) by releasing sodium chloride. Our study of HA-Pt NPs formed through anionic polymer-metal complexation between cisplatin and HA, has suggested no significant difference between cisplatin and HA-Pt cytotoxicity *in vitro*. However, HA-Pt maintains its cytotoxicity in conjugate form, although HA alone did not show any cytotoxic effects up to concentration as high as 10 mg/mL. The noncytotoxic nature of HA could remarkably reduce the systemic cytotoxicity and also enhance the accumulation of cisplatin towards the site of tumor via HA-CD44 receptor-ligand uptake mechanism.

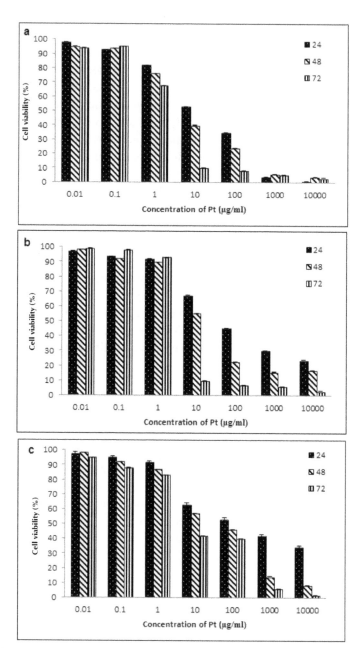

FIGURE 1.10 Cytotoxic effects of cisplatin on the viability of lung (A549) (a), breast (MCF-7) (b), colorectal (HT-29) cancer cell lines (c) after 24, 48 and 72 hour incubation (Gul-e-Saba *et al.*, 2012c).

1.8.2 HA-PACLITAXL (PTX)

PTX is an anticancer drug that promotes stabilization of tubulin polymerization (Spratlin and Sawyer, 2007). This stable condition of microtubule consequently promotes cell cycle arrest at the G2/M phase and results in inhibition of mitosis (Gligorov and Lotz, 2004). Although PTX has shown remarkable potential as anticancer compound, it causes toxicity, drug resistance in some tumors, poor stability and aqueous solubility. Taxol™, the clinically used formulation of PTX, contains polyoxyethylated castor oil (Cremophor™ EL) and ethanol. Cremophor, is considered responsible for hypersensitivity reactions observed along with PTX infusions. Side effects of Taxol™ include nausea, vomiting, diarrhea, mucositis, myelosuppression, cardiotoxicity, and neurotoxicity. Zhao *et al.* (2012) have reported that PTX-loaded folic acid modified lipid-shell and polymer-core nanoparticles (FLPNPs) have higher cytotoxicity than the commercial PTX formulation (taxol).

In our study (Fig. 1.11), there is a significant increase in PTX cytotoxicity in the range of 0.1–1.0 µg/mL after 48 hours as compared to 24 hours (data not shown), suggesting the time dependent increase in cytotoxicity of PTX in lung and breast cancer cells. Cytotoxicity of PTX on lung cancer cells (A549) (IC_{50}=0.7±0.012 µg/mL) is similar to that reported earlier (Liu *et al.*, 2011); while the breast (MCF-7) cancer cell line with IC_{50}=0.6±0.015 µg/mL, is similar to that reported by Caliceti *et al.* (2010). These are also much lower than the IC_{50} of cisplatin.

HA-PTX nanoparticles show IC_{50}=0.2±0.015 µg/mL in lung cells, which is comparable to breast cells (IC50=0.3±0.012 µg/mL). This is, however, 2–3 fold higher cytotoxicity than PTX alone in both cells. The increase in cytotoxicity could be a result of HA-CD44 binding, as HA has strong affinity to receptor molecule CD44, being overexpressed in various types of tumor cells (Hua *et al.*, 1993). HA-PTX nanoparticles possibly allow greater interaction with the receptor molecules and internalized through CD44 receptor mediated endocytosis. It has been reported that HA-PTX conjugate is well suited to locoregional application in the treatment of ovarian cancer, leading to enhanced therapeutic effectiveness (Rosato *et al.*, 2006).

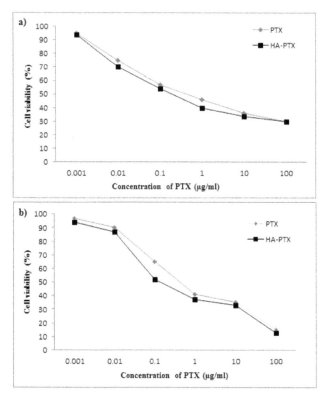

FIGURE 1.11   Cytotoxic effects of PTX and HA-PTX on lung (A549) (a) and breast (MCF-7) (b) cancer cell-lines, after 48-hour incubation (Gul-e-Saba *et al.*, 2012c).

HA-PTX cytotoxicity may occur by the activation of caspase 8 in A549, MCF7 and HT 29 cancer cell lines (Fig. 1.12). Caspases are proteases that act as essential initiators and executioners of the apoptosis (Fulda, 2009). They are single chain inactive enzymes containing 3 domains: an N-terminal prodomain, a p20 large subunit, and a p10 small subunit. Activation of caspases occurs through cleavage after an aspartate residue forms large and small subunits, which together constitute the catalytic form of the enzyme (Li and Yuan, 2008). Caspase-8 is involved in the apoptotic pathway and in the activation of several effector procaspases involving caspase-3 (Medema *et al.*, 1997). However, caspase-8 activation and the involvement of CD95/CD95-L interaction induced by anticancer drugs, including PTX, are subject to speculation and debate (Fulda *et al.*, 1997, 1998; Micheau *et al.*, 1999; Wesselborg *et al.*, 1999; Ferreira *et al.*, 2004).

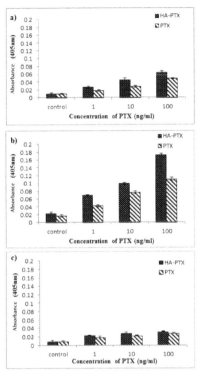

FIGURE 1.12 HA-PTX NPs and PTX induce activation of caspase 8 on A549 (a), MCF-7 (b) and HT-29 (c) cancer cell lines after 24h and 48h incubation (Gul-e-Saba *et al.*, 2012d).

1.8.3 HA-CURCUMIN

Curcumin (diferuloylmethane) is a low molecular weight natural polyphenolic compound found in the Indian turmeric (*Curcuma longa*) spice. Curcuminoids give tumeric its yellow color, with curcumin as the principal curcuminoid. Curcumin exists predominantly in two tautomeric forms- keto and enol, and the enol form is more energetically stable in solid phase and solution. Curcumin has been used for centuries as a therapy for many disorders such as cardiovascular, autoimmune, pulmonary, neurodegenerative and neoplastic diseases (Aggarwal and Harikumar, 2009). Curcumin mechanism of action shows that it inhibits several important cellular targets such as NF-*kb* and results in induction of apoptosis and also blocks the function of protein kinase C, epidermal growth factor receptor tyrosine kinase, and HER-2. As a potent anticancer agent, curcumin has potential use in DDS but it faces problem of poor aqueous solubility and low bioavailability that limits its functionality as anticancer drug. Manju and Sreenivasan (2011) have shown that curcumin undergoes rapid hydrolysis at physiological pH and HA-curcumin could prevent curcumin degradation, along with enhanced cytotoxicity (IC_{40}=3.25 μg/mL).

1.9 CONCLUSION AND FUTURE PERSPECTIVES

Various types of polymeric formulations in drug delivery nanosphere and capsules are already well established as the uni-targeted drug. The now advanced multitargeted formulation of polymeric micelles/nanoparticles may be the perfect cocktail combination for commercialization due to their acquired dual power to attack and ligate with cancer tissue site. The effect of various solvents in preparations of nanoparticles and size-morphology changes may be critical for effective formulation. Rapid expansion of supercritical solution method, which is environmentally safe, is hampered by poor solubility. By using the pH and temperature changes in various compartments of cell and tissue site, responsive polymers may be key players of controlled release mechanism. PLGA and HA are important polymers in drug delivery, with several studies have shown enhanced cytotoxicity as compared to naked drug. Future research will focus more on elucidating signaling pathways to understand whether the route taken by commercially available nanoparticles follow the same drug pathways. Utilization of multitargeted polymeric nanoparticles for effective uptake by cancer cells with reduced desirable side effects, may involve combinations of ligand-receptor and antibody-antigen in formulation with novel polymeric nanoparticles for targeted anticancer drug delivery to cancer cells.

KEYWORDS

- Cancer cells
- Cancer pathways
- Dendrimers
- Drug delivery system
- HA-Cisplatin
- HA-Curcumin
- HA-Paclitaxl
- Liposome
- Nanomedicine
- Paclitaxl
- Polylactic coglycolic acid
- Polymeric micelles
- Polymeric nanoparticles

REFERENCES

Aggarwal, B. B.; Harikumar, K. B. Potential therapeutic effects of curcumin, the antiinflammatory agent against neurodegenerative, cardiovascular, pulmonary, metabolic, autoimmune and neoplastic diseases. *Int J Biochem Cell Biol* 2009, 41, 40–59.

Akagi, T.; Kaneko, T.; Kida, T.; Akashi, M. Preparation and characterization of biodegradable nanoparticles based on poly(glutamic acid) with l-phenylalanine as a protein carrier. *J Control Rel* 2005, 108, 226–236.

Alarcon, C. H.; Pennadam, S.; Alexander, C. Stimuli-responsive polymers for biomedical applications. *Chem Soc Rev* 2005, 34, 276–285.

Algeciras-Schimnich, A.; Barnhart, B. C.; Peter, M. E. Apoptosis independent functions of killer caspases. *Curr Opin Cell Biol* 2002, 14, 721–726.

Allemann, E.; Gurnay, R.; Doelker, E. Preparation of aqueous polymeric nanodispersions by a reversible salting-out process: influence of process parameters on particle size. *Int J Pharm* 1992, 87, 247–253.

Allemann, E.; Leroux, J. C.; Gurnay, R.; Doelker, E. *In vitro* extended-release properties of drug-loaded poly(D, L-lactic) acid nanoparticles produced by a salting-out procedure. *Pharm Res* 1993, 10, 1732–1737.

Alonso, M. J. Nanoparticulate drug carrier technology. In: *Microparticulte systems for the delivery of proteins and vaccines;* Cohen, C., Bernstein, H., Eds.; Marcel Dekker: New York, 1996, 203–242.

Alper, J. Rising chemical "stars" could play many roles. *Science* 1991, 251, 1562.

Amiya, T.; Hirokawa, Y.; Hirose, Y.; Li, Y.; Tanaka, T. Reentrant phase transition of N-isopropylacrylamide gels in mixed solvents. *J Chem Phys* 1987, 86, 2375.

Anand, P.; Nai, H. B.; Sung, B.; Kunnumakkara, A. B.; Yadav, V. R.; Tekmal, R. R. Design of curcumin-loaded PLGA nanoparticles formulation with enhanced cellular uptake, and increased bioactivity *in vitro* and superior bioavailability *in vivo*. *Biochem Pharmacol* 2010, 79(3), 330–338.

Aoki, T.; Kawashima, M.; Katono, H.; Sanui, K.; Ogata, N.; Okano, T.; Sakurai, Y. Temperature responsive interpenetrating polymer networks constructed with poly(acrylic acid) and poly(N, N-dimethylacrylamide). *Macromolecules* 1994, 27, 947–952.

Avgoustakis, K.; Beletsi, A.; Panagi, Z.; Klepetsanis, P.; Karydas, A. G.; Ithakissios, D. S. PLGA-mPEG nanoparticles of cisplatin: *in vitro* nanoparticle degradation, *in vitro* drug release and *in vivo* drug residence in blood properties. *J Control Rel* 2002, 79, 123–135.

Bassett, E. A.; Wang, W.; Rastinejad, F.; El-Deiry, W. S. Structural and functional basis for therapeutic modulation of p53 signaling. *Clin Cancer Res* 2008, 14, 6376–6386.

Benard, J.; Douc-Rasy, S.; Ahomadegbe, J. C. TP53 family members and human cancers. *Hum Mutat* 2003, 21, 182–191.

Betancourt, T.; Brown, B.; BrannonPeppas, L. Doxorubicin-loaded PLGA nanoparticles by nanoprecipitation: preparation, characterization and *in vitro* evaluation. *Nanomed.* 2007, 2, 219–232.

Birnbaum, D. T.; Kosmala, J. D.; Henthorn, D. B.; Peppas, L. B. Controlled release of β-estradiol from PLGA microparticles: the effect of organic phase solvent on encapsulation and release. *J Control Rel* 2000, 65, 375–387.

Blasig, A.; Shi, C.; Enick, R. M.; Thies, M. C. Effect of concentration and degree of saturation on RESS of a CO_2 soluble fluoropolymer. *Ind Eng Chem Res* 2002, 41, 4976–4983.

Bouchet, B. P.; De-Fromentel, C. C.; Puisieux, A. p53 as a target for anticancer drug development. *Crit Rev Oncol Hematol* 2006, 58, 190–207.

BrannonPeppas, L.; Blanchette, J. O. Nanoparticle and targeted systems for cancer therapy. *Advanced Drug Delivery Reviews* 2004, 56, 1649–1659.

Broome, A. M.; Ramamurthy, G.; Lavik, K.; Verma, A.; Pinter, M.; Basilion, J. P. Molecular imaging of the cancer signature using targeted-split enzyme complementation. *Cancer Res* 2012, 72, 8.

Bykov, V. J.; Selivanova, G.; Wiman, K. G. Small molecules that reactivate mutant p53. *Eur J Cancer* 2003, 39, 1828–1834.

Cafaggi, S.; Russo, E.; Stefani, R.; Leardi, R.; Caviglioli, G.; Parodi, B.; Bignardi, G.; De Totero, D.; Aiello, C.; Viale, M. Preparation and evaluation of nanoparticles made of chitosan or n-trimethyl chitosan and a cisplatin-alginate complex. *J Control Rel* 2007, 121, 110–123.

Caliceti, P.; Salmaso, S.; Bersani, S. Polysaccharide-based anticancer prodrugs macromolecular anticancer therapeutics. *Cancer Drug Discovery and Development* 2007, 2, 163–219.

Chen, G.; Hoffman, A. S. Graft copolymers that exhibit temperature induced phase transition over a wide range of pH. *Nature* 1995, 373, 49–52.

Chernyak, Y.; Henon, F.; Harris, R. B.; Gould, R. D.; Franklin, R. K.; Edwards, J. R.; De Simone, J. M.; Carbonell, R. G. Formation of perfluoropolyether coatings by the rapid expansion of supercritical solutions (RESS) process, Part 1: experimental results. *Ind Eng Chem Res.* 2001, 40, 6118–6126.

Choi, K. Y.; Yoon, H. Y.; Kim, J. H.; Bae, S. M.; Park, R. W.; Kang, Y. M.; Kim, I. S.; Kwon, I. C.; Choi, K.; Jeong, S. Y.; Kim, K.; Park, J. H. Smart nanocarrier based on PEGylated hyaluronic acid for cancer therapy. *ACS Nano.* 22; 2011, 5(11), 8591–8599.

Couvreur, P. Polyalkylcyanoacrylates as colloidal drug carriers. *Crit Rev Ther Drug Carr Syst.* 1988, 5, 1–20.

Couvreur, P.; Dubernet, C.; Puisieux, F. Controlled drug delivery with nanoparticles: current possibilities and future trends. *Eur J Pharm* 1995, 41, 2–13.

Danhier, F.; Feron, O.; Preat, V. To exploit the tumor microenvironment: passive and active tumor targeting of nanocarriers for anticancer drug delivery. *J Control Rel* 2010, 148, 135–146.

De-Jaeghere, F.; Allemann, E.; Leroux, J. C.; Stevels, W.; Feijen, J.; Doelker, E.; Gurny, R. Formulation and lyoprotection of poly(lactic acid-coethylene oxide) nanoparticles: influence on physical stability and *in vitro* cell uptake. *Pharm Res* 1999, 16, 859–866.

Deyoung, M. P.; Ellisen, L. W. p63 and p73 in human cancer: defining the network. *Oncogene* 2007, 26, 5169–5183.

Dhar, S.; Kolishetti, N.; Lippard, S. J.; Farokhzad, O. C. Targeted delivery of a cisplatin prodrug for safer and more effective prostate cancer therapy *in vivo*. *Proc Natl Acad Sci USA* 2011, 108(5), 1850–1855.

Dhar, S.; Lipparda, S. J. Mitaplatin, a potent fusion of cisplatin and the orphan drug dichloroacetate. *Proc Natl Acad Sci USA* 2009, 106(52), 22199–22204.

Dinarvand, R.; Sepehri, N.; Manoochehri, S.; Rouhani, H.; Atyabi, F. Polylactide-coglycolide nanoparticles for controlled delivery of anticancer agents. *Int J Nanomed* 2011, 87, 78–95.

Dissemond, J.; Witthoff, M.; Brauns, T. C.; Harberer, D.; Gros, M. pH values on chronic wounds: evaluation during modern wound therapy. *Hautarzt* 2001, 54, 959–965.

Dobson, P. D.; Kell, D. B. Carrier-mediated cellular uptake of pharmaceutical drugs: an exception or the rule? *Nat. Rev. Drug Discov* 2008, 7, 205–220.

Dong, L. C.; Hoffman, A. S. A novel approach for preparation of pH-sensitive hydrogels for enteric drug delivery. *J. Control Rel* 1991, 15, 141–152.

Dreher, M. R.; Raucher, D.; Balu, N., Colvin, O. M.; Ludeman, S. M.; Chilkoti, A. A evaluation of an elastin-like polypeptide-doxorubicin conjugate for cancer therapy. *J Control Rel* 2003, 91, 31–43.

Duncan, R. The dawning era of polymer therapeutics. *Nat Rev Drug Discov* 2003, 2, 347–360.

Essa, S.; Rabanel, J. M.; Hildgen, P. Characterization of rhodamine loaded PEGg-PLA nanoparticles (NPs): effect of poly(ethylene glycol) grafting density. *Int J Pharm* 2011, 411, 178–187.

Feng, S. S.; Chien, S. Chemotherapeutic engineering: application and further development of chemical engineering principles for chemotherapy of cancer and other diseases. *Chem Eng Sci* 2003, 58, 4087–4114.

Ferrari, M. Cancer nanotechnology: opportunities and challenges. *Nat Rev Cancer* 2005, 5, 161–171.

Ferreira, C. G.; Tolis, C.; Span, S. W.; Peters, G. J.; van Lopik, T.; Kummer, A. J.; Pinedo, H. M.; Giaccone, G. Drug-induced apoptosis in lung cancer cells is not mediated by the Fas/FasL (CD95/APO1) signaling pathway. *Clin Cancer Res* 2000, 6, 203–212.

Fessi, H.; Puisieux, F.; Devissaguet, J. P.; Ammoury, N.; Benita, S. Nanocapsule formation by interfacial polymer deposition following solvent displacement. *Int J Pharm* 1989, 55, 1–4.

Florence, A. T.; Attwood, D. *Physicochemical Principles of Pharmacy,* 3rd ed.; Macmillan Press: London, 1998, p 380.

Folkman, J. Tumor angiogenesis: therapeutic implications. *N Engl J Med* 1971, 285, 1182–1186.

Frechet, J. M. J. Functional polymers and dendrimers: reactivity, molecular architecture and interfacial energy. *Science* 1994, 263, 1710.

Fulda, S. Caspase-8 in cancer biology and therapy. *Cancer Lett* 2009, 281, 128–133.

Fulda, S.; Sieverts, H.; Friesen, C.; Herr, I.; Debatin, K. M. The CD95 (Apo-1/Fas) system mediates drug-induced apoptosis in neuroblastoma cells. *Cancer Res* 1997, 57, 3823–3829.

Fulda, S.; Susin, S. A.; Kroemer, G.; Debatin, K. M. Molecular ordering of apoptosis induced by anticancer drugs in neuroblastoma cells. *Cancer Res* 1998, 58, 4453–4460.

Furgeson, D. Y.; Dreher, M. R.; Chilkoti, A. A structural optimization of a "smart" doxorubicin-polypeptide conjugate for thermally targeted delivery to solid tumors. *J Control Rel* 2006, 110, 362–369.

Galindo-Rodriguez, S. A.; Allemann, E.; Fessi, H.; Doelker, E. Physicochemical parameters associated with nanoparticle formation in the salting-out, emulsification-diffusion, and nano-precipitation methods. *Pharm Res.* 2004, 21, 1428–1439.

Galindo-Rodriguez, S. A.; Puel, F.; Briancon, S.; Allemann, E.; Doelker, E.; Fessi, H. Comparative scale-up of three methods for producing ibuprofen-loaded nanoparticles. *Eur J Pharm Sci* 2005, 25, 357–367.

Ganachaud, F.; Katz, J. L. Nanoparticles and nanocapsules created using the ouzo effect: spontaneous emulsification as an alternative to ultrasonic and high-shear devices. *Chem Phys Chem* 2005, 6, 209–216.

Giovanella, B. C.; Stehlin, J. S.; Wall, M. E.; Wani, M. C.; Nicholas, A. W.; Liu, L. F.; Silber, R.; Potmesil, M. DNA topoisomerase I targeted chemotherapy of human colon cancer in xenografts. *Science* 1989, 246, 1046–1048.

Gligorov, J.; Lotz, J. P. Preclinical pharmacology of the taxanes: implications of the differences. *Oncologist* 2004, 9(2), 3–8.

Godwin, A. K.; Meister, A.; O'Dwyer, P. J.; Huang, C. S.; Hamilton, T. C.; Anderson, M. E. High resistance to cisplatin in human ovarian cancer cell lines is associated with marked increase of glutathione synthesis. *Proc Natl Acad Sci USA* 1992, 89, 3070–3074.

Gomez-Gaete, C.; Tsapis, N.; Besnard, M.; Bochot, A.; Fattal, E. Encapsulation of dexamethasone into biodegradable polymeric nanoparticles. *Int J Pharm* 2007, 331, 153–159.

Grabe, M.; Oster, G. Regulation of organelle acidity. *J Gen Physiol* 2001, 117, 329–344.

Grabovac, V.; Bernkop-Schnurch, A. Development and *in vitro* evaluation of surface modified poly(lactide-coglycolide) nanoparticles with chitosan-4-thiobutylamidine. *Drug Dev Ind Pharm* 2007, 33, 767–774.

Gringauz, A. *Introduction to medicinal chemistry: how drugs act and why,* Wiley-VCH: New York, 1997.

Gul-e-Saba.; Abdah, A.; Abdullah, M. A. Hyaluronan-mediated CD44 receptor cancer cells progression and the application of controlled drug delivery system. *Int J Cur Chem.* 2010, 1(4), 195–215.

Guo, Y.; Xu, F.; Lu, T.; Duan, Z.; Zhang, Z. Interleukin-6 signaling pathway in targeted therapy for cancer. *Cancer Treat Rev* 2012, 38(7), 904–910.

Hao, J.; Fang, X.; Zhou, Y.; Wang, J.; Guo, F.; Li, F.; Peng, X. Development and optimization of solid lipid nanoparticle formulation for ophthalmic of chloramphenicol using a Box-Behnken design. *Int. J. Nanomed.* 2011, 6, 683–692.

Hirokawa, Y.; Tanaka, T. Volume phase transition in nonionic gel. *J Chem Phys* 1984, 81, 6379–6380.

Hoffman, A. S. Conventional and environmentally sensitive hydrogels for medical and industrial uses: a review paper. *Polymer Gels* 1991, 268, 82–87.

Horne, R.; Almeida, J. P.; Day, A. F.; Yu, N. Macromolecule hydration and the effect of solutes on the cloud point of aqueous solutions of polyvinyl methyl ether: a possible model for protein denaturation and temperature control in homeothermic animals. *J Colloid Interface Sci* 1971, 35, 77–84.

Hua, Q.; Knudson, C. B.; Knudson, W. J. Internalization of hyaluronan by chondrocytes occurs via receptor-mediated endocytosis. *J Cell Sci* 1993, 106, 365–375.

Hussain, S. P.; Harris, C. C. Molecular epidemiology and carcinogenesis: endogenous and exogenous carcinogens. *Mutat Res* 2000, 462, 311–322.

Hutchenson, K. W. Organic chemical reactions and catalysis in supercritical fluid media. In *Supercritical fluid technology in materials science and engineering: synthesis, properties, and applications.*, Sun, Y. P., Ed.; Marcel Dekker: New York, 2002, pp 87–188.

Idziak, I.; Avoce, D.; Lessard, D.; Gravel, D.; Zhu, X. X. Copolymers of N-alkylacrylamides as thermosensitive hydrogels. *Macromolecules* 1999, 32, 1260.

Irie, M. Stimuli-responsive poly(N-isopropyl acrylamide) photo- and chemical-induced phase transitions. *Adv Polym Sci* 1993, 110, 49.

Jain, S.; Kaur, A.; Puri, R.; Utreja, P.; Jain, A.; Ratnam, R.; Singh, V.; Patil, A. S.; Jayaraman, N.; Kaushik, G.; Yadav, S.; Khanduja, K. L. Poly propyl ether imine (PETIM) dendrimer: a novel nontoxic dendrimer for sustained drug delivery. *Eur J Med Chem* 2010, 45, 4997–5005.

Jain, R.; Dandekar, P.; Loretz, B.; Melero, A.; Stauner, T.; Wenz, G.; Koch, M.; Lehr, C. M. Enhanced cellular delivery of idarubicin by surface modification of propyl starch nanoparticles employing pteroic acid conjugated polyvinyl alcohol. *Int J Pharm* 2011, 420, 147–155.

Jeong, Y. I.; Cho, C. S.; Kim, S. H.; Ko, K. S.; Kim, S. I.; Shim, Y. H.; Nah, J. W. Preparation of poly(dl-lactide-coglycolide) nanoparticles without surfactant. *J Appl Polym Sci* 2011, 80, 2228–2236.

Jin, C.; Bai, L.; Wu, H.; Song, W.; Guo, G.; Dou, K. Cytotoxicity of paclitaxl incorporated in PLGA nanoparticles on hypoxic human tumor cells. *Pharm Res* 2009, 26, 1776–1784.

Johnson, S. W.; Laub, P. B.; Beesley, J. S.; Ozols, R. F.; Hamilton, T. C. Increased platinum-DNA damage tolerance is associated with cisplatin resistance and cross-resistance to various chemotherapeutic agents in unrelated human ovarian cancer cell lines. *Cancer Res* 1997, 57, 850–856.

Jung, J.; Perrut, M. Particle design using supercritical fluids: literature and patent survey. *J Supercrit Fluids* 2001, 20, 179–219.

Kawashima, Y. Nanoparticulate systems for improved drug delivery. *Adv Drug Deliv Rev* 2001, 47, 1–2.

Kell, D. B.; Dobson, P. D.; Oliver, S. G. Pharmaceutical drug transport: the issues and the implications that it is essentially carrier-mediated only. *Drug Discov Today* 2011, 16, 704–714.

Kelland, L. R.; Mistry, P.; Abel, G.; Loh, S. Y.; O'Neill, C. F.; Murrer, V. A.; Harrap, K. R. Mechanism-related circumvention of acquired cis-diamminedichloroplatinum(II) resistance using two pairs of human ovarian carcinoma cell lines by ammine/amine platinum(II) dicarboxylates. *Cancer Res* 1992, 52, 3857–3864.

Kingsley, J. D.; Dou, H.; Morehead, J.; Rabinow, B.; Gendelman, H. E.; Destache, C. J. Nanotechnology: a focus on nanoparticles as a drug delivery system. *J Neuroimmunol Pharmacol* 2006, 1, 340–350.

Kirpotin, D. B.; Drummond, D. C.; Shao, Y.; Shalaby, M. R.; Hong, K.; Nielsen, U. B.; Marks, J. D.; Benz, C. C.; Park, J. W. Antibody targeting of long-circulating lipidic nanoparticles does not increase tumor localization but does increase internalization in animal models. *Cancer Res* 2006, 66, 6732–6740.

Konan, Y. N.; Gurny, R.; Allemann, E. Preparation and characterization of sterile and freeze-dried sub 200 nm nanoparticles. *Int J Pharm* 2002, 233, 239–252.

Kreuter, J. Nanoparticles. In *Colloidal drug delivery systems*, Kreuter, J., Ed.; Marcel Dekker: New York, 1994, Vol. 66; pp 219–342.

Kumari, A.; Yadav, S. K.; Yadav, S. C. Biodegradable polymeric nanoparticles based drug delivery systems. *Colloids Surf B Biointerfaces* 2009, 75, 1–18.

Kunstfeld, R.; Wickenhauser, G.; Michaelis, U.; Teifel, M.; Umek, W.; Naujoks, K.; Wolff, K.; Petzelbauer, P. Paclitaxl encapsulated in cationic liposomes diminishes tumor angiogenesis and melanoma growth in a "humanized" SCID mouse model. *J Invest Dermatol* 2003, 120, 476–482.

Kwon, I. K.; Jeong, S. H.; Kang, E.; Park, K. *Nanoparticulate drug delivery systems for cancer therapy*, American Scientific Publishers: New York, 2006.

Laga, R.; Carlisle, R.; Tangney, M.; Ulbrich, K.; Seymour, L. W. Polymer coatings for delivery of nucleic acid therapeutics. *J Control Rel* 2012, 161, 537–553.

Launay, S.; Hermine, O.; Fontenay, M.; Kroemer, G.; Solary, E.; Garrido, C. Vital functions for lethal caspases. *Oncogene* 2005, 24, 5137–5148.

Lee, J.; Cho, E. C.; Cho, K. Incorporation and release behavior of hydrophobic drug in functionalized poly(d, l-lactide)-blockpoly(ethylene oxide) micelles. *J Control Rel* 2004, 94, 323–335.

Leroux, J. C.; Allemann, E.; Doelker, E.; Gurnay, R. New approach for the preparation of nanoparticles by an emulsification-diffusion method. *Eur J Pharm Biopharm* 1995, 41, 14–18.

Li, J.; Yuan, J. Caspases in apoptosis and beyond. *Oncogene* 2008, 27, 6194–6206.

Liu, Y.; Sun, J.; Cao, W.; Yang, J.; Lian, H.; Li, X.; Sun, Y.; Wang, Y.; Wang, S.; He, Z. Dual targeting folate-conjugated hyaluronic acid polymeric micelles for paclitaxl delivery. *Int J Pharm* 2011, 421, 160–169.

Machado-Silva, A.; Stéphane, P.; Jean-Christophe, B. p53 family members in cancer diagnosis and treatment. *Seminars in Cancer Biology* 2010, 20, 57–62.

Makhaeva, E. E.; Tenhu, H.; Khokhlov, A. R. Conformational changes of poly(vinylcaprolactam). *Macromolecules* 1998, 31, 6112.

Mallery, S. R.; Shenderova, A.; Pei, P.; Begum, S.; Ciminieri, J. R.; Wilson, R. F.; Casto, B. C, Schuller, D. E.; Morse, M. A. Effects of 10-hydroxycamptothecin, delivered from locally injectable poly(lactide-coglycolide) microspheres, in a murine human oral squamous cell carcinoma regression model. *Anticancer Res.* 2001, 21, 1713–1722.

Manju, S.; Sreenivasan, K. Conjugation of curcumin onto hyaluronic acid enhances its aqueous solubility and stability. *J Colloid Interface Sci* 2011, 359(1), 318–325.

Matsumura, Y.; Maeda, H. A new concept for macromolecular therapeutics in cancer chemotherapy: mechanism of tumoritropic accumulation of proteins and the antitumor agent. *Cancer Res* 1986, 46, 6387–6392.

Medema, J. P.; Scaffidi, C.; Kischkel, F. C.; Shevchenko, A.; Mann, M.; Krammer, P. H.; Peter, M. E. FLICE is activated by association with the CD95 death-inducing signaling complex (DISC). *EMBO J* 1997, 16, 2794–2804.

Meyer, O.; Kirpotin, D.; Hong, K.; Sternberg, B.; Park, J. W.; Woodle, M. C.; Papahadjopoulos, D. Cationic liposomes coated with polyethylene glycol as carriers for oligonucleotides. *J Biol Chem* 1998, 273, 15621–15627.

Micheau, O.; Solary, E.; Hammann, A.; Dimanche-Boitrel, M. T. Fas ligand-independent, FADD-mediated activation of the Fas death pathway by anticancer drugs. *J Biol Chem* 1999, 274, 7987–7992.

Mikheeva, L. M.; Grinberg, N. V.; Mashkevich, A. Y.; Grinberg, V. Y.; Thanh, L. T. M.; Makhaeva, E. E.; Khokhlov, A. R. Microcalorimetric study of thermal cooperative transitions in poly(N-vinylcaprolactam) hydrogels. *Macromolecules* 1997, 30, 2693.

Mishima, K. Biodegradable particle formation for drug and gene delivery using supercritical fluid and dense gas. *Adv Drug Deliv Rev* 2008, 60, 411–432.

Misra, R.; Sahoo, S. K. Intracellular trafficking of nuclear localization signal conjugated nanoparticles for cancer therapy. *Eur J Pharm Sci* 2010, 39, 152–163.

Moerkerke, R.; Meeussen, F.; Koningsveld, R.; Berghmans, H.; Mondelaers, W.; Schacht, E.; Dusek, K.; Solc, K. Phase transitions in swollen networks. 3. swelling behavior of radiation cross-linked poly(vinyl methyl ether) in water. *Macromolecules* 1998, 31, 2223–2229.

Moreno, D.; Zalba, S.; Navarro, I.; De Ilarduya, C. T.; Garrido, M. J. Pharmacodynamics of cisplatin-loaded PLGA nanoparticles administered to tumor-bearing mice. *Eur J Pharm Biopharm* 2010, 74, 265–274.

Mukerjee, A.; Vishwanatha, J. K. Formulation, characterization and evaluation of curcumin-loaded PLGA nanospheres for cancer therapy. *Anticancer Res* 2009, 29, 3867–3875.

Murakami, H.; Yoshino, H.; Mizobe, M.; Kobayashi, M.; Takeuchi, H.; Kawashima, Y. Preparation of poly(D, L-lactide-coglycolide) latex for surface modifying material by a doublecoacervation method. *Proc Intl Symp Control Rel Bioact Mater* 1996, 23, 361–362.

Na, K.; Lee, K. H.; Lee, D. H.; Bae, Y. H. Biodegradable thermo-sensitive nanoparticles from poly(l-lactic acid)/poly(ethylene glycol) alternating multiblock copolymer for potential anticancer drug carrier. *Eur J Pharm Sci* 2006, 27, 115–122.

Nguyen, C. A.; Allemann, E.; Schwach, G.; Doelker, E.; Gurny, R. Synthesis of a novel fluorescent poly(d, l-lactide) end-capped with1-pyrenebutanol used for the preparation of nanoparticles. *Eur J Pharm Sci* 2003, 20, 217–222.

Niwa, T.; Takeuchi, H.; Hino, T.; Kunou, N.; Kawashima, Y. Preparations of biodegradable nanospheres of water-soluble and insoluble drugs with D, L-lactide/glycolide copolymer by a novel spontaneous emulsification solvent diffusion methodand the drug release behavior. *J Control Rel* 1993, 25, 89–98.

Oh, I.; Lee, K.; Kwon, H. Y.; Lee, Y. B.; Shin, S. C.; Cho, C. S.; Kim, C. K. Release of adriamycin from poly(benzyl-glutamate)/poly(ethylene oxide) nanoparticles. *Int J Pharm* 1999, 181, 107–115.

Ong, B. Y.; Ranganath, S. H.; Lee, L. Y.; Lu, F.; Lee, H. S.; Sahinidis, N. V.; Wang, C. H. Paclitaxl delivery from PLGA foams for controlled release in postsurgical chemotherapy against glioblastoma multiforme. *Biomater* 2009, 30, 3189–3196.

Orive, G.; Hernandez, R. M.; Rodriguez, G. A.; Dominguez-Gil, A.; Pedraz, J. L. Drug delivery in biotechnology: present and future. *Curr Opin Biotechnol* 2003, 14, 659–664.

Osada, Y.; Okuzaki, H.; Hori, H. A polymer gel with electrically driven motility. *Nature* 1992, 355, 242–243.

Parker, R. J.; Eastman, A.; Bostick-Bruton, F.; Reed, E. Acquired cisplatin resistance in human ovarian cancer cells is associated with enhanced repair of cisplatin-DNA lesions and reduced drug accumulation. *J Clin Invest* 1991, 87, 772–777.

Perugini, P.; Simeoni, S.; Scalia, S.; Genta, I.; Modena, T.; Conti, B.; Pavanetto, F. Effect of nanoparticle encapsulation on the photostability of the sunscreen agent, 2-ethylhexyl-p-methoxycinnamate. *Int J Pharm* 2002, 246, 37–45.

Quintanar, G. D.; Ganem, Q. A.; Allemann, E.; Fessi, H.; Doelker, E. Influence of the stabilizer coating layer on the purification and freeze drying of poly(DL-lactic acid) nanoparticles prepared by emulsification-diffusion technique. *J Microencap* 1998, 15, 107–119.

Ramkissoon-Ganorkar, C.; Liu, F.; Baudys, M.; Kim, W. Effect of molecular weight and polydispersity on kinetics of issolution and release from pH/temperature sensitive polymers. *J Biomater Sci Polym* 1999, 10, 1149–1161.

Reverchon, E. Supercritical antisolvent precipitation of microand nano-particles *J Supercrit Fluids.* 1999, 15, 1–21.

Rodríguez-Berriguete, G.; Galvis, L.; Fraile, B.; De-Bethencourt, FR.; Martínez-Onsurbe, P.; Olmedilla, G.; Paniagua, R.; Royuela, M. Immunoreactivity to caspase-3, caspase-7, caspase-8, and caspase-9 forms is frequently lost in human prostate tumors. *Human Pathol* 2012, 43, 229–237.

Rodriguez-Cabello, J. C.; Reguera, J.; Girotti, A.; Arias, F. J.; Alonso, M. Genetic engineering of protein-based polymers: the example of elastin like polymers. *Adv Polym Sci* 2006, 200, 119–167.

Rofstad, E. K.; Mathiesen, B.; Kindem, K.; Galappathi, K. Acidic extracellular pH promotes experimental metastasis of human melanoma cells in athymic nude mice. *Cancer Res* 2006, 66, 6699–6707.

Rosato, A.; Banzato, A.; De-Luca, G. HYTAD1–20: a new paclitaxl-hyaluronic acid hydrosoluble bioconjugate for treatment of superficial bladder cancer. *Urol Oncol* 2006, 24, 207–215.

Rosenberg, B. Fundamental studies with cisplatin. *Cancer* 1985, 55, 2303–2306.

Rueda, J.; Zschoche, S.; Komber, H.; Schmaljohann, D.; Voit, B. Synthesis and characterization of thermoresponsive graft copolymers of NIPAAm and 2-alkyl-2-oxazolines by the "grafting from" method. *Macromolecules* 2005, 38, 7330–7336.

Schild, H. G. Poly(N-isopropylacrylamide) experiment, theory and application. *Prog Polym Sci* 1992, 17, 163–249.

Serova, M.; Calvo, F.; Lokiec, F.; Koeppel, F.; Poindessous, V.; Larsen, A. K.; Van Laar, E. S.; Waters, S. J.; Cvitkovic, E.; Raymond, E. Characterizations of irofulven cytotoxicity in combination with cisplatin and oxaliplatin in human colon, breast, and ovarian cancer cells. *Cancer Chemother Pharmacol* 2006, 57, 491–499.

Serres, A.; Baudys, M.; Kim, S. W. Temperature and pH-sensitive polymers for human calcitonin delivery. *Pharm Res* 1996, 13, 196–201.

Shahani, K.; Swaminathan, S. K.; Freeman, D.; Blum, A.; Ma, L.; Panyamb, J. Injectable sustained release microparticles of curcumin: a new concept for cancer chemoprevention. *Cancer Res* 2010, 70(11), 4443–4452.

Shariati, A.; Peters, C. J. Recent developments in particle design using supercritical fluids. *Curr Opin Solid State Mater Sci* 2003, 7, 371–383.

Shibayama, M.; Norisuye, T.; Nomura, S. Cross-link density dependence of spatial inhomogeneities and dynamic fluctuations of poly(*N*-isopropylacrylamide) gels. *Macromolecules* 1996, 29, 8746–8750.

Shukla, R.; Thomas, T. P.; Desai, A. M.; Kotlyar, A.; Park, S. J.; Baker, J. R. HER2 specific delivery of methotrexate by dendrimer conjugated antiHER2 mAb. *Nanotech* 2008, 19, 295102.

Siegel, R.; Naishadham, D.; Jemal, A. Cancer Statistics, 2012. *Cancer J Clin* 2012, 62, 10–29.

Spratlin, J.; Sawyer, M. B. Pharmacogenetics of paclitaxl metabolism. *Crit Rev Oncol Hematol* 2007, 61, 222–229.

St-Germain, C.; Niknejad, N.; Ma, L.; Garbuio, K.; Hai, T. Cisplatin induces cytotoxicity through the mitogen-activated protein kinase pathways and activating transcription factor 3. *Neoplasia* 2010, 12, 527–538.

Su, J.; Kim, C. J.; Ciftci, K. Characterization of poly [(N-trimethylammonium) ethyl methacrylate]-based gene delivery systems. *Gene Ther* 2002, 9, 1031–1036.

Sugano, K.; Kansy, M.; Artursson, P.; Avdeef, A.; Bendels, S.; Di Lecker, G. F.; Faller, B.; Fischer, H.; Gerebtzoff, G.; Lennernaes, H.; Senner, F. Coexistence of passive and carrier-mediated processes in drug transport. *Nat Rev Drug Discov* 2010, 9, 597–614.

Sun, Y. P.; Rolling, H. W.; Bandara, J.; Meziani, J. M.; Bunker, C. E. Preparation and processing of nanoscale materials by supercritical fluid technology. In: *Supercritical fluid technology in materials science and engineering: synthesis, properties, and applications*, Sun, Y. P., Ed.; Marcel Dekker: New York, 2002; Vol. 9, pp 491–576.

Suzuki, A.; Tanaka, T. Phase transition in polymer gels induced by visible light. *Nature* 1990, 346, 345–347.

Szabo, D.; Szeghy, G.; Zrínyi, M. Shape transition of magnetic field sensitive polymer gels. *Macromolecules* 1998, 31, 6541–6548.

Tanaka, T. Collapse of gels and the critical endpoint. *Phys Rev Lett* 1978, 40, 820.

Tanaka, T.; Fillmore, D.; Sun, S. T.; Nishio, I.; Swislow, G.; Shah, A. Phase transition in ionic gel. *Phys Rev Lett* 1980, 45, 1636–1639.

Tanaka, T.; Nishio, I.; Sun, S. T.; Ueno-Nishio, S. Collapse of gels in an electric field. *Science* 1982, 218, 467–469.

Torchilin, V. P. Targeted polymeric micelles for delivery of poorly soluble drugs. *Cell Mol Life Sci* 2004, 61, 2549–2559.

Torchilin, V. P. Recent advances with liposomes as pharmaceutical carriers. *Nat Rev Drug Discov* 2005, 4, 145–160.

Tsai, Y. M.; Chien, C. F.; Lin, L. C.; Tsai, T. H. Curcumin and its nano-formulation: the kinetics of tissue distribution and blood-brain barrier penetration. *Int. J. Pharm* 2011, 416, 331–338.

Twaites, B.; Alarcon, C. H.; Alexander, C. Synthetic polymers as drugs and therapeutics. *J Mater Chem* 2005, 15, 441–455.

Varkouhi, A. K.; Mountrichas, G.; Mountrichas, R. M.; Lammers, T.; Storm, G. Polyplexes based on cationic polymers with strong nucleic acid binding properties. *Eur J Pharm Sci* 2012, 45, 459–466.

Vaupel, P.; Kallinowski, F.; Okunieff, P. Blood flow, oxygen and nutrient supply, and metabolic microenvironment of human tumors: a review. *Cancer Res* 1989, 49, 6449–6465.

Vauthier, C.; Beanabbou, S.; Spenlehauer, G.; Veillard, M. P.; Couvreur, P. Methodology of ultra-dispersed polymer system. *STP Pharm Sci* 1991, 1, 109–116.

Vauthier, C.; Couvreur, P. Development of nanoparticles made of polysaccharides as novel drug carrier systems. In: *Handbook of pharmaceutical controlled release technology*, Marcel Dekker; New York, 2000; pp 13–429.

Vemavarapu, C.; Mollan, . .; Lodaya, M.; Needham, T. E. Design and process aspects of laboratory scale SCF particle formation systems. *Int J Pharm* 2005, 292, 1–16.

Wang, H.; Zhao, P.; Su, W.; Wang, S.; Liao, Z.; Niu, R.; Chang, J. PLGA/polymeric liposome for targeted drug and gene codelivery. *Biomater.* 2010, 31, 8741–8748.

Watson, P.; Jones, A. T.; Stephens, D. J. Intracellular trafficking pathways and drug delivery: fluorescence imaging of living and fixed cells. *Adv Drug Deliv Rev* 2005, 57, 43–61.

Wehrle, P.; Magenheim, B.; Benita, S. The Influence of process parameters on the PLA nanoparticle size distribution evaluated by means of factorial design. *J Pharm Biopharm* 1995, 41, 19–26.

Weissleder, R. Scaling down imaging: molecular mapping of cancer in mice. *Nature Rev Cancer.* 2002, 2, 1–8.

Wesselborg, S.; Engels, H.; Rossmann, E.; Los, M.; Schulze-Osthoff, K. M. Anticancer drugs induce caspase-8/FLICE activation and apoptosis in the absence of CD95 receptor/ligand interaction. *Blood* 1999, 93, 3053–3063.

Xin, D.; Wang, Y.; Xiang, J. The use of amino acid linkers in the conjugation of paclitaxl with hyaluronic acid as drug delivery system: synthesis, self-assembled property, drug release, and *in vitro* efficiency. *Pharm Res* 2010, 27(2), 380–389.

Yadav, A. K.; Agarwal, A.; Rai, G.; Mishra, P.; Jain, S.; Mishra, A. K.; Agrawal, H.; Agrawal, G. P. Development and characterization of hyaluronic acid decorated PLGA nanoparticles for delivery of 5-fluorouracil. *Drug Deliv* 2010, 17(8), 561–572.

Yallapu, M. M.; Gupta, B. K.; Jaggi, M.; Chauhan, S. C. Fabrication of curcumin encapsulated PLGA nanoparticles for improved therapeutic effects in metastatic cancer cells. *J Colloid Interface Sci* 2010, 351, 19–29.

Yong-Hee, K.; Bae, Y. H.; Kim, S. W. pH/temperature sensitive polymers for macromolecular drug loading and release. *J Control Rel* 1994, 28, 143–152.

Yoo, H. S.; Oh, J. E.; Lee, K. H.; Park, T. G. Biodegradable nanoparticles containing doxorubicin-PLGA conjugate for sustained release. *Pharm Res* 1999, 16, 1114–1118.

York, P. Strategies for particle design using supercritical fluid technologies. *Pharm Sci Technol Today* 1999, 2, 430–440.

Yu, W.; Pirollo, K. F.; Yu, B.; Rait, A.; Xiang, L.; Huang, W. Enhanced transfection efficiency of a systemically delivered tumor-targeting immunolipoplex by inclusion of a pH-sensitive histidylated oligolysine peptide. *Nucl Acids Res* 2004, 32, 48.

Zambaux, M. F.; Bonneaux, F.; Gref, R.; Maincent, P.; Dellacherie, E.; Alonso, M. J.; Labrude, P.; Vigneron, C. Influence of experimental parameters on the characteristics of poly(lac-tic acid) nanoparticles prepared by double emulsion method. *J Control Rel* 1998, 50, 31–40.

Zambito, Y.; Pedreschi, E.; Di-Colo, G. Is dialysis a reliable method for studying drug release from nanoparticulate systems? A case study. *Int J Pharm* 2012, 434, 28–34.

Zhao, P.; Wang, H.; Yu, M.; Liao, Z.; Wang, X.; Zhang, F.; Han, J.; Zhang, H.; Wang, H.; Chang, J.; Niu, R.; Ji, W.; Wu, B. Paclitaxl loaded folic acid targeted nanoparticles of mixed lipid-shell and polymer-core: *in vitro* and *in vivo* evaluation. *Eur J Pharm Biopharm* 2012, 81, 248–256.

Zhen, W.; Link, C. J. Jr.; O'Connor, P. M.; Reed, E.; Parker, R.; Howell, S. B.; Bohr, V. A. Increased gene-specific repair of cisplatin interstrand cross-links in cisplatin-resistant human ovarian cancer cell lines. *Mol Cell Biol* 1992, 12, 3689–3698.

Zweers, M. L. T.; Engbers, G. H. M.; Grijpma, D. W.; Feijen, J. *In vitro* degradation of nanoparticles prepared from polymers based on dl-lactide, glycolide and poly(ethylene oxide). *J Control Rel* 2004, 100, 347–356.

Zweers, M. L. T.; Engbers, G. H. M.; Grijpma, D. W.; Feijen, J. Release of antirestenosis drugs from poly(ethylene oxide)-poly(dl-lactic-coglycolic acid) nanoparticles. *J Control Rel* 2006, 114, 317–324.

Zweers, M. L. T.; Grijpma, D. W.; Engbers, G. H. M.; Feijen, J. The preparation of monodisperse biodegradable polyester nanoparticles with a controlled size. *J Biomed Mater Res* B 2003, 66, 559–566.

CHAPTER 2

PHARMACOGENETICS AND PHARMACOGENOMICS IN PERSONALIZED MEDICINE: ROLE OF GENE POLYMORPHISM IN DRUG RESPONSE

ROWNOCK AFRUZA, FUMIAKI SUZUKI,
and A. H. M. NURUN NABI

CONTENTS

2.1   INTRODUCTION

Magnitude of the information regarding pharmacogenetics and pharmacogenomics with increasing knowledge in the field of molecular biology has profoundly augmented day-by-day for giving the best-tailored medication to the patients in a most effective way with less involvement of cost and time. It is well understood and recognized that with the same kind of medication, different patients predisposed to a same kind of disease would respond in a different manner. Human genome project has been completed in 2003, there are only 20,000–25,000 genes, which are much less in number than it was expected or hypothesized. More than 1.4 million single-nucleotide polymorphisms were identified in the initial sequencing of the human genome (Sachidanandam *et al.*, 2001) for genotyping and phenotyping studies. Of them, over 60,000 have been identified in the coding region of genes. Some of these single-nucleotide polymorphisms (SNPs or also called "snips") have been reported to be associated with substantial changes in the metabolism or effects of medications and some are being used to predict clinical response (Evans *et al.*, 1960; Evans and Relling, 1999; Evans and Johnson, 2001; McLeod and Evans, 2001). Depending on the nature of the protein required to perform a specific biological functions, the exons of the assigned genes are tagged together by splicing and then expressed. Each codon within the gene determines the nature of the amino acid and any change in a single base in the codon that may change the open reading frame of the codon followed by a change in amino acid, which could ultimately alter whole structure of the three dimensional structure of protein. As a result, response of the expressed protein towards its substrate or ligand will also be altered that ultimately cause reduction or unnecessary augmentation of function. Thus, genes determine the structural and functional properties of all body proteins, and the efficacy of medicines, as medicines work on body proteins. So, the response to a particular medicine is different in different people due to variations in genetic makeup. It is estimated that genetics can account for 20 to 95% of variability in drug disposition and effects (Kalow *et al.*, 1998; Evans and McLeod, 2003) although many nongenetic factors such as age, organ function, concomitant therapy, drug interactions, and the nature of the disease influence the effects of medications (Evans and Relling, 1999). Interindividual differences in drug response are due to sequence variants in genes encoding drug-metabolizing enzymes, drug transporters, or drug targets (Evans and Johnson, 2001; McLeod and Evans, 2001). That's why personalized medicine is a concerned topic of recent days. Although many people are in favor of personalized medicine, few of them understand exactly what it is, how to invent it and how to use it. Personalized medical treatment expects the best possible outcomes by providing the right treatment to the right person in the right amount at the right time. Thus, personalized medicine is the use of genetic or other molecular biomarker information to improve the safety, effectiveness and health outcomes of patients via more efficiently targeted risk stratification, prevention and tailored treatment management approaches.

Based on trial and error or test-and-treat approach, a good physician has always customized treatment to ameliorate symptoms and reduce probable side effects of a drug on the treated patient. However, selective treatments using personal information regarding the patient's variation due to environmental, genetic and immunological history would certainly help to alleviate the symptoms of a disease and its adverse effects more potently by reducing the possibility of being partially less effective. This approach defines disease subtypes and biomarkers used to identify the particular disease. Thus, collective information of these various histories would change the way of diagnosing a disease, developing a drug and best-fitted therapeutic dose of a drug for the affected tissues or cells. Knowledge of pharmacogenetics and pharmacogenomics are the major base line prerequisites for designing personalized medicine. Variety of prospective products for the pharmaceutical industries should be the strategic outcome for discovering and developing personalized medicine. It is important to have risk-assessing and disease-monitoring diagnostic tools for accurately enumerating load of disease in patients. Clinical observations of inherited differences in drug effects were first documented in the 1950s (Kalow, 1956; Vesell, 1989; Yates *et al.*, 1997; Sachidanandam *et al.*, 2001; Weinshilboum, 2003) giving rise to the field of pharmacogenetics, later pharmacogenomics and now, personalized medicine. Therefore, acquiring knowledge in the areas of pharmacogenetics and pharmacogenomics is a demand of time for managing diseases at the individual level.

2.2   RESEARCH AREAS IN PERSONALIZED MEDICINE

There are five major types of research activities that are currently under practice in the area of personalized medicine. (i) Pharmacogenetics, a field of science that explains how different people responds in different ways to same drug by testing candidate genes for drug-patient interactions. This area also promotes drugs that have a favorable effect on any gene that is responsible for some or all of the disease phenotype (Klein *et al.*, 2001). (ii) Pharmacogenomics deals with the levels of gene expression over time (Bozkurt *et al.*, 2007). The content of genes within DNA of a person does not change over time, rather its expression levels may change, i.e., the RNA content may change over time. The measurement of gene products over time adds great complexity to the process for identifying genes that are integral to the disease state. Ultimately the relationships that are investigated using pharmacogenomics are more robust than those from pharmacogenetics alone. (iii) Nutrigenomics is the approach that uses identifying genetically mediated responses to foods and then adjusting the diet to take advantage of these responses (El-Sohemy, 2007; Kaput and Dawson, 2007). (iv) Biomarkers can be used to predict, diagnose, or monitor diseases. For example, autoantibodies can be measured to predict type 1 diabetes (In't Veld *et al.*, 2007) and adipokines (Mojiminiyi and Abdella, 2007; Shaibi *et al.*, 2007) can be measured to predict type 2 diabetes. (v) Finally, systems biology (Petrasek, 2008; Teixeira and Malin, 2008) measures interactions between the components of biological systems and how these interactions give rise to the function and behavior of that system. Systems biology analyzes complex data from multiple sources by using such tools as transcriptomics (which assesses gene expression measurements), proteomics (which completely identifies proteins and protein expression patterns of a cell or tissue), metabolomics (which identifies and measures all the small molecule metabolites within a cell or tissue), and glycomics (which identifies all carbohydrates in a cell or tissue).

Precise diagnosis of a disease determines the success rate of any personalized medicine. Indeed, the specificity and reliability of diagnostic tests limits the degree to which a treatment can be personalized. Clinicians generally use physical symptoms to identify an illness. However, two patients with identical symptoms might be suffering from very different conditions. This is important if the genes or biological pathways involved require different treatments. Diagnostic tests of certain biomarkers are useful to determine the specific genes or pathways responsible for the onset of the diseases. Biomarkers are measurable material indicators of current health status, or predictors of susceptibility to disease and likely effectiveness of treatment. Genetic information is one of many such indicators. Other biomarkers include the products of genes or metabolic activity, such as proteins, hormones, RNA or other signaling molecules used by cells. Biomarkers are used in clinical practice to predict or identify risk of disease; diagnose and assess severity of existing disease; stratify patients to potentially tailor treatment.

RNA expression profiling and pharmacogenomics as well as DNA sequence profiling (pharmacogenetics and SNPs analyzes) are the most reliable and used toolkit for the diagnosis of diseases where personalized medicine could be applied to get most effective outcomes (Overdevest *et al.*, 2009).

RNA expression profiling technology relies heavily on nucleic acid amplification technologies to accurately boost the level of target RNA species to be probed by the complimentary capture sequences on arrays. The strength of this technology includes target patient subpopulations directly with specific molecular characteristics, limits toxicity and troublesome side effects and cost-effective and well-established technical methodology that could be applied in targeted therapeutics, disease subclassification and RNA expression for diagnosis/response classification. Strength of the protein expression profiling, pharmacoproteomics, is that it indicates the direct functional interaction of drugs with molecular pathways, which could be applied in testing and patient stratification for drug sensitivity. The strength of pharmacogenetics (SNP testing and differential quantitative PCR or qPCR, haplotyping) includes cost-effective large-scale genetic analysis, variation screening in patients, low error rate and well-established analysis tools, which could be applied in many disease applications, prediction of therapeutic response, metabolic potential and high-risk variations. High throughput gene sequencing helps in tumor classification and assessment of epigenetics (Overdevest *et al.*, 2009).

2.3   PHARMACOGENETICS, PHARMACOGENOMICS AND PHARMACOPROTEOMICS

Three discoveries in the 1950s gave rise to the discipline of pharmacogenetics (Motulsky, 1957) and these are sensitivity to primquine (glucose-6-phosphate dehydrogenase deficiency), the slow metabolism of isoniazid (acetylation polymorphism and tuberculosis) and prolonged effects of succinylcholine due to an atypical plasma cholinesterase (respiratory apnea). The term pharmacogenetics is generally associated with inheritance, which can be defined as "*the study of the role of inheritance in inter-individual variation in drug response*" by Weilshboum and Wong (2004). Thus, pharmacogenetics deals with the genetic differences that give rise to interpatient variation in drug absorption, distribution, biotransformation, and elimination. The way a person responds to a drug (this includes both positive and negative reactions) is a complex trait that is influenced by many different genes and genetic variations. Knowing whether a patient carries any of the genetic variations can help medical practitioners to prescribe tailored medicine, decrease the chance for adverse drug events, and increase the effectiveness of drugs. Study of pharmacogenetics helps to improve drug choices by predicting individualized negative or positive reaction to drug, boost safer dosing options by means of testing genomic variation that ultimately leads to correct dose for each individual, improve drug development for determining efficacy of a specific drug for a specific group of population, and decrease health care costs by reducing number of deaths due to adverse drug reactions and reducing procurement of such drugs, which are ineffective in certain individuals due to genetic variations. Thus, knowing whether a patient carries any of the genetic variations can help medical practitioners to prescribe tailored medicine, decrease the chance for adverse drug events, and increase the effectiveness of drugs.

Pharmacogenomics is the whole genome association of pharmacogenetics. The term pharmacogenomics defined as the study of information regarding genomics and proteomics (Fig. 2.1) for identifying new drug targets and their mechanism of action that has emerged in the late 1990s. This area of science is often associated with the industrial application of discovery and development of new drugs. Pharmacogenomic markers of efficacy and side effects will be used in conjunction with specific drugs to target drug therapy to those patients who will have an optimal response. With the availability of more trustworthy biomarkers as diagnostic tool, more advanced molecular utensils for detecting SNPs or more complex gene polymorphisms, advances in bioinformatics and functional genomics, and the bunch of new data after completion of human genome projects, characterization and effects of a particular drug on the basis genetic determinants of an individual or a specific population have been rapidly elucidated, and these data are being used in rational drug therapy that has given rise to a new era of translational medicine (Evans and Relling, 1999; McLeod and Evans, 2001).

Pharmacoproteomics is the use of high-resolution proteomic technologies such as high resolution electrophoresis and mass spectroscopy to identify serum-, fluid-, or tissue-based markers of drug for the discovery and development of new drugs. Along with pharmacogenomics and pharmacogenetics, pharmacoproteomics will play an important role in the development of personalized medicines to facilitate tailored therapy (Fig. 2.1). Thus, proteomic technologies are contributing to molecular diagnostics, which is a basis of personalized medicine. Pharmacoproteomics is a more functional representation of patient-to-patient variation than that provided by genotyping. Proteomics-based characterization of multifactorial diseases may help to match a particular target-based therapy to a particular marker in a subgroup of patients. Individualized therapy may be based on differential protein expression rather than a genetic polymorphism. Protein chips will be used increasingly in clinical diagnostics in the coming years, particularly in the point-of-care diagnostics that would allow the practice of personalized medicine in the clinic by the end of this decade.

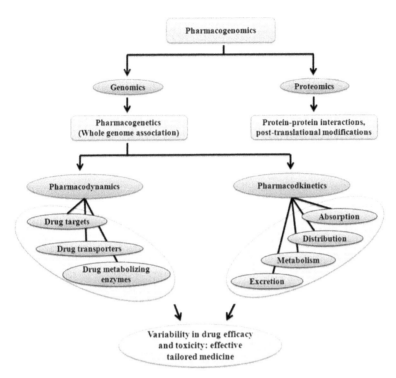

FIGURE 2.1 Key components in pharmacogenomics and pharmacogenetics. Pharmacogenomics deals with genomics and proteomics. There are two broad areas of research in pharmacogenetics: pharmacokinetics and pharmacodynamics. Drug targets, metabolizing enzymes and transporters frequently contribute to variable pharmacokinetics and thus, could be candidate genes in pharmacogenetics studies.

Decoding the human genome has made sense of the complex genetic code that directs the fate of the 100 trillion cells in the human body. This has provided hope of elucidating the genetic architecture of disease. Genetic information has enabled the identification of potential therapeutic targets. However, it is clear that sequence information alone cannot resolve the complex patterns of gene activities that ultimately determines the physiology and pathology of an organism. Hence, experimental approaches aimed to analyze gene function and the interactions of their products on a genome-wide scale. These functions all are collectively termed as 'functional genomics.' Validated drug-target discovery is pacing-up with the acceleration of availability of sequence data that has enabled a fundamental paradigm shift in functional genomics. This new biological perception, genome analysis, has conveyed unique advances in the understanding of gene function and their causative role in disease. Thus, postgenomics era has brought a comprehensive change in the perception of the researchers in the context of biological system and improvement of more reliable diagnostic biomarkers for the identification as well as health care prognosis of a particular disease, of the investors in pharmaceutical industries and policy makers for encouraging instrumentation by accepting and putting money on more automation system for developing new drugs. The evolution of more advanced genome sequencing technologies over the years has resulted in affluence of sequence information. Analysis of the primary sequence data is possibly the most important feature of the postgenomics era (i.e., after the completion of DNA sequencing). The large number of prospective drug targets rising from genome sequencing claims new approaches to discover drug. Application of mathematical, statistical and computational knowledge in the primary DNA sequence to obtain more functional biological information is the main aim of bioinformatics. A major part of this discipline is not only to identify single genes or multiple genes responsible for the onset of various diseases but also to recognize genes for the elucidation of probable novel drug targets (Broder and Venter, 2000). However, it is indeed a challenge to identify the most likely therapeutic targets from many new sequences encode gene products. Various methods and technologies are currently being used to validate these genes as drug targets, ranging from hypothesis-driven studies of single genes in model physiological systems, to global scans for genes underlying disease processes.

2.4   FACTORS INFLUENCING INDIVIDUALIZED DRUG RESPONSE

The processes of absorption, distribution into tissues, metabolism and elimination determine the amount of drug and metabolites that are delivered to target sites. Group of patients even diagnosed with a same disease may have different response to the same kind and dose of drug (Fig. 2.2). Two issues should be taken into consideration for evaluating responses to drugs –most notable one is the genetic issue and the equally important other one is nongenetic issue including environmental aspects such as occupation, smoking, diet, age and sex. Diet can influence the expression and activity of hepatic drug metabolizing enzymes (Ioannides, 1999). Charcoal-broiled foods and cruciferous vegetables are known to induce CYP1A enzymes, whereas grapefruit juice is known to inhibit the CYP3A metabolism of co-administered drug substrates (Conney, 1967). Some children have matured CYP enzymes within their 6 months of age while it takes up to 12 months for others (Stewart and Hampton, 1987). For example, the half-life of a bronchodilator called theophylline, primarily metabolized by CYP1A2 and CYP3A4, in serum is significantly longer in neonates and younger infants than in adults. Similarly, N-demethylation of diazepam, a CYP2C19-mediated pathway, is also significantly slower in infants. Cigarette smokers metabolize some drugs more rapidly than non-smokers because of enzyme induction. Industrial workers exposed to some pesticides metabolize certain drugs more rapidly than unexposed individuals. Such differences make it difficult to determine effective and safe doses of drugs that have narrow therapeutic indices. Thus, genetic diversity, most conspicuously through single nucleotide polymorphisms and copy-number variation, together with specific environmental exposures, contributes to both disease susceptibility and drug response variability. It has proved difficult to isolate disease genes that confer susceptibility to complex disorders, and as a consequence, even fewer genetic variants that influence clinical drug responsiveness have been uncovered. Following sections will cover drug response on the basis of variations of different genes involved in drug disposition.

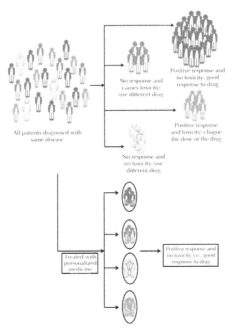

FIGURE 2.2 A group of patients diagnosed with same disease may show different responses while treated with same kind and dose of drug. However, this can be overcome by personalized medicine.

2.4.1 GENETIC VARIATION AND DRUG RESPONSE

Gene is the functional unit of heredity. Each gene is a sequence within the genome that functions by giving rise to a discrete product; which may be a polypeptide or RNA. There are approximately 25,000 genes, which give more than millions of proteins in one person and there are so many varieties in protein combinations as well as genetic variations within different individuals too. These genetic variations are often greater among members of a population than they are within the same person at different times (or between monozygotic twins) (Vesell, 1989; Evans and McLeod, 2003). At present, there is a race to catalog as many of the genetic variations found within the human genome as possible. Highly variable drug response and toxicity exclude clinical use of a drug. Even with some of the most advanced current drugs, favorable response occurs in only 30–70% of patients, with a significant portion showing adverse effects, yielding a poor risk/benefit ratio for a diverse patient population. Pharmacokinetics and pharmacodynamics provide quantitative measures of drug exposure and effect. Genetic variation in drug metabolism influences pharmacokinetics that focuses on absorption, distribution, metabolism and excretion (ADME), while pharmacodynamics deals with drug targets that include receptors (beta2-adrenergic receptors), enzymes (Phase I and II enzyme system functions affect drug metabolism; phase I includes CYP450 enzymes and phase II includes N-acetyl transferase, thiopurine S-methyltransferase), ion channels and transporters (that affect drug absorption), second messenger and/or downstream signaling pathways, immune system. Polymorphic genes relevant to PK-PD are listed in Table 2.1. Since ADME directs drug exposure, drug level monitoring yields phenotypic markers useful for individualizing therapy. Large interindividual variability is often associated with frequent mutations in cytochrome P450s and in conjugating enzymes such as glucuronyl transferases (Table 2.1). Through a meta analysis, a high level of association between polymorphic CYP450 genes and the lead drugs causing adverse reactions have been revealed (Sachidanandam et al., 2001). Several hundred genes are involved to encode drug transporters, playing all-encompassing role in ADME and drug targeting (Weinshilboum, 2003). Drug responses have been found to be affected by numerous functional polymorphisms, while their impact is poorly understood. Similarly, the effect of polymorphisms in genes encoding drug receptors is difficult to assess. Activating mutations are a possible exception, particularly those of tyrosine kinases involved in cancer progression. For example, constitutive activity of the fusion protein BCR/ABL (generated through chromosomal translocation

TABLE 2.1   Survey of polymorphic genes relevant to drug response.

Gene symbol	Polymorphisms	Clinical phenotype
Phase I		
CYP2D6[a] CYP2C9[a] CYP2C19 CYP3A5 CYP2B6[a]	Numerous and frequent functional polymorphisms (non-synonymous SNPs, splice variants, regulatory, gene amplification) (http://www.imm.ki.se/cypalleles/)	Variable metabolism of CYP substrates depending on enzyme specificity. Allele frequencies vary among ethnic groups
CYP2E1		
DPYD	Multiple polymorphims including splice variant IVS14 þ 1G.A (DYPD_2A)	2A increases 5FU neurotoxicity
Phase II		
TPMT[a]	Multiple polymorphisms	Null genotype associated with hematopoietic thiopurine toxicity, homozygous frequency 1/300
COMT	Val158Met, haplotypes	Met158 associated with higher daily neuroleptic dosage and poor response; haplotye affects mRNA levels
NAT2	Arg197Gln, Ile114The, Lys268Arg, G191GA, G857A	Affect acetylation rates for drugs (e.g. isoniazid)
GSTM1	M1 Deletion causing null genotype	Decreased glutathione conjugation GSTT1 Glutathione-S-transferase T1 Deletion causing null genotype
UGT1A1[a]	Multiple polymorphisms in the promoter and coding regions	UGT1A1_28 variants associated with increased irinotecan toxicity
Drug transport		
ABCB1[a]	Several polymorphisms including C3435T	C3435 associated with higher drug transport activity
BCRP (ABCG2)	G34A, C421A and 944–949 deletion	Minor alleles with lower BRCP expression, enhanced drug sensitivity

TABLE 2.1   *(Continued)*

Gene symbol	Polymorphisms	Clinical phenotype
Drug targets/pathway protein		
MTHFR	C(677)T and A(1298)	CT677 increases toxicity to methotrexate, drug resistance
TYMS	Two to nine 28 bp repeats in the 50 enhancer	TSER_3 associated with drug resistance
ADRB2	Arg16Gly, Gln27Glu and Thr164Ile	16Gly decreases response to altuterol and salbutamol
ADRB1	Arg389Gly	Gly389 with decreased cardiovascular response to b1-antagonists
ALOX5	Variant promoter SP1 binding site	Decreased drug response with the mutant genotype
HTR2A	His452Tyr	Tyr452 decreases response to the antipsychotic Clozapine
ACE	Ins/del in intron 16	Del/Del decreases efficacy of the reduction of proteinuria by ACE inhibitors

[a]Genes or proteins for prospective genotyping tests are in clinical use or under consideration.

in leukemia) conveys high sensitivity to imatinib (Gleevec) for the treatment of chronic myelogenous leukemia and gastrointestinal stromal tumors, whereas activating mutations of EGFR appear to correlate with responsiveness to gefitinib (Iressa) (Meyer, 2000). Over-expression of ErbB2 is a requisite for successful herceptin treatment of breast cancer (Kuehl *et al.*, 2001). These have brought to the diagnostic laboratory an expanding role for the testing of patients genetic variations to determine their eligibility to receive these new therapies (O'Dwyer and Druker, 2000; Vogel *et al.*, 2002). For cardiology, inflammatory conditions, neurodegenerative diseases, and psychiatric disorders, similar expectations have emerged among patients and physicians concerning the ability of diagnostic tests to customize and individualize therapy (Ross and Ginsburg, 2002). The clinical application of molecular diagnostics is expected to continue to greatly impact the drug discovery and development process; customize the selection, dosing, and route of administration of existing and new therapeutic agents; and facilitate further development of personalized medical care (Ross and Ginsburg, 2002; Goldstein *et al.*, 2003; Johnson, 2003; Ulrich *et al.*, 2003).

2.4.2   GENETIC POLYMORPHISMS INFLUENCING DRUG DISPOSITION

The field of pharmacogenetics began with a focus on drug metabolism (Weinshilboum, 2003), but it has been extended to cover the full spectrum of drug disposition including a growing list of transporters that influence drug absorption, distribution, and excretion (Evans and Relling, 1999; Meyer, 2000; Evans and Johnson, 2001; McLeod and Evans, 2001). Single nucleotide variations in a specific gene are supposed to yield crucial information for new drug development and individualized drug prescription. Synonymous and nonsynonymous SNPs should be taken into consideration to validate the association of a candidate gene with a disease, that is, the clinical consequences of SNPs. Thus, it should be kept in mind that localization of the SNPs is just the beginning of the challenging work. Its clinical utility being a confirmed SNP, pharmacological as well as toxicological impacts should equally be assessed and evaluated. The SNPs are associated with a number of diseases such as Alzheimer's disease (Martin *et al.*, 2000), osteoporosis (Grant *et al.*, 1996), Crohn's disease (Hugot *et al.*, 2001) and obesity (Esterbauer *et al.*, 2001). The SNP scoring technologies, either direct or indirect (Grant *et al.*, 2002), are required for SNP detection and confirmation. DNA sequencing and use of DNA polymerase technologies are the prerequisites to detect the polymorphic nucleotide directly. Indirect scoring technologies use hybridization to identify base composition of the SNP. The hybridization approach is more accurate than the direct one though it is harder to optimize. However, combination of the two methodologies involving hybridization coupled to an enzymatic reaction, a time consuming method, offers both advantages.

Numerous studies have demonstrated the role of SNPs and microsatellite instability in disease susceptibility (Bowater and Wells, 2001; Peltomaki, 2001) by influencing the disease progression and clinical management. Interindividual variations in the genes of liver enzymes that affect the biochemistry of drug metabolism (Wolf *et al.*, 2000) have been identified. An individual with less metabolic activity due to such less effective enzymes would have greater possibility of over dosing of drugs. On the contrary, a person with a genetic sequence, which can express more potent enzyme with efficient metabolizing ability would need a higher dose of the drug for a successful therapy.

Congestive heart failure represents the final stage of a continuous disease process. It originates from prevalent cardiovascular diseases, for example, dilated cardiomyopathy or ischemic cardiomyopathy based on coronary artery disease. Although pharmacotherapy has provided significant improvement of overall survival, the prognosis of advanced heart failure is still rather poor.

Guidelines for pharmacological treatment of heart failure have been demonstrated (Hunt, 2002). ACE-inhibitors (Hall *et al.*, 1997) and/or AT_1-blockers (Cohn and Tognoni, 2001), antiadrenergic drugs (The Cardiac Insufficiency Bisoprolol Study II, 1999), diuretics (no prospective mortality trial available), digitalis (The Digitalis Investigation Group, 1997), and lately aldosterone antagonists (Pitt *et al.*, 1999) comprise standard treatment regimens, while others such as endothelin antagonists are still rather experimental (Rich and McLaughlin, 2003). Such standardized pharmacotherapy, however, leads to variable clinical outcome. One factor possibly contributing to this variability is the modification of drug disposition and action by genetic traits. Highly variable plasma concentrations and hence effects following administration of standard doses have been described for many cardiovascular drugs. This phenomenon can be attributed to variable expression of drug metabolizing enzymes (in particular cytochrome P450) and, less well explored, to variable expression of drug transporters. Some drugs used in treatment of heart failure patients are metabolized by polymorphic enzymes (e.g., beta-adrenergic blockers and AT_1-antagonists) or subject to polymorphic transport (e.g., digitalis). The large prospective trials such as MERIT-HF (Hjalmarson *et al.*, 2000), and CIBIS (Lechat *et al.*, 1997) gave clear evidence that treatment with selective beta$_1$-adrenoreceptor blockers like metoprolol and bisoprolol improves survival of patients with mild to moderate heart failure. Lately, the COPERNICUS-Study (Packer *et al.*, 2002) demonstrated beneficial effects of carvedilol also for NYHA IV patients. In contrast to β_1-selective blockers, carvedilol inhibits not only β_1- and β_2 but also α_1-receptors. Metoprolol and carvedilol, but not bisoprolol, are metabolized by cytochrome *P*4502D6. This enzyme has been studied in detail and 7–10% of the Caucasian population exhibits no CYP2D6 activity due to well characterized hereditary polymorphisms, which has a major impact on CYP2D6 dependent clearance of metoprolol and carvedilol (Pepper *et al.*, 1991; Oldham and Clarke, 1997). For example, CYP2D6 poor metabolizers provide a six-fold elevated metabolic ratio of metoprolol (Rau *et al.*, 2002). From a survey on adverse events after metoprolol treatment, the same group concluded that side effects may be due to CYP2D6 genotype, since the frequency of poor metabolizers was elevated in the group of patients, who reported pronounced adverse symptoms (Wuttke *et al.*, 2002). The number of poor metabolizer patients studied in this trial, however, was rather limited. In contrast, an unexpected low drug effect may arise from the phenomenon of gene duplications of CYP2D6 leading to high protein expression. The frequency of the CYP2D6 duplications is given with 1–3% in Middle Europeans but up to 29% in Ethiopia (Aklillu *et al.*, 1996). Aside selected betablockers, angiotensin-II type-1 (AT1) receptor antagonists such as losartan and irbesartan are subject to polymorphic metabolism (Yasar *et al.*, 2001; Hallberg *et al.*, 2002; Lee *et al.*, 2003). The prodrug losartan is activated in the liver by the drug metabolizing enzyme cytochrome P450–2C9 (CYP2C9) to the active metabolite EXP3174, and Sekino *et al.* (2003) demonstrated in healthy Japanese subjects that CYP2C9 wild type carriers have lower systolic blood pressure after losartan therapy than poor metabolizers. Irbesartan, in contrast, is inactivated by CYP2C9. The two major hereditary polymorphisms of this enzyme lead to amino acid replacements causing diminished enzyme activity. Consequently, hypertensive patients being low active CYP2C9 carriers showed a stronger effect on reduction of the diastolic and – less significant, of the systolic blood pressure – following treatment with irbesartan (Hallberg *et al.*, 2002). The impact of the variable pharmacokinetics of AT1-antagonists to the outcome on heart failure, however, has not yet been verified, since prospective studies considering this aspect are lacking.

One of the major obstacles of cancer chemotherapy is the development of drug resistance, which prevents the application of sufficient high doses to eradicate less-sensitive tumor cell populations. Inter-individual differences in response to xenobiotics, which include many clinically used drugs, are extensive and represent a major problem in rational therapeutics. Such differences in many cases may be caused by inherited differences in enzymes and transporters, which function in drug elimination (Evans and Relling, 1999). Owing to its possible effect on gene expression, it was anticipated that polymorphisms of drug metabolizing enzymes, genes and drug transporters genes may influence tumor response to platinum-based chemotherapy. The identification of molecular variables that predict either sensitivity or resistance to chemotherapy is of major interest in selecting the most likely effective first-line treatment. Some studies suggested that there is no difference whether cisplatin- or carboplatin-based chemotherapy regimens in the clinical efficacy (Schiller *et al.*, 2002; Zatloukal *et al.*, 2003). Moreover, the results tend to be similar whether the partner drug is paclitaxl, docetaxl or gemcitabine. Similar results are generally obtained with carboplatin (Scagliotti *et al.*, 2002; Schiller *et al.*, 2002; Zatloukal *et al.*, 2003).

Glutathione *S*-transferases are crucial for the cell defense system. These phase II detoxification enzymes are involved in the detoxification of a variety of chemotherapeutics including platinum. *In vitro* analyzes revealed a significant association between high GSTP1 expression of tumor cells and decreased sensitivity to platinum agents (Oguri *et al.*, 2000). Cisplatin is detoxified by glutathione through adduct formation (Rudin *et al.*, 2003). GSTP1 interacts with platinum-based compounds (Goto *et al.*, 1999), and glutathione-conjugated platinum can be quickly effluxed from cells (Harpole *et al.*, 2001). Thus, it is plausible that high GST activity may result in more rapid drug metabolism that diminishes the cytotoxic effects of chemotherapy on tumor cells (Beeghly *et al.*, 2006). Multidrug resistance-associated protein 2 (MRP2) is responsible for the intracellularly formed glucuronide and glutathione (GSH) conjugates of clinically important drugs. MRP2 is expressed in many tumor tissues, and the tumor cells over expressing MRP2 might acquire the multidrug resistance (Borst *et al.*, 2000). A significant correlation has been observed between MRP2 mRNA levels and cisplatin resistance in colorectal carcinoma (Hinoshita *et al.*, 2000) suggesting that MRP2 contributes to resistance against treatment with the chemotherapeutic drugs.

Board *et al.* (1989) first identified the Glutathione S-transferease P1 (GSTP1) polymorphisms. Alterations in the structure, function, or expression levels of GSTP1 due to genetic polymorphisms could alter the ability to detoxify chemotherapeutic agents and modulate drug response. The Ile105 Val GSTP1 polymorphism, an A/G SNP located within the substrate binding domain of GSTP1 at position +313 within exon 5 results in an amino acid substitution of isoleucine by valine at codon 105 of the enzyme. This substitution has been shown to significantly influence the catalytic activity and thermal stability of the enzyme and to affect the conjugation capacity of GSTP1 for certain substrates, including platinum agents (Watson *et al.*, 1998; Allan *et al.*, 2001). Rather than being present or absent, the GSTP1 gene has alleles that encode enzymes with different activities. The highest level of GSTP1 activity is seen in individuals with homozygous wild genotype (Ile/Ile). The activity is somewhat reduced in heterozygotes (Ile/Val) and further diminished for those with homozygous mutant genotype (Val/Val) (Allan *et al.*, 2001). It is estimated that approximately 50% of people have either one or two valine alleles (Ile/Val or Val/Val) (Garte *et al.*, 2001). Thus, this polymorphism is linked to clinical outcome of patients who received platinum-based chemotherapy (Stoehlmacher *et al.*, 2002, 2004). Individuals with these variant GSTP1 genotypes that result in reduced GST enzymatic activity may be good responders due to decreased detoxification of chemotherapeutic agents.

Because the important factors influencing interindividual differences in the drug disposition, many analyzes of SNPs of drug metabolizing enzymes and drug transporters have been performed.

2.4.3 ASSOCIATION OF GENE POLYMORPHISMS WITH DRUG METABOLIZING ENZYMES

Many drug metabolizing enzymes have been identified and studied extensively so far, and there are >30 families of drug-metabolizing enzymes in humans (Evans and Relling, 1999). Fundamentally all of them have genetic dissimilarities and some of them are translated into proteins with altered functions. With the advanced technological penetration in molecular biology, significant progress has been made in understanding the role of genetic polymorphisms in drug metabolism. The cytochrome P450 superfamily (CYP) consists of a large and diverse group of major enzymes, involved in drug metabolism, bioactivation and clearance of greater part of currently used medicines accounting for about 75% of the total number of different metabolic reactions (Guengerich, 2008). In humans up to 21 families, 20 subfamilies and 57 genes have been recognized and described. Out of these, CYP 1, 2 and 3 accounts for 70% of total hepatic CYPs content and are responsible for 94% of drug metabolism in liver (Chang and Kam, 1999). Thus, polymorphisms in any of the genes of this family may affect patient's response to drug and dose requirement. The major polymorphisms with clinical implications are related to the oxidation of drugs by CYP2D6 and CYP2C19 (Wilkinson et al., 1989; Alvan et al., 1990; Meyer et al., 1990, 1992; Broly and Meyer, 1993; Meyer, 1994), acetylation by N-acetyltransferase (Evans, 1992), and S-methylation by thiopurine methyltransferase (Weinshiboum, 1992; Creveling and Thakker, 1994). Individuals who inherit an impaired ability to catalyze one or more of these enzymatic reactions may be at an increased risk of concentration-related adverse effects and toxicity. For example, variation in the gene of cytochrome P450 CYP2C9 affects clearance of warfarin compounds, while CYP2D6 gene polymorphism, can inhibit conversion of prodrug codeine to analgesic drug morphine (Wolf et al., 2000). In some patient, any variation in the CYP2C9 enzyme may cause slower metabolism of warfarin which could generate dose toxicity, thus putting the patient at an increased risk of bleeding (Higashi et al., 2002). Two main variant alleles of CYP2C9, the *2 allele and the *3 allele have been revealed (Anderson et al., 2007). These alleles cause reductions in enzymatic activity by approximately 30% and 80%, respectively that increase the chance of bleeding (Higashi et al., 2002). Perhaps, CYP2D6 polymorphism is the most studied genetic polymorphism in drug metabolism (Mahgoub *et al.*, 1977) and many drugs are known to be catalyzed primarily by CYP2D6 and of them antidepressants, antipsychotics and cardiovascular drugs are notable (Parkinson, 1996). Drug metabolism is also affected by another form of CYP enzyme known as CYP2C19 which has been studied far less than those of CYP2D6 and CYP2C9 with respect to its polymorphic effects on the clinical implications (Parkinson, 1996). Demethylation of diazepam in humans is carried out by CYP2C19 (Anderson *et al.*, 1990). The disposition of diazepam has been studied in 13 Caucasians of the extensive metabolizers (persons with two normally functioning alleles that cause efficient metabolism of probe drug) phenotype and three Caucasians of the poor metabolizers (persons who metabolize a probe-drug at a slower rate than others) (Bertilsson *et al.*, 1989). The plasma clearance of diazepam in the extensive metabolizers was more than 2 times that in the poor metabolizers (11.0 and 5.0 mL/min, respectively), whereas the half life in the extensive metabolizers was shorter compared to the poor metabolizers (59 and 128 h, respectively). The difference in the plasma clearance appeared to be related to formation of the desmethyl metabolite. The incidences of rate of metabolism mediated by CYP2C19 also vary in populations with different racial origins. For example, due to variations in CYP2C19 encoding genes, approximately 2 to 6% of individuals in the Caucasian populations have been found to be poor metabolizer, while in the Asian population this percentage varies from 14 to 22% (Wilkinson *et al.*, 1992; Kalow and Bertilsson, 1994). On the other hand, the lack of functional CYP3A5 may not be readily noticeable, because many drugs metabolized equally by CYP3A5 and universally expressed CYP3A4. Drugs that are equally metabolized by these two enzymes is the net rate of metabolism due to CYP3A4 and CYP3A5. Consequently, clinical effects of genetic polymorphism of CYP3A5 are partially hindered due to the existence of this dual pathway but contribute largely to the activity of CYP3A in humans. The CYP3A pathway of drug elimination is further confounded by the presence of single-nucleotide polymorphisms in the *CYP3A4* gene that alter the activity of this enzyme for some substrates but not for others (Sata *et al.*, 2000). The genetic basis of CYP3A5 deficiency is predominantly a single-nucleotide polymorphism in intron 3 that creates a cryptic splice site causing 131 nucleotides of the intronic sequence to be inserted into the RNA, introducing a termination codon that prematurely truncates the CYP3A5 protein (Kuehl *et al.*, 2001).

2.4.4 ASSOCIATION OF GENE POLYMORPHISMS WITH DRUG TRANSPORTERS

Transport proteins are one of the important and determining factors directing the pharmacokinetics and pharmacodynamic profile of drugs as these proteins have an important role in regulating the absorption, distribution, and excretion of many medications. While the majority of transporter-related pharmacogenetic research has been in regards to classic genes encoding the outward-directed ATP-binding cassette (ABC) transporters, such as ABCB1 (P-glycoprotein), ABCC2 (MRP2), and ABCG2 (BCRP), more studies have been conducted in recent years evaluating genes encoding solute carriers (SLC) that mediate the cellular uptake of drugs, such as, SLCO1B1 (OATP1B1) and SLC22A1 (OCT1). The distribution of ABC and SLC transporters in tissues key to pharmacokinetics, such as intestine (absorption), blood-brain-barrier (distribution), liver (metabolism), and kidneys (excretion), strongly suggests that variations in these transporter encoding genes may have a substantial impact on systemic drug exposure and toxicity. Members of the adenosine triphosphate (ATP)-binding cassette (ABC) family of membrane transporters (Borst et al., 2000) are among the most extensively studied transporters involved in drug disposition and response. A member of the ABC family, P-glycoprotein, is encoded by the human MDR1 gene (ABCB1). It plays a significant role in ADME (absorption, distribution, metabolism, and excretion) processes (Gottesmann et al., 2002) and drug-drug interaction (Callen et al., 1987; Mizuno et al., 2003; Sun et al., 2004). It is of great clinical interest because of its broad substrate specificity, including a variety of structurally divergent drugs in clinical use. A principal function of P-glycoprotein is the energy-dependent cellular efflux of substrates, including bilirubin, several anticancer drugs, cardiac glycosides, immunosuppressive agents, glucocorticoids, human immunodeficiency virus (HIV) type 1 protease inhibitors, and many other medications. The expression of P-glycoprotein in many normal tissues suggests that it has a role in the excretion of xenobiotics and metabolites into urine, bile, and the intestinal lumen (Thiebaut et al., 1987; Rao et al., 1999). At the blood-brain barrier, P-glycoprotein in the choroid plexus limits the accumulation of many drugs in the brain, including digoxin, ivermectin, vinblastine, dexamethasone, cyclosporine, domperidone, and loperamide (Thiebaut et al., 1987; Schinkel et al., 1996; Rao et al., 1999).

The MDR1 gene is located on the long arm of 7th chromosome at q21.1 and consists of 28 exons, which encodes for polypeptide of 1280 amino acids (Chen *et al.*, 1990). MDR1 genes are highly polymorphic and till to date 50 single nucleotide polymorphisms have been identified (Chinn and Kroetz, 2007). Genetic polymorphisms of the MDR1 gene have been reported to be associated with alteration in disposition kinetics and interaction profiles of various drugs including statins (atravostatin, stevastatin, simvastatin, mevas-tatin) (Kajinami *et al.*, 2004). Genetic polymorphisms were first identified by Kioka *et al.* (1989) from *in vitro* studies with cancer cells. Subsequently, other groups, including Hoffmeyer *et al.* (2000) have screened the entire MDR1 coding region (Kioka *et al.*, 1989; Gerloff *et al.*, 2002; Goto *et al.*, 2002; Siegmund *et al.*, 2002; Yamauchi *et al.*, 2002). A synonymous SNP (i.e., a single-nucleotide polymorphism that does not alter the amino acid encoded) in exon 26 (3435C→T) has been associated with variable expression of P-gly-coprotein in the duodenum; in patients homozygous for the T allele, duodenal expression of P-glycoprotein was less than half that in patients with the CC genotype (Hoffmeyer *et al.*, 2000). CD56+ natural killer cells from subjects homozygous for 3435C demonstrated significantly lower retention of the P-glycoprotein substrate rhodamine (i.e., higher P-glycoprotein function) (Hitzl *et al.*, 2001). Digoxin, another P-glycoprotein substrate, has significantly higher bioavailability in subjects with the 3435TT genotype (Hoffmeyer *et al.*, 2000; Sakaeda *et al.*, 2001). As is typical for many pharmacogenetic traits, there is considerable racial variation in the frequency of the 3435C→T sin-gle-nucleotide polymorphism (Ameyaw *et al.*, 2001; Schaeffeler *et al.*, 2001; McLeod, 2002). The 3435C→T single-nucleotide polymorphism is in linkage disequilibrium with a nonsynonymous single-nucleotide polymorphism (i.e., one causing an amino acid change) in exon 21 (2677G→T, leading to Ala-893Ser) that alters P-glycoprotein function (Kim *et al.*, 2001). Because these two single-nucleotide polymorphisms travel together, it is unclear whether the 3435C→T polymorphism is of functional importance or is simply linked with the causative polymorphism in exon 21. The 2677G→T single-nucleotide polymorphism has been associated with enhanced P-glycoprotein function *in vitro* and lower plasma fexofenadine concentrations in humans (Kim *et al.*, 2001), effects opposite to those reported with digoxin (Sakaeda *et al.*, 2001). Also, pharmacogenomic studies have reported three main SNPs C3435T (exon 26, synonymous), G2677T/A (exon 21, nonsynonymous) and C1236T (exon 16, synonymous), responsible for the reduced or decreased bioavail-ability of statins (Bercovich *et al.*, 2006).

Members of the organic anion transporting polypeptide (OATP/SLCO) family are responsible for the sodium-independent transport cellular uptake of a broad range of structurally diverse endogenous compounds and xenobiotics in multiple tissues including the gastrointestinal tract, liver, kidney, lung, brain, testis, placenta and ciliary body. Rat Oatp1a1, the first cloned member of the OATP/Oatp gene family, was first expressed in liver and kidney (Jacquemin *et al.*, 1994) and since then a total of 52 members comprising 12 families have been identified across eight species (Konig *et al.*, 2000). OATP1A2 was the first human family member described (Kullak-Ublick *et al.*, 1995), followed by the discovery of the second family member OATP1B1 (formerly termed OATP2 or OATP-C). The OATP/Oatp family consists of 20 different members, 11 of them belonging to the human OATP family, including 10 OATPs and the prostaglandin transporter OATP2A1 (formerly termed as PGT). All OATP proteins have several amino acids highly conserved throughout the family and they share a very similar transmembrane organization. They have 12 predicted transmembrane helices and a large fifth extracellular loop (Hagenbuch *et al.*, 2003). This loop contains many conserved cysteine residues that resemble the zinc-finger domains of transcription factors (Hakes *et al.*, 1991). Additional conserved attributes are the OATP family signature at the border between extracellular loop 3 and transmembrane helix 6, as well as conserved *N*-glycozylation sites in the extracellular loops 2 and 5. Highly conserved amino acids are preferentially found in transmembrane helices 2, 3, 4, 5 and 6, in intracellular loops 1, 2, 4 and 5 and in extracellular loops 1, 3 and 5 (Hagenbuch *et al.*, 2003).

2.4.5 ASSOCIATION OF DRUG RESPONSE WITH VARIATIONS IN GENES EXPRESSING DRUG TARGETS

Identifying protein targets of bioactive compounds is an effective approach to discover unknown protein functions, identify molecular mechanisms of drug action, and obtain information for optimization of lead compounds. Genetic variation in drug targets has profound association with drug efficacy and toxicity. In the literature, there are now various examples of relationships between drug target polymorphisms and drug effect. For example, sequence variants with a direct effect on response occur in the gene for the β_2-adrenoreceptor, affecting the response to β_2-agonists (Liggett, 2000; Dishy *et al.*, 2001). Drug targets can be broken into three main categorized as the direct protein target of the drug, signal transduction cascades or downstream proteins and disease pathogenesis proteins. Protein kinases are one of the most investigated classes of drug targets (Daub *et al.*, 2004). Imatinib (Gleevec), a tyrosine kinase inhibitor, is such a drug approved by FDA to treat certain cancers like Chronic myelogenous leukemia, Gastrointestinal stromal tumors (Druker *et al.*, 1996).

Signal transduction cascades are mediated by hormones, growth factors, neurotransmitters, and cytokines to exhibit a specific internal cellular response like gene expression, cell division, or even cell suicide begins at the cell membrane where an external stimulus initiates a cascade of enzymatic reactions inside the cell that typically includes phosphorylation of proteins as mediators of downstream processes. The p38 MAPK regulates intracellular signaling pathways in response to environmental stress. Its inhibition blocks the production of inflammatory cytokines, such as IL-1β and TNF-α, and inhibitors are effective in animal models of arthritis and bone resorption. Numbers of Pharmaceutical companies are using a rational design approach to create novel p38 MAP kinase inhibitors that will have antiinflammatory activity.

Genetic differences may also have indirect effects on drug response that are unrelated to drug metabolism or transport, such as methylation of the methylguanine methyltransferase (MGMT) gene promoter, which alters the response of gliomas to treatment with carmustine (Esteller *et al.*, 2000). The mechanism of this effect is related to a decrease in the efficiency of repair of alkylated DNA in patients with methylated O(6)-methylguanine-DNA methyltransferase (MGMT). It is critical to distinguish this target mechanism from genetic polymorphisms in drug-metabolizing enzymes that affect response by altering drug concentrations, such as the thiopurine methyltransferase polymorphism associated with the hematopoietic toxicity of mercaptopurine (Black *et al.*, 1998; Relling *et al.*, 1999a; Evans *et al.*, 2001) and susceptibility to radiation-induced brain tumors (Relling *et al.*, 1999b).

The link between genetic polymorphisms in drug targets and clinical responses has also been demonstrated by the β_2-adrenoreceptor (coded by the *ADRB2* gene). Genetic polymorphism of the β_2-adrenoreceptor can alter the process of signal transduction by these receptors (Liggett, 2000; Dishy *et al.*, 2001). Three SNPs have been detected in *ADRB2*, which have been found to be associated with altered expression, down-regulation, or coupling of the receptor in response to β_2-adrenoreceptor agonists (Liggett, 2000). Polymorphisms resulting in an Arg-to-Gly amino acid change at codon 16 and a Gln-to-Glu change at codon 27 are relatively common, with allele frequencies of 0.4 to 0.6, and are under intensive investigation for their clinical relevance. An agonist-mediated vasodilatation and desensitization study revealed that patients who were homozygous for Arg at *ADRB2* codon 16 and patients who were homozygous for Gly at *ADRB2* codon 16 showed completely different response when treated with isoproterenol (Dishy *et al.*, 2001). Homozygotic patients with Arg at codon 16 had nearly complete desensitization after continuous infusion of isoproterenol, with venodilatation decreasing from 44% at

base line to 8% after 90 minutes of infusion. On the contrary, patients homozygous for Gly at codon 16 had no significant change in venodilatation, regardless of their codon 27 status. Polymorphism at codon 27 was also of functional relevance; subjects homozygous for the Glu allele had higher maximal
venodilatation in response to isoproterenol than those with the codon 27 Gln
genotype, regardless of their codon 16 status (Dishy *et al.*, 2001).

At least 13 distinct single-nucleotide polymorphisms have been identified in *ADRB2* (Drysdale *et al.*, 2000). This finding has led to evaluation of
the importance of haplotype structure as compared with individual single-
nucleotide polymorphisms in determining receptor function and pharmaco-
logic response. Among 77 white, black, Asian, and Hispanic subjects, only
12 distinct haplotypes of the 8192 possible *ADRB2* haplotypes were actually observed (Drysdale *et al.*, 2000). The bronchodilator response to inhaled
β-agonist therapy in patients with asthma revealed a stronger association between bronchodilator response and haplotype than between bronchodilator response and any single-nucleotide polymorphism alone (Drysdale *et al.*, 2000).
This is not surprising, because haplotype structure is often a better predictor
of phenotypic consequences than are individual polymorphisms. This result
suggests that it would be desirable to develop simple but robust molecular
methods to determine the haplotype structure of patients (McDonald *et al.*,
2002). Neurotransmitter receptors are involved in the efficacy and side effects
of antipsychotics. A considerable number of pharmacogenetic studies have
been performed to define the association of antipsychotic medication response
with dopamine receptor polymorphisms. In the dopamine D2, subjects with
−141C Ins allele in −141C Ins/Del polymorphism and subjects with A1 allele
in Taq1A have shown better response to dopamine antagonists. Association of
Ser allele with typical antipsychotics and of Gly allele with atypical antipsychotics has also been extensively investigated in case of D3 receptor (Saito
et al., 2005). A pattern of association is seen between the Ser9 allele and a
response to typical antipsychotics and between the Gly9 allele and a response
to atypical antipsychotics has been revealed. The Gly9-allele of the dopamine
D3 receptor has been found to be associated with the response to clozapine
(Wilffert *et al.*, 2005). For the D4 receptor, no convincing association results
have been reported to date.

2.5   PERSONALIZED MEDICINE IN DEVELOPING COUNTRIES

Genomics, postgenomics and biotechnology related diagnoses as well as
medical treatments have been familiarized in developing countries with few

scattered research groups involved in such works. Whereas 90% of the total health dollars are used up for 10% of the total population (Global Forum for Health Research) and developing countries currently account for the 80% of the world's population (Cohen, 2006), isolated research activities and efforts would not create equal benefits for the health care receivers. Moreover, to avoid being set aside as passive benefiters, it will be important for developing countries to develop their own local research and development capacities. However, access to the specific problem-based applied education programs relevant to the reality of developing countries will be important determinants in this respect.

The World Health Organization (WHO) released a report entitled "Genomics and World Health" by emphasizing the potential of genomics to improve global health. It identifies limited resources devoted to health research in developing countries and that there is an urgent need to focus attention on the most promising technologies. The report recommended that WHO "*should develop the capacity to evaluate advances in genomics, to anticipate their potential for research and clinical application... and to assess their effectiveness and cost in comparison to current practice*" (WHO, 2002). To address the recommendation by WHO, the first step is to identify priority technologies. Daar *et al*. (2002) have identified the 10 most promising biotechnologies for improving health in developing countries using a study that began with an open-ended question "*What do you think are the major biotechnologies that can help improve health in developing countries in the next 5 to 10 years?*" provided to the panelists comprising of 28 scientists (who are at the forefront in their field) from all over the world, half of them were from the developing countries. To assess the importance of the technologies, the panelists took several factors into their consideration that include impact, appropriateness, feasibility, burden, knowledge gap and indirect benefits. On the basis of these considerations, using Delphi Method to achieve consensus, Daar *et al*. (2002) scored top priority technologies and concluded that biotechnology, especially molecular diagnostics (rated first), can be made affordable for the developing world. However, to date most of the people of the developing countries are the receivers or passive benefiters of the most of the modern technologies built up in the developed worlds. But it is indeed the necessity of time to address problems of developing countries by their own, bringing these problems to the research field using own resources and capabilities. It can be initiated with the advantage of open innovation in the field of biotechnology (sometimes referred to as open biotechnology) (Jefferson, 2006) by involving sharing data, expertise and resources to promote collaboration, transparency and cumulative public knowledge (Masum and Harris, 2011). Open biotechnology is already producing significant results through open genomic database projects and open publications (Hrynaszkiewicz, 2011; Gitter, 2010). Complete personalized medicine textbooks and other educational resources relevant to developing countries could be made openly available at little cost via dedicated internet storehouses. Open genomic databases and bioinformatics research tools should present students in developing countries with unique low-cost training opportunities, as well as the opportunity to learn from and collaborate on common goals with scientists in other developed and developing countries (Joly, 2011).

2.6   CONCLUSION

For personalized medicine to become established it will be necessary for people's genotypes to be analyzed completely, which is not currently possible on a large scale in developing countries because of the great expense. The cost of performing such an analysis is coming down and it is expected that this type of test will soon be available for as little as $1000 (Hutchison, 2007) within the next decade. To popularize personal medicine, researchers, physicians, medical trainers and educators, geneticists, policy makers, patient advocates, clinical laboratories, pharmaceutical companies, diagnostics companies, information technology managers, payers, and government regulators should work together. Personalized medicine will be used to treat cases of diseases with specific approaches that will be effective for a given patient but not necessarily effective for another patient with similar height, weight and age. Personalized medicine would not only help to treat disease but also would help to prevent onset of specific diseases by reducing the cost and efforts of a particular individual.

Pharmacogenomics serves as an increasingly powerful tool in understanding interindividual variability in drug response and toxicity. Although pharmacogenomics continues to improve understanding of drug response, progress is gradual, with clinical implementation lagging far behind. Several obstacles need to be overcome for successful application of pharmacogenomics to drug therapy. Multiple processes contribute to the response to drugs and drug combinations, with drug-drug interactions leading to unexpected outcomes linked to polymorphic genes. Genetic analysis of overall drug response requires a systems analysis using medical informatics for integration of all relevant information. It is essential to understand the contribution of genetic factors to the target phenotype in quantitative terms. In addition to molecular genetic analysis of polymorphisms affecting protein primary structure, we propose the systematic use of allelic expression imbalance, for quantitative assessment of cis-acting factors in transcription and mRNA processing. We must assess the role of epigenetic factors, and of small regulatory RNAs, in determining interindividual variability. A regulatory framework is needed to assure that pharmacogenomic data are incorporated into drug development and postapproval surveillance. Because the impact of genetic and genomic data is still poorly understood, the FDA has implemented a 'safe haven policy' by, which pharmaceutical companies are encouraged to include genomic data for the New Drug Approval process without risking delays or other regulatory actions. In time, our knowledge will progress to a point where such data will become a cornerstone of the drug approval process. The inclusion of pharmacogenetic data (e.g., on CYP polymorphisms) on drug package inserts has already been implemented, providing genetic information accessible to patients and physicians alike. Our knowledge of genetic variations has been so profoundly influenced by Mendelian genetics that it is difficult to speculate about the ways in which our thinking will need to change with further insights into genomics. We have far to go in teasing apart the multiple variables of complex traits and diseases, the relationships between hereditary, somatic and environmental factors, and in making the transition from focusing on monogenic diseases with high penetrance to polygenic conditions with greatly varying degrees of penetrance.

KEYWORDS

- β_2-Adrenoreceptor
- Biomarkers
- CYP1A enzymes
- Cytochrome P450s
- DNA sequence profiling
- Drug metabolizing enzymes
- Drug response
- Gene polymorphism
- Haplotyping
- Metabolomics
- Nutrigenomics

- O(6)-methylguanine-DNA methyltransferase
- Personalized medicine
- Pharmacodynamics
- Pharmacogenetics
- Pharmacogenomics
- Pharmacoproteomics
- RNA expression profiling
- SNP testing
- Translational medicine

REFERENCES

Aklillu, E.; Persson, I.; Bertilsson, L.; Johansson, I.; Rodrigues, F.; Ingelman-Sundberg, M. Frequent distribution of ultrarapid metabolizers of debrisoquine in an ethiopian population carrying duplicated and multiduplicated functional CYP2D6 alleles. *J Pharmacol Exp Ther* 1996, 278, 441–446.

Allan, J. M.; Wild, C. P.; Rollinson, S.; Willett, E. V.; Moorman, A. V.; Dovey, G. J.; Roddam, P. L.; Roman, E.; Cartwright, R. A.; Morgan, G. J. Polymorphism in glutathione S-transferase P1 is associated with susceptibility to chemotherapy induced leukemia. *Proc Natl Acad Sci USA* 2001, 98, 11592–11597.

Alvan, G.; Bechtel, P.; Iselius, L.; Gundert-Remy, U. Hydroxylation polymorphisms of debrisoquine and mephenytoin in European populations. *Eur J Clin Pharmacol* 1990, 39, 533–537.

Ameyaw, M. M.; Regateiro, F.; Li, T.; Liu, X.; Tariq, M.; Mobarek, A.; Thornton, N.; Folayan, G. O.; Githanga, J.; Indalo, A.; Ofori-Adjei, D.; Price-Evans, D. A.; McLeod, H. L. MDR1 pharmacogenetics: frequency of the C3435T mutation in exon 26 is significantly influenced by ethnicity. *Pharmacogenetics* 2001, 11, 217–221.

Anderson, J. L.; Horne, B. D.; Stevens, S. M., Grove, A. S.; Barton, S.; Nicholas, Z. P.; Kahn, S. F.; May, H. T.; Samuelson, K. M.; Muhlestein, J. B.; Carlquist, J. F. Randomized trial of genotype-guided versus standard warfarin dosing in patients initiating oral anticoagulation. *Circulation* 2007, 116(22), 2563–2570.

Beeghly, A.; Katsaros, D.; Chen, H.; Fracchioli, S.; Zhang, Y.; Massobrio, M.; Risch, H.; Jones, B.; Yu, H. Glutathione S-transferase polymorphisms and ovarian cancer treatment and survival. *Gynecol Oncol* 2006, 100(2), 330–337.

Bercovich, D.; Friedlander, Y.; Korem, S.; Houminer, A.; Hoffman, A.; Kleinberg, L.; Shochat, C.; Leitersdorf, E.; Meiner, V. The association of common SNPs and haplotypes in the CETP and MDR1 genes with lipids response to fluvastatin in familial hypercholesterolemia. *Atherosclerosis* 2006, 185(1), 97–107.

Bertilsson, L.; Henthorn, T. K.; Sanz, A.; Tybring, G.; Sawe, J.; Villén, T. Importance of genetic factors in the regulation of diazepam metabolism: relationship to S-mephenytoin, but not debrisoquin hydroxylation phenotype. *Clin Pharmacol Ther* 1989, 45, 348–355.

Black, A. J.; McLeod, H. L.; Capell, H. A.; Powrie, R. H.; Matowe, L. K.; Pritchard, S. C.; Collie-Duguid, E. S.; Reid, D. M. Thiopurine methyltransferase genotype predicts therapy-limiting severe toxicity from azathioprine. *Ann Intern Med* 1998, 129, 716–718.

Board, P. G.; Weber, G. C.; Coggan, M. Isolation of a cDNA clone and localization of the human glutathione S-transferase 3 genes to chromosome bands 11q13 and 12q13–14. *Ann Hum Genet* 1989, 53, 205–213.

Borst, P.; Evers, R.; Kool, M.; Wijnholds, J. A family of drug transporters: the multidrug resistance-associated proteins. *J Natl Cancer Inst* 2000, 92, 1295–1302.

Bowater, R. P.; Wells, R. D. The intrinsically unstable life of DNA triplet repeats associated with human hereditary disorders. *Prog Nucleic Acid Res Mol Biol* 2001, 66, 159–202.

Bozkurt, O.; de Boer, A.; Grobbee, D. E.; Heerdink, E. R.; Burger, H.; Klungel, O. H. Pharmacogenetics of glucose-lowering drug treatment: a systematic review. *Mol Diagn Ther* 2007, 11, 291–302.

Broder, S.; Venter, J. C. Sequencing the entire genomes of free-living organisms: the foundation of pharmacology in the new millennium. *Annu Rev Pharmacol Toxicol* 2000, 40, 97–132.

Broly, F.; Meyer, U. A. Debrisoquine oxidation polymorphism: phenotypic consequences of a 3-base-pair deletion in exon 5 of the CYP2D6 gene. *Pharmacogenetics* 1993, 3, 123–130.

Callen, D. F.; Baker, E.; Simmers, R. N.; Seshadri, R.; Roninson, I. B. Localization of the human multiple drug resistance gene mdr1 to 7q21.1. *Human Genetics* 1987, 77, 142–144.

Chang, G. W.; Kam, P. C. The physiological and pharmacological role of CYP450 isoenzymes. *Anesthesia* 1999, 54, 42–50.

Chen, C. J.; Clark, D.; Ueda, K.; Pastan, I.; Gottesman, M. M.; Roninson, I. B. Genomic organization of the human multidrug resistance (MDR1) gene and origin of P-glycoproteins. *J Biol Chem* 1990, 265, 506–514.

Chinn, L. W.; Kroetz, D. L. ABCB1 pharmacogenetics: progress, pitfalls and promise. *Clin Pharmacol Ther* 2007, 81, 265–269.

Cohen, B. Urbanization in developing countries: current trends, future projections, and key challenges for sustainability. *Technology in Society* 2006, 28, 63–80.

Cohn, J. N.; Tognoni, G. A randomized trial of the angiotensin-receptor blocker valsartan in chronic heart failure. *N Engl J Med* 2001, 345, 1667–1675.

Conney, A. H. Pharmacological implications of microsomal enzyme induction. *Pharmacol Res* 1967, 19, 317–366.

Creveling, C. R.; Thakker, D. R. O-, N- and S-methyltransferase. In *Conjugation-deconjugation reactions in drug metabolism and toxicity*, Kauffman, F. C., Ed.; Springer-Verlag: Berlin, 1994, pp. 189–216.

Daar, S. A.; Thorsteinsdóttir, H.; Martin, D. K.; Smith, A. C.; Nast, S.; Singer, P. A. Top ten biotechnologies for improving health in developing countries. *Nature Genetics* 2002, 32, 229–232.

Daub, H.; Godl, K.; Brehmer, D.; Klebl, B.; Muller, G. Evaluation of kinase inhibitor selectivity by chemical proteomics. *Assay Drug Dev Technol* 2004, 2, 215–224.

Dishy, V.; Sofowora, G. G.; Xie, H. G.; Kim, R. B.; Byrne, D. W.; Stein, C. M.; Wood, A. J. The effect of common polymorphisms of the β_2-adrenergic receptor on agonist-mediated vascular desensitization. *N Engl J Med* 2001, 345, 1030–1035.

Druker, B. J.; Tamura, S.; Buchdunger, E.; Ohno, S.; Segal, G. M.; Fanning, S.; Zimmermann, J.; Lydon, N. B. Effects of a selective inhibitor of the Abl tyrosine kinase on the growth of Bcr-Abl positive cells. *Nat Med* 1996, 2, 561–566.

Drysdale, C. M.; McGraw, D. W.; Stack, C. B.; Stephens, J. C.; Judson, R. S.; Nandabalan, K.; Arnold, K.; Ruano, G.; Liggett, S. B. Complex promoter and coding region beta 2-adrenergic receptor haplotypes alter receptor expression and predict *in vivo* responsiveness. *Proc Natl Acad Sci USA* 2000, 97, 10483–10488.

El-Sohemy, A. Nutrigenetics. *Forum Nutr.* 2007, 60, 25–30.

Esteller, M.; Garcia-Foncillas, J.; Andion, E.; Goodman, S. N.; Hidalgo, O. F.; Vanaclocha, V.; Baylin, S. B.; Herman, J. G. Inactivation of the DNA-repair gene MGMT and the clinical response of gliomas to alkylating agents. *N Engl J Med* 2000, 343, 1350–1354.

Esterbauer, H.; Schneitler, C.; Oberkofler, H.; Ebenbichler, C.; Paulweber, B.; Sandhofer, F.; Ladurner, G.; Hell, E.; Strosberg, A. D.; Patsch, J. R.; Krempler, F.; Patsch, W. A common polymorphism in the promoter of UCP2 is associated with decreased risk of obesity in middle-aged humans. *Nature Genet* 2001, 28, 178–183.

Evans, D. A. P. N-acetylesterase. In *Pharmacogenetics of drug metabolism*, Kalow, W., Ed.; Pergamon Press: New York, 1992, pp 95–178.

Evans, D. A. P.; Manley, K. A.; McKusick, V. A. Genetic control of isoniazid metabolism in man. *Br Med J* 1960, 2, 485–491.

Evans, W. E.; Hon, Y. Y.; Bomgaars, L.; Coutre, S.; Holdsworth, M.; Janco, R.; Kalwinsky, D.; Keller, F.; Khatib, Z.; Margolin, J.; Murray, J.; Quinn, J.; Ravindranath, Y.; Ritchey, K.; Roberts, W.; Rogers, Z. R.; Schiff, D.; Steuber, C.; Tucci, F.; Kornegay, N.; Krynetski, E. Y.; Relling, M. V. Preponderance of thiopurine S-methyltransferase deficiency and heterozygosity among patients intolerant to mercaptopurine or azathioprine. *J Clin Oncol* 2001, 19, 2293–2301.

Evans.; W. E.; Johnson, J. A. Pharmacogenomics: the inherited basis for interindividual differences in drug response. *Annu Rev Genomics Hum Genet* 2001, 2, 9–39.

Evans, W. E.; McLeod, H. L. Pharmacogenomics – drug disposition, drug targets and side effects. *N Engl J Med* 2003, 348, 538–549.

Evans, W. E.; Relling, M. V. Pharmacogenomics: translating functional genomics into rational therapeutics. *Science* 1999, 286, 487–491.

Garte, S.; Gaspari, L.; Alexandrie, A. K.; Ambrosone, C.; Autrup, H.; Autrup, J. L.; Baranova, H.; Bathum, L.; Benhamou, S.; Boffetta, P.; Bouchardy, C.; Breskvar, K.; Brockmoller, J.; Cascorbi, I.; Clapper, M. L.; Coutelle, C.; Daly, A.; Dellomo, M.; Dolzan, V.; Dresler, C. M.; Fryer, A.; Haugen, A.; Hein, D. W.; Hildesheim, A.; Hirvonen, A.; Hsieh, L. L.; Ingelman-Sundberg, M.; Kalina, I.; Kang, D.; Kihara, M.; Kiyohara, C.; Kremers, P.; Lazarus, P.; Le Marchand, L.; Lechner, M. C.; van Lieshout, E. M.; London, S.; Manni, J. J.; Maugard, C. M.; Morita, S.; Nazar-Stewart, V.; Noda, K.; Oda, Y.; Parl, F. F.; Pastorelli, R.; Persson, I.; Peters, W. H.; Rannug, A.; Rebbeck, T.; Risch, A.; Roelandt, L.; Romkes, M.; Ryberg, D.; Salagovic, J.; Schoket, B.; Seidegard, J.; Shields, P. G.; Sim, E.; Sinnet, D.; Strange, R. C.; Stücker, I.; Sugimura, H.; To-Figueras, J.; Vineis, P.; Yu, M. C.; Taioli, E. Metabolic gene polymorphism frequencies in control populations. *Cancer Epidemiol Biomarkers Prev* 2001, 10(12), 1239–1248.

Gerloff, T.; Schaefer, M.; Johne, A.; Oselin, K.; Meisel, C.; Cascorbi, I.; Roots, I. MDR1 genotypes do not influence the absorption of a single oral dose of 1 mg digoxin in healthy white males. *Br J Clin Pharmacol* 2002, 54, 610–616.

Gitter, D. M. The challenges of achieving open-source sharing of biobank data. *Biotechnology Law Report* 2010, 29(6), 623–635.

Goldstein, D. B.; Tate, S. K.; Sisodiya, S. M. Pharmacogenetics goes genomic. *Nat Rev Genet* 2003, 4, 937–947.

Goto, M.; Masuda, S.; Saito, H.; Uemoto, S.; Kiuchi, T.; Tanaka, K.; Inui, K. C3435T polymorphism in the MDR1 gene affects the enterocyte expression level of CYP3A4 rather than Pgp in recipients of living-donor liver transplantation. *Pharmacogenetics* 2002, 12, 451–457.

Goto, S.; Iida, T.; Cho, S.; Oka, M.; Kohno, S.; Kondo, T. Overexpression of glutathione S-transferase pi enhances the adduct formation of cisplatin with glutathione in human cancer cells. *Free Radic Res* 1999, 31(6), 549–558.

Gottesman, M. M. Mechanisms of cancer drug resistance. *Annu Rev Med* 2002, 53, 615–627.

Grant, S. F.; Reid, D. M.; Blake, G.; Herd, R.; Fogelman, I.; Ralston, S. H. Reduced bone density and osteoporosis associated with a polymorphic Sp1 binding site in the collagen type I alpha 1 gene. *Nature Genet* 1996, 14, 203–205.

Grant, S. F.; Steinlicht, S.; Nentwich, U.; Kern, R.; Burwinkel, B.; Tolle, R. SNP genotyping on a genome-wide amplified DOP-PCR template. *Nucleic Acids Research* 2002, 30, 22–25.

Guengerich, F. P. Cytochrome p450 and chemical toxicology. *Chem Res Toxicol* 2008, 21(1), 70–83.

Hagenbuch, B.; Meier, P. J. The superfamily of organic anion transporting polypeptides. *Biochim Biophys Acta – Biomembranes* 2003, 1609(1), 1–18.

Hakes, D.; Berezney, R. Molecular cloning of matrin F/G: a DNA binding protein of the nuclear matrix that contains putative zinc finger motifs. *Proc Natl Acad Sci USA* 1991, 88(14), 6186.

Hall, A. S.; Murray, G. D.; Ball, S. G. Follow-up study of patients randomly allocated ramipril or placebo for heart failure after acute myocardial infarction: AIRE Extension (AIREX) study, acute infarction ramipril efficacy. *Lancet* 1997, 349, 1493–1497.

Hallberg, P.; Karlsson, J.; Kurland, L.; Lind, L.; Kahan, T.; Malmqvist, K.; Ohman, K. P.; Nystrom, F.; Melhus, H. The CYP2C9 genotype predicts the blood pressure response to irbesartan: results from the Swedish irbesartan left ventricular hypertrophy investigation vs. atenolol (SILVHIA) trial. *J. Hypertens* 2002, 20, 2089–2093.

Harpole, D. H. Jr.; Moore, M. B.; Herndon, J. E. I. I.; Aloia, T.; D'Amico, T. A.; Sporn, T. The prognostic value of molecular marker analysis in patients treated with trimodality therapy for esophageal cancer. *Clin Cancer Res* 2001, 7(3), 562–569.

Higashi, M. K.; Veenstra, D. L.; Kondo, L. M.; Wittkowsky, A. K.; Srinouanprachanh, S. L.; Farin, F. M.; Rettie, A. E. Association between CYP2C9 genetic variants and anticoagulation-related outcomes during warfarin therapy. *JAMA* 2002, 287(13), 1690–1698.

Hitzl, M.; Drescher, S.; van der Kuip, H.; Schäffeler, E.; Fischer, J.; Schwab, M.; Eichelbaum, M.; Fromm, M. F. The C3435T mutation in the human MDR1 gene is associated with altered efflux of the P-glycoprotein substrate rhodamine 123 from CD56+ natural killer cells. *Pharmacogenetics* 2001, 11, 293–298.

Hinoshita, E.; Uchiumi, T.; Taguchi, K.; Kinukawa, N.; Tsuneyoshi, M.; Maehara, Y.; Sugimachi, K.; Kuwano, M. Increased expression of an ATP-binding cassette superfamily transporter, multidrug resistance protein 2, in human colorectal carcinomas. *Clin Cancer Res* 2000, 6, 2401–2407.

Hjalmarson, A.; Goldstein, S.; Fagerberg, B.; Wedel, H.; Waagstein, F.; Kjekshus, J.; Wikstrand, J.; El Allaf, D.; Vítovec, J.; Aldershvile, J.; Halinen, M.; Dietz, R.; Neuhaus, K. L.; Jánosi, A.; Thorgeirsson, G.; Dunselman, P. H.; Gullestad, L.; Kuch, J.; Herlitz, J.; Rickenbacher, P.; Ball, S.; Gottlieb, S.; Deedwania, P. Effects of controlled-release metoprolol on total mortality, hospitalizations, and well-being in patients with heart failure: the Metoprolol CR/XL Randomized Intervention Trial in congestive heart failure (MERIT-HF). *JAMA* 2000, 283, 1295–1302.

Hoffmeyer, S.; Burk, O.; von Richter, O.; Arnold, H. P.; Brockmöller, J.; Johne, A.; Cascorbi, I.; Gerloff, T.; Roots, I.; Eichelbaum, M.; Brinkmann, U. Functional polymorphisms of the human multidrug-resistance gene: multiple sequence variations and correlation of one allele with P-glycoprotein expression and activity *in vivo*. *Proc Natl Acad Sci USA* 2000, 97, 3473–3478.

Hrynaszkiewicz, I. The need and drive for open data in biomedical publishing. *Serials* 2011, 24(1), 31–37.

Hugot, J. P.; Chamaillard, M.; Zouali, H.; Lesage, S.; Cézard, J. P.; Belaiche, J.; Almer, S.; Tysk, C.; O'Morain, C. A.; Gassull, M.; Binder, V.; Finkel, Y.; Cortot, A.; Modigliani, R.; Laurent-Puig, P.; Gower-Rousseau, C.; Macry, J.; Colombel, J. F.; Sahbatou, M.; Thomas, G. Association of NOD2 leucine-rich repeat variants with susceptibility to Crohn's disease. *Nature* 2001, 411, 599–603.

Hunt, S. A.; Baker, D. W.; Chin, M. H.; Cinquegrani, M. P.; Feldman, A. M.; Francis, G. S.; Ganiats, T. G.; Goldstein, S.; Gregoratos, G.; Jessup, M. L.; Noble, R. J.; Packer, M.; Silver, M. A.; Stevenson, L. W.; Gibbons, R. J.; Antman, E. M.; Alpert, J. S.; Faxon, D. P.; Fuster, V.; Gregoratos, G.; Jacobs, A. K.; Hiratzka, L. F.; Russell, R. O.; Smith, S. C. ACC/AHA guidelines for the evaluation and management of chronic heart failure in the adult: executive summary. *J Heart Lung Transplant* 2002, 21, 189–203.

Hutchison, C. A. DNA sequencing: bench to bedside and beyond. *Nucleic Acids Res* 2007, 35, 6227–6237.

In't Veld, P.; Lievens, D.; De Grijse, J.; Ling, Z.; Van der Auwera, B.; Pipeleers-Marichal, M.; Gorus, F.; Pipeleers, D. Screening for insulitis in adult autoantibody-positive organ donors. *Diabetes* 2007, 56, 2400–2404.

Ioannides, C. Effect of diet and nutrition on the expression of cytochrome P450. *Xenobiotica* 1999, 29, 109–154.

Jacquemin, E.; Hagenbuch, B.; Stieger, B.; Wolkoff, A. W.; Meier P. J. Expression cloning of a rat liver Na(+)-independent organic anion transporter. *Proc Natl Acad Sci USA* 1994, 91(1), 133–137.

Jefferson, R. Science as social enterprise: the CAMBIA BiOS initiative. *Innovations* 2006, 1(4), 13–44.

Johnson, J. A. Pharmacogenetics: potential for individualized drug therapy through genetics. *Trends Genet* 2003, 19, 660–666.

Joly, Y. Personalized medicine in developing countries: a roadmap to personalized innovation. *Current Pharmacogenomics and Personalized Medicine* 2011, 9, 156–158.

Kajinami, K.; Brousseau, M. E.; Ordovas, J. M.; Schaefer, E. J. *Am. J. Cardiol.* 2004, 93(8), 1046–1050.

Kalow, W. Familial incidence of low pseudocholinesterase level. *Lancet* 1956, 2, 576.

Kalow, W.; Bertilsson, L. Interethnic factors affecting drug response. In *Advances in Drug Research*, Testa, B., Meyer, U. A., Eds.; Academic Press: New York, 1994, pp 1–53.

Kalow, W.; Tang, B. K.; Endrenyi, I. Hypothesis: comparisons of inter and intraindividual variations can substitute for twin studies in drug research. *Pharmacogenetics* 1998, 8, 283–289.

Kaput, J.; Dawson, K. Complexity of type 2 diabetes mellitus datasets emerging from nutrigenomic research: a case for dimensionality reduction? *Mutat Res* 2007, 622, 19–32.

Kim, R. B.; Leake, B. F.; Choo, E. F.; Dresser, G. K.; Kubba, S. V.; Schwarz, U. I.; Taylor, A.; Xie, H. G.; McKinsey, J.; Zhou, S.; Lan, L. B.; Schuetz, J. D.; Schuetz, E. G.; Wilkinson, G.R. Identification of functionally variant MDR1 alleles among European Americans and African Americans. *Clinical Pharmacology and Therapeutics* 2001, 70(2), 189–199.

Kioka, N.; Tsubota, J.; Kakehi, Y.; Komano, T.; Gottesman, M. M.; Pastan, I.; Ueda, K. P-glycoprotein gene (MDR1) cDNA from human adrenal: normal P-glycoprotein carries Gly185 with an altered pattern of multidrug resistance. *Biochem Biophys Res Commun* 1989, 162, 224–231.

Klein, T. E.; Chang, J. T.; Cho, M. K.; Easton, K. L.; Fergerson, R.; Hewett, M.; Lin, Z.; Liu, Y.; Liu, S.; Oliver, D. E.; Rubin, D. L.; Shafa, F.; Stuart, J. M.; Altman, R. B. Integrating genotype and phenotype information: an overview of the PharmGKB project. Pharmacogenetics Research Network and Knowledge Base. *Pharmacogenomics J* 2001, 1, 167–170.

Konig, J.; Cui, Y.; Nies, A. T.; Keppler, D. Localization and genomic organization of a new hepatocellular organic anion transporting polypeptide. *J Biol Chem* 2000, 275(30), 23161–23168.

Kuehl, P.; Zhang, J.; Lin, Y.; Lamba, J.; Assem, M.; Schuetz, J.; Watkins, P. B.; Daly, A.; Wrighton, S. A.; Hall, S. D.; Maurel, P.; Relling, M.; Brimer, C.; Yasuda, K.; Venkataramanan, R.; Strom, S.; Thummel, K.; Boguski, M. S.; Schuetz, E. Sequence diversity in CYP3A promoters and characterization of the genetic basis of polymorphic CYP3A5 expression. *Nat Genet* 2001, 27, 383–391.

Kullak-Ublick, G. A.; Hagenbuch, B.; Stieger, B.; Schteingart, C. D.; Hofmann, A. F.; Wolkoff, A. W.; Meier, P. J. Molecular and functional characterization of an organic anion transporting polypeptide cloned from human liver. *Gastroenterology* 1995, 109(4), 1274–1282.

Lechat, P.; Escolano, S.; Golmard, J. L.; Lardoux, H.; Witchitz, S.; Henneman, J. A.; Maisch, B.; Hetzel, M.; Jaillon, P.; Boissel, J. P.; Mallet, A. Prognostic value of bisoprolol-induced hemodynamic effects in heart failure during the cardiac insufficiency bisoprolol study (CIBIS). *Circulation* 1997, 96, 2197–2205.

Lee, C. R.; Pieper, J. A.; Hinderliter, A. L.; Blaisdell, J. A.; Goldstein, J. A. Losartan and E3174 pharmacokinetics in cytochrome P450–2C9*1/*1, *1/*2, and *1/*3 individuals. *Pharmacotherapy* 2003, 23, 720–725.

Liggett, S. B. Beta (2)-adrenergic receptor pharmacogenetics. *Am J Respir Crit Care Med* 2000, 161, S197-S201.

Mahgoub, A.; Idle, J. R.; Dring, L. G.; Lancaster, R.; Smith, R. L. Polymorphic hydroxylation of debrisoquine in man. *Lancet* 1977, 2, 584–586.

Martin, E. R.; Lai, E. H.; Gilbert, J. R.; Rogala, A. R.; Afshari, A. J.; Riley, J.; Finch, K. L.; Stevens, J. F.; Livak, K. J.; Slotterbeck, B. D.; Slifer, S. H.; Warren, L. L.; Conneally, P. M.; Schmechel, D. E.; Purvis, I.; Pericak-Vance, M. A.; Roses, A. D.; Vance, J. M. SNPing away at complex diseases: analysis of single-nucleotide polymorphisms around APOE in Alzheimer disease. *Am J Hum Genet* 2000, 67, 383–394.

Masum, H.; Harris, R. *Open source for neglected disease: magic bullet or mirage?* Results for Development Institute: Washington D.C., 2011.

McDonald, O. G.; Krynetski, E. Y.; Evans, W. E. Molecular haplotyping of genomic DNA for multiple single-nucleotide polymorphisms located kilobases apart using long-range polymerase chain reaction and intramolecular ligation. *Pharmacogenetics* 2002, 12, 93–99.

McLeod, H. Pharmacokinetic differences between ethnic groups. *Lancet* 2002, 359, 78.

McLeod, H. L.; Evans, W. E. Pharmacogenomics: unlocking the human genome for better drug therapy. *Annu Rev Pharmacol Toxicol* 2001, 41, 101–121.

Meyer, U. A. The molecular basis of genetic polymorphisms of drug metabolism. *J Pharm Pharmacol* 1994, 46(S1), 409–415.

Meyer, U. A. Pharmacogenetics and adverse drug reactions. *Lancet* 2000, 356, 1667–1671.

Meyer, U. A.; Skoda, R. C.; Zanger, U. M.; Heim, M.; Broly, F. The genetic polymorphism of debrisoquine/sparteine metabolism: molecular mechanisms. In *Pharmacogenetics of Drug Metabolism*, Kallow, W., Ed.; Pergamon Press: New York, 1992, pp 609–623.

Meyer, U. A.; Zanger, U. M.; Grant, D.; Blum, M. Genetic polymorphisms of drug metabolism. *Adv Drug Res* 1990, 19, 197–241.

Mizuno, N.; Niwa, T.; Yotsumoto, Y.; Sugiyama, Y. Impact of drug transporter studies on drug discovery and development. *Pharmacological Reviews* 2003, 55, 425–461.

Mojiminiyi, O. A.; Abdella, N. A. Associations of resist in with inflammation and insulin resistance in patients with type 2 diabetes mellitus. *Scand J Clin Lab Invest* 2007, 67, 215–225.

Motulsky, A. G. Drug reactions, enzymes and biochemical genetics. *J Am Med Assoc* 1957, 165, 835–837.

O'Dwyer, M. E.; Druker, B. J. STI571: an inhibitor of the bcr-abl tyrosine kinase for the treatment of chronic myelogenous leukemia. *Lancet Oncol* 2000, 1, 207–211.

Oguri, T.; Fujiwara, Y.; Katoh, O.; Daga, H.; Ishikawa, N.; Fujitaka, K.; Yamasaki, M.; Yokozaki, M.; Isobe, T.; Ishioka, S.; Yamakido, M. Glutathione S transferase-pi gene expression and platinum drug exposure in human lung cancer. *Cancer Lett* 2000, 156, 93–99.

Oldham, H. G.; Clarke, S. E. *In vitro* identification of the human cytochrome P450 enzymes involved in the metabolism of R(+)- and S(−)-carvedilol. *Drug Metab Dispos* 1997, 25, 970–977.

Overdevest, J. B.; Theodorescu, D.; Lee, J. K. Utilizing the molecular gateway: the path to personalized cancer management. *Clin Chem* 2009, 55, 684–697.

Packer, M.; Fowler, M. B.; Roecker, E. B.; Coats, A. J.; Katus, H. A.; Krum, H.; Mohacsi, P.; Rouleau, J. L.; Tendera, M.; Staiger, C.; Holcslaw, T. L.; Amann-Zalan, I.; DeMets, D. L.

Effect of carvedilol on the morbidity of patients with severe chronic heart failure: results of the carvedilol prospective randomized cumulative survival (COPERNICUS) study. *Circulation* 2002, 106, 2194–2199.

Parkinson, A. An overview of current cytochrome P-450 technology for assessing the safety and efficacy of new materials. *Toxicol Pathol* 1996, 24, 45–57.

Peltomaki, P. Deficient DNA mismatch repair: a common etiologic factor for colon cancer. Deficient DNA mismatch repair: a common etiologic factor for colon cancer. *Hum Mol Genet* 2001, 10, 735–740.

Pepper, J. M.; Lennard, M. S.; Tucker, G. T.; Woods, H. F. Effect of steroids on the cytochrome P4502D6-catalyzed metabolism of metoprolol metabolism persists during long-term treatment. *Pharmacogenetics* 1991, 1, 119–122.

Petrasek, D. Systems biology: the case for a systems science approach to diabetes. *J Diabetes Sci Technol* 2008, 2, 131–134.

Pitt, B.; Zannad, F.; Remme, W. J.; Cody, R.; Castaigne, A.; Perez, A.; Palensky, J.; Wittes, J. The effect of spironolactone on morbidity and mortality in patients with severe heart failure. *N Engl J Med* 1999, 341, 709–717.

Rao, V. V.; Dahlheimer, J. L.; Bardgett, M. E.; Snyder, A. Z.; Finch, R. A.; Sartorelli, A. C.; Piwnica-Worms, D. Choroid plexus epithelial expression of MDR1 P glycoprotein and multidrug resistance-associated protein contribute to the blood-cerebrospinal-fluid drug-permeability barrier. *Proc Natl Acad Sci USA* 1999, 96, 3900–3905.

Rau, T.; Heide, R.; Bergmann, K.; Wuttke, H.; Werner, U.; Feifel, N.; Eschenhagen, T. Effect of the CYP2D6 genotype on metoprolol. *Pharmacogenetics* 2002, 12, 465–472.

Relling, M. V.; Hancock, M. L.; Rivera, G. K.; Sandlund, J. T.; Ribeiro, R. C.; Krynetski, E. Y.; Pui, C. H.; Evans, W. E. Mercaptopurine therapy intolerance and heterozygosity at the thiopurine S-methyltransferase gene locus. *J Natl Cancer Inst* 1999a, 91, 2001–2008.

Relling, M. V.; Rubnitz, J. E.; Rivera, G. K.; Boyett, J. M.; Hancock, M. L.; Felix, C. A.; Kun, L. E.; Walter, A. W.; Evans, W. E.; Pui, C. H. High incidence of secondary brain tumors after radiotherapy and antimetabolites. *Lancet* 1999b, 354, 34–39.

Rich, S. McLaughlin, V. V. Endothelin receptor blockers in cardiovascular disease. *Circulation* 2003, 108, 2184–2190.

Ross, J. S.; Ginsburg, G. S. Integrating diagnostics and therapeutics: revolutionizing drug discovery and patient care. *Drug Discov Today* 2002, 7, 859–864.

Rudin, C. M.; Yang, Z.; Schumaker, L. M.; Vander Weele, D. J.; Newkirk, K.; Egorin, M. J.; Zuhowski, E. G.; Cullen, K. J. Inhibition of glutathione synthesis reverses Bcl-2-mediated cisplatin resistance. *Cancer Res* 2003, 63(2), 312–318.

Sachidanandam, R.; Weissman, D.; Schmidt, S. C.; Kakol, J. M.; Stein, L. D.; Marth, G.; Sherry, S.; Mullikin, J. C.; Mortimore, B. J.; Willey, D. L.; Hunt, S. E.; Cole, C. G.; Coggill, P. C.; Rice, C. M.; Ning, Z.; Rogers, J.; Bentley, D. R.; Kwok, P. Y.; Mardis, E. R.; Yeh, R. T.; Schultz, B.; Cook, L.; Davenport, R.; Dante, M.; Fulton, L.; Hillier, L.; Waterston, R. H.; McPherson, J. D.; Gilman, B.; Schaffner, S.; Van Etten, W. J.; Reich, D.; Higgins, J.; Daly, M. J.; Blumenstiel, B.; Baldwin, J.; Stange-Thomann, N.; Zody, M. C.; Linton, L.; Lander, E. S.; Altshuler, D. A map of human genome sequence variation containing 1.42 million single nucleotide polymorphisms. *Nature* 2001, 409, 928–933.

Saito, M.; Yasui-Furukori, N.; Kaneko, S. Clinical pharmacogenetics in the treatment of schizophrenia. *Nihon Shinkei Seishin Yakurigaku Zasshi* 2005, 25(3), 129–135.

Sakaeda, T.; Nakamura, T.; Horinouchi, M.; Kakumoto, M.; Ohmoto, N.; Sakai, T.; Morita, Y.; Tamura, T.; Aoyama, N.; Hirai, M.; Kasuga, M.; Okumura, K. MDR1 genotype-related pharmacokinetics of digoxin after single oral administration in healthy Japanese subjects. *Pharm Res* 2001, 18, 1400–1404.

Sata, F.; Sapone, A.; Elizondo, G.; Stocker, P.; Miller, V. P.; Zheng, W.; Raunio, H.; Crespi, C. L.; Gonzalez, F. J. CYP3A4 allelic variants with amino acid substitutions in exons 7 and 12, evidence for an allelic variant with altered catalytic activity. *Clin Pharmacol Ther* 2000, 67, 48–56.

Scagliotti, G. V.; De Marinis, F.; Rinaldi, M.; Crinò, L.; Gridelli, C.; Ricci, S.; Matano, E.; Boni, C.; Marangolo, M.; Failla, G.; Altavilla, G.; Adamo, V.; Ceribelli, A.; Clerici, M.; Di Costanzo, F.; Frontini, L.; Tonato, M. Italian Lung Cancer Project: phase III randomized trial comparing three platinum-based doublets in advanced nonsmall-cell lung cancer. *J Clin Oncol* 2002, 20(21), 4285–4291.

Schaeffeler, E.; Eichelbaum, M.; Brinkmann, U.; Penger, A.; Asante-Poku, S.; Zanger, U. M.; Schwab, M. Frequency of C3435T polymorphism of MDR1 gene in African people. *Lancet* 2001, 358, 383–384.

Schiller, J. H.; Harrington, D.; Belani, C. P.; Langer, C.; Sandler, A.; Krook, J.; Zhu, J.; Johnson, D. H. Comparison of four chemotherapy regimens for advanced nonsmall-cell lung cancer. *New Engl J Med* 2002, 346, 92–98.

Schinkel, A. H.; Wagenaar, E.; Mol, C. A.; van Deemter, L. P-glycoprotein in the blood-brain barrier of mice influences the brain penetration and pharmacological activity of many drugs. *J Clin Invest* 1996, 97, 2517–2524.

Sekino, K.; Kubota, T.; Okada, Y.; Yamada, Y.; Yamamoto, K.; Horiuchi, R.; Kimura, K.; Iga, T. Effect of the single CYP2C9*3 allele on pharmacokinetics and pharmacodynamics of losartan in healthy Japanese subjects. *Eur J Clin Pharmacol* 2003, 59, 589–592.

Shaibi, G. Q.; Cruz, M. L.; Weigensberg, M. J.; Toledo-Corral, C. M.; Lane, C. J.; Kelly, L. A.; Davis, J. N.; Koebnick, C.; Ventura, E. E.; Roberts, C. K.; Goran, M. I. Adiponectin independently predicts metabolic syndrome in overweight Latino youth. *J Clin Endocrinol Metab* 2007, 92, 1809–1813.

Siegmund, W.; Ludwig, K.; Giessmann, T.; Dazert, P.; Schroeder, E.; Sperker, B.; Warzok, R.; Kroemer, H. K.; Cascorbi, I. The effects of the human MDR1 genotype on the expression of duodenal P-glycoprotein and disposition of the probe drug talinolol. *Clin Pharmacol Ther* 2002, 72, 572–583.

Stewart, C. F.; Hampton, E. M. Effect of maturation on drug disposition in pediatric patients. *Clin Pharm* 1987, 6, 548–564.

Stoehlmacher, J.; Park, D. J.; Zhang, W.; Groshen, S.; Tsao-Wei, D. D.; Yu, M. C.; Lenz, H. J. Association between glutathione S-transferase P1, T1, and M1 genetic polymorphism and survival of patients with metastatic colorectal cancer. *J Natl Cancer Inst* 2002, 94, 936–942.

Stoehlmacher, J.; Park, D. J.; Zhang, W.; Yang, D.; Groshen, S.; Zahedy, S.; Lenz, H. J. A multivariate analysis of genomic polymorphisms: prediction of clinical outcome to 5-FU/oxaliplatin combination chemotherapy in refractory colorectal cancer. *Br J Cancer* 2004, 91, 344–354.

Sun, J.; He, Z. G.; Cheng, G.; Wang, S. J.; Hao, X. H.; Zou, M. J. Multidrug resistance P-gly-coprotein: crucial significance in drug disposition and interaction. *Medical Science Monitor* 2004, 10(1), 14.

Teixeira, R. E.; Malin, S. The next generation of artificial pancreas control algorithms. *J Diabetes Sci Technol* 2008, 2, 105–112.

The Digitalis Investigation Group. The effect of digoxin on mortality and morbidity in patients with heart failure. *N. Engl. J. Med* 1997, 336, 525–533.

Thiebaut, F.; Tsuruo, T.; Hamada, H.; Gottesman, M. M.; Pastan, I.; Willingham, M. C. Cellular localization of the multidrug-resistance gene product P-glycoprotein in normal human tissues. *Proc Natl Acad Sci USA* 1987, 84, 7735–7738.

Ulrich, C. M.; Robien, K.; McLeod, H. L. Cancer pharmacogenetics: polymorphisms, pathways and beyond. *Nat Rev Cancer* 2003, 3, 912–920.

Vesell, E. S. Pharmacogenetic perspectives gained from twin and family studies. *Pharmacol Ther* 1989, 41, 535–552.

Vogel, C. L.; Cobleigh, M. A.; Tripathy, D.; Gutheil, J. C.; Harris, L. N.; Fehrenbacher, L.; Slamon, D. J.; Murphy, M.; Novotny, W. F.; Burchmore, M.; Shak, S.; Stewart, S. J.; Press, M. Efficacy and safety of trastuzumab as a single agent in first-line treatment of HER2-overexpressing metastatic breast cancer. *J Clin Oncol* 2002, 20, 719–726.

Watson, M. A.; Stewart, R. K.; Smith, G. B.; Massey, T. E.; Bell, D. A. Human glutathione S-transferase P1 polymorphisms: relationship to lung tissue enzyme activity and population frequency distribution. *Carcinogenesis* 1998, 19, 275–280.

Weilshboum, R.; Wong, L. Pharmacogenomics: bench to bedside. *Nat Rev Drug Discov* 2004, 3, 739–748.

Weinshilboum, R. Inheritance and drug response. *N Engl J Med* 2003, 348, 529–537.

Weinshilboum, R. M. Methyltransferase pharmacogenetics. In *Pharmacogenetics of Drug Metabolism*, Kalow, W., Ed.; Pergamon Press: New York, 1992, pp 179–194.

Wilffert, B.; Zaal, R.; Brouwers, J. R. Pharmacogenetics as a tool in the therapy of schizophrenia. *Pharm World Sci.* 2005, 27, 20–30.

Wilkinson, G. R.; Guengerich, F. P.; Branch, R. A. Genetic polymorphism of S-mephenytoin hydroxylation. *Pharmacol Ther* 1989, 43, 53–76.

Wilkinson, G. R.; Guengerich, F. P.; Branch, R. A. Genetic polymorphism of S-mephenytoin hydroxylation. In *Pharmacogenetics of Drug Metabolism*, Kalow, W., Ed.; Pergamon Press: New York, 1992, pp 657–685.

Wolf, C. R.; Smith, G.; Smith, R. L. Science, medicine and the future: pharmacogenetics. *Brit Med J* 2000, 320, 987–990.

World Health Organization. *Genomics and World Health*, WHO: Geneva, 2002.

Wuttke, H.; Rau, T.; Heide, R.; Bergmann, K.; Böhm, M.; Weil, J.; Werner, D.; Eschenhagen, T. Increased frequency of cytochrome P450–2D6 poor metabolizers among patients with metoprolol-associated adverse effects. *Clin Pharmacol Ther* 2002, 72, 429–437.

Yamauchi, A.; Ieiri, I.; Kataoka, Y.; Tanabe, M.; Nishizaki, T.; Oishi, R.; Higuchi, S.; Otsubo, K.; Sugimachi, K. Neurotoxicity induced by tacrolimus after liver transplantation: relation to genetic polymorphisms of the ABCB1 (MDR1) gene. *Transplantation* 2002, 74, 571–588.

Yasar, U.; Tybring, G.; Hidestrand, M.; Oscarson, M.; Ingelman-Sundberg, M.; Dahl, M. L.; Eliasson, E. Role of CYP2C9 polymorphism in losartan oxidation. *Drug Metab Dispos* 2001, 29, 1051–1056.

Yates, C. R.; Krynetski, E. Y.; Loennechen, T.; Fessing, M. Y.; Tai, H. L.; Pui, C. H.; Relling, M. V.; Evans, W. E. Molecular diagnosis of thiopurine S-methyltransferase deficiency: genetic basis for azathioprine and mercaptopurine intolerance. *Ann Intern Med* 1997, 126, 608–614.

Zatloukal, P.; Petruzelka, L.; Zemanová, M.; Kolek, V.; Skricková, J.; Pesek, M.; Fojtů, H.; Grygárková, I.; Sixtová, D.; Roubec, J.; Horenková, E.; Havel, L.; Průsa, P.; Nováková, L.; Skácel, T.; Kůta, M. Gemcitabine plus cisplatin vs. gemcitabine plus carboplatin in stage IIIb and IV nonsmall cell lung cancer: a phase III randomized trial. *Lung Cancer* 2003, 41, 321–331.

CHAPTER 3

NATURAL HISTORY AND MOLECULAR BIOLOGY OF HEPATITIS B VIRUS

ESSAM MOHAMMED JANAHI

CONTENTS

3.1   INTRODUCTION

Viral hepatitis means liver inflammation caused by a viral infection. MacCallum was the first to introduce the terms hepatitis A and hepatitis B, in 1947, to sort out infectious (epidemic) from serum hepatitis (Fields *et al.*, 2001; Thomas *et al.*, 2005). In 1965, Blumberg found an antigen in patients with serum type hepatitis (type B hepatitis) and was initially named "Australian Antigen" (now known as hepatitis B surface antigen) (Fields *et al.*, 2001; Thomas *et al.*, 2005). The virion was first seen in the electron microscope by Dane in 1970, who clearly described its structure and correctly interpreted the nature of the associated smaller particles and tubular forms (Monjardino, 1998). It was hence known as the Dane particle. Soon after, the core of the virion was identified in infected hepatocytes and also isolated from complete virions following detergent removal of the envelope. Major advances during the early seventies, to which Robinson's group at Stanford was a main contributor, led to the characterization of the virus genome, virion-associated proteins (including DNA polymerase) and the major HBV antigens and antiH-BV antibodies present during HBV infections. As techniques for cloning and amplification of DNA became available, the virus genome was sequenced and putative genes identified. In 1982, the elegant experiments of Summers and Mason characterized the mechanism of replication and opened up a new phase in HBV research (Monjardino, 1998; Fields *et al.*, 2001; Thomas *et al.*, 2005).

3.2   CLASSIFICATION

Human HBV and animal hepatitis viruses share common characteristics, which are the basis for their classification as hepadnaviruses (HBV being the prototype of the *Hepadnaviridae* family). The Woodchuck hepatitis virus (WHV, 1978), the Ground squirrel hepatitis virus (GSHV, 1980), and the Tree squirrel hepatitis virus (TSHV, 1986) are mammalian hepadnaviruses (*Orthohepadnavirus* genus), while the Duck hepatitis B virus (DHBV, 1980) and the Heron hepatitis B virus (HHBV, 1988) are avian hepadnaviruses (*Avihepadnavirus* genus) (Monjardino, 1998; Fields *et al.*, 2001; Thomas *et al.*, 2005). Similarities among them include virion and DNA size, structure, and genetic organization as well as their replication mechanism that involves reverse transcription of a pregenomic RNA transcript of more than genome length. Phylogenetically, hepadnaviruses, due to similarities in gene number, function, and organization, are related to retroviruses. Hepadnaviruses also show similarities in genetic organization and replication strategy with members of the *Caulimoviridae*, for example, Cauliflower mosaic virus (CaMV), and viruses from both of these families are sometimes referred to as pararetroviruses (Fields *et al.*, 2001).

3.3 EPIDEMIOLOGY

The prevalence of HBV and its modes of transmission vary geographically, and it can be classified into three endemic patterns. Around 45% of the world's population live in regions of high endemicity, defined as areas where 8% of the population is positive for HBsAg such as South-east Asia and Sub-Saharan Africa. The moderately endemic areas, such as in Mediterranean countries and Japan, are defined as those areas where 2–7% of the population are HBsAg positive, and around 43% of the world's population live in regions of moderate endemicity. Western Europe and North America are considered as areas with low endemicity (<2% of the population is HBsAg positive) and it constitutes 12% of the world's population (Fig. 3.1) (Fields *et al.*, 2001; Previsani *et al.*, 2002).

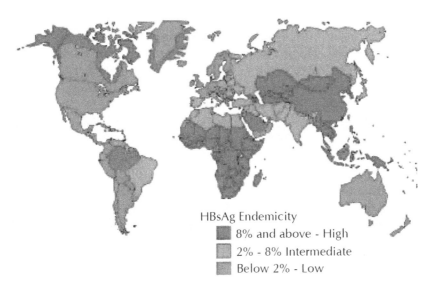

FIGURE 3.1 Geographic distribution of chronic hepatitis B infection. High levels of infection are more prevalent in the third world than the developed countries (Previsani *et al.*, 2002).

In Western Europe and the United States of America, HBV is usually transmitted horizontally by blood products or mucosal contact. In highly endemic areas like South-east Asia or Equatorial Africa, the most common mode of transmission is vertical transmission perinatally from an HBV-infected mother to the newborn child (Fields *et al.*, 2001; Previsani *et al.*, 2002; Thomas *et al.*, 2005). Certain types of behaviors increase the risk for contracting HBV such as use of contaminated needle during acupuncture, intravenous drug abuse, ear piercing and tattooing, sexually active heterosexuals or homosexuals (having more than one sexual partner in the last 6 months), infants/children in highly endemic areas, infants born to infected mothers, health care workers, haemodialysis patients, blood receivers prior to 1975 (blood transfusion), hemophiliacs, prisoners with long-term sentences as well as visitors to highly endemic regions (Previsani *et al.*, 2002).

Hepatitis B infections are either acute or chronic. Acute Hepatitis B virus infection is a short-term illness with an incubation period that ranges from 45 to 120 days and about 50–67% of older children and adults are asymptomatic. Chronic Hepatitis B virus infection is a long-term illness that occurs when the person has HBsAg present in his/her serum for more than 6 months (Fields *et al.*, 2001; Thomas *et al.*, 2005). Up to 85% of chronic hepatitis B cases displaying severe histological changes progress to cirrhosis; 2 to 15% of these will develop hepatocellular carcinoma (HCC), depending on regional factors (Koshy *et al.*, 1998). The probability of becoming a chronic carrier is inversely related to age at time of infection. In infants, 90–95% of those who contract hepatitis B become chronic carriers, as opposed to 5–10% of adults (Fig. 3.2). Therefore, the younger the age that infection occurs, the more likely that HBV infection will become chronic and thus developing complications of severe liver disease and hepatocellular carcinoma. Immuno-compromised patients are more likely to become chronic carriers than their healthier counterparts (Fields *et al.*, 2001; Koshy *et al.*, 1998; Thomas *et al.*, 2005). The World Health Organization (WHO) has estimated that more than 2 billion people worldwide have been infected with HBV and about 350 million of these are chronic carriers. About a quarter of the chronic carriers develop serious liver diseases such as chronic hepatitis, liver cirrhosis and HCC, and more than a million of them die annually from the complications of HBV-associated liver disease (Previsani *et al.*, 2002).

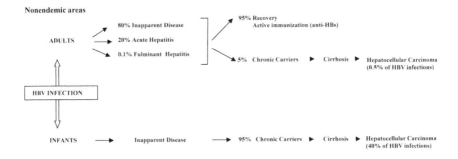

FIGURE 3.2 The relationship between HBV infection and age. The outcome of HBV infection varies substantially depending on the age at, which infection occurs. In infants, 95% of them would become chronic carriers, as opposed to 5% of adults (Thomas *et al.*, 2005).

3.4 NATURAL HISTORY OF THE DISEASE

3.4.1 ACUTE HEPATITIS B INFECTION

The course of acute viral hepatitis can be divided into four clinical phases: (a) the incubation period; (b) a prodromal or preicteric stage; (c) the icteric phase; and (d) the convalescent period. The severity of symptoms varies from mild and anicteric to severe and associated with jaundice. Regardless of the clinical course of infection, major increment of the serum transaminases (ALT, AST) occur within 2–4 weeks (Fig. 3.3), and then start to drop. The mean incubation period of acute HBV infection is around 60–90 days (range 45–120 days) (Fields *et al.*, 2001; Previsani *et al.*, 2002; Thomas *et al.*, 2005). The differences in the incubation period are associated with the amount of virus in the infecting inoculum, the mode of transmission and host factors (Previsani *et al.*, 2002). Fever, RUQ abdominal pain, nausea, vomiting, anorexia, repulsion to tobacco, headache and malaise may occur 1 to 2 weeks prior to the preicteric stage (onset of jaundice). The icteric phase of acute hepatitis B is characterized by the appearance of dark, golden-brown urine (bilirubinuria), followed one to several days later by pale stools and yellowish discoloration of the mucous membranes, conjunctivae, sclerae and skin (Fields *et al.*, 2001). In over 85% of HBV cases, theicteric phase starts within 10 days of the initial symptoms. The convalescent phase, that is, recovery from clinical symptoms, occurs over 4 to 12 weeks from onset of the icteric phase (Previsani *et al.*,

2002). There is no supporting data that early treatment of acute HBV with antiviral agents has an effect on chronicity (Thomas *et al.*, 2005).

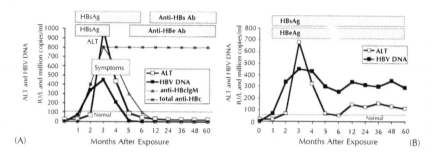

(A)   Months After Exposure   Months After Exposure   (B)

FIGURE 3.3   (A) Typical serological profile of acute hepatitis B infection; (B) Typical serological profile of chronic hepatitis B infection (Fields *et al.*, 2001).

In acute infection, HBsAg becomes detectable in the serum several weeks after infection and 3–5 weeks before the inception of clinical symptoms (Thomas *et al.*, 2005). It reaches highest concentration during the acute stage of the illness, then gradually decreases to undetectable levels within 4 to 6 months. HBV DNA is detected 3 to 5 weeks after infection and up to 6–15 days before the appearance of HBsAg. HBeAg and antiHBc IgM are usually detected few weeks after the appearance of HBsAg and simultaneous with the beginning of ALT elevations. Because IgM-specific antiHBc is produced in high titre in response to the synthesis of the nucleocapsid protein of the virion, its appearance in the serum is indicative of ongoing viral replication. However, elevated titres decline to undetectable levels regardless of whether the disease resolves or becomes chronic and consequently IgG antibodies against core are produced (Koshy *et al.*, 1998; Previsani *et al.*, 2002; Thomas *et al.*, 2005). Even before HBsAg disappears, antiHBe becomes detectable and decreases HBeAg, which signals a decline in HBV replication and the beginning of disease resolution. HBeAg and HBV DNA are markers of active HBV replication in hepatocytes. These markers can be detected in both the early course of acute infection, and in chronically infected patients (Thomas *et al.*, 2005). The resolution of acute HBV infection is characterized by the disappearance of HBsAg and the appearance of antiHBs. Anti HBs antibodies appear weeks to months following disappearance of serum HBsAg (Fig. 3.3). Anti-HBs is associated with neutralization of the virus and resolution of infection as shown by the accompanying normalization of all other biochemical markers of hepatic dysfunction such as billirubin, transaminases, etc. (Fields *et al.*, 2001; Previsani *et al.*, 2002). Patients who recover from acute infection develop antibodies against HBsAg for their lifetime. Such antibodies are also produced by HBV vaccine; 10 mIU/mL of antiHBs is considered to indicate a protective level of immunity. Some individuals continue to be immune from HBV by having anti HBs, but others may lose these antibodies and become susceptible to future infection. Most individuals who have recovered from an acute infection continue to have anti HBc IgG antibodies for life. Acutely infected individuals, who fail to clear HBV, will persistently have serum HBsAg. HBsAg is present during both acute and chronic infection and its presence indicates active viral replication (Fields *et al.*, 2001; Previsani *et al.*, 2002; Thomas *et al.*, 2005). Table 3.1 provides a description of HBV serologic markers in patients with hepatitis B.

TABLE 3.1 Interpretation of serologic markers in HBV patients with hepatitis (Fields *et al.*, 2001).

HBsAg	Anti-HBs	Anti-HBc	Interpretation
Positive	Negative	Negative	Early acute HBV infection.
Positive	Positive or Negative	Positive	HBV infection, either acute or chronic; differentiate with IgM anti-HBc; determine level of replicative activity (infectivity) with HBeAg or HBV DNA.
Negative	Positive	Positive	Indicates previous HBV infection and immunity to hepatitis B.
Negative	Negative	Positive	Possibilities include HBV infection in remote past; "low-level" HBV carrier; "window" between disappearance of HBsAg and appearance of anti-HBs; or false-positive or non-specific reaction.
Negative	Negative	Negative	Another infectious agent, toxic injury to the liver, disorder of immunity, hereditary disease of the liver, or disease of the biliary tract.
Negative	Positive	Negative	Vaccine-type response.

Fulminant hepatitis is a severe form of acute hepatitis and individuals with fulminant hepatitis usually die of liver failure within a short period of time. Hepatitis B is responsible for more than 50% of fulminant hepatitis, an extremely serious disease. Over half of the victims of fulminant hepatitis B die from it. Fulminant hepatitis is usually fatal in adults, whereas children may survive. This condition results from a sudden breakdown of liver function, which can lead to fatal liver failure. Liver damage is often accompanied by renal failure. Those who do survive may make a full recovery without liver damage (Koshy *et al.*, 1998; Thomas *et al.*, 2005).

3.4.2 *CHRONIC HEPATITIS B INFECTION*

In about 5–10% of cases, infection does not resolve but instead progresses to chronicity during, which time the virus persists for many years, associated with either no detectable clinical disease, mild disease or very active disease. Chronic hepatitis B is serologically characterized as HBsAg positivity for more than 6 months (Koshy *et al.*, 1998; Previsani *et al.*, 2002; Thomas *et al.*, 2005). Some chronically infected patients may have no clinical or biochemical evidence of liver disease. To distinguish this group from patients with chronic hepatitis, they are often categorized as asymptomatic hepatitis B carriers, or simply HBsAg carriers. There are two types of chronic HBV infection, chronic persistent hepatitis (CPH) and chronic active hepatitis (CAH). Of HBV chronic patients, about 70% have CPH whereas 30% have CAH. Viral replication during the course of HBV infection, especially in chronic patients, can be divided into three phases: a high replicative phase, low replicative phase and nonreplicative phase. In the high replicative phase, HBsAg, HBeAg, and HBV DNA are present and detectable in the sera. Aminotransferase levels may increase, and mild inflammation is histologically apparent. The risk of developing cirrhosis is high. The loss of HBeAg, or decrease or loss of the HBV DNA, and the appearance of antiHBe antibodies are all characteristics of the low replicative phase. Also a decrease in inflammatory activity is histologically apparent. The loss of HBV DNA and HBeAg are described as seroconversion. In the nonreplicative phase, markers of viral replication are either not present or beneath detection level, and inflammation is significantly reduced. However, if cirrhosis has already started, it persists indefinitely. Up to 20% of chronic persistent hepatitis cases develop cirrhosis. Cirrhosis is the end result of chronic liver damage, which occurs over a period of several years. It is characterized by diffuse hepatic fibrosis and the conversion of normal liver architecture into abnormal nodules. This leads to the impediment of blood flow and reduced liver function. The presence of HBV virus stimulates recurrent immune attacks on the hepatocytes, which ultimately leads to liver cirrhosis (Previsani *et al.*, 2002; Thomas *et al.*, 2005).

A further complication of chronic HBV hepatitis is the development of hepatocellular carcinoma. HBV is the cause of 60–80% of the world's primary liver cancers. The lifetime risk of death from chronic HBV in men is around 40–50% placing it among the top 10 causes of death in men (Previsani *et al.*, 2002). While in women the risk is only about 15%. In particular groups of patients where HBV is vertically transmitted (from mother to baby), the incidence of chronicity is much higher, reaching about 90–95%, an effect thought to be due to partial tolerance. The virus is nonetheless associated with liver injury and these patients normally evolve to cirrhosis and, in a significant number of cases, to hepatocellular carcinoma (Fields *et al.*, 2001; Thomas *et al.*, 2005). Chronic HBV infection is serologically identified as HBsAg (+), HBeAg (+), HBV-DNA (+), antiHBc (+), and antiHBs (–) (Table 3.1.).

3.5   PATHOGENESIS

HBV is normally noncytopathic to hepatocytes as illustrated by a number of situations where active viral replication is associated with very minor or no cellular injury. Such examples include 'healthy' carriers of the virus, cases of persistent infection where, in spite of a localized mild inflammatory infiltrate and active viral replication, there is no histological evidence of cellular necrosis, and transgenic mice expressing high levels of replicating virus, where again no cell injury is observed (Chisari *et al.*, 1995, 1995b). Current consensus on the mechanisms leading to the necroinflammatory lesions of hepatitis is that they result from the cell-mediated immune clearance of infected hepatocytes and that progression to chronicity results from inadequacy of such a response (Chisari *et al.*, 1995a, 1995b). This view has received support from the induction of a histologically acute hepatitis-like condition in transgenic mice expressing HBsAg as a result of the adoptive transfer of CD8$^+$ HBsAg specific cytotoxic T lymphocytes (CTL) (Moriyama *et al.*, 1990; Chisari, 1995). These studies also suggest that although the changes are initiated by CTLs, other cells, particularly macrophages and antigen nonspecific cytokines, such as interferon-γ (IFN-γ) and tumor necrosis factor-α (TNF-α), are subsequently recruited to the initial focus of inflammation and are ultimately responsible for most of the hepatocyte damage (Moriyama *et al.*, 1990; Chisari, 1995; Chisari *et al.*, 1995a, 1995b).

Progression of HBV infection beyond the acute stage, that is, to chronicity, is the result of immune response inadequacy. In successfully resolved acute infection, a strong, polyclonal, T-cell immune response against core, envelope and polymerase epitopes has been documented as accompanied by effectively neutralizing antiHBs circulating antibodies. In those cases, which become chronically infected, a weak and narrowly targeted T-cell response, unaccompanied by the production of neutralizing antibodies, appears to be the norm (Chisari *et al.*, 1995a, 1995b). Such chronic patients often display weak responses of helper (CD4$^+$) T-cells to viral antigens, as well (Fields *et al.*, 2001). Amongst other mechanisms leading to viral persistence, the escape from T-cell recognition as a result of high mutation rate in the sequence of a dominant epitope has been recognized as a possible mechanism. However, its contribution is unlikely to be significant during acute infection in the presence of a strong response targeted at a variety of epitopes in the envelope, nucleocapsid and polymerase (Chisari *et al.*, 1995a, 1995b).

3.6 DIAGNOSIS

HBV infection is generally diagnosed by using serological markers. The two assays used commercially for HBV diagnosis are radioimmunoassays (RIA) and enzyme-linked immunosorbent assays (ELISA). Both of, which are very sensitive and specific assays based on using specific antibodies against various HBV proteins and can detect low HBsAg concentrations as low as 0.25 ng/mL (Fields *et al.*, 2001; Thomas *et al.*, 2005). All HBV patients, whether acutely or chronically infected, will have detectable HBsAg in their serum. The diagnosis of acute HBV infection is most accurately made by the detection of antiHBc IgM, which develops few weeks after the presence of HBsAg in the blood. Anti-HBc IgM is rapidly converted to antiHBc IgG upon recovery. Chronic HBV infection is diagnosed by a negative test for antiHBc IgM, plus a positive test for HBsAg in the same blood sample (Fields *et al.*, 2001; Previsani *et al.*, 2002). In the majority of cases, the chronic infection becomes "nonreplicative" or "low replicative" which is characterized by the loss of serum HBeAg and replacement with anti HBe antibodies, but in few cases "high replicative" infection continues without losing serum HBeAg. In such individuals, infection can switch from "nonreplicative" to "replicative" and *vice-versa* (Fields *et al.*, 2001). Beside serology, polymerase chain reaction (PCR) has been used as well in detecting low levels of HBV DNA in both blood and liver tissue samples (Previsani *et al.*, 2002). Liver biopsy is essential to confirm hepatitis B diagnosis and assess its prognosis. In general, most chronic carriers (who are asymptomatic, HBsAg positive and have a normal serum aminotransferase levels) have minor or no inflammation on biopsy. In such chronic carriers, "ground glass cells" are often observed in liver biopsy, representing liver cells in which a large amount of HBsAg is being produced. Other chronic patients will have different degrees of liver inflammation on biopsy, and some will have fibrosis or cirrhosis. Bad prognosis is usually associated with higher level of inflammation, and with the presence of fibrosis or cirrhosis (Fields *et al.*, 2001; Thomas *et al.*, 2005).

3.7   TREATMENT

There is currently no specific treatment for acute viral hepatitis. Current therapy is mostly supportive to maintain comfort. Patients are strongly advised to minimize their alcohol consumption if not eliminate it altogether. This helps the liver to regenerate (Koshy *et al.*, 1998). Treatment of chronic HBV involves the use of interferon and/or nucleoside analogs. The goal of therapy is to suppress HBV replication to reduce symptoms, minimize chronic inflammation, and prevent progression to cirrhosis and HCC. Treatment is aimed at decreasing persistent inflammation through suppressing viral replication, which results in lowered level of HBV antigens' expressions on the surfaces of infected hepatocytes. Patients who are persistently HBeAg (+), HBV DNA (+), have constantly high transaminase levels, and show histological evidence of moderate-to-severe inflammation or fibrosis are considered good candidates for treatment (Koshy *et al.*, 1998; Previsani *et al.*, 2002; Thomas *et al.*, 2005).

3.7.1   INTERFERON ALPHA-2B

About 35–40% of HBV chronic patients, in contrast to 12% in placebo controls, respond positively to treatment with Interferon alpha-2b at a dosage of either 5 million units per day or 10 million units three times weekly for 4–6 months. Up to 6 months after therapy completion, HBV replication markers (HBeAg and HBV DNA) are either decreased or lost, and ALT levels are also normalized (Koshy *et al.*, 1998; Thomas *et al.*, 2005; Wong *et al.*, 1993). Persistence of covalently closed circular HBV DNA (cccDNA) molecules in the nucleus, in spite of a favorable response to therapy, demonstrates failure to eradicate the virus and a risk of recurrence. Relapse rates range from 10 to 15%, and even in responders to therapy HBsAg may remain present for years. Favorable response to interferon alpha-2b therapy are seen among patients with ALT level > 200 IU/L, HBV DNA < 100 copies/mL, infection of short duration, presence of necroinflammatory activity in liver biopsy, and no underlying immunosuppressive disease or treatment (Koshy *et al.*, 1998; Wong *et al.*, 1993). Many individuals experience flu-like symptoms following the first administration of interferon. Other side effects of long-term treatment are poor appetite, weight loss, fatigue, muscle aches, bone marrow suppression, autoimmune phenomena, bacterial infections, alopecia and psychological side effects (Koshy *et al.*, 1998). During therapy, patients should be monitored frequently using hematological and biochemical tests. Serologic assays for HBsAg, HBeAg, and HBV DNA level are usually reserved for the end of treatment (Fields *et al.*, 2001). HBV DNA estimation by PCR, as a measure of viral activity, is also used to monitor the therapy response. Detectable DNA should disappear within 4–6 months followed by the elimination of HBeAg. HBe antibodies found in the serum indicates successful seroconversion (Koshy *et al.*, 1998; Wong *et al.*, 1993).

3.7.2   LAMIVUDINE

Lamivudine monotherapy (3TC, Epivir-HBV) was approved for the treatment of chronic HBV infection in 1998, at an oral dosage of 100 mg per day. Positive response after one year of therapy included suppression of HBV-DNA in nearly all patients; normalization of ALT levels in 40–50%; loss of HBeAg in 17–33%; appearance of HBeAb in 17–20%; and reduction in liver inflammation in 50–60% of patients (Dienstag *et al.*, 1995). Lamivudine is a 2,'3'-dideoxy cytosine analog, which inhibits HBV polymerase and, hence, inhibits HBV replication. Its inhibitory effect was proven both *in vitro* and *in vivo*. The drug is well tolerated and relatively few serious side-effects are observed. After lamivudine is stopped, HBV replication may restart again and relapse occurs (Thomas *et al.*, 2005). After one year of therapy, HBV resistance to lamivudine was found in 14–39% of patients and appeared more in persistently HBeAg(+) patients (Dienstag *et al.*, 1995; Thomas *et al.*, 2005). In spite of the persistent HBeAg in these patients, histological recovery may result alongside decease in HBV DNA and transaminase levels (Dienstag *et al.*, 1995). Lamivudine-resistant HBV strains contain mutations in the YMDD motif, which is part of the catalytic site of reverse transcriptase. Results from a pilot study and a multicenter placebo-controlled trial showed that the combination of lamivudine and interferon may confer a marginally better response than lamivudine alone. Therefore, the more promising approach probably involves the use of direct antivirals, such as nucleoside analogs, with interferon in combination therapy (Fields *et al.*, 2001; Thomas *et al.*, 2005).

3.8   PREVENTION

3.8.1   PRE-EXPOSURE HBV PROPHYLAXIS

Three doses of the hepatitis B vaccine induces protective levels of antibody in greater than 90% of immunocompetent adults (Koshy *et al.*, 1998; Previsani *et al.*, 2002; Thomas *et al.*, 2005). The immunogenicity of the two commercially available vaccines, Recombivax-HB and Energix-B, are equivalent (Previsani *et al.*, 2002). These vaccines are recombinant yeast derived, where the small hepatitis B surface antigen (SHBs) is synthesized in the yeast cells and this leads to the formation of SHBs particles (Koshy *et al.*, 1998; Previsani *et al.*, 2002). The timings of the three vaccine doses are 0, 1–2 months, and 6 months or one year. The doses of vaccine used vary with vacinees's age, and produce antibody titres of more than 100 mIU/mL in 90-95% of vaccinees. The levels drop markedly over a period of 5 to 7 years in a significant number of cases and a booster is normally required (Koshy *et al.*, 1998; Previsani *et al.*, 2002; Thomas *et al.*, 2005). One to two months following taking the third vaccine dose, a test for antiHBs antibody is advised. Hepatitis B vaccine has been introduced into the national immunization programs of infants in more than 110 countries (Previsani *et al.*, 2002). However, the need for three injections and a possible booster, and cost, still make the vaccine less than optimal for mass vaccination in rural areas of the world where HBV is endemic (Monjardino, 1998).

3.8.2   POST-EXPOSURE HBV PROPHYLAXIS

Percutaneous exposure of a nonimmune person to HBsAg(+) blood carries an infection risk of 2.5% in HBeAg(–) blood samples and 19% in HBeAg(+) blood samples (Fields *et al.*, 2001). Both active and passive immunization are administered subsequent to percutaneous or mucosal membrane exposure. In this case, hepatitis B vaccine (active immunization) and hepatitis B immunoglobulin (HBIG) (passive immunization) are usually given (Koshy *et al.*, 1998; Thomas *et al.*, 2005). Management of patients sexually exposed to an HBV infected person is similar. HBIG given within 14 days of sexual exposure decreases the transmission risk by around 75% (Thomas *et al.*, 2005). The perinatal transmission risk vary considerably in relation to HBeAg, being 70–90% in neonates born to mothers who are both HBsAg (+) and HBeAg (+) and only 10–20% in neonates born to mothers who are HBsAg (+) and

HBeAg (–). The majority (90–95%) of these infants will become chronic carriers if not given any treatment. If the neonate was given a single dose of HBIG within 24 hours of birth plus vaccination beginning within 12 hours of birth, this would result in reducing the transmission risk by 85–95% (Koshy *et al.*, 1998; Thomas *et al.*, 2005).

3.9   VIRION MORPHOLOGY

The complete infectious hepatitis B virus particle is spherical, with an overall diameter of 42–47 nm and is sometimes referred to as a Dane particle (Fig. 3.4). It consists of a 25–27 nm diameter core or nucleocapsid, which is mainly comprised of HBcAg surrounded by a 7 nm lipoprotein bilayer derived from the endoplasmic reticulum (ER) membrane of the host. Three differently glycozylated HBV surface proteins of varying sizes, LHBs (large hepatitis B surface protein), MHBs (middle hepatitis B surface protein), and SHBs (small (major) hepatitis B surface protein), are inserted in this lipoprotein bilayer. The viral DNA, a virus-encoded RNA/DNA dependent DNA polymerase, a genome bound protein covalently linked to the 5'-end of the negative (-) strand of HBV DNA, and a capped RNA primer covalently linked to the 5'-end of the positive (+) strand of HBV DNA are contained in the nucleocapsid (Shafritz *et al.*, 1984; Monjardino, 1998; Seeger *et al.*, 2000; Fields *et al.*, 2001). Purified cores also contain a protein kinase activity detected by its ability to phosphorylate HBc protein *in vitro*; because recombinant C and P proteins do not possess kinase activity, this enzyme is thought to be of host origin (Fields *et al.*, 2001).

In the initial phase of infection, up to 10^{10} infectious particles/mL maybe detectable in serum (Fields *et al.*, 2001). Besides virion particles, 20 nm spheres and filaments, 20 nm in diameter and of variable length, are also detectable (Fig. 3.4). These particles consist exclusively of hepatitis B surface antigen (HBsAg) and host-derived lipids (phospholipids, cholesterol, cholesterol esters, and triglycerides). However, they do not contain viral DNA and are therefore noninfectious. The 20 nm spherical and filamentous structures are found in sera of infected patients, sometimes reaching titres of 10^{12}/mL, and far outnumbering the virion particles (10,000-1000,000:1). The presence of such high concentration of these noninfectious particles acts as an immunological decoy that will protect the infectious viral particles from antibodies neutralization in the blood stream (Shafritz *et al.*, 1984; Fields *et al.*, 2001). Subviral 20 nm particles are made up largely of SHBs protein, with variable quantities of MHBs polypeptides and few or no LHBs chains. By contrast,

Dane particles are substantially enriched for LHBs chains. Because LHBs chains are thought to carry the receptor recognition domain, this enrichment may prevent the more numerous 20 nm particles from competing effectively with virions for cell surface receptors (Fields *et al.*, 2001).

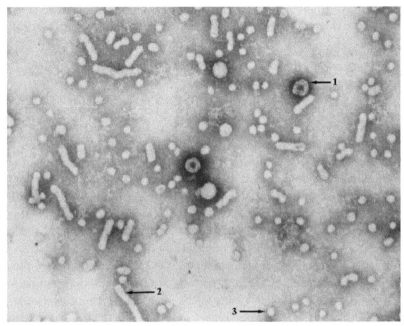

FIGURE 3.4   Hepatitis B virus morphology. HBV is a 42 nm, double-shelled DNA virus that possesses an icosahedral nucleocapsid or "core," 27 nm in diameter, surrounded by an outer envelope nearly 7 nm thick (arrow 1). The outer envelope contains hepatitis B surface antigen (HBsAg), which is synthesized in large quantities, and usually found in the blood stream in the form of 20 nm filamentous (arrow 2) and spherical particles (arrow 3). The filamentous particles vary in length and sometimes display regular, nonhelical transverse striations (Fields *et al.*, 2001).

Electron cryomicroscopy with computer tri-dimensional reconstitution has shown recombinant core particles to have icosahedral T4 symmetry with 240 constitutive subunits, with a less-abundant population showing T3 symmetry and 180 subunits (Crowther *et al.*, 1994; Monjardino, 1998; Fields *et al.*, 2001). The proposed structures contain holes or channels of up to 10 nm diameter, previously described in other transcriptionally active viral cores, which are thought to allow the entry of small molecules (Crowther *et al.*, 1994). The structure of the core particle has been resolved further by electron cryomicroscopy to reveal spikes on the surface of the shell made up of dimers of core polypeptide in the form of radial bundles of four α-helices (Bottcher *et al.*, 1997). The N-termini of the two core proteins in the dimer are located at the tip of the spike. A more recent study has confirmed these results and showed that the capsid spikes do not penetrate the inner leaflet of the lipid bilayer and that LHBs seems to play an important role in maintaining stable, noncovalent interactions between the capsid and envelope (Dryden *et al.*, 2006).

3.10   GENOMIC ORGANIZATION

3.10.1   OPEN READING FRAMES

The hepatitis B virus genome is a circular partly double-stranded DNA molecule consisting of one complete strand about 3,200 nucleotides long (negative strand) with a single, unique interruption and a complementary incomplete (positive strand) of variable length (Shafritz *et al.*, 1984; Monjardino, 1998; Seeger *et al.*, 2000; Fields *et al.*, 2001) (Fig. 3.5). The overall lengths of positive strands among different molecules range between 50% and 70% of full size (Monjardino, 1998). The large [L (-)] strand of the partially double-stranded genome has four overlapping ORFs: preS/S (surface), preC/C (core), P (polymerase), and X, that encode at least seven different proteins (Fig. 3.5). By the use of overlapping ORFs the HBV genome can encode one and a half times the information content of an equivalent DNA molecule that only has nonoverlapping ORFs. Circularization of the genome is achieved by base pairing of the free 5' end of the S(+) strand with the nicked 5' end of the L(-) strand in a region of approximately 220 nt flanked by two 11-bp direct repeats, DR1 and DR2. The 5' end of the L(-) strand begins within the direct repeat termed DR1, whereas the S(+) strand begins within DR2. A polypeptide encoded by the amino-terminal region of the HBV P gene is covalently linked to the 5' end of the L(-) strand and serves as a protein primer for L(-) strand DNA synthesis during replication. A 19-nt capped ribonucleotide, covalently attached to the 5' end of the S(+) strand, most likely serves as a primer for S(+) strand DNA synthesis (Shafritz *et al.*, 1984; Monjardino, 1998; Seeger *et al.*, 2000; Fields *et al.*, 2001).

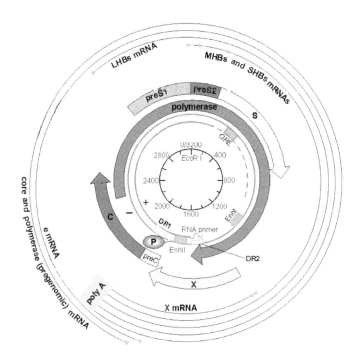

FIGURE 3.5 Organization of HBV Genome (subtype ayw). The partially double-stranded circle is comprised of L(–) and S(+) strands with cohesive complementary 5′ ends. Locations of the direct repeats 1 and 2 (DR1 and DR2), the enhancers (EnhI and EnhII respectively) as well as glucocorticoid-responsive element (GRE) indicated. The genome bound protein and the oligoribonucleotide at the 5′ end of the L(–) strand and at the 5′ end of the S(+) strand, respectively are indicated. The inner arrows denote viral ORFs (pre-S1/S2/S, pre-C/C, P, and X). The outer circle depicts unspliced viral transcripts with heterologous initiation sites and the common polyadenylation site (Seeger *et al.*, 2000).

Four ORFs (preS/S, preC/C, P and X) encode for the expression of seven different HBV proteins via the use of different in-frame start codons (Shafritz *et al.*, 1984; Monjardino, 1998; Seeger *et al.*, 2000; Fields *et al.*, 2001). The pre-S/S gene is divided into pre-S1, pre-S2 and S domains by three in-frame start codons and thus encodes three different forms of the envelope protein: the small (major) (SHBs), middle (MHBs) and large (LHBs) hepatitis B surface proteins. The SHBs protein is synthesized when a ribosome begins translation at an AUG at position 157 of the preS/S gene. It is comprised of 226 amino acids and contains the group (a) and subtype (d/y and r/w) antigenic determinants of HBsAg. The MHBs protein is produced from an upstream AUG at position 3174, which adds 55 amino acids onto the small protein, and the LHBs surface protein is generated from the AUG at position 2850, adding 108 amino acids onto the middle protein (Shafritz *et al.*, 1984; Monjardino, 1998; Seeger *et al.*, 2000; Fields *et al.*, 2001). Based on the presence or absence of S-domain glycozylation, there are two isomeric forms for each of the three HBV surface proteins. Furthermore, the MHBs protein carries a further N-linked glycozylation within the pre-S2-specific domain (Stibbe *et al.*, 1983; Heermann *et al.*, 1984). The LHBs protein contains a further posttranslational modification that involves a myristic fatty acid group linked by an amide bond to its amino-terminal glycine residue (Persing *et al.*, 1987).

A similar alternative use of initiation codons is seen in the preC/C ORF where the core protein (HBcAg, the structural protein of the nucleocapsid) initiates at the AUG at position 1903 and terminates at position 2451, but may be preceded by a 29 amino acid precore sequence in the same reading frame starting from AUG at position 1816 to form the 'e' antigen (HBeAg). HBc encapsulates the HBV nucleic acid and has nucleic acid binding capacity (Petit *et al.*, 1985). The pre-C polypeptide (HBeAg precursor) has a signal peptide in its first 19 N-terminus amino acids. This signal directs the pre-C polypeptide to the endoplasmic reticulum where the N- and C-terminus amino acids are cleaved by host proteases. After further processing by the Golgi apparatus, the secretory protein is released in to extracellular medium (or serum) as the 17-kDa protein known serologically as 'e' antigen (HBeAg) (Ou *et al.*, 1986; Standring *et al.*, 1988). The physiological function of HBeAg is less clear than that of the structural proteins, as it is not required for viral replication, nor acute infection *in vivo* (Schlicht *et al.*, 1987; Standring *et al.*, 1988; Chen *et al.*, 1992). HBeAg is conveniently used clinically as an index of viral replication, infectivity, and severity of disease. It may have an immunomodulating function in the course of infection, playing an important role in the natural history of disease (Standring *et al.*, 1988; Milich *et al.*, 2003). It has been proposed that one function of HBeAg is to maintain chronicity (Milich *et al.*, 1990), which is suggested by the induction of fulminant hepatitis following primary infection in animal models with HBeAg negative variants of hepadnaviruses (Ogata *et al.*, 1993; Cote *et al.*, 2000). In the precore region at the position 1896, a mutation from guanosine (G) to adenosine (A), results in a stop codon (TGG to TAG). A second G to A mutation occurs, accompanying the mutation at 1896, is at position 1899 (Carman *et al.*, 1989; Brunetto *et al.*, 1993). These two mutations together stabilize the stem loop structure, which is a secondary structure with a pregenome encapsidation sequence of the precore/core region (Ogata *et al.*, 1993). These mutations are relatively common in patients in Mediterranean countries, particularly in Greece and

Italy. Such mutated viruses continue to replicate, but do not secrete HBeAg. In a clinical setting, these patients are antiHBe positive with high HBV DNA levels. The mutations are believed to occur during seroconversion to antiHBe under immunological pressure during the immune clearance phase (Carman *et al.*, 1991).

Of the two other reading frames, P extends over more than three quarters of the genome from the AUG at position 2309 to the termination codon at position 1622 (equivalent to 832 amino acids). Four domains are found in the P ORF. The amino-terminal domain encodes a terminal protein (primase), required for priming (–) strand synthesis. The next domain has no known function. The third domain encodes a RNA/DNA-dependent DNA polymerase, which serves as a reverse transcriptase and DNA polymerase. The fourth domain in the carboxyl terminal, encodes an RNase H, since the RNA template has to be removed in order to allow second-strand DNA synthesis of the (+) strand (Schlicht *et al.*, 1989). Finally, the X ORF which initiates at position 1374 and terminates at position 1836, encodes a complex transactivating protein (HBx) of 154 amino acids. This protein is conserved among all mammalian hepadnaviruses but not in avian hepadnaviruses and it is essential for the life cycle of HBV *in vivo* but not *in vitro* (Monjardino, 1998; Fields *et al.*, 2001).

3.10.2 REGULATORY ELEMENTS

Regulatory sequences act as binding sites for various proteins and are important for viral replication and gene regulation. Several regulatory sequences have been identified in different regions of the HBV genome. These include 4 promoters, 2 enhancers, the glucocorticoid-responsive element (GRE), 2 short direct repeats (DR1 and DR2), the Epsilon-Stem loop, the polyadenylation signal sequences and a posttranscriptional regulatory element.

3.10.2.1 PROMOTERS

A promoter functions as a binding site for RNA polymerase and the initiation of RNA transcription. Four promoters have been identified in the HBV genome. The three different hepatitis B surface proteins are expressed from three different in-frame start sites within the S ORF. Two promoters (pre-S1 promoter and pre-S2 promoter) control the expression of these proteins. The pre-S1promoter gives rise to a single 2.4 kb transcript, which contains the full S ORF. While the pre-S2 promoter gives rise to various transcripts of 2.1 kb in length (Eble *et al.*, 1986; Fields *et al.*, 2001). The pre-S2 promoter is located within the coding region of the pre-S1 domain, and lacks a classic TATA box sequence (Antonucci *et al.*, 1989; Chang *et al.*, 1989). The pre-S2 promoter, is stronger than the pre-S1 promoter, consequently more MHBs and SHBs are expressed, compared to LHBs protein (Eble *et al.*, 1986; Antonucci *et al.*, 1989). Transcription factors such as Oct-1, HNF-1 and HNF-3 regulate the activation of the pre-S1 promoter (Eble *et al.*, 1986), as do HBV enhancer elements (Antonucci *et al.*, 1989; Su and Yee, 1992). The pre-S1 promoter can be down-regulated by a poorly understood mechanism, which is dependent on the 3160 to 3221 nucleotide sequence within the pre-S2 promoter (Eble *et al.*, 1986; Fields *et al.*, 2001). The pre-S1 promoter can also be down-regulated by a CCAAT element, which is also involved in the up-regulation of the pre-S2 promoter. The pre-S2 promoter is activated by NF-Y, the CCAAT-binding factor (Lu *et al.*, 1996), as well as by both of HBV enhancers (Su and Yee, 1992).

The core promoter gives rise to multiple RNAs, with different sequences at the 5' end. These transcripts are the core antigen, '*e*' antigen, polymerase, and pregenomic RNA (pgRNA) transcripts. The core promoter is regulated by both HBV enhancer elements and by a negative regulatory element (NRE) (Shaul *et al.*, 1985). However, its location has yet to be accurately defined, as it has been roughly mapped in the 1591 to 1851 nucleotide region (Lopez-Cabrera *et al.*, 1990). Transcription from the C promoter is tightly restricted to liver (Honigwachs *et al.*, 1989; Raney *et al.*, 1997). The liver-specific expression of C gene transcription is thought to result from the requirement of several liver-enriched transcription factors to bind to the C promoter. These include factors *c*/EBP, HBF1, HNF3, and HNF4 (Dikstein *et al.*, 1990; Lopez-Cabrera *et al.*, 1990; Garcia *et al.*, 1993; Chen *et al.*, 1994). The C promoter is also susceptible to both positive and negative regulation (Raney *et al.*, 1997; Yu *et al.*, 1997). Part of the hepatotropism of HBV is likely to be attributable to liver-specific transcription of the C promoter, which is required for synthesis of pregenomic RNA.

The polymerase gene does not have a promoter element directly upstream. Its expression appears to result from ribosome scanning of the pgRNA transcript and initiation of translation at the P ORF start codon (Monjardino, 1998; Fields *et al.*, 2001). ORF X has its own promoter that gives rise to the 0.9 kb RNA transcript, which is in turn translated to a 17-kd protein (HBx). The promoter is thought to be located within the 1230 to 1376 nucleotide sequence (Siddiqui *et al.*, 1987; Fields *et al.*, 2001). However, the exact location remain contentious, because of the proximity of the X promoter to enhancer I region.

3.10.2.2   ENHANCERS

Enhancer I (EnhI, nt 1135-1254) spans the region between ORF S and ORF X, and has binding sites for various liver-specific and ubiquitous DNA-binding transcription factors. These include c/EBP, HBF4, and HBLF (Levy *et al.*, 1989; Guo *et al.*, 1991; Trujillo *et al.*, 1991). EnhI stimulates gene transcription from the pre-S1, pre-S2 and C promoters as well as from the X promoter via an X-responsive element, XRE (Guo *et al.*, 1991). EnhI functions in both hepatic and nonhepatic cells, although it more strongly activates transcription in hepatocytes (Shaul *et al.*, 1985; Antonucci *et al.*, 1989). Enhancer II (EnhII, nt 1627-1774) is situated in ORF X, partially overlaps the C promoter, and has an activity largely restricted to hepatic cells (Yuh *et al.*, 1993). It consists of two functional subunits, element A (nt 1627-1687) and element B (nt 1688–1774). Element A seems mainly to be responsible for the liver specificity of EnhII, by binding to liver-specific transcription factors including the c/EBP family of proteins, HBF1, HNF3, HNF4; element B is essential for its transcriptional stimulatory function (Lopez-Cabrera *et al.*, 1990; Lopez-Cabrera *et al.*, 1991; Li *et al.*, 1995; Li *et al.*, 1998). EnhII strongly promotes liver-specific transcription of the C promoter, and possibly X and S promoters (Lopez-Cabrera *et al.*, 1990; Su and Yee, 1992). A negative regulatory element, NRE (nt 1613-1636) represses EnhII activity in differentiated cells, possibly by masking transcription factor binding sites. When isolated, it has a marginal inhibitory effect but show strong repressive potential in the presence of a functional EnhII (Lo *et al.*, 1994).

3.10.2.3   GLUCOCORTICOID-RESPONSIVE ELEMENT

A glucocorticoid-responsive element (GRE) is located in the region where ORF P and ORF S overlap, upstream of the enhancers (Tur-Kaspa *et al.*, 1986). It is a specific DNA binding site for the glucocorticoid receptor. By interacting with the glucocorticoid receptor, transcription stimulation activity of the enhancers is increased. In a transgenic mice model, HBsAg gene expression is increased by sex steroids and glucocorticoids (Farza *et al.*, 1987).

3.10.2.4   POLYADENYLATION SIGNAL

A polyadenylation signal is located in ORF C (Simonsen *et al.*, 1983). It is important for the termination of transcription and the polyadenylation of the mRNA transcripts 3' end (Miller *et al.*, 1989). It is a well-conserved sequence, consisting of 13 nucleotides: ATAAAGAATTTGG.

3.10.2.5   DIRECT REPEAT 1(DR1) AND DIRECT REPEAT 2 (DR2)

The (−) and (+) strands of HBV DNA overlap over a region of 220 nucleotides and form short cohesive ends through the base-pairing between the 5' termini of the two strands (Summers *et al.*, 1975). At both sides of the cohesive ends, there is a conserved region of an 11 base pair direct repeat, 5' TTCACCTCT-GC. These direct repeats are designated as DR1 and DR2, and start at nucleotides sequence 1824 and 1590. DR1 is responsible for (−) strand synthesis, and DR2 for (+) strand synthesis (Summers *et al.*, 1975; Fields *et al.*, 2001).

3.10.2.6   EPSILON-STEM LOOP

A stem-loop structure (the Epsilon (ε) stem loop) is formed from nucleotides 1847–1907 in the HBV genome and has an essential role in HBV DNA encapsidation (Chiang *et al.*, 1992; Fields *et al.*, 2001). By fusing foreign genes to different regions of the HBV genome and testing for their encapsidation, the position of the stem-loop was determined (Chiang *et al.*, 1992). Because of the terminal redundancy in pgRNA, the ε element is present at both ends of the pgRNA. Interestingly, however, only the 5' copy is functional for RNA packaging; deletion of the 3' element has no impact on viral replication (Hirsch *et al.*, 1991). For this reason, only pgRNA is encapsidated, even though all HBV transcripts have the stem loop sequence at their 3' termini (Hirsch *et al.*, 1991; Chiang *et al.*, 1992). The stem loop sequence contains a series of inverted repeats and on the basis of their sequence is thought to fold into a three-dimensional stem-loop structure (Fig. 3.6) In spite of the differences in the primary sequence, the stem-loop structure is highly conserved among all hepadnaviruses (Chiang *et al.*, 1992). The HBV polymerase enzyme identifies and interacts with the stem-loop structure, which is necessary for initiating both encapsidation and reverse transcription of the HBV pgRNA (Pollack *et al.*, 1994).

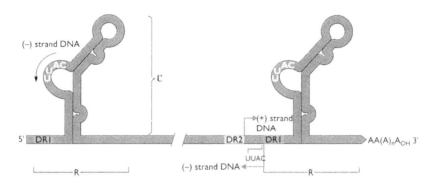

FIGURE 3.6 HBV Epsilon-Stem Loop. HBV pgRNA bears terminal repetitions of ca. 200 nt [®] that contain copies of the ε stem-loop and a short sequence of 11 nt termed Direct repeats (DR). Only the 5' copy of ε stem-loop is functional for RNA packaging, and for this reason, only pgRNA is encapsidated, even though all HBV transcripts have the stem loop sequence at their 3' termini. The internal bulge in the stem-loop structure contains the 5'-UUAC-3' motif, the binding site for HBV polymerase for the initiation of (−) strand synthesis. Pregenomic RNA contains a 5' copy of DR1, a copy of DR2, and an extra copy of DR1 at the 3' end (Flint, 2000)

3.10.2.7 POST-TRANSCRIPTIONAL REGULATORY ELEMENT

RNA splicing in eukaryotes is tightly coupled to 3' end formation and export of the RNA from the nucleus to the cytoplasm (Nigg, 1997). Consequently, intronless RNAs are poorly exported to the cytoplasm. A number of animal viruses encode intronless mRNAs that, nevertheless, efficiently accumulate in the cytoplasm. To overcome the export restriction to intronless RNAs, hepadnaviruses use a large RNA element known as the posttranscriptional regulatory element (PRE) (Huang and Yen, 1995). The PRE is located partially within the X promoter and ORF, and downstream of ORF S, indicating that it is contained in all viral transcripts (Huang and Yen, 1995). The PRE was shown to promote directly the export of unspliced transcripts. The HBV PRE is conserved in mammalian hepadnaviruses, but its mechanism of action is not well understood.

3.11 HBV LIFE CYCLE

Figure 3.7 represents different phases in the HBV replication cycle. The mechanism by which HBV binds and enters hepatocytes is poorly understood. It is thought that an interaction takes place between a cellular receptor and the pre-S1 region of the virus surface protein (Ryu *et al.*, 2000). Binding of HBV virions to human liver plasma membrane fractions can be blocked by monoclonal antibodies to the HBV pre-S1 domain and antipeptide antibodies to pre-S1 block adherence of HBV particles to HepG2 cells, indicating that the pre-S1 domain is important for HBV receptor binding (Neurath *et al.*, 1986; Pontisso *et al.*, 1989). Much less is known of the identities of the cellular receptors for hepadnaviruses. Mehdi and co-workers (Mehdi *et al.*, 1994) have produced good evidence that HBV S determinants bind the serum protein apolipoprotein H, but this molecule is not an integral transmembrane protein of the hepatocyte and its role in HBV infection is uncertain. Like other members of Hepadnaviridae, HBV displays hepatotropism. Hepatocytes, the major cell type in the liver, are the major target of infection, although infections in other extrahepatic sites, such as, the kidney (Dejean *et al.*, 1984), peripheral blood mononuclear cells (PBMCs) (Kock *et al.*, 1996), pancreas (Shimoda *et al.*, 1981) and spleen (Lieberman *et al.*, 1987), have been reported.

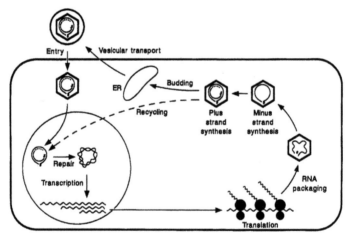

FIGURE 3.7   Schematic diagram of the HBV replication life cycle (Fields *et al.*, 2001).

Thus the HBV virion attaches to the hepatocytes via the pre-S1 domain of LHBs and enters by receptor mediated endocytosis, however, the events following entry into the host cell and the subsequent release of the capsid from the endosome are still unclear. It is still unclear, whether capsids then enter the nucleus, directed by putative nuclear targeting sequences in the carboxy terminus of the core protein, or directly release their DNA into the nucleus. The mechanism of DNA replication however, is well characterized. When inside the nucleus, HBV DNA molecule is repaired to form the covalently closed circular form (cccDNA). A number of steps are involved in generation of cccDNA. These include: a) completion of the plus strand of the viral genome; b) removal of the 5' oligoribonucleotide cap from the plus strand; c) removal of the 5' terminal protein and the 7–9 base terminal redundancy from the minus strand; d) ligation of the 5' and 3' ends of both DNA strands; and e) supercoiling of the covalently closed DNA by gyrase and DNA topoisomerase activities (Shafritz et al., 1984; Monjardino, 1998; Seeger et al., 2000; Fields et al., 2001). The minus strand of cccDNA serves as the template for transcription by cellular RNA polymerase II. Transcription is highly regulated by the four promoters, the two enhancers and the glucocorticoid-responsive element. All HBV mRNAs possess a 5' cap and are 3' polyadenylated. The transcripts generated can be divided into two categories: subgenomic and genomic. The smaller, subgenomic transcripts serve as mRNA for the expression of the X and surface proteins [LHBs mRNA (2.4 kb), MHBs and SHBs mRNA (2.1 kb) and X mRNA (0.9 kb)]. The two larger (3.5 kb) genomic transcripts are longer than one genome in length and serve as mRNAs for 'e,' core and polymerase proteins by the varying activities of the appropriate promoter and enhancer sequences. One 3.5 kb transcript, lacking the ATG start codon for the 'e' protein, is designated the pgRNA (pregenomic RNA), since it is selected specifically by the HBV polymerase protein for packaging and reverse transcription. This involves recognition of the ε-stem loop at the 5' end of the pgRNA by the polymerase and the subsequent binding of the polymerase to the stem-loop. This complex interacts with dimerised core protein to form the viral capsid (Shafritz et al., 1984; Monjardino, 1998; Seeger et al., 2000; Fields et al., 2001). The assembly of the polymerase and pgRNA (RNP, ribonucleoprotein) has been found to require host molecular chaperones (heat shock protein 90, hsp90) and ATP (Hu et al., 1996, 1997). One proposed function of the stably complexed chaperones in the RNP is to maintain the P protein in a conformation that is competent for noncovalent interactions with assembling C subunits.

Reverse transcription/DNA replication is initiated once pgRNA has been encapsidated (Monjardino, 1998). Pregenomic RNA is terminally redundant; it's approximately 200-nucleotide redundancies (termed R) include the ε stem-loop and a short sequence of 11 nucleotides termed Direct repeats (DR). Pregenomic RNA contains a 5' copy of DR1, a copy of DR2, and an extra copy of DR1 at the 3' end. Fig. 3.8 shows the general steps in HBV reverse transcription/DNA replication. Here, the priming hydroxyl group is supplied, not by a 3'OH of DNA or RNA, but by the side chain of tyrosine 63 residue

in the P protein (terminal protein). After the binding of the polymerase priming domain to the bulge of ε, a 5'-UUAC-3' motif in the bulge is used as the
template for the initiation of reverse transcription (Weber et al., 1994; Rieger
et al., 1995). Both DR1 and ε are represented twice in the pgRNA template.
The fact that the 3' copy of ε can be deleted from pgRNA without blocking
viral replication (Seeger et al., 1990, 1991) suggests that it is the 5' copy of
ε, which functions in initiation. Due to the sequence homology between the
first four nucleotides of the minus strand and the DR1 element, the nascent
'polymerase-oligonucleotide complex is then translocated to the 3' copy of
DR1. Reverse transcription continues from the 3' DR1 of pgRNA, synthesizing a full length minus strand DNA in a 5' to 3' manner. As minus strand
DNA synthesis proceeds, the pgRNA template in the RNA-DNA hybrid is
degraded by the RNase H activity of the polymerase protein; this is necessary
to make negative-strand DNA available to act as template for positive-strand
synthesis. Probably due to the spatial distance between the reverse transcriptase domain and the RNase domain, all but the 18 nucleotides at the 5' end of
pgRNA is degraded (Loeb et al., 1991). This 5' capped 18 nucleotide RNA is
used as a template for plus strand synthesis. The final product of this elongation reaction is a negative-strand DNA that is actually terminally redundant by
about 8 nucleotides (Lien et al., 1986; Seeger et al., 1986; Will et al., 1987)
(known as r).

Synthesis of the HBV plus strand begins with another primer translocation
event. The 18 nucleotide capped RNA harbors the DR1 sequence. Because of
the homology between DR1 and DR2, the DR1 sequence on the capped RNA
translocates and primes to the DR2 sequence of the newly synthesized minus
strand (Lien et al., 1986; Seeger et al., 1986). Synthesis of the plus strand
DNA then initiates from this priming site and proceeds toward the 5' end of
the minus strand. Due to the linearity of the minus strand DNA, synthesis of
the plus strand has to stop at the 5' end of the template (Ganem et al., 1994).
However, because both the 5' and 3' ends of the minus strand contain the
8 bases terminal redundancy sequence, the polymerase can undergo another
template switch, often called circularization, in which the nascent positive
strand DNA (by its r homology) anneals to the 3' end of the negative strand
and circularizes the genome, which allows the synthesis of the positive-strand
to continue (Havert et al., 1997). Synthesis of the plus strand, however, does
not go to completion, resulting in an incomplete and variable lengths plus
strand. The reason for this premature termination is unknown. One suggestion
is that steric hindrance within the nucleocapsid prevents further DNA replication events (Monjardino, 1998). Another possible explanation is that there

may be insufficient quantities of deoxyribonucleotides in the capsid, which prevents further genome replication (Ganem *et al.*, 1994). As a result of this series of template switches, a characteristic partially doubled-stranded rcDNA genome, containing a full length negative strand with a terminal protein covalently attached to its 5′ end and a variable length positive strand with a 5′ capped RNA primer, is formed within the mature nucleocapsid (Shafritz *et al.*, 1984; Monjardino, 1998; Seeger *et al.*, 2000; Fields *et al.*, 2001).

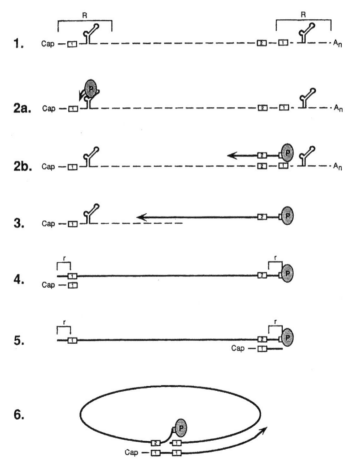

FIGURE 3.8 Mechanism of HBV DNA replication. The HBV polymerase is denoted (P) and HBV pregenomic RNA shown as discontinuous line. Hairpin structures at both ends of the pregenomic RNA represent the encapsidation signal ε. DR1 and DR2 elements are shown as boxed numbers "1" and "2." Details of replication steps are described in the text (Fields *et al.*, 2001).

Mature nucleocapsids can be either become enveloped and leave the host cell or be transported back to the nucleus. Hepatitis B surface proteins are synthesized in the ER as integral membrane proteins. These transmembrane monomers accumulate in the ER and simultaneously exclude host proteins from those parts of the ER. The mature nucleocapsid associates with areas of the ER containing high concentrations of HBV surface proteins. LHBs have been suggested to play an essential role in the envelopment of viral capsids, as demonstrated by their affinity to mature nucleocapsids (Bruss and Ganem, 1991; Bruss et al., 1994; Gerelsaikhan et al., 1996). The large surface protein is believed to interact with core protein at the cytoplasmic face of the ER membrane, pulling the nucleocapsid into the forming vesicle, and ultimately resulting in the 42 nm enveloped virion particle. The enveloped nucleocapsids are then translocated to the Golgi apparatus where further processing of the envelope proteins occurs and are secreted from the cell as mature virions via the constitutive cellular secretory pathway. However, not all nucleocapsids are secreted, some are recycled back to the host cell nucleus, where the rcDNA genome is released and more cccDNA is synthesized. As HBV DNA is incapable of semiconservative DNA-to-DNA replication, a stable pool of cccDNA molecules, and hence persistent infection, is maintained by this indirect, RNA-mediated replication. The two distinct destinies of the mature nucleocapsids are regulated by the level of LHBs (Summers et al., 1990; Lenhoff et al., 1994b). This control is of particular importance, as mutations in DHBV LHBs result in the accumulation of cccDNA in infected cells, which in turn is cytopathic (Lenhoff et al., 1994a; Lenhoff et al., 1999).

3.12 CONCLUSION

Thus, at an early phase of infection where the amount of surface proteins in infected cells is low, most of the nucleocapsids enter the recycle pathway. It is not until a later stage, when there are sufficient amounts of surface proteins, that the nucleocapsids are enveloped. In this way, the viral components are efficiently used during the early stage of infection. HBc is also suggested to contribute to this regulation, as the recycling of mature nucleocapsids is shown to be dependent on the phosphorylation of HBc NLS domain (Rabe et al., 2003).

KEYWORDS

- **Acute hepatitis B infection**
- **Chronic hepatitis B infection**
- **Epsilon-stem loop**
- **Glucocorticoid-responsive element**
- **HBV life cycle**
- **Hepatitis B surface antigen**
- **Hepatitis B virus**
- **Interferon alpha-2b**
- **Lamivudine**
- **Middle hepatitis B surface protein**
- **Peripheral blood mononuclear cells**
- **Pregenomic RNA**
- **Specific cytotoxic T lymphocytes**
- **Viral hepatitis**
- **X-responsive element**

REFERENCES

Antonucci, T. K.; Rutter, W. J. Hepatitis B virus (HBV) promoters are regulated by the HBV enhancer in a tissue-specific manner. *J. Virol.* 1989, 63(2), 579–583.

Ben Levy, R.; Faktor, O.; Berger, I.; Shaul, Y. Cellular factors that interact with the hepatitis B virus enhancer. *Mol. Cell Biol.* 1989, 9(4), 1804–1809.

Bottcher, B.; Wynne, S. A.; Crowther, R. A. Determination of the fold of the core protein of hepatitis B virus by electron cryomicroscopy. *Nature* 1997, 386(6620), 88–91.

Brunetto, M. R.; Giarin, M.; Saracco, G.; Oliveri, F.; Calvo, P.; Capra, G.; Randone, A.; Abate, M. L.; Manzini, P.; Capalbo, M. Hepatitis B virus unable to secrete e antigen and response to interferon in chronic hepatitis B. *Gastroenterology* 1993, 105(3), 845–850.

Bruss, V.; Ganem, D. The role of envelope proteins in hepatitis B virus assembly. *Proc. Natl. Acad. Sci. USA* 1991, 88(3), 1059–1063.

Bruss, V.; Lu, X.; Thomssen, R.; Gerlich, W. H. Post-translational alterations in transmembrane topology of the hepatitis B virus large envelope protein. *EMBO J.* 1994, 13(10), 2273–2279.

Carman, W. F.; Fagan, E. A.; Hadziyannis, S.; Karayiannis, P.; Tassopoulos, N. C.; Williams, R.; Thomas, H. C. Association of a precore genomic variant of hepatitis B virus with fulminant hepatitis. *Hepatology* 1991, 14(2), 219–222.

Carman, W. F.; Jacyna, M. R.; Hadziyannis, S.; Karayiannis, P.; McGarvey, M. J.; Makris, A.; Thomas, H. C. Mutation preventing formation of hepatitis B e antigen in patients with chronic hepatitis B infection. *Lancet* 1989, 2(8663), 588–591.

Chang, H. K.; Ting, L. P. The surface gene promoter of the human hepatitis B virus displays a preference for differentiated hepatocytes. *Virology* 1989, 170(1), 176–183.

Chen, H. S.; Kew, M. C.; Hornbuckle, W. E.; Tennant, B. C.; Cote, P. J.; Gerin, J. L.; Purcell, R. H.; Miller, R. H. The precore gene of the woodchuck hepatitis virus genome is not essential for viral replication in the natural host. *J. Virol.* 1992, 66(9), 5682–5684.

Chen, M.; Hieng, S.; Qian, X.; Costa, R.; Ou, J. H. Regulation of hepatitis B virus ENI enhancer activity by hepatocyte-enriched transcription factor HNF3. *Virology* 1994, 205(1), 127–132.

Chiang, P. W.; Jeng, K. S.; Hu, C. P.; Chang, C. M. Characterization of a cis element required for packaging and replication of the human hepatitis B virus. *Virology* 1992, 186(2), 701–711.

Chisari, F. V. Hepatitis B virus transgenic mice: insights into the virus and the disease. *Hepatology* 1995, 22, 1316–1325.

Chisari, F. V.; Ferrari, C. Hepatitis B virus immunopathogenesis. *Annu. Rev. Immunol.* 1995a, 13, 29–60.

Chisari, F. V.; Ferrari, C. Hepatitis B virus immunopathology. *Springer Semin. Immunopathol.* 1995b, 17(2–3), 261–281.

Cote, P. J.; Korba, B. E.; Miller, R. H.; Jacob, J. R.; Baldwin, B. H.; Hornbuckle, W. E.; Purcell, R. H.; Tennant, B. C.; Gerin, J. L. Effects of age and viral determinants on chronicity as an outcome of experimental woodchuck hepatitis virus infection. *Hepatology* 2000, 31(1), 190–200.

Crowther, R. A.; Kiselev, N. A.; Bottcher, B.; Berriman, J. A.; Borisova, G. P.; Ose, V.; Pumpens, P. Three-dimensional structure of hepatitis B virus core particles determined by electron cryomicroscopy. *Cell* 1994, 77(6), 943–950.

Dejean, A.; Lugassy, C.; Zafrani, S.; Tiollais, P.; Brechot, C. Detection of hepatitis B virus DNA in pancreas, kidney and skin of two human carriers of the virus. *J. Gen. Virol.* 1984, 65(3), 651–655.

Dienstag, J. L.; Perrillo, R. P.; Schiff, E. R.; Bartholomew, M.; Vicary, C.; Rubin, M. A preliminary trial of lamivudine for chronic hepatitis B infection. *N. Engl. J. Med.* 1995, 333(25), 1657–1661.

Dikstein, R.; Faktor, O.; Shaul, Y. Hierarchic and cooperative binding of the rat liver nuclear protein C/EBP at the hepatitis B virus enhancer. *Mol. Cell Biol.* 1990, 10(8), 4427–4430.

Dryden, K. A.; Wieland, S. F.; Whitten-Bauer, C.; Gerin, J. L.; Chisari, F. V.; Yeager, M. Native hepatitis B virions and capsids visualized by electron cryomicroscopy. *Mol. Cell* 2006, 22(6), 843–850.

Eble, B. E.; Lingappa, V. R.; Ganem, D. Hepatitis B surface antigen: an unusual secreted protein initially synthesized as a transmembrane polypeptide. *Mol. Cell Biol.* 1986, 6(5), 1454–1463.

Farza, H.; Salmon, A. M.; Hadchouel, M.; Moreau, J. L.; Babinet, C.; Tiollais, P.; Pourcel, C. Hepatitis B surface antigen gene expression is regulated by sex steroids and glucocorticoids in transgenic mice. *Proc. Natl. Acad. Sci. USA* 1987, 84(5), 1187–1191.

Fields, B. N.; Knipe, D. M.; Howley, P. M.; Griffin, D. E. *Fields virology*, 4th ed.; Lippincott Williams & Wilkins: Philadelphia, 2001.

Flint, S. J. *Principles of virology: molecular biology, pathogenesis and control*, ASM Press: Washington, D.C., 2000.

Ganem, D.; Pollack, J. R.; Tavis, J. Hepatitis B virus reverse transcriptase and its many roles in hepadnaviral genomic replication. *Infect. Agents Dis.* 1994, 3(2–3), 85–93.

Garcia, A. D.; Ostapchuk, P.; Hearing, P. Functional interaction of nuclear factors EF-C, HNF-4, and RXR alpha with hepatitis B virus enhancer I. *J. Virol.* 1993, 67(7), 3940–3950.

Gerelsaikhan, T.; Tavis, J. E.; Bruss, V. Hepatitis B virus nucleocapsid envelopment does not occur without genomic DNA synthesis. *J. Virol.* 1996, 70(7), 4269–4274.

Guo, W. T.; Bell, K. D.; Ou, J. H. Characterization of the hepatitis B virus EnhI enhancer and X promoter complex. *J. Virol.* 1991, 65(12), 6686–6692.

Havert, M. B.; Loeb, D. D. *Cis*-acting sequences in addition to donor and acceptor sites are required for template switching during synthesis of plus-strand DNA for duck hepatitis B virus. *J. Virol.* 1997, 71(7), 5336–5344.

Heermann, K. H.; Goldmann, U.; Schwartz, W.; Seyffarth, T.; Baumgarten, H.; Gerlich, W. H. Large surface proteins of hepatitis B virus containing the pres sequence. *J. Virol.* 1984, 52(2), 396–402.

Hirsch, R. C.; Loeb, D. D.; Pollack, J. R.; Ganem, D. *Cis*-acting sequences required for encapsidation of duck hepatitis B virus pregenomic RNA. *J. Virol.* 1991, 65(6), 3309–3316.

Honigwachs, J.; Faktor, O.; Dikstein, R.; Shaul, Y.; Laub, O. Liver-specific expression of hepatitis B virus is determined by the combined action of the core gene promoter and the enhancer. *J. Virol.* 1989, 63(2), 919–924.

Hu, J. Seeger, C. Hsp90 is required for the activity of a hepatitis B virus reverse transcriptase. *Proc. Natl. Acad. Sci. USA* 1996, 93(3), 1060–1064.

Hu, J.; Toft, D. O.; Seeger, C. Hepadnavirus assembly and reverse transcription require a multicomponent chaperone complex, which is incorporated into nucleocapsids. *EMBO J.* 1997, 16(1), 59–68.

Huang, Z. M.; Yen, T. S. Role of the hepatitis B virus posttranscriptional regulatory element in export of intronless transcripts. *Mol. Cell Biol.* 1995, 15(7), 3864–3869.

Kock, J.; Theilmann, L.; Gallo, P.; Schlicht, H. J. Hepatitis B virus nucleic acids associated with human peripheral blood mononuclear cells do not originate from replicating virus. *Hepatology* 1996, 23(3), 405–413.

Koshy, R.; Caselmann, W. H. *Hepatitis B Virus: molecular mechanisms in disease and novel strategies for therapy*, Imperial College Press: London, 1998.

Lenhoff, R. J.; Luscombe, C. A.; Summers, J. Acute liver injury following infection with a cytopathic strain of duck hepatitis B virus. *Hepatology* 1999, 29(2), 563–571.

Lenhoff, R. J. Summers, J. Construction of avian hepadnavirus variants with enhanced replication and cytopathicity in primary hepatocytes. *J. Virol.* 1994a, 68(9), 5706–5713.

Lenhoff, R. J.; Summers, J. Coordinate regulation of replication and virus assembly by the large envelope protein of an avian hepadnavirus. *J. Virol.* 1994b, 68(7), 4565–4571.

Li, M.; Xie, Y.; Wu, X.; Kong, Y.; Wang, Y. HNF3 binds and activates the second enhancer, ENII, of hepatitis B virus. *Virology* 1995, 214(2), 371–378.

Li, M.; Xie, Y. H.; Kong, Y. Y.; Wu, X.; Zhu, L.; Wang, Y. Cloning and characterization of a novel human hepatocyte transcription factor, hB1F, which binds and activates enhancer II of hepatitis B virus. *J. Biol. Chem.* 1998, 273(44), 29022–29031.

Lieberman, H. M.; Tung, W. W.; Shafritz, D. A. Splenic replication of hepatitis B virus in the chimpanzee chronic carrier. *J. Med. Virol.* 1987, 21(4), 347–359.

Lien, J. M.; Aldrich, C. E.; Mason, W. S. Evidence that a capped oligoribonucleotide is the primer for duck hepatitis B virus plus-strand DNA synthesis. *J. Virol.* 1986, 57(1), 229–236.

Lo, W. Y.; Ting, L. P. Repression of enhancer II activity by a negative regulatory element in the hepatitis B virus genome. *J. Virol.* 1994, 68(3), 1758–1764.

Loeb, D. D.; Hirsch, R. C.; Ganem, D. Sequence-independent RNA cleavages generate the primers for plus strand DNA synthesis in hepatitis B viruses: implications for other reverse transcribing elements. *EMBO J.* 1991, 10(11), 3533–3540.

Lopez-Cabrera, M.; Letovsky, J.; Hu, K. Q.; Siddiqui, A. Multiple liver-specific factors bind to the hepatitis B virus core/pregenomic promoter: transactivation and repression by CCAAT/ enhancer binding protein. *Proc. Natl. Acad. Sci. USA* 1990, 87(13), 5069–5073.

Lopez-Cabrera, M.; Letovsky, J.; Hu, K. Q.; Siddiqui, A. Transcriptional factor C/EBP binds to and transactivates the enhancer element II of the hepatitis B virus. *Virology* 1991, 183(2), 825–829.

Lu, C. C.; Yen, T. S. Activation of the hepatitis B virus S promoter by transcription factor NF-Y via a CCAAT element. *Virology* 1996, 225(2), 387–394.

Mehdi, H.; Kaplan, M. J.; Anlar, F. Y.; Yang, X.; Bayer, R.; Sutherland, K.; Peeples, M. E. Hepatitis B virus surface antigen binds to apolipoprotein H. *J. Virol.* 1994, 68(4), 2415–2424.

Milich, D.; Liang, T. J. Exploring the biological basis of hepatitis B e antigen in hepatitis B virus infection. *Hepatology* 2003, 38(5), 1075–1086.

Milich, D. R.; Jones, J. E.; Hughes, J. L.; Price, J.; Raney, A. K.; McLachlan, A. Is a function of the secreted hepatitis B e antigen to induce immunologic tolerance in utero?. *Proc. Natl. Acad. Sci. USA* 1990, 87(17), 6599–6603.

Miller, R. H.; Kaneko, S.; Chung, C. T.; Girones, R.; Purcell, R. H. Compact organization of the hepatitis B virus genome. *Hepatology* 1989, 9(2), 322–327.

Monjardino, J. *Molecular biology of human hepatitis viruses*, Imperial College Press: London, 1998.

Moriyama, T.; Guilhot, S.; Klopchin, K.; Moss, B.; Pinkert, C. A.; Palmiter, R. D.; Brinster, R. L.; Kanagawa, O.; Chisari, F. V. Immunobiology and pathogenesis of hepatocellular injury in hepatitis B virus transgenic mice. *Science* 1990, 248(4953), 361–364.

Neurath, A. R.; Kent, S. B.; Strick, N.; Parker, K. Identification and chemical synthesis of a host cell receptor binding site on hepatitis B virus. *Cell* 1986, 46(3), 429–436.

Previsani, N.; Lavanchy, D. *Hepatitis B*, World Health Organization: Geneva, 2002.

Nigg, E. A. Nucleocytoplasmic transport: signals, mechanisms and regulation. *Nature* 1997, 386(6627), 779–787.

Ogata, N.; Miller, R. H.; Ishak, K. G.; Purcell, R. H. The complete nucleotide sequence of a precore mutant of hepatitis B virus implicated in fulminant hepatitis and its biological characterization in chimpanzees. *Virology* 1993, 194(1), 263–276.

Ou, J. H.; Laub, O.; Rutter, W. J. Hepatitis B virus gene function: the precore region targets the core antigen to cellular membranes and causes the secretion of the e antigen. *Proc. Natl. Acad. Sci. USA* 1986, 83(6), 1578–1582.

Persing, D. H.; Varmus, H. E.; Ganem, D. The pre-S1 protein of hepatitis B virus is acylated at its amino terminus with myristic acid. *J. Virol.* 1987, 61(5), 1672–1677.

Petit, M. A.; Pillot, J. HBc and HBe antigenicity and DNA-binding activity of major core protein P22 in hepatitis B virus core particles isolated from the cytoplasm of human liver cells. *J. Virol.* 1985, 53(2), 543–551.

Pollack, J. R.; Ganem, D. Site-specific RNA binding by a hepatitis B virus reverse transcriptase initiates two distinct reactions: RNA packaging and DNA synthesis. *J. Virol.* 1994, 68(9), 5579–5587.

Pontisso, P.; Petit, M. A.; Bankowski, M. J.; Peeples, M. E. Human liver plasma membranes contain receptors for the hepatitis B virus pre-S1 region and via polymerized human serum albumin, for the pre-S2 region. *J. Virol.* 1989, 63(5), 1981–1988.

Rabe, B.; Vlachou, A.; Pante, N.; Helenius, A.; Kann, M. Nuclear import of hepatitis B virus capsids and release of the viral genome. *Proc. Natl. Acad. Sci. USA* 2003, 100(17), 9849–9854.

Raney, A. K.; Johnson, J. L.; Palmer, C. N.; McLachlan, A. Members of the nuclear receptor superfamily regulate transcription from the hepatitis B virus nucleocapsid promoter. *J. Virol.* 1997, 71(2), 1058–1071.

Rieger, A.; Nassal, M. Distinct requirements for primary sequence in the 5'- and 3'- part of a bulge in the hepatitis B virus RNA encapsidation signal revealed by a combined *in vivo* selection/*in vitro* amplification system. *Nucleic Acids Res.* 1995, 23(19), 3909–3915.

Ryu, C. J.; Cho, D. Y.; Gripon, P.; Kim, H. S.; Guguen-Guillouzo, C.; Hong, H. J. An 80-kilodalton protein that binds to the pre-S1 domain of hepatitis B virus. *J. Virol.* 2000, 74(1), 110–116.

Schlicht, H. J.; Salfeld, J.; Schaller, H. The duck hepatitis B virus preC region encodes a signal sequence, which is essential for synthesis and secretion of processed core proteins but not for virus formation. *J. Virol.* 1987, 61(12), 3701–3709.

Schlicht, H. J.; Schaller, H. Analysis of hepatitis B virus gene functions in tissue culture and *in vivo*. *Curr. Top. Microbiol. Immunol.* 1989, 144, 253–263.

Seeger, C.; Ganem, D.; Varmus, H. E. Biochemical and genetic evidence for the hepatitis B virus replication strategy. *Science* 1986, 232(4749), 477–484.

Seeger, C.; Maragos, J. Identification and characterization of the woodchuck hepatitis virus origin of DNA replication. *J. Virol.* 1990, 64(1), 16–23.

Seeger, C.; Maragos, J. Identification of a signal necessary for initiation of reverse transcription of the hepadnavirus genome. *J. Virol.* 1991, 65(10), 5190–5195.

Seeger, C.; Mason, W. S. Hepatitis B virus biology. *Microbiol. Mol. Biol. Rev.* 2000, 64(1), 51–68.

Shafritz, D. A.; Lieberman, H. M. The molecular biology of hepatitis B virus. *Annu. Rev. Med.* 1984, 35, 219–232.

Shaul, Y.; Rutter, W. J.; Laub, O. A human hepatitis B viral enhancer element. *EMBO J.* 1985, 4(2), 427–430.

Shimoda, T.; Shikata, T.; Karasawa, T.; Tsukagoshi, S.; Yoshimura, M.; Sakurai, I. Light microscopic localization of hepatitis B virus antigens in the human pancreas: possibility of multiplication of hepatitis B virus in the human pancreas. *Gastroenterology* 1981, 81(6), 998–1005.

Siddiqui, A.; Jameel, S.; Mapoles, J. Expression of the hepatitis B virus X gene in mammalian cells. *Proc. Natl. Acad. Sci. USA* 1987, 84(8), 2513–2517.

Simonsen, C. C.; Levinson, A. D. Analysis of processing and polyadenylation signals of the hepatitis B virus surface antigen gene by using simian virus 40-hepatitis B virus chimeric plasmids. *Mol. Cell Biol.* 1983, 3(12), 2250–2258.

Standring, D. N.; Ou, J. H.; Masiarz, F. R.; Rutter, W. J. A signal peptide encoded within the precore region of hepatitis B virus directs the secretion of a heterogeneous population of e antigens in Xenopus oocytes. *Proc. Natl. Acad. Sci. USA* 1988, 85(22), 8405–8409.

Stibbe, W.; Gerlich, W. H. Structural relationships between minor and major proteins of hepatitis B surface antigen. *J. Virol.* 1983, 46(2), 626–628.

Su, H.; Yee, J. K. Regulation of hepatitis B virus gene expression by its two enhancers. *Proc. Natl. Acad. Sci. USA* 1992, 89(7), 2708–2712.

Summers, J.; O'Connell, A.; Millman, I. Genome of hepatitis B virus: restriction enzyme cleavage and structure of DNA extracted from Dane particles. *Proc. Natl. Acad. Sci. USA* 1975, 72(11), 4597–4601.

Summers, J.; Smith, P. M.; Horwich, A. L. Hepadnavirus envelope proteins regulate covalently closed circular DNA amplification. *J. Virol.* 1990, 64(6), 2819–2824.

Thomas, H. C.; Lemon, S. M.; Zuckerman, A. J. *Viral hepatitis*, 3rd ed.; Blackwell Publishing: Malden, 2005.

Trujillo, M. A.; Letovsky, J.; Maguire, H. F.; Lopez-Cabrera, M.; Siddiqui, A. Functional analysis of a liver-specific enhancer of the hepatitis B virus. *Proc. Natl. Acad. Sci. USA* 1991, 88(9), 3797–3801.

Tur-Kaspa, R.; Burk, R. D.; Shaul, Y.; Shafritz, D. A. Hepatitis B virus DNA contains a glucocorticoid-responsive element. *Proc. Natl. Acad. Sci. USA* 1986, 83(6), 1627–1631.

Weber, M.; Bronsema, V.; Bartos, H.; Bosserhoff, A.; Bartenschlager, R.; Schaller, H. Hepadnavirus P protein uses a tyrosine residue in the TP domain to prime reverse transcription. *J. Virol.* 1994, 68(5), 2994–2999.

Will, H.; Reiser, W.; Weimer, T.; Pfaff, E.; Buscher, M.; Sprengel, R.; Cattaneo, R.; Schaller, H. Replication strategy of human hepatitis B virus. *J. Virol.* 1987, 61(3), 904–911.

Wong, D. K.; Cheung, A. M.; O'Rourke, K.; Naylor, C. D.; Detsky, A. S.; Heathcote, J. Effect of alpha-interferon treatment in patients with hepatitis B e antigen-positive chronic hepatitis B. A meta-analysis. *Ann. Intern. Med.* 1993, 119(4), 312–323.

Yu, X. Mertz, J. E. Differential regulation of the preC and pregenomic promoters of human hepatitis B virus by members of the nuclear receptor superfamily. *J. Virol.* 1997, 71(12), 9366–9374.

Yuh, C. H.; Ting, L. P. Differentiated liver cell specificity of the second enhancer of hepatitis B virus. *J. Virol.* 1993, 67(1), 142–149.

CHAPTER 4

RECONFIGURABLE PARALLEL PLATFORMS FOR DEPENDABLE PERSONALIZED HEALTH MONITORING

OLUFEMI ADELUYI and JEONG-A LEE

CONTENTS

4.1   INTRODUCTION

Health care delivery efforts have evolved over the years to reflect the available technology, medication, approaches to treatment and global economic realities. Medical diagnostics is a platform upon which many other sectors of healthcare delivery are built and its level of development dictates the level of development in the other sectors as well. A fundamental aspect of medical diagnostics is the acquisition of information upon which a clinical decision can be based on (Splinter, 2010). The overarching theme of the twenty-first century healthcare delivery evolution has been the paradigm shift from curative to preventive medicine and from a doctor-centered approach to a patient-centered one. This paradigm shift allowed diagnostic medicine to assume an even greater level of importance and is responsible for the growing trend of Personalized Health Monitoring (PHM). Health monitoring refers to a systematic approach for keeping track of the health status of any given entity. It can refer to both animate and inanimate objects. For example, the process of monitoring the degradation of infrastructure like bridges and buildings falls under this category, as does the process of monitoring the health of a human being. In this chapter, our definition of PHM refers to the process of monitoring the vital and nonvital health signals of a human being. PHM is useful for monitoring signals that change over time. It is not suited to readings that rarely change or those that do not change, like a patient's genotype.

4.1.1   PHM SYSTEMS: COVERAGE AND STRUCTURE

PHM refers to long-term health monitoring that is performed by relatively novice patients in uncontrolled environments, such as their homes (Pärkkä, 2011). These systems use sensors to capture bio-signals from the patient and process the signals in order to enable the patients to track, predict and maintain their health. The signals consist of a succession of analog or digital values (mainly analog). There are four main health vital signs that are used to monitor a patient's health status. They are the heart rate, body temperature, blood pressure and respiratory rate. Many times, other signals outside this group are also monitored. In some cases, especially where observed bio-signals exceed some predefined safety threshold, the results can also be sent to a remote monitoring physician.

Three key technologies have aided the development of PHM: advances in wireless technologies, advances in sensor technologies, and advances in embedded computing technologies. A standard PHM system consists of a suite of

relevant sensors to receive the patient's physiological signals, some front end
electronics for analog-to-digital conversion, the PHM processing hub, output
device(s) and, optionally, devices for remote communication. These systems
are usually designed to support mobility, with key requirements that include
intuitive interface, lightweight PHM processing hub (low resource overhead,
low performance overhead and low area overhead), portable power supply,
efficient analysis and decision algorithm, and, intelligent communication interface. The PHM devices are made up of embedded systems on chip and the
environment is usually resource-constrained. This increases the attraction for
lightweight solutions. In theory, within the acceptable limits of processing
requirements, a PHM system can be used to monitor a large percentage of all
the physiological signals that can be monitored in a standard clinical environment. The key is having the appropriate sensor and running the relevant algorithm on the PHM hub. Figure 4.1 shows the approximate location of PHM
systems (used for personalized healthcare) in the healthcare delivery research
space; toward the top ends of preventive and automated medicine.

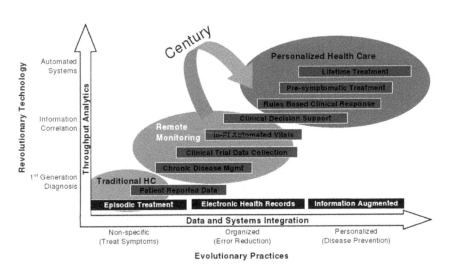

FIGURE 4.1   Location of PHM systems on the Healthcare Delivery Research Space.

4.1.2   RECONFIGURATION IN PHM SYSTEMS

In all spheres of endeavor, technology vendors have increased the level of
device customization in order to satisfy the specific needs of individual users.

For a system to be customizable, it needs to be able to undergo reconfiguration in response to some stimuli fed back into the system. In the case of PHM systems, this reconfiguration can be initiated when certain vital sign thresholds have been exceeded or even just as a process of adapting the features of a device to a new user (Mouttham *et al.*, 2009). Reconfigurable systems have an architecture that takes advantage of the high performance of hardware as well as the high flexibility of software and the Field Programmable Gate Arrays (FPGAs) represent the best example of such systems. For example, in Huang and Lee (2009), the FPGA is reconfigured to conserve power used in implementing video coding algorithms while in Alemzadeh *et al.* (2011) and Jiang *et al.* (2005) the reconfiguration is used to change the functionality of the system. Run-time reconfiguration is defined as the ability of a system to use either hardware or software approaches to modify or change the functional configuration of the device during operation. It enables system flexibility, system reuse, power reduction and the avoidance of system obsolescence, among other things (http://goo.gl/wfxbK). Traditional reconfiguration usually requires the system to be offline for a minimal duration. This requirement can be avoided through the choice of an alternative partial reconfiguration approach. In this approach, the system is stratified into static and reconfigurable modules. The core system functionality is programmed into the static module and is thus unaffected by reconfiguration of the reconfigurable modules at runtime (Huang and Lee, 2009).

4.1.3 DEPENDABILITY IN PHM SYSTEMS

The *dependability* of a system is a measure of its availability, reliability and maintenance. It is the ability of a system to deliver a service that can justifiably be trusted and the ability to avoid failures that are more frequent and more severe than is acceptable (Avizienis *et al.*, 2004). As a result of the sensitive nature of healthcare, the ultimate decision of the medical choices that are made lies with the physician. However, a lot will depend on the reliability of the data made available to them by the PHM systems. As such, the accuracy of the data provided by these systems goes a long way in determining the type of treatment administered and, by extension, the recovery of the patient, aggravation of the disease or, in dire cases, the death of the patient. Thus, PHM systems must be dependable enough for caregivers to base their decisions on. A standard PHM system comprises a number of constituent parts, each potentially susceptible to faults that can undermine the reliability of the full system. Sabry *et al.* (2012) discussed the vulnerabilities of such constituent parts, especially in the area of scaled voltages, and how they can undermine the reliability of the entire system. Taking such challenges into account will enable the design to accommodate the strict levels of tolerance that are usually associated with medical devices (Milenković *et al.*, 2006). The focus of many dependability research efforts in contemporary PHM systems has been on ensuring reliability with respect to the wireless transfer of the signals between the patient-end and the physician-end.

4.1.4 PARALLELIZATION IN PHM SYSTEMS

Computers are generally designed to accept data, carry out some processing and produce a given output. Traditional processing techniques have involved a single Central Processing Unit (CPU) designed to run software that are written for serial computation, allowing instructions to be processed one after the other. However, the increasing complexity and resource demands of applications in fields such as medicine, geology, entertainment and financial modeling far outstrip the capacity of regular serial computing systems. An alternative approach is to split a given computation among multiple processing platforms, a process known as parallelization. Current parallelization approaches in PHM systems enable these systems to process multiple vital parameters simultaneously (Rousselot and Decotignie, 2009; Panagiotako-poulos *et al.*, 2010; Zhang *et al.*, 2013). The processing of multiple parameters is only a basic use of the potential of parallelization and PHM systems can well benefit from the other big advantages of parallelization.

Common advantages of FPGAs include system speed-up, reduced implementation costs, increased flexibility and increased computational capacity of systems. A few examples of typical speed-up values of Digital Signal Processing (DSP) algorithms implemented on FPGAs, as compared to CPU implementations, include DES data encryption algorithm (100x), video applications (20–100x) and the Smith-Waterman pairwise protein sequence alignment (120–200x) (Schmidt, 2010). These days it has become routine to implement a system that combines the high performance of hardware with the high flexibility of software in a hardware–software codesigned system. Many times the processor representing the software part of the system is implemented as a soft-core processor on the FPGA. Some proprietary softcore processors include: MicroBlaze (Xilinx), NiOS (Altera) and FPGA-optimized Cortex M1 (ARM). Embedded processors form the core of a great number of systems across the technology spectrum.

4.2 TAXONOMY

The term "personalized health monitoring" has been described in many research papers using different naming schemes. A brief list of terms that have some connotation to PHM system are mobile Health (mHealth), ubiquitous healthcare, pervasive healthcare, care-in-home, edge care, personal health system, ambient assisted living, aging well and aging-in-place. The taxonomy presented here classifies PHM systems based on the different salient features that characterize them. We have chosen "3As" – Architecture, Application and Adaptation – to categorize the PHM systems as shown in Fig. 4.2. Some intersection may occur between some of the groupings.

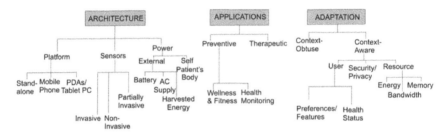

FIGURE 4.2   Taxonomy of PHM systems.

4.2.1   ARCHITECTURE

In classifying the architecture of PHM systems, we have considered the hardware and software components that exist between the sensing device that measures the relevant signal from the patient and the remote healthcare professional that makes the medical decisions based on those signals. The platform comprises components at the patient-end, those at the physician's end and the pathway in between. As a result of their growing ubiquity, mobile phones are now used in a large proportion of patient-end PHM systems. Apart from the mobile phone, other devices like Personal Digital Assistants (PDAs) and portable devices like the laptop are also used. Stand-alone devices refer to devices that are neither mobile phones nor PDAs/laptops. They are devices that are usually specifically designed for the PHM system. Desktop computers and higher end servers that may be used at the physician-end also fall under this category. They are used to provide more compute power for analysis and for simultaneous management of multiple remote patients. Other types of stand-alone systems can include nonconventional options like vehicle-based or home-based platforms.

It is also worth noting that many PHM systems use wireless networks as pathways to communicate between the patient-end and the physician-end devices. The most common wireless technologies used for this purpose are Bluetooth, IrDA, ZigBee, GSM/CDMA and WiFi. The sensing technology arguably represents the most important technology for the adoption and growth of PHM. There are several sensors that are used in PHM systems to measure different types of signals at the patient-end and they can be categorized based on the level of invasiveness. Invasiveness refers to a need for a puncture of a patient's skin or the introduction of a foreign body into a patient in order to monitor some signal. Thus, PHM systems can be classified based on access as invasive, noninvasive or partially invasive systems. Invasive systems include hemodynamic monitors (Øyri et al., 2010) and partially invasive systems include glucose meters (Shahriyar et al., 2009).

The sensed signals include temperature, ECG, EEG, Galvanic Skin Response (GSR), Electromyogram (EMG), blood pressure, glucose level, gait and oxygen saturation. Table 4.1 shows a list of common sensors used in PHM, the types of signals that they sense and the general disease domain where they are used. The PHM systems are either self-powered or they draw power from external sources, usually from a battery. These externally powered systems may also rely on energy harvesting techniques to power the system. Self powered systems rely on energy generated in the body of the patient; energy from body heat and motion are common.

TABLE 4.1 Type of sensors, sensing agents and disease domains.

Sensor	Sensed signal	Disease domain
ECG/EKG	Electrical activity of the heart	Cardiovascular health
GSR	Skin conductance	Stress response, polygraphy, psychotherapy
Thermometer	Temperature	General health
Chemical bio-sensor	Toxicity	Septicaemia
Accelerometer	Motion	Elderly fall detection
Saturation of Peripheral Oxygen (SPO$_2$)	Oxygen saturation	Cardio-respiratory health

TABLE 4.1   *(Continued)*

Sensor	Sensed signal	Disease domain
EEG	Electrical activity along the scalp	Brain health
Blood pressure	Blood induced pressure on the arteries	General health, kidney health
Piezoresistive sensor	Respiration rate	Respiratory health
EMG	Electrical activity in the muscle	Neuromuscular health
Strip base glucose meter	Glucose level	Diabetes
Photoplethysmograph (PPG)	Blood volume pulse	Cardiovascular health

4.2.2   APPLICATIONS

PHM systems are used for a wide range of applications that are targeted at a broad range of users from the fitness-conscious all the way up to the critically ill. We have categorized these applications into two broad classes, namely preventive and therapeutic. The preventive applications focus on wellness and fitness, as well as, traditional monitoring, while the therapeutic applications include the ability to contribute to the patient's healing process. Examples of therapeutic actions include automatic drug administration (Malaguttiy *et al.*, 2012) and automated defibrillation (Amann *et al.*, 2007). The other two subclassifications are based on the number of parameters analyzed and the hardware/software component of the application. Based on the number of parameters that are being sensed, the system can be classified as a unitary-parameter or multiparameter system. The unitary parameter class refers to PHM systems that have been designed to monitor a single bio-signal. This is the approach used by many standard equipment used in the clinical environment and it is also used by a few PHM systems. A much higher percentage of PHM systems fall under the multiparameter class. For the hardware/software based classification, we have software-based, hardware-based and hardware/software codesigned systems. The software-based systems have every part of the system is implemented as a software engine that runs on an off-the-shelf commodity device such as a mobile phone. The hardware based systems are wholly hardware driven. A more common approach is to have applications that run on a hardware–software codesigned platform.

4.2.3   ADAPTATION

This classification is based on the ability of the system to change its internal configuration and/or its action in response to changes in the context. The context-aware class refers to systems that have been designed to effect such changes while the context-obtuse systems are those that are invariant to changes in context. There are many types of contexts that can be used to trigger reconfiguration in the PHM system, depending on the priority of the system designer. The main ones used are based on awareness of the user status, level of privacy/security and status of the available resources. For the user status, this can include functionality to customize the readings for a specific individual and it usually requires an initial system training process. It can also include the ability of the system to change its configuration when the measured signal exceeds some predetermined threshold. For example, the systems may start out as a simple health monitor that does not include any remote communication states. It can then include these states after the threshold is exceeded in order to enable the system to alert a remote caregiver. Many times, medical information is classified as sensitive and they require some level of security to protect them from unauthorized access. However, increasing the level of security tends to increase the level of system complexity and, by extension, the power expended and the overall cost of the system. It is worth noting that not all the readings are accorded the same level of sensitivity. As an example, the temperature readings do not require the same level of privacy as mental test results. Such types of variations can be exploited in context-aware systems. In resource-aware systems, the level of resource utilization (such as power, memory or bandwidth) can be used to trigger a change in the system in a way that further preserves the resource.

4.3   PHM RESEARCH: TRENDS, DIRECTIONS AND CHALLENGES

PHM research is entering a level of maturity, following its introduction a little over a decade ago. Research teams around the world have focused on the various aspects of research we described in the taxonomy section and will use the taxonomy to organize our discussions on PHM research. Europe was ahead of the curve in PHM research, buoyed by the funds provided for ICT for aging research in the EU-FP7 project, along with the synergy and visibility it provides.

4.3.1 ARCHITECTURE: PROJECTS AND TRENDS

The portable and simple-to-use Tinke device by Zensorium (http://goo.gl/Hesfl) uses readings from a finger placed on the camera lens of an iPhone's dongle. It uses this to measure the multiparameters of heart rate, respiratory rate, blood oxygen level and heart rate variability. These values are then used to calculate two indices, namely vita index (for fitness score) and zen index (for stress level). The subject needs to attach the *Tinke* to an iPhone.

The mHealth Management and Rescue system (http://goo.gl/WJWcx) of IVT Corporation, China is another example of a custom mobile device that uses a parallel multiparameter approach to collect vital signs (ECG, blood pressure, glucose level, pulse oximeter, blood gas analyzer and cardiac troponim). The signals (including realtime image signals) are then sent via a 3G, bluetooth or satellite enabled tablet PC to the online E-Health record on IVT's cloud platform. Positive ID's verichip-*in-vivo* glucose sensing is an example of an invasive sensor. It is embedded in the patient and can be used to measure the glucose level in an approach that does not require a patient to draw his blood. It works with an iglucose device, which seamlessly communicates the readings to a diabetes management portal by sending an sms through an an embedded AirPrime GSM module. Microsoft's HealthVault (http://goo.gl/lNfcd) provides patient's with a free portal to their health information. It supports multiple industry standard formats like Continuity of Care Document and Continuity of Care Record and is compatible with 170 industry standard sensors. Textile sensors are viewed as the sensors of the future and several research teams are exploring the use of textiles to sense patient signals. They have the advantages of low cost, less invasiveness and have better esthetic value, among other things. Tom Martins and Mark Jones of the University of Virginia designed e-Textile Attached Gadgets (e-TAGs) as a cheap and reliable method for connecting electronic components to fabrics. The prototype uses a system architecture that consists of detachable PCB and a network of data and power lines woven into the fabric. Each of the e-TAGs has a microcontroller and uses the 1-wire system to support low communication bandwidth applications (Lehn *et al.*, 2004). These e-TAGs have been used with gyroscopes and accelerometers to identify the individuals with a higher risk of falling.

A recent research at the University of Boras, Sweden, explored the use of textile electrodes (also known as textrodes) in the measurement of electrical bioimpedance (Marquez *et al.*, 2009) for the assessment of total body composition and fluid distribution. The authors used the 4-Electrode method for wrist-to-ankle Electrical Bioimpedance Spectroscopy (EBIS) measurements. The Cole Parameter was estimated from the Modulus of the electrical bioimpeadance for assessment of body composition. The results closely matched those of the traditional Ag/AgCl electrode. Apart from sensors, these smart textiles may also include features like data processing, actuators, communication and storage. Platforms of PHM systems are usually mobile-based or PDA/tablet PC based. An example of the smartphone-based medical monitoring device was developed by the laboratory of Professor Gilwon Yoon at the Seoul National University. It uses a low power biosignal monitoring unit (BMU) as its hub and like a number of other projects, it incorporates the parallel processing of multiparameters like ECG, photoplethysmogram (PPG), temperature, oxygen saturation, energy expenditure and location information. The BMU is a separate unit that communicates with the smartphone via Bluetooth and uses the smartphone's 3G or WiFi network to communicate with a remote healthcare server. To increase system dependability, the researchers implemented a data compression and lightweight error correction scheme based on differential Huffman coding. This approach improves the efficiency of the transmission.

Power supply has always been an issue for mobile devices and it assumes an even greater level of importance for mobile medical devices, especially those that are implanted in the body. Interoperability between different vendors has also emerged as an important theme. The Open architecture for Accessible Services Integration and Standardization (OASIS) (Castro *et al.*, 2008) is an EU-FP7 project that addresses this need for an open, modular and holistic approach to the platform design of PHM systems, as well as the provision of a standard reference architecture for PHM systems. OASIS provides seamless connectivity between hardware, from hardware to service and from service to service. It provides a platform for monitoring multiple vital sign (including temperature, pulse rate, respiration rate and blood pressure). It is a context-aware platform that adjusts itself in response to the user's health status monitored over a wireless sensor network. Other EU FP7 projects involved in the provision of open platforms include the UNIVERsal open platform and reference Specification for Ambient Assisted Living (UNIVERSAAL) (Hanke *et al.*, 2011) and the Common Platform Services for Ageing Well in Europe (CommonWell) (http://goo.gl/8ufaa) projects. The success of any technology lies not just in its level of innovativeness, but much more in the willing adoption and usage by the intended user. With this in mind, a lot of innovativeness now goes into the process of getting the users to adopt the products. One of such products is MIT's health monitoring mirror. It takes advantage of the fact that most people stand before the mirror everyday to simultaneously provide them with health monitoring information on the mirror. It determines the heart rate from the light that reflects off the subject's body, since the level of reflected light is inversely proportional to the blood flow rate, which is proportional to the heart rate.

4.3.2 ARCHITECTURE: DIRECTION AND CHALLENGES

In terms of the platform, the research direction aims to increase the level of portability, while trying to integrate all the units into one device. Smartphones are emerging as the leading platform for that "one device." For sensors, non-invasive options are the overwhelming choice and the smart textiles are seen as an important area in this regard. Unfortunately, these smart textiles need circuits that are extremely rugged enough to remain dependable in the presence of mechanical stress and strain (Cherenack and Pieterson, 2011) as well as when immersed in fluids (like during washing). Another challenge is that they also require portable and durable power. This challenge is not limited to e-textiles alone, but is of relevance in the choice of sensors and of PHM systems. The power conversion efficiency, which is the ratio of the useful output of an energy conversion machine to its input, tends to be low for harvested energy. Increasing the amount of harvested energy that can actually be used to power a device is an ongoing research theme. Highly parallel PHM systems may be required to simultaneously process multiple signals from the patient. These systems are likely to generate very large datasets and transferring these wirelessly can create a strain on the already scarce bandwidth resource. This, along with the fact that wireless networks are inherently unreliable, can reduce the level of dependability of such PHM systems. Storing large datasets is rather challenging due to the limited memory resources of PHM systems and this has led to many designers migrating the datasets to less constrained cloud resources.

4.3.3 APPLICATIONS: PROJECTS AND TRENDS

PHM systems are increasingly being used for analyzing stress and the state of a subject's psychological health. One of such projects is StressSense (Lu *et al.*, 2012). It is a joint project between groups at Intel Laboratory, Cornell University, École Polytechnique Fédérale de Lausanne (EPFL), University of Neuchatel, Dartmouth College, Idiap Research Institute. StressSense is a software-based solution that inputs the subject's voice to a commodity Android phone and uses it to noninvasively infer his stress level in diverse acoustic environments in real-time. They use Gaussian Mixture Models (GMMs) with diagonal covariance matrix to classify the stress levels. They also ensure privacy by using only nonvoice regions for their analysis and the system adapts to the individual characteristics of the user. The Ubicomp laboratory of Professor Shwetak Patel (University of Washington) also has some interesting PHM projects based on the use of mobile phones. These projects include SpiroSmart and CoughSense (Larson *et al.*, 2011). SpiroSmart is completely software-based, performs spirometry sensing using the built-in microphone of the mobile phone after analyzing lip reverberations and it can be used to screen for common lung function measures like Forced Vital Capacity (FVC), Peak Expiratory Flow (PEF) and Forced Expiratory Volume in one second (FEV1). They reported an accuracy of 81% and 76% for indoor and outdoor environments respectively. Coughsense also uses a mobile phone to detect coughing patterns from an audio stream and it is based on the use of a novel algorithm that incorporates Principal Component Analysis (PCA) and a random forest classifier. The authors used a user context-aware approach to generate a cough model from random annotated coughs of a given user and it is used to train the cough detection system. It retains the fidelity of the voice to enable the physician use it for diagnosis and has been reported to have an average true positive accuracy of 92%.

In both SpiroSmart and CoughSense, the data is received on the mobile phone and transferred to a server for processing as the phone does not have the required computational power. CoughSense requires the mobile phone to be placed close to the neck for reliable detection, even though this position is not convenient for the patient. The researchers also identified the robustness to external noise as the most limiting factor for PHM systems based on audio signals. The need for the analysis of mental/psychological state seems to have assumed a new level of importance, especially in Europe, judging by the number of projects that focus on this application domain. The Online Predictive Tools for Intervention In Mental Illness (OPTIMI) is a 3-year EU-FP7 project that uses EEG, ECG and activity monitoring, along with voice analysis and cortisol sampling to detect early signs of stress and depression. IMEC International, located in Belgium, also uses a body area network of sensors to detect the level of stress of the wearer. It uses an innovative concept where the readings of the sensors are connected to an emotionally responsive photo whose level of illumination is inversely proportional to the level of stress (Brown and Penders, 2011). They hope that a gloomy looking picture will encourage the subject to cheer up, leading to a reduction in stress levels.

PredictCAD (Antilla *et al*., 2013) (an EU-FP7 project) and another EU project, Interreality in the Management and Treatment of Stress Related Disorders (INTERSTRESS) (Riva *et al*., 2010) also use PHM to determine the psychological state of the patient. PredictCAD is used in the early diagnosis of Alzheimer disease. The design follows a dependability-aware approach as portrayed in its primary technical objective, which is the development of an efficient and reliable software for diagnosing Alzheimer's disease using heterogeneous data. INTERSTRESS uses an innovative approach that combines the use of biosensors and virtual reality simulations to predict and manage stress. A remote controlled Sensorized ARTificial heart (SensorART) (Lee *et al*., 2011) is an example of a PHM system in the therapeutic application domain. It is a 4-year EU-FP7 project that supports telemonitoring and telecontrol and is able to interact with cardiac support systems in order to check and influence parameters like the pumping action of cardiovascular implanted assist devices. Another important contribution of SensorART is open interoperable device-independent platform for supporting personalized care. With the gamut of available PHM related devices in the market, efforts aimed at standardization and interoperability are truly welcome. The Integrated Cognitive Assistive and Domotic Companion Robotic Systems for Ability and Security (COMPANIONABLE) project (Badii *et al*., 2009) takes PHM to another level through the use of a companion robot called Hector. It monitors patients in order to detect falls, initiates emergency calls and provides a platform for a two-way video communication between the patient and the care giver. The Combining social interaction and long-term monitoring for promoting independent living (GIRAFF+) project (Frennert *et al*., 2012) is another EU-FP7 project that supports health monitoring and initiates emergency calls through the aid of a robot, Giraff. Giraff is monitored and controlled over the Internet.

4.3.4 APPLICATIONS: DIRECTION AND CHALLENGES

A great number of the current projects are for preventive purposes; many of which focus on discerning the psychological state of the patient of the occurrence of falls in the seniors. However, a growing number of researchers are trying to design reconfigurable platforms that can be modified to support therapeutic ends. This is nice in theory, but the route towards translating the laboratory design to a clinical design makes it quite challenging. The strict dependability issues involved in this process makes it difficult to get the required approval from bodies like the Food and Drug Administration (FDA). PHM systems operate in surroundings that may be affected by noise. Many processing techniques use a significant amount of resources to eliminate or reduce the level of noise. Recent research efforts aim at reducing this noise levels in a highly efficient manner. Noise also affects the appropriate placement of sensors and has caused researchers to resort to placing sensors in less convenient locations. Real-time processing is more resource-hungry than their non-real-time counterparts. Parallel processing can handle the level of complexity better and increase the level of system performance. Many of the mobile phone based devices require signals to be processed on the mobile phone or on another device in close proximity to the phone (such as a server) and others still require a transfer of the raw signals to a remote location for processing. In the future, we expect that it would be easier to efficiently and reliably process the signals both on the phone and a closely connected server without having to spend time to transfer and process on remote servers.

4.3.5   ADAPTATION: PROJECTS AND TRENDS

A lot of the PHM systems are only used as monitors and, at best, also used to trigger alarms when necessary. However, these context-obtuse systems require some user intervention to be able to alter their functionality. An increasing number of designers are making their devices to be context-aware and are providing for autonomic reconfiguration of functionality. The *FALL Repository for the design of Smart and sElf-adaptive Environments prolonging INdependent livinG* (FARSEEING) project (Fleming *et al.*, 2008) collects data from smartphone, wearable and environmental sensors and analyzes this long-term data. Armed with this knowledge, the FARSEEING system will initiate self adaptive responses in order to promote better prediction and prevention of falls in seniors. FPGAs are reconfigurable and this makes them well suited to systems that require adaptation. A research at the Center for Reliable and High-Performance Computing is one of such research efforts that have taken advantage of the FPGA's capacity for reconfigurability in the personalized health monitoring domain (Alemzadeh *et al.*, 2011). This research employed the use of a Xilinx Virtex-5 XC5 VFX70T FPGA that reconfigures its processing elements and communication blocks in response to a "Health Index." This health index is an aggregation of the physiological signals (including blood pressure, heart rate and ECG) of the patient. Active research is ongoing on the use of Brain Computer Interfaces (BCI) in adaptive context-aware PHM systems. For example, the Mind Controlled Orthosis And Virtual Reality Training Environment for Walk Empowering (Mindwalker) is an EU-FP7 project based on the use of smart dry EEG biosensors for non-invasive Brain Computer Interfacing (BCI). Their system controls purpose-designed lower limbs orthosis enabling different types of gaits (Duvinage *et al.*, 2012).

4.3.6   *ADAPTATION: DIRECTION AND CHALLENGES*

Designers now focus on the need to reduce the complexity of the system, while increasing the level of intuitiveness. Brain-Computer-Interfaces (BCIs) or the so-called Mind-over Matter interfaces have been used as exciting gaming interfaces but they are now being integrated into PHM systems. They also provide an interesting user input for triggering system reconfiguration/adaptation. System adaptation can be used to take advantage of the features that make each person unique. To enable the adaptation it is usually necessary for the PHM systems to undergo a level of "training," where the system is tuned to enable it to effectively adapt to the specific user features or preferences. Sometimes short training periods can have noise-induced errors and thus not be able to provide accurate results. This may affect the level of system dependability. At other times, depending on the complexity of the system and training approach, these sessions may tend to be too long. This increases the overall processing time and can affect the system's ability to process real time data. Researchers are working to strike a balance between the complexity, duration and performance of device training techniques.

4.4   CONCLUSION AND FUTURE PERSPECTIVES

In the future, the technology requirements of PHM systems will likely go beyond just a failure-free reporting of the measured vital signs, to a therapeutic role in intensive care units, as well as a stabilizing role in life support systems. The bioinformatics research scope covers the extraction, processing and storage of biological signals. The advances in this research area will combine well with PHM systems to deliver the platform required to make affordable personalized medicine a reality. Personalized medicine is a type of medicine that is customized to the needs an individual patient by the use of his genetic information. Portable, reconfigurable, parallel and dependable PHM systems could well be an interesting option for driving down the costs, shrinking down the size and accelerating the transition into the personalized medicine era. In this chapter, we have discussed the emerging field of personalized health monitoring systems. A description of the current direction and challenges in the field in terms of a new taxonomy has been provided. We have also identified the benefits of reconfiguration, parallelization and dependability in medical systems. Key PHM issues like the need for portability, robustness to noise, noninvasiveness, therapeutic systems, among others, have been discussed. We then identified potentials for future development in the PHM area.

KEYWORDS

- Biosignal monitoring unit
- Brain computer interfacing
- Electromyogram
- Field programmable gate arrays
- Non-invasiveness
- Parallelization
- Personal digital assistants
- Personalized health monitoring
- Run-time reconfiguration

REFERENCES

Alemzadeh, H.; Jin, Z.; Kalbarczyk, Z. T.; Iyer, R. K. *An embedded reconfigurable architecture for patient-specific multiparameter medical monitoring*, 33rd Annual International IEEE EMBS Conference, Boston, USA, 1896–1900.

Amann, A.; Tratnig, R.; Unterkofler, K. Detecting ventricular fibrillation by time-delay methods. *IEEE Transaction on Biomedical Engineering* 2007, 54(1), 174–177.

Avizienis, A.; Laprie, J. -C.; Randell, B.; Landwehr, C. Basic concepts and taxonomy of dependable and secure computing. *IEEE Transaction on Dependable and Secure Computing* 2004, 1(1), 11–33.

Badii, A.; Etxeberria, I.; Huijnen, C.; Maseda, M.; Dittenberger, S.; Hochgatterer, A.; Thiemert, D.; Rigaud, A. -S. Companionable-mobile robot companion and smart home system for people with mild cognitive impairment. *Journal of Nutrition, Health and Aging* 2009, 13(S1), S113.

Brown, L.; Penders, J. Wearable health companions. *Cyber Therapy and Rehabilitation* 2011, 1, 16–17.

Castro, A. G.; Normann, I.; Hois, J.; Kutz, O. *Ontologizing metadata for assistive technologies – the OASIS repository*, International Workshop on Ontologies in Interactive Systems, Liverpool, UK, 2008, pp 57–62.

Cherenack, K.; van Pieterson, L. Smart textiles: challenges and opportunities. J. *Appl. Phys.* 2011, 112, 91301, http://dx.doi.org/10.1063/1.4742728.

Duvinage, M.; Castermans, T.; Jimenez-Fabian, R.; Petieau, M.; Seetharaman K.; Hoellinger, T.; De Saedeleer, C.; Cheron, G.; Verlinden, O.; Dutoit, T. *A five-state P300-based foot lifter orthosis: proof of concept*, 3rd IEEE Biosignals and Biorobotics conference, Manaus, Brazil, 2012.

Fleming, J.; Matthews, F.; Brayne, C. The Cambridge City over75 s Cohort (CC75C) study collaboration. *BMC Geriatr.* 2008, 8, 6.

Frennert, S.; Östlund, B.; Eftring, H. *Would granny let an assistive robot into her home?* 4th International Conference on Social Robotics, Chengdu, China, 2012, pp 128–137.

Hanke, S.; Mayer, C.; Hoeftberger, O.; Boos, H.; Wichert, R.; Tazari, M. -R.; Wolf, P.; Furfari, F. (2011) universAAL – an open and consolidated AAL platform. In: *Ambient assisted living;* Wichert, R., Eberhardt, B., Eds.; Springer: Heidelberg, 2011, pp 127–140.

Huang, J.; Lee, J. A self-reconfigurable platform for scalable DCT computation using compressed partial bitstreams and blockRAM prefetching. *IEEE Transaction on Circuits and Systems for Video Technology* 2009, 19(11), 1623–1632.

Jiang, W.; Kong, S.G.; Peterson, G. D. *ECG signal classification using block-based neural networks.* IEEE International Joint Conference on Neural Networks, Montreal, Canada, 2005; Vol. 1; pp 326–331.

Larson, E.; Tien Jui, L.; Liu, S.; Rosenfeld, M.; Patel, S. N. *Accurate and privacy preserving cough sensing using low-cost microphone.* 13th International Conference on Ubiquitous Computing, Beijing, China, 2011; pp 375–384.

Lee, M.; Zine, N.; Baraket, A.; Zabala, M.; Campabadal, F.; Trivella, M. G.; Jaffrezic-Renault, N.; Errachid, A. Novel capacitance biosensor based on hafnium oxide for interleukin-10 protein detection. *Procedia Eng.* 2011, 25, 972–975.

Lee, M. -S.; Flammer, A. J.; Lerman, L. O.; Lerman, A. Personalized medicine in cardiovascular diseases. *Korean Circulation J.* 2012, 42(9), 583–591.

Lehn, D.; Neely, C.; Schoonover, K.; Martin, T.; Jones, M. *e-TAGS: e-Textile Attached Gadgets.* Communication Networks and Distributed Systems Modeling and Simulation Conference, San Diego, USA, 2004.

Lu, H.; Rabbi, M.; Chittaranjan, G. T.; Frauendorfer, D.; Mast, M. S.; Campbell, A. T.; Gatica-Perez, D.; Choudhury, T. *StressSense: detecting stress in unconstrained acoustic environments using smartphones.* The 14th International Conference on Ubiquitous Computing, Pittsburg, USA, 2012, pp 351–360.

Malaguttiy, N.; Dehghaniz, A.; Kennedy, R. A. *An approach to controlled drug infusion via tracking of the time-varying dose-response.* International Conference of the IEEE Engineering in Medicine and Biology Society, San Diego, USA, 2012, pp 3539–3542.

Marquez, J. C.; Seoane, F.; Valimaki, E.; Lindecrantz, K. *Textile electrodes in electrical bioimpedance measurements – a comparison with conventional Ag/AgCl electrodes.* 31st Annual International Conference of the IEEE/EMBS, Minneapolis, USA, 2009, pp 4816–4819.

Milenković, A.; Otto, C.; Jovanov, E. Wireless sensor networks for personal health monitoring: issues and an implementation. *Comput. Commun.* 2006, 29(13–14), 2521–2533.

Mouttham, A.; Peyton, L.; Eze, B.; El Saddik, A. Event-driven data integration for personal health monitoring. *J. Emerg. Technol. Web Intell.* 2009, 1(2), 110–118.

Øyri, K.; Støa, S.; Fosse, E. *A biomedical wireless sensor network for hemodynamic monitoring.* 5th International Conference on Body Area Networks, Corfu Island, Greece, 2010, pp 175–180.

Panagiotakopoulos, T. C.; Lyras, D. P.; Livaditis, M.; Sgarbas, K. N.; Anastassopoulos, G. C.; Lymberopoulos, D. K. A contextual data mining approach toward assisting the treatment of anxiety disorders. IEEE *Transaction on Information Technology in Biomedicine* 2010, 14(3), 567–581.

Pärkkä, J. *Analysis of personal health monitoring data for physical activity recognition and assessment of energy expenditure, mental load and stress.* PhD Thesis, Tampere University of Technology, Finland, 2011.

Riva, G.; Raspelli, S.; Pallavicini, F.; Grassi, A.; Algeri, D.; Wiederhold, B. K.; Gaggioli, A. Interreality in the management of psychological stress: a clinical scenario. *Studies in Health Technology and Informatics* 2010, 154, 21–25.

Rousselot, J.; Decotignie, J. -D. *Wireless communication systems for continuous multiparameter health monitoring.* IEEE International Conference on Ultra-Wideband, Vancouver, Canada, 2009, pp. 480–484.

Schmidt, B. *Bioinformatics: high performance parallel computer architectures.* CRC Press: Florida, 2009, pp 242–250.

Shahriyar, R.; Bari, M. F.; Kundu, G.; Ahamed, S. I.; Akbar, M. M. Intelligent mobile health monitoring system (IMHMS). *Int. J. Control Autom.* 2009, 2(3), 13–28.

Splinter, R. *Handbook of Physics in Medicine and Biology.* CRC Press: Florida, 2010.

Zhang, Y. -T.; Zheng, Y. -L.; Lin, W. -H.; Zhang, H. -Y.; Zhou, X. -L. Challenges and opportunities in cardiovascular health informatics. *IEEE Transaction on Biomedical Engineering* 2013, 60(3), 633–642.

CHAPTER 5

HIGH THROUGHPUT EVALUATION OF GENE EXPRESSION FROM FORMALIN-FIXED PARAFFIN-EMBEDDED TISSUES

PARASKEVI A. FARAZI

CONTENTS

5.1   INTRODUCTION

In order to understand disease pathogenesis, the signaling networks governing normal and disease tissues must be delineated. The most abundant source of disease tissue material is in the form of formalin-fixed paraffin embedded (FFPE) samples. Unfortunately, the fixation procedure causes damage to the most important biomolecules (DNA, RNA, proteins) with the exception of microRNAs, thus making their extraction and analysis difficult. However, several methods of isolation of these biomolecules have been developed to allow reasonable extraction of DNA, RNA, and proteins, which can subsequently be used for downstream assays. Traditional low-throughput assays, such as immunohistochemistry and reverse transcription polymerase chain reaction (RT-PCR) have been used extensively to study gene expression in FFPE tissue samples. With the advancements of science, several high-throughput methods of gene expression analysis have also been optimized for FFPE material including reverse phase protein microarrays, DNA methylation assays, microarrays, cDNA-mediated Annealing, Selection, Extension and Ligation assay, quantitative RT-PCR. These developments allow the design of explorative studies to study whole genome gene expression changes in disease tissues, which can be related to patient treatment sensitivity or survival. The latter allows the generation of prediction models that are able to predict a patient's disease progression and response to treatment based on gene expression data.

5.2   FORMALIN-FIXED PARAFFIN-EMBEDDED TISSUES AS A MEANS OF STUDYING GENE EXPRESSION IN HUMAN DISEASE

To understand the molecular mechanisms driving human disease, one has to create a picture of the signaling pathways that are engaged during the development and progression of disease. Therefore, studying gene expression profiles in human disease tissue is of paramount importance in enabling scientists to understand the mechanisms of disease pathogenesis and thus generate effective therapeutics. One source of a disease sample is fresh frozen tissue from a patient. However, such a sample is not always available, especially in certain hospitals where there might not be systematic tissue collection and storage of such tissues. Storage of fresh/frozen tissue is expensive and requires appropriate facilities to maintain the tissue (–80°C freezers and liquid nitrogen storage for long-term storage). In addition, fresh frozen tissue might be more readily available for certain stages of disease (usually the most advanced stages when surgical removal of disease tissue is common practice).

On the other hand, all hospitals store disease tissue fixed in formalin and embedded in paraffin since this is the way to obtain thin sections for histological analysis and histopathological evaluation of the disease. Both biopsy and surgical material are stored in this manner thus broadening the number of available disease tissues and stages. In the case of cancer for example, biopsy material of suspected masses are a good source of tissue from different cancer stages. The conservation of morphology during the process of fixation allows scientists to have an accurate picture of the tissue structure they are studying. Laser capture microdissection (LCM) allows the isolation of nucleic acids and proteins from the exact tissue location that needs to be investigated (i.e., one can isolate an area with a tumor and exclude normal surrounding cells) (Malinowsky *et al.*, 2010). FFPE tissues also allow the design of retrospective studies since one can find a large collection of disease tissues already collected at a certain hospital during many years, thus increasing sample size. When the number of cases for a particular disease are limited (e.g., in small populations), being able to obtain disease material collected for many years makes the sample size large enough to conduct a study with potentially statistically significant results. Retrospective studies using archived FFPE tissue material with known clinical follow-up make the identification of predictive and prognostic molecular markers of disease possible (Kotorashvili *et al.*, 2012). In summary, low cost, easy storage and availability of FFPE tissues makes them an excellent source of study material (Kroll *et al.*, 2008).

To illustrate an example where FFPE tissues are an important source of study material, one can consider the example of nonsmall cell lung cancer where certain patients with a specific genetic mutation in EGFR respond to treatment with EGFR inhibitors Gefitinib (Iressa) and Erlotinib (Tarceva) more effectively than those who lack the mutation (Dietel and Sers, 2006). Studying the molecular profile of the tumors of patients who respond to a particular treatment and those who do not might enable the design of prediction models that would allow medical doctors to direct patients to the appropriate treatment based on the patient's tumor molecular profile. This would reduce the cost for treatment and improve patients' quality of life since the appropriate treatment will be used for each patient. In other words, gene expression data obtained from FFPE tissues would enable more effective personalized medicine.

To generate FFPE tissue the first step is to fix the tissue in formalin which cross-links macromolecules and maintains the tissue in good condition for histopathological evaluation (Becker *et al.*, 2007). In addition, fixation maintains protein modifications such as phosphorylation, which can be analyzed even decades after the sample has been collected (Liotta and Petricoin, 2000). Formalin is composed of 37–40% formaldehyde in water and 10% methanol. Formaldehyde leads to the formation of highly reactive methylol adducts through a reaction with the amino groups of basic amino acids. A condensation reaction of adducts then occurs via Schiff base formation, resulting in methylene bridges with amine, guanidyl, phenol imidazol and indole groups of other amino acids. The result is inter- and intramolecular cross-linking of proteins (Berg *et al.*, 2010). These cross-links make the morphology of the tissue stable for years (Fox *et al.*, 1985). Autolysis of the tissue components is prevented by inhibition of proteinolytic/nucleic acid degrading enzymes and heterolysis by bacterial enzymes is also inhibited through the microbicidal properties of the fixative. DNA fragmentation occurs due to the low pH of the fixative (formaldehyde-based) solutions, which results in hydrolysis of the glycosidic bonds in the purine bases (Bonin *et al.*, 2003; Klopfleisch *et al.*, 2011). Cross-linking of proteins with nucleic acids also leads to fragmentation of DNA and RNA and hampers their extraction from FFPE tissues (Lehmann and Kreipe, 2001). RNA is also modified and degraded when tissues are fixed in formalin. In the case of RNA, there is covalent bonding of monomethylol groups to purines as well as loss of the polyA tail, which makes binding to oligo (dT) primers for reverse transcription impossible (McGhee and von Hippel, 1977; Srinivasan *et al.*, 2002). microRNAs, which are short in length (on average 22 nucleotides long), remain largely unaffected by formalin fixation (Liu *et al.*, 2009). The paraffin used in the process of making FFPE tissues also affects extraction of nucleic acids, since some types of paraffin may contain contaminants (Fergenbaum *et al.*, 2004).

Considering the detrimental effects of formalin on the quality of nucleic acids, other methods of fixation have also been investigated. Formalin is cheap and represents an easy method of fixation that preserves the morphology of the tissue, however, it is also a hazardous chemical (Klopfleisch *et al.*, 2011). Alternative fixatives include gluteraldehyde, alcohols, hepes-glutamic acid buffer-mediated organic solvent protection effect (HOPE), Weigner's solution (nitrite pickling salt – NPS), methacarn, mercurials, Bouin's solution and several others (Klopfleisch *et al.*, 2011). Gluteraldehyde actually has a worse effect compared to formalin on the quality of nucleic acids (O'Leary *et al.*, 1994), whereas alcohols actually allow for extraction of better quality nucleic acids from FFPE tissues comparable to those obtained from fresh frozen tissue (Linke *et al.*, 2010; Soukup *et al.*, 2003). However, alcohols have the disadvantage of being expensive, evaporate easily, and incompletely fixing thicker tissue samples thus limiting their use (Barnes *et al.*, 2000). HOPE fixative has actually been shown to retain the features of proteins as well as result in better yield and quality of nucleic acids, however, its microbicidal, fungicidal and virucidal properties are not clear (Kahler *et al.*, 2010; Vollmer *et al.*, 2006; Klopfleisch *et al.*, 2011). NPS has also been investigated for its fixative properties and has been shown to retain the morphology of the tissue as well as nucleic acid quality (Janczyk *et al.*, 2010; Klopfleisch *et al.*, 2011).

Preanalytics and tissue quality of FFPE material is an important consideration in their use. The specimens undergo processing during the preanalytical phase of tissue collection and preservation. Effects of temperature, duration of transportation and duration of fixation on FFPE tissue quality must be considered. Along these lines, there are two international initiatives that are trying to address these issues and create guidelines for tissue handling and processing of FFPE tissues (EU initiative: www.spidia.edu; US initiative: http://biospecimens.cancer.gov) (Malinowsky *et al.*, 2010). Of note, formalin fixation for less than 48 hours at 4°C has been suggested to give the best results in terms of nucleic acid quality (Abrahamsen *et al.*, 2003).

5.3   METHODS TO INVESTIGATE GENE EXPRESSION AND GENE/CHROMOSOME STRUCTURE FROM FFPE TISSUES

Gene expression can be studied at the level of mRNA and proteins (including posttranslationally modified proteins). The most common method used for analysis of gene expression in FFPE tissue is immunohistochemistry (IHC). IHC gives researchers information about the expression of a particular protein in a tissue, however, the quantification is rather limited especially since it depends at least partially on the observer. In addition, it does not provide many opportunities for multiplexing and thus analysis of multiple proteins simultaneously is not possible (Paweletz *et al.*, 2001). Since one protein needs to be analyzed at a time, researchers have to select the protein to be analyzed and therefore IHC depends on the generation of hypotheses prior to the experiment. Another aspect to consider is that formalin fixation causes many protein-protein cross links, which often make antigen retrieval difficult; one might have to try out several methods of antigen retrieval such as proteinase K digestion, heat treatment in acidic or basic buffers, or heat/pressure treatment (Ikeda *et al.*, 1998).

To achieve evaluation of protein concentration in FFPE tissues, successful isolation of proteins from FFPE tissues must be in place. To this end, several groups have worked out protocols to isolate proteins from FFPE tissues some using their own reagents and others using commercially available reagents suitable for the specific method (Berg *et al.*, 2010). High SDS concentrations and high temperature are used to successfully extract full-length proteins in all these protocols (Ikeda *et al.*, 1998; Chu *et al.*, 2005; Shi *et al.*, 2006; Becker *et al.*, 2007; Becker *et al.*, 2008; Chung *et al.*, 2008; Nirmalan *et al.*, 2009; Addis *et al.*, 2009). Different protocols have different advantages and disadvantages (time consuming, cheap, low yield, etc. (Berg *et al.*, 2010). It is not the scope of this book chapter to go over these different methods, but rather just show that many protocols have been developed for protein isolation from FFPE tissues.

In an effort to move away from hypothesis-driven research (e.g., analysis of protein expression using IHC) and move towards explorative studies, proteome analysis has been put to work. This involves separation of proteins according to their chemical properties and size by two-dimensional gel electrophoresis, followed by quantification of proteins using fluorescence differential gel electrophoresis, isotope-coded affinity tags, and identification of differentially expressed proteins by mass spectrometry (Klopfleisch *et al.*, 2010). However, such methods have not been found to work on FFPE material initially due to extensive protein cross-linking (Ahram *et al.*, 2003). Optimized antigen retrieval methods, as for example boiling of FFPE tissues have made proteome analyzes on FFPE material possible (Prieto *et al.*, 2005; Shi *et al.*, 2006). Imaging mass spectrometry (IMS) has also been used on FFPE tissues by digesting proteins *in situ* with trypsin and then identifying proteins directly on the slide, however, resolution and sensitivity are low (Groseclose *et al.*, 2008).

A method that has been used for the evaluation of proteins in FFPE tissues is reverse phase protein array (RPPA). It can detect changes in protein phosphorylation and allows for analysis of multiple samples at the same time (Wulfkuhle *et al.*, 2006; Paweletz *et al.*, 2001; Grubb *et al.*, 2003). Purified recombinant proteins of known concentration are used as internal standards in RPPA, which allows for more precise protein expression quantification than IHC (Berg *et al.*, 2010). In addition, since denatured lysates can be used there are no significant issues with antigen retrieval and antibody accessibility as encountered in IHC. However, it should be noted that in RPPA a correlation between histology and protein expression is no longer feasible as it is with IHC. RPPA though offers many advantages compared to other assays. For example, it is much more sensitive than ELISA – it can detect proteins at the level of attograms (10^{-18} g) and variances less than 10%. The limiting factor in using this technology is the availability of antibodies that recognize posttranslationally modified proteins as well as the variability of staining between arrays (Berg *et al.*, 2010). The latter makes it difficult to compare results from different groups or different experiments among the same group. An interesting development in this technology is that it could be used to predict drug sensitivity of lung cancer cell lines using a Bayesian network classifier (Kim *et al.*, 2012). One can envision the application of this to FFPE samples.

To study gene expression at the mRNA level, effective isolation of RNA from FFPE tissues must be possible. Such protocols are in place and there are many commercially available kits for RNA isolation from FFPE samples (Bonin *et al.*, 2010; Okello *et al.*, 2010). However, the degradation observed in RNA isolated from FFPE specimens poses limitations in their use in subsequent molecular analyzes. Quantitative reverse transcription PCR (qRT-PCR) has been conducted using RNA isolated from FFPE samples to study gene expression (Farragher *et al.*, 2008). For more consistent results the target size of the sequence to be amplified should be less than 200 bp (for best results target size should be up to 120 bp) (Jackson *et al.*, 1990; Farragher *et al.*, 2008). Due to loss of the polyA tails during fixation, reverse transcription on RNA from FFPE tissue is achieved using random hexamers or specific antisense primers (Lewis *et al.*, 2001; Xiang *et al.*, 2003). Using this technology, Genomic Health Inc. along with National Surgical Adjuvant Breast and Bowel Project researchers developed an assay (Oncotype DX) that is able to predict breast cancer recurrence based on a recurrence score algorithm that uses data from the expression of 21 genes on archival FFPE-derived material (Paik *et al.*, 2004). High throughput gene expression/microarray analysis has also been successfully conducted from RNA derived from FFPE tissues and some studies have shown good correlation with results obtained using RNA from fresh frozen tissue, especially when short prefixation time is achieved and mRNA preamplification before reverse transcription is performed (April *et al.*, 2009; Coudry *et al.*, 2007; Frank *et al.*, 2007; Ravo *et al.*, 2008). Problems have been reported by these studies as well: cross-hybridization due to mRNA degeneration and consequent presence of high numbers of genes on the microarrays (Coudry *et al.*, 2007; Farragher *et al.*, 2008). Others, however, report low quality and insufficient results of microarray analysis of FFPE-derived RNA due to formalin fixation and the various fixation protocols used (Penland *et al.*, 2007). Massive parallel sequencing has also been performed on RNA derived from FFPE tissues and yielded transcriptome profiles (Qu *et al.*, 2010; Weng *et al.*, 2010).

To overcome the extensive degradation often seen in FFPE-derived RNA, the DASL (cDNA-mediated annealing, selection, extension and ligation) assay was developed by Illumina Inc. (Bibicova *et al.*, 2004; Fan *et al.*, 2004). In this assay, a biotinylated cDNA is created using the total FFPE-derived RNA as the template and then two primers are designed for each specific sequence to be amplified. The upstream primer contains a universal PCR primer sequence at the 5′ end along with the gene specific sequence and the downstream primer contains a gene-specific sequence and a universal primer sequence at the 3′ end. The idea is that the upstream oligo binds to its corresponding cDNA, extends and ligates to the appropriate 3′ oligo. This product serves as a template for PCR using the primer sequences on the aforementioned oligos. The assay has been shown to yield comparable expression profiling data in FFPE tissue material compared to matched fresh/frozen material (Mittempergher et al., 2011). DASL has been used successfully to investigate gene expression in breast cancer using a custom 512-gene breast cancer bead array-based platform, which allowed investigators to identify a set of markers that can be used to classify breast cancer into specific subtypes (Abramovitz et al., 2011).

MicroRNAs (miRNAs) represent a type of RNA found in cells, which are usually 20–22 nucleotides long and play a role in the control of gene expression. miRNA gene expression profiles are actually the easiest to obtain from FFPE tissues since the small size of microRNAs and their association with large proteins allows them to remain intact during formalin fixation and RNA extraction and unaffected by formalin fixation times (Streichert et al., 2011; Giricz et al., 2011; Liu et al., 2009; Xi et al., 2007). Reverse transcription of miRNAs does not depend on the poly A tail, which tends to be lost during formalin fixation, but rather uses direct end-labeling thus making the assay work much more efficiently on miRNAs isolated from FFPE tissues (Castoldi et al., 2006). A good correlation between miRNA expression studies using fresh frozen and FFPE tissues has been found (Glud et al., 2009; Liu et al., 2009). miRNA works much better than mRNA derived from FFPE tissues in several assays, including RT-PCR (Li et al., 2007; Doleshal et al., 2008), microarrays (Glud et al., 2009; Liu et al., 2009), and deep sequencing (Weng et al., 2010). The biggest limitation of miRNAs is their low abundance in tissues, however, methods have been developed to quantify miRNAs in samples with low miRNA levels using locked nucleic acid (LNA)-enhanced primers in quantitative RT-PCR (Andreasen et al., 2010). In situ hybridization and in situ RT-PCR allows for detection of miRNAs with as low as one copy per cell (Nuovo, 2008).

As mentioned earlier, gene expression is affected by posttranslational modifications such as methylation. DNA methylation involves covalent addition of a methyl group to the carbon-5 position of cytosine bases (mostly in the context of cytosine-guanine nucleotides, CpG), a process that is reversible

(Bird, 2002). Hypermethylation of promoters can result in decreased gene expression, whereas hypomethylation of promoters can result in increased gene expression (Sato *et al.*, 2003). Approaches to study DNA methylation in FFPE tissues include methylation-specific PCR (MSP), MethyLight and combined bisulphite restriction analysis (COBRA) (Beck and Rakyan, 2008). These are mostly candidate locus approaches and not high throughput. GoldenGate (which involves multiplexed methylation-specific primer extension of bisulphite converted DNA at 1536 CpG sites in promoter regions) and MassArray were the first genome-wide DNA methylation promoter assays used to assess methylation in FFPE-derived material (Bibikova *et al.*, 2009; Jurinke *et al.*, 2005). These approaches rely on a bisulphite conversion reaction, which involves deamination of unmethylated cytosines to uracils (Shapiro *et al.*, 1973; Frommer *et al.*, 1992). A novel genome-wide DNA methylation platform was recently developed (Illumina Infinium HumanMethylation27 BeadChip), which analyzes specific promoter regions for DNA methylation, however, it was designed for good quality DNA and not for FFPE samples (Bibikova and Fan, 2009). The reason it does not work well for FFPE-derived DNA is that the assay requires the template DNA to be >1 kb in length and FFPE-derived DNA is usually fragmented to sizes below 1 kb. Another group introduced an extra DNA ligation step in the protocol to increase the length of the DNA, thus making the modified assay suitable for use with FFPE material (Thirlwell *et al.*, 2010). Even archival tissue up to 10 years old worked well in this assay as well as DNA derived from different tissue types. This development provides the opportunity to analyze the DNA methylation status in samples from different diseases in order to understand how gene expression is altered in disease pathogenesis. The fact that it works in FFPE material it greatly increases the sample size to be analyzed since FFPE disease material is abundant.

Array-comparative Genomic Hybridization (array-CGH) is a method used for studying DNA copy number alterations that are a result of genomic instability and can ultimately affect gene expression. Genomic instability is a common feature of diseases like cancer and results in gene amplifications and deletions (described herein as DNA copy number alterations – CNAs). Increasing number of CNAs is an indicator of poor prognosis in certain cancer types and different CNA profiles correspond to different tumor subtypes. Thus, it may be possible to classify cancers based on their CNA profile and direct patients to the appropriate types of treatment based on their genomic profiles (Abramovitz *et al.*, 2007). Array-CGH was introduced in the late 90s and basically involves hybridization of fluorescently labeled genomic DNA (from disease and corresponding normal tissue) onto an array with probes (either large insert clones such as bacterial artificial chromosomes or oligo-nucleotides) (Pinkel *et al.*, 1998). Array-CGH has actually been successfully performed on DNA isolated from FFPE tumor tissues (Johnson *et al.*, 2006; Joosse *et al.*, 2007; Little *et al.*, 2006).

Single nucleotide polymorphisms (SNPs) often account for genetic differences among individuals in relation to their disease susceptibility, reaction to drugs, interaction with the environment, and response to treatment. SNPs can also affect the expression of genes. SNP analysis is performed to assess the allelic constitution of an individual's genotype and is often conducted using conventional PCR or real-time PCR and melting curve analysis (Song *et al.*, 2005). SNP analysis has been successfully conducted in DNA obtained from FFPE tissues and gave results that were comparable to those from analysis of fresh frozen material (Lyons-Weiler *et al.*, 2008). However, other studies have questioned the results obtained from mutational analysis of FFPE tissue material, which might be influenced by cross-linking of cytosine nucleotides during fixation and the generation of a C-T or G-A exchange during DNA synthesis by DNA polymerase in a PCR (Verhoest *et al.*, 2010; Williams *et al.*, 1999).

An important development in the analysis of nucleic acids derived from FFPE tissues was the efficient concurrent isolation of DNA and RNA from the same FFPE sample, which broadens the spectrum of analyzes one could perform to study the molecular mechanisms of disease pathogenesis (Koto-rashvili *et al.*, 2012). To achieve this, the FFPE tissue sections are first depa-raffinized in a nonpolar solvent (such as xylene, δ-limonene or citrisolv) and the tissue is digested with proteinase K. The incubation in proteinase K ranges from 15 minutes to overnight for successful isolation of RNA and for DNA this is extended to 48 hours. For DNA, heating at 95–98°C in alkaline buffer helps to remove DNA-protein cross-links (this is not done for RNA) (Shi *et al.*, 2004). Trizol or phenol/chloroform or silica-based column purification are used for DNA/RNA separation from proteins (Loudig *et al.*, 2007; Okello *et al.*, 2010). For best RNA quality (i.e., removal of chemical modifications such as methylol groups acquired during formalin fixation) heating at 70°C for 1 hour in Tris-EDTA or citrate-based buffer is performed (Masuda *et al.*, 1999; Hamatani *et al.*, 2006). Effective DNA/RNA coisolation allows for re-trieval of matched nucleic acids from the same cells, especially helpful in validation and integrative studies. The quality of nucleic acids isolated from such copurification protocols was assessed using qRT-PCR and global gene expression analysis using whole-genome DASL (for mRNA), qRT-PCR and expression profiling (for microRNAs), and methylation assays (for genomic

DNA) (Kotorashvili *et al.*, 2012). Being able to isolate DNA and RNA from the same sample allows one to investigate the mutational status of genes and relate such mutations to gene expression data (i.e., understand how mutations alter the gene expression profile in a tissue and drive disease pathogenesis).

FFPE material can be used directly for gene expression analysis by spotting FFPE samples as small as 0.6 mm on microarray slides and then subjecting them to microarray analysis (called tissue microarrays, TMAs, Fig. 5.1) An antibody is added to each array to examine the expression of a particular protein across many disease and normal samples (Giltnane and Rimm, 2004). One can thus analyze the expression of multiple proteins in a set of disease and normal samples in order to understand the signaling pathways that are driving disease pathogenesis. There was skepticism whether TMAs are equivalent to whole tissue sections, however, this was addressed in breast carcinoma samples (Camp *et al.*, 2000). TMAs have actually been used to study gene expression in various cancer types (Schraml *et al.*, 1999; Richter *et al.*, 2000; Nocito *et al.*, 2001; Callagy *et al.*, 2003).

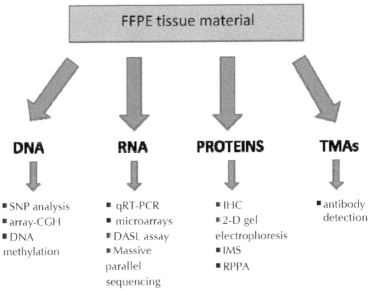

FIGURE 5.1 Summary of the most commonly used assays to study gene expression from FFPE tissues (SNP = single nucleotide polymorphism; array-CGH = array comparative genomic hybridization; qRT-PCR = quantitative reverse transcription PCR; DASL = cDNA-mediated annealing, selection, extension and ligation; IHC = immunohistochemistry; IMS = Imaging mass spectrometry; RPPA = reverse phase protein array; TMAs = tissue microarrays).

5.4   CONCLUSION

Even though FFPE-derived biomolecules pose challenges in regards to their analysis, FFPE tissues are a significant source of disease and normal material that can be used to better understand the molecular processes governing disease. Given that one can obtain FFPE tissues from decades back, there is a wealth of information to be obtained from such tissues. A lot of progress has been made to isolate good quality biomolecules from FFPE samples, which consequently allows for the execution of numerous assays to study the signaling networks governing normal and disease cells in the context of the tissue. Assays such as *in situ* hybridization and IHC allow for low throughput evaluation of gene expression in FFPE samples based on hypotheses. High throughput assays have also been tested on FFPE tissues allowing researchers to do explorative gene expression studies in order to better understand disease pathogenesis. These include two dimensional gel electrophoresis followed by mass spectrometry, RPPA, qRT-PCR, microarrays, the DASL assay, TMAs and DNA methylation assays. Gene and chromosome structure, which can affect gene expression, can also be studied through SNP assays and array-CGH. microRNAs are well preserved in FFPE samples, thus allowing analysis of their expression relatively easily.

KEYWORDS

- Array-comparative genomic hybridization
- cDNA-mediated annealing
- Formalin-fixed paraffin embedded tissues
- Gene expression
- Human disease
- Imaging mass spectrometry
- Immunohistochemistry
- Laser capture microdissection
- MicroRNAs
- Nitrite pickling salt
- Reverse phase protein array
- Reverse transcription polymerase chain reaction
- Tissue microarrays

REFERENCES

Abrahamsen, H. N.; Steiniche, T.; Nexo, E.; Hamilton-Dutoit, S. J.; Sorensen, B. S. Towards quantitative mRNA analysis in paraffin-embedded tissues using real-time reverse transcriptase-polymerase chain reaction: a methodological study on lymph nodes from melanoma patients. *J Mol Diagn* 2003, 5, 34–41.

Abramovitz, M.; Leyland-Jones, B. Application of array-based genomic and epigenomic technologies to unraveling the heteregenous nature of breast tumors: on the road to individualized treatment. *Cancer Genomics and Proteomics* 2007, 4, 135–146.

Abramovitz, M.; Barwick, B. G.; Willis, S.; Young, B.; Catzavelos, C.; Li, Z.; Kodani, M.; Tang, W.; Bouzyk, M.; Moreno, C. S.; Leyland-Jones, B. Molecular characterization of formalin-fixed paraffin-embedded (FFPE) breast tumor specimens using a custom 512-gene breast cancer bead array-based platform. *British Journal of Cancer* 2011, 105, 1574–1581.

Addis, M. F.; Tanca, A.; Pagnozzi, D.; Crobu, S.; Fanciulli, G.; Cossu Rocca, P.; Uzzau, S. Generation of high-quality protein extracts from formalin-fixed, paraffin-embedded tissues. *Proteomics* 2009, 9, 3815–3823.

Ahram, M.; Flaig, M. J.; Gillespie, J. W.; Duray, P. H.; Linehan, W. M.; Ornstein, D. K.; Niu, S.; Zhao, Y.; Petricoin, E. F. Evaluation of ethanol-fixed, paraffin-embedded tissues for proteomic applications. *Proteomics* 2003, 3, 413–421.

Andreasen, D.; Fog, J. U.; Biggs, W.; Salomon, J.; Dahslveen, I. K.; Baker, A.; Mouritzen P. Improved microRNA quantification in total RNA from clinical samples. *Methods* 2010, 50, S6–9.

April, C.; Klotzle, B.; Royce, T.; Wickham-Garcia, E.; Boyaniwsky, T.; Izzo, J.; Cox, D.; Jones, W.; Rubio, R.; Holton, K.; Matulonis, U.; Quackenbush, J.; Fan, J. -B. Whole-genome gene expression profiling of formalin-fixed, paraffin-embedded tissue samples. *PLoS One* 2009, 4, e8162.

Barnes, I.; Holton, J.; Vaira, D.; Spigelman, M.; Thomas, M. G. An assessment of the long-term preservation of the DNA of a bacterial pathogen in ethanol-preserved archival material. *J Pathol* 2000, 192, 554–559.

Beck, S.; Rakyan, V. K. The methylome: approaches for global DNA methylation profiling. *Trends Genet* 2008, 24(5), 231–327.

Becker, K. F.; Schott, C.; Hipp, S.; Metzger, V.; Porschewski, P.; Beck, R.; Nährig, J.; Becker, I.; Höfler, H. Quantitative protein analysis from formalin-fixed tissues: implications for translational clinical research and nanoscale molecular diagnosis. *Journal of Pathology* 2007, 211, 370–378.

Becker, K. F.; Mack, H.; Schott, C.; Hipp, S.; Rappl, A.; Piontek, G.; Hofler, H. Extraction of phosphorylated proteins from formalin-fixed cancer cells and tissues. *TOPAT J.* 2008, 2, 46–52.

Becker, K. F.; Schott, C.; Becker, I.; Hofler, H. Guided protein extraction from formalin-fixed tissues for quantitative multiplex analysis avoids detrimental effects of histological stains. *Proteomics Clin Appl* 2008, 2, 737–743.

Berg, D.; Hipp, S.; Malinowsky, K.; Bollner, C.; Becker, K. -F. Molecular profiling of signaling pathways in formalin-fixed and paraffin-embedded cancer tissues. *European Journal of Cancer* 2010, 46, 47–55.

Bibikova, M.; Talantov, D.; Chudin, E.; Yeakly, J. M.; Chen, J.; Doucet, D.; Wickham, E.; Atkins, D.; Barker, D.; Chee, M.; Wang, Y.; Fan, J. B. Quantitative gene expression profiling in formalin-fixed, paraffin-embedded tissues using universal bead arrays. *Am J Pathol* 2004, 165, 1799–1807.

Bibikova, M.; Fan, J. B. GoldenGate® assay for DNA methylation profiling. *Methods Mol Biol* 2009, 507, 149–163.

Bibikova, M. L.; Barnes, B.; Gunderson, K. Genome-wide DNA methylation profiling using Infinium® assay. *Epigenomics* 2009, 1(1), 177–200.

Bird, A. DNA methylation patterns and epigenetic memory. *Genes Dev* 2002, 16(1), 6–21.

Bonin, S.; Petrera, F.; Niccolini, B.; Stanta, G. PCR analysis in archival postmortem tissues. *Mol Pathol* 2003, 56, 184–186.

Bonin, S.; Hlubek, F.; Benhattar, J.; Denkert, C.; Dietel, M.; Fernandez, P. L.; Höfler, G.;, H.; Kruslin, B.; Mazzanti, C.M.; Perren, A.; Popper, H.; Scarpa, A.; Soares, P.; Stanta, G.; Groenen, P. Multicenter validation study of nucleic acids extraction of FFPE tissues. *Virchows Arch* 2010, 457, 309–317.

Callagy, G.; Cattaneo, E.; Daigo, Y.; Haperfield, L.; Bobrow, L. G.; Pharoah, P. D.; Caldas, C. Molecular classification of breast carcinomas using tissue microarrays. *Diagn Mol Pathol* 2003, 12, 27–34.

Camp, R. L.; Charette, L. A.; Rimm, D. L. Validation of tissue microarray technology in breast carcinoma. *Lab Invest* 2000, 80, 1943–1949.

Castoldi, M.; Schmidt, S.; Benes, V.; Noerholm, M.; Kulozik, A. E.; Hentze, M. W.; Muckenthaler, M. U. A sensitive array for microRNA expression profiling (miChip) based on locked nucleic acids (LNA). *RNA* 2006, 12, 913–920.

Chu, W. S.; Liang, Q.; Liu, J.; Wei, MQ.; Winters, M.; Liotta, L.; Sandberg, G.; Gong, M. A nondestructive molecule extraction method allowing morphological and molecular analyzes using a single tissue section. *Lab Invest* 2005, 85, 1416–1428.

Chung, J. Y.; Lee, S. J.; Kris, Y.; Braunschweig, T.; Traicoff, J. L.; Hewitt, S. M. A well-based reverse-phase protein array applicable to extracts from formalin-fixed paraffin-embedded tissue. *Proteomics Clin Appl* 2008, 2, 1539–1547.

Coudry, R. A.; Meireles, S. I.; Stoyanova, R.; Cooper, H. S.; Carpino, A.; Wang, X.; Engstrom, P. F.; Clapper, M. L. Successful application of microarray technology to microdissected formalin-fixed paraffin-embedded tissue. *J Mol Diagn* 2007, 9, 70–79.

Dietel, M.; Sers, C. Personalized medicine and development of targeted therapies: the upcoming challenge for diagnostic molecular pathology: a review. *Virchows Arch* 2006, 448, 744–755.

Doleshal, M.; Magotra, A. A.; Choudhury, B.; Cannon, B. D.; Labourier, E.; Szafranska, A. E. Evaluation and validation of total RNA extraction methods for microRNA expression analyzes in formalin-fixed, paraffin-embedded tissues. *J Mol Diagn* 2008, 10, 203–211.

Entze, M. W.; Muckenthaler, M. U. A sensitive assay for microRNA expression profiling (miChip) based on locked nucleci acids (LNA). *RNA* 2006, 12, 913–920.

Fan, J. B.; Yeakley, J. M.; Bibikova, M.; Chudin, E.; Wickham, E.; Chen, J.; Doucet, D.; Rifgault, P.; Zhang, B.; Shen, R.; McBride, C.; Li, H. R.; Fu, X. D.; Oliphant, A.; Barker, D. L.; Chee, M. S. A versatile assay for high-throughput gene expression profiling on universal array matrices. *Genome Res* 2004, 14, 878–885.

Farragher, S. M.; Tanney, A.; Kennedy, R. D.; Harkin, P. D. RNA expression analysis from formalin fixed paraffin embedded tissues. *Histochem Cell Biol* 2008, 130, 435–445.

Fergenbaum, J. H.; Garcia-Closas, M.; Hewitt, S. M.; Lissowska, J.; Sakoda, L. C.; Sherman, M. E. Loss of antigenicity in stored sections of breast cancer tissue microarrays. *Cancer Epidemolo Biomarkers Prev* 2004, 13, 667–672.

Fox, C. H.; Johnson, F. B.; Whiting, J.; Roller, P. P. Formaldehyde fixation. *J Histochem Cytochem* 1985, 33, 845–853.

Frank, M.; Doring, C.; Metzler, D.; Eckerle, S.; Hansmann, M. L. Global gene expression profiling of formalin-fixed paraffin-embedded tumor samples: a comparison to snap-frozen material using oligonucleotide microarrays. *Virchows Arch* 2007, 450, 699–711.

Frommer, M.; McDonald, L. E.; Millar, D. S.; Collis, C. M.; Watt, F.; Grigg, G. W.; Molloy, P. L.; Paul, C. L. A genomic sequencing protocol that yields a positive display of 5-methylcytosine residues in individual DNA strands. *Proc Natl Acad Sci USA* 1992, 89(5), 1827–1831.

Giltnane, J. M.; Rimm, D. L. Technology insight: identification of biomarkers with tissue microarray technology. *Nat Clin Pract Oncol* 2004, 1, 104–111.

Giricz, O.; Reynolds, P. A.; Ramnauth, A.; Liu, C.; Wang, T.; Stead, L.; Childs, G.; Rohan, T.; Shapiro, N.; Fineberg, S.; Kenny, P. A.; Loudig, O. Hsa-miR-375 is differentially expressed during breast lobular neoplasia and promotes loss of mammary acinar polarity. *J Pathol* 2011, 226(1), 108–119.

Glud, M.; Klausen, M.; Gniadecki, R.; Rossing, M.; Hastrup, N.; Nielsen, F. C.; Drzewiecki, K. T. Micro-RNA expression in melanicytic nevi: the usefulness of formalin-fixed, paraffin-embedded material for miRNA microarray profiling. *J Invest Dermatol* 2009, 129, 1219–1224.

Grosenclose, M. R.; Massion, P. P.; Chaurand, P.; Caprioli, R. M. High-throughput proteomic analysis of formalin-fixed paraffin-embedded tissue microarrays using MALDI imaging mass spectrometry. *Proteomics* 2008, 8, 3715–3724.

Grubb, R. L.; Calvert, V. S.; Wulkuhle, J. D.; Paweletz, C. P.; Linehan, W. M.; Phillips, J. L., Chuaqui, R.; Valasco, A.; Gillespie, J.; Emmert-Buck, M.; Liotta, L. A.; Petricoin, E. F. Signal pathway profiling of prostate cancer using reverse phase protein arrays. *Proteomics* 2003, 3, 2142–2146.

Hamatani, K.; Eguchi, H.; Takahashi, K.; Koyama, K.; Mukai, M.; Ito, R.; Taga, M.; Yasui, W.; Nakachi, K. Improved RT-PCR amplification for molecular analyzes with long-term preserved formalin-fixed, paraffin-embedded tissue specimens. *J Histochem Cytochem* 2006, 54, 773–780.

Ikeda, K.; Monden, T.; Kanoh, T.; Tsujie, M.; Izawa, H.; Haba, A.; Ohnishi, T.; Sekimoto, M.; Tomita, N.; Shiozaki, H.; Monden, M. Extraction and analysis of diagnostically useful proteins from formalin-fixed, paraffin-embedded tissue sections. *J Histochem Cytochem* 1998, 46, 397–403.

Jackson, D. P.; Lewis, F. A.; Taylor, G. R.; Boylston, A. W.; Quirke, P. Tissue extraction of DNA and RNA and analysis by the polymerase chain reaction. *J Clin Pathol* 1990, 43, 499–504.

Janczyk, P.; Weigner, J.; Luenke-Becker, A.; Kaessmeyer, S.; Plendl, J. Nitrite pickling salt as an alternative to formaldehyde for embalming in veterinary anatomy – a study based on histo- and microbiological analyzes. *Ann. Anat* 2010, 193, 71–75.

Johnson, N. A.; Hamoudi, R. A.; Ichimuera, K.; Liu, L.; Pearson, D. M.; Collins, V. P.; Du, M. Q. Application of array CGH on archival formalin-fixed paraffin-embedded tissues including small numbers of microdissected cells. *Lab Invest* 2006, 86, 968–978.

Joosse, S. A.; van Beers, E. H.; Nederlof, P. M. Automated array-CGH optimized for archival formalin-fixed, paraffin-embedded tumor material. *BMC Cancer* 2007, 7, 43.

Jurinke, C.; Denissenko, M.; Oeth, P.; Ehrich, M.; van den Boom, D.; Cantor, C. A single nucleotide polymorphism based approach for the identification and characterization of gene expression modulation using MassARRAY. *Mutat Res* 2005, 573(1–2), 83–95.

Kahler, D.; Alexander, C.; Schultz, H.; Abdullah, M.; Branscheid, D.; Lindner, B.; Zabel, P.; Vollmer, E.; Goldmann, T. Proteomics out of the archive: two-dimensional electrophoresis and mass spectrometry using HOPE-fixed, paraffin-embedded tissues. *J Histochem Cytochem* 2010, 58, 221–228.

Kim, D. C.; Wang, X.; Yang, C. -R.; Gao, J. X. A framework for personalized medicine: prediction of drug sensitivity in cancer by proteomic profiling. *Proteome Science* 2012, 10(S1), S13.

Klopfleisch, R.; Weiss, A. T. A.; Gruber, A. D. Excavation of a buried treasure – DNA, mRNA, miRNA and protein analysis in formalin fixed, paraffin embedded tissues. *Histol Histopathol* 2011, 26, 797–810.

Klopfleisch, R.; Klose, P.; Weise, C.; Bondzio, A.; Multhaup, G.; Einspanier, R.; Gruber, A. D. Proteome of metastatic canine mammary carcinomas: similarities to and differences from human breast cancer. *J Proteome Res* 2010, 9(12), 6380–6391.

Kotorashvili, A.; Ramnauth, A.; Liu, C.; Lin, J.; Ye, K.; Kim, R.; Hazan, R.; Rohan, T.; Fineberg, S.; Loudig, O. Effective DNA/RNA coextraction for analysis of microRNAs, mRNAs and genomic DNA from formalin-fixed paraffin-embedded specimens. *PLOS One* 2012, 7(4), e34683.

Kroll, J.; Becker, K. F.; Kuphal, S.; Hein, R.; Hofstadter, F.; Bosserhoff, A. K. Isolation of high quality protein samples from punches of formalin-fixed and paraffin-embedded tissues blocks. *Histology and Histopathology* 2008, 23, 391–395.

Lehmann, U.; Kreipe, H. Real-time PCR analysis of DNA and RNA extracted from formalin-fixed and paraffin-embedded biopsies. *Methods* 2001, 25, 409–418.

Lewis, F.; Maughan, N. J.; Smith, V.; Hillan, K.; Quirke, P. Unlocking the archive-gene expression in paraffin-embedded tissue. *J Pathol* 2001, 195, 66–71.

Li, J.; Smyth, P.; Flavin, R.; Cahill, S.; Denning, K.; Ahefne, S.; Guenther, S. M.; O'Leary, J. J.; Shells, O. Comparison of miRNA expression patterns using total RNA extracted from matched samples of formalin-fixed paraffin-embedded (FFPE) cells and snap frozen cells. *BMC Biotechnol* 2007, 7, 36.

Linke, B.; Schroder, K.; Arter, J.; Gasperazzo, T.; Woehlecke, H.; Ehwald, R. Extraction of nucleic acids from yeast cells and plant tissues using ethanol as medium for sample preservation and cell disruption. *Biotechniques* 2010, 49, 655–657.

Liotta, L.; Petricoin, E. Molecular profiling of human cancer. *Nature Reviews Genetics* 2000, 1, 48–56.

Little, S. E.; Vuononvirta, R.; Reis-Filho, J. S.; Natrajan, R.; Iravani, M.; Fenwick, K.; Mackay, A.; Ashworth, A.; Pritchard-Jones, K.; Jones, C. Array-CGH using whole genome amplification of fresh-frozen and formalin-fixed, paraffin-embedded tumor DNA. *Genomics* 2006, 87, 298–306.

Liu, A.; Tetzlaff, M. T.; Vanbelle, P.; Elder, D.; Feldman, M.; Tobias, J. W.; Sepulveda, A. R.; Xu, X. MicroRNA expression profiling outperforms mRNA expression profiling in formalin-fixed paraffin-embedded tissues. *Int J Clin Exp Pathol* 2009, 2, 519–527.

Loudig, O.; Milova, E.; Brandwein-Gensler, M.; Massimi, A.; Belbin, T. J.; Childs, G.; Singer, R.; Rohan, T.; Prystowski, M. Molecular restoration of archived transcriptional profiles by complementary-template reverse transcription (CT-RT). *Nucleic Acids Res* 2007, 35, e94.

Lyons-Weiler, M.; Hagenkord, J.; Sciulli, C.; Dhir, R.; Monzon, F. A. Optimization of the Affymetrix GeneChip Mapping 10K 2.0 Assay for routine clinical use on formalin-fixed paraffin-embedded tissues. *Diagn Mol Pathol* 2008, 17, 3–13.

Malinowsky, K.; Wolff, C.; Ergin, B.; Berg, D.; Becker, K. D. Deciphering signaling pathways in clinical tissues for personalized medicine using protein microarrays. *Journal of Cellular Physiology* 2010, 225, 364–370.

Masuda, N.; Ohnishi, T.; Kawamoto, S.; Monden, M.; Okubo, K. Analysis of chemical modification of RNA from formalin-fixed samples and optimization of molecular biology applications for such samples. *Nucleic Acids Res* 1999, 27, 4436–4443.

McGhee, J. D.; von Hippel, P. H. Formaldehyde as a probe of DNA structure. 3. Equilibrium denaturation of DNA and synthetic polynucleotides. *Biochemistry* 1977, 16, 3267–3276.

Mittempergher, L.; de Ronde, J. J.; Nieuwland, M.; Kerkhoven, R. M.; Simon, I.; Rutgers, E. J.; Wessels, L. F.; Van't Veer, L. J. Gene expression profiles from formalin fixed paraffin embedded breast cancer tissue are largely comparable to fresh frozen matched tissue. *PLoS One* 2011, 6(2), e17163.

Nirmalan, N. J.; Harnden, P.; Selby, P. J.; Banks, R. E. Development and validation of a novel protein extraction methodology for quantitation of protein expression in formalin-fixed paraffin-embedded tissues using western blotting. *J Pathol* 2009, 217, 497–506.

Nocito, A.; Bubendorf, L.; Maria Tinner, E.; Suess, K.; Wagner, U.; Forster, T.; Kononen, J.; Fijan, A.; Bruderer, J.; Schmid, U.; Ackermann, D.; Maurer, R.; Alund, G.; Knonagel, H.; Rist, M.; Anabitarte, M.; Hering, F.; Hardmeier, T.; Schoenenberger, A. J.; Flury, R.; Jager, P.; Luc Fehr, J.; Schraml, P.; Moch, H.; Mihatsch, M. J.; Gasser, T.; Sauter, G. Microarrays of bladder cancer tissue are highly representative of proliferation index and histological grade. *J Pathol* 2001, 194, 349–357.

Nuovo, G. J. *In situ* detection of precursor and mature microRNAs in paraffin embedded, formalim fixed tissues and cell preparations. *Methods* 2008, 44, 39–46.

Schraml, P.; Moch, H.; Mihatsch, M. J.; Gasser, T.; Sauter, G. Microarrays of bladder cancer tissue are highly representative of proliferation index and histological grade. *J Pathol* 2001, 194, 349–357.

O'Leary, J. J.; Browne, G.; Landers, R. J.; Crowley, M.; Healy, I. B.; Street, J. T.; Pollock, A. M.; Murphy, J.; Johnson, M. I.; Lewis, F. A.; Mohamdee, O.; Cullinane, C.; Doyle, C. T. The importance of fixation procedures on DNA template and its suitability for solution-phase polymerase chain reaction and PCR *in situ* hybridization. *Histochem J* 1994, 26, 337–346.

Okello, J. B.; Zurek, J.; Devault, A. M.; Kuch, M.; Okwi, A. L.; Sewankambo, N. K.; Bimenya, G. S.; Poinar, D.; Poinar, H. N. Comparison of methods in the recovery of nucleic acids from archival formalin-fixed paraffin-embedded autopsy tissues. *Anal Biochem* 2010, 400, 110–117.

Paik, S.; Shak, S.; Tang, G.; Kim, C.; Baker, J.; Cronin, M.; Baehner, F. L.; Walker, M. G.; Watson, D.; Park, T.; Hiller, W.; Fisher, E. R.; Wickerham, D. L.; Bryant, J.; Wolmark, N. A multi-gene assay to predict recurrence of tamoxifen-treated, node-negative breast cancer. *N Engl J Med* 2004, 351, 2817–2826.

Paweletz, C. P.; Charnonaeu, L.; Bichsel, V. E.; Simone, N. L.; Chen, T.; Gillespie, J. W.; Emmert-Buck, M. R.; Roth, M. J.; Petricoin III, E. F.; Liotta, L. A. Reverse phase protein microarrays, which capture disease progression show activation of prosurvival pathways at the cancer invasion front. *Oncogene* 2001, 20, 1981–1989.

Penland, S. K.; Keku, T. O.; Torrice, C.; He, X.; Krishnamurthy, J.; Hoadley, K. A.; Woosley, J. T.; Thomas, N. E.; Perou, C. M.; Sandler, R. S.; Sharpless, N. E. RNA expression analysis of formalin-fixed paraffin-embedded tumors. *Lab Invest* 2007, 87, 383–391.

Pinkel, D.; Segraves, R.; Sudar, D.; Clark, S.; Poole, I.; Kowbel, D.; Collins, C.; Kuo, W. L.; Chen, C.; Zhai, Y.; Dairkee, S. H.; Ljung, B. M.; Gray, J. W.; Albertson, D. G. High resolution analysis of DNA copy number variation using comparative genomic hybridization to microarrays. *Nat Genet* 1998, 20, 207–211.

Prieto, D. A.; Hood, B. L.; Darfler, M. M.; Guiel, T. G.; Lucas, D. A.; Conrads, T. P.; Veenstra, T. D.; Krizman, D. B. Liquid tissue: proteomic profiling of formalin-fixed tissues. *Biotechniques Suppl* 2005, 32–35.

Qu, K.; Morlan, J.; Stephans, J.; Li, X.; Baker, J.; Sinicropi, D. Transcriptome profiling from formalin-fixed paraffin-embedded tumor specimens by RNA-seq. *Genome Biology* 2010, 11(S1), 31.

Ravo, M.; Mutarelli, M.; Ferraro, L.; Grober, O. M.; Paris, O.; Tarallo, R.; Vigilante, A.; Cimino, D.; De Bortoli, M.; Nola, E.; Cicatiello, L.; Weisz, A. Quantitative expression profiling of highly degraded RNA from formalin-fixed, paraffin-embedded breast tumor biopsies by oligonucleotide microarrays. *Lab Invest* 2008, 88, 430–440.

Richter, J.; Wagner, U.; Kononen, J.; Fijan, A.; Bruderer, J.; Schmid, U.; Ackermann, D.; Maurer, R.; Alund, G.; Knonagel, H.; Rist, M.; Wilber, K.; Anabitarte, M.; Hering, F.; Hardmeier, T.; Kallioniemi, O. P.; Sauter, G. High-throughput tissue microarray analysis of cyclin E gene amplification and overexpression in urinary bladder cancer. *Am J Pathol* 2000, 157, 787–794.

Sato, N.; Maitra, A.; Fukushima, N.; van Heek, T. N.; Matsubayashi, N.; Iacobuzio-Donahue, C. A.; Rosty, C.; Goggins, M. Frequent hypomethylation of multiple genes overexpressed in pancreatic ductal adenocarcinoma. *Cancer Res* 2003, 63(14), 4158–4166.

Schraml, P.; Kononen, J.; Bubendorf, L.; Moch, H.; Bissig, H.; Nocito, A.; Mihatsch, M. J.; Kallioniemi, O. P.; Sauter, G. Tissue microarrays for gene amplification surveys in many different tumor types. *Clin Cancer Res* 1999, 5, 1966–1975.

Shapiro, R.; Braverman, B.; Louis, J. B.; Servis, R. E. Nucleic acid reactivity and conformation II. Reaction of cytosine and uracil with sodium bisulfite. *J Biol Chem* 1973, 248(11), 4060–4064.

Shi, S. R.; Datar, R.; Liu, C.; Wu, L.; Zhang, Z.; Cote, R. J.; Taylor, C. R. DNA extraction from archival formalin-fixed, paraffin-embedded tissues: heat-induced retrieval in alkaline solution. *Histochem Cell Biol* 2004, 122, 211–218.

Shi, S. R.; Liu, C.; Balgley, B. M.; Lee, C.; Taylor, C. R. Protein extraction from formalin-fixed, paraffin-embedded tissue sections: quality evaluation by mass spectrometry. *J Histochem Cytochem* 2006, 54, 739–743.

Song, Y.; Araki, J.; Zhang, L.; Froehlich, T.; Sawabe, M.; Arai, T.; Shirasawa, T.; Muramatsu, M. Haplotyping of TNFa gene promoter using melting temperature analysis: detection of a novel 856 (G/A) mutation. *Tissue Antigens* 2005, 66(4), 284–290.

Soukup, R. W.; Krskova, L.; Hilska, I.; Kodet, R. Ethanol fixation of lymphoma samples as an alternative approach for preservation of the nucleic acids. *Neoplasma* 2003, 50, 300–304.

Srinivasan, M.; Sedmak, D.; Jewell, S. Effect of fixatives and tissue processing on the content and integrity of nucleic acids. *Am J Pathol* 2002, 161, 1961–1971.

Streichert, T.; Otto, B.; Lehmann, U. microRNA expression profiling in archival tissue specimens: methods and data processing. *Mol Biotechnol* 2011, 50(2), 159–169.

Thirlwell, C.; Eymard, M.; Feber, A.; Teschendorff, A.; Pearce, K.; Lechner, M.; Widschwendter, M.; Beck, S. Genome-wide DANN methylation analysis of archival formalin-fixed paraggin-embedded tissue using the Illumina Infinium HumanMethylation27 BeadChip. *Methods* 2010, 52, 248–254.

Verhoest, G.; Patard, J. J.; Fergelot, P.; Jouan, F.; Zerrouki, S.; Dreano, S.; Mottier, S.; Rioux-Leclercq, N.; Denis, M. G. Paraffin-embedded tissue is less accurate than frozen section analysis for determining VHL mutational status in sporadic renal cell carcinoma. *Urol Oncol* 2010, 30(4), 469–475.

Vollmer, E.; Galle, J.; Lang, D. S.; Loeschke, S.; Schultz, H.; Goldmann, T. The HOPE technique opens up a multitude of new possibilities in pathology. *Rom J Morphol Embryol* 2006, 47, 15–19.

Weng, L.; Wu, X.; Gao, H.; Mu, B.; Li, X.; Wang, J. H.; Guo, C.; Jin, J. M.; Chen, Z.; Covarrubias, M.; Yuan, Y. C.; Weiss, L. M.; Wu, H. microRNA profiling of clear cell renal cell carcinoma by whole-genome small RNA deep sequencing of paired frozen and formalin-fixed, paraffin-embedded tissue specimens. *J Pathol* 2010, 222, 41–51.

Williams, C.; Ponten, F.; Moberg, C.; Soderkvist, P.; Uhlen, M.; Ponten, J.; Sitbon, G.; Lundeberg, J. A high frequency of sequence alterations is due to formalin fixation of archival specimens. *Am J Pathol* 1999, 155, 1467–1471.

Wulfkuhle, J. D.; Edminston, K. H.; Liotta, L. A. Petricoin III, E. F. Technology insight: pharmacoproteomics for cancer-promises of patient-tailored medicine using protein microarrays. *Nat Clin Pract Oncol* 2006, 3, 256–268.

Xi, Y.; Nakajima, G.; Gavin, E.; Morris, C. G.; Kudo, K.; Hayashi, K.; Ju, J. Systematic analysis of microRNA expression of RNA extracted from fresh frozen and formalin-fixed paraffin-embedded samples. *RNA* 2007, 13, 1668–1674.

Xiang, C. C.; Chen, M.; Ma, L.; Pahn, Q. N.; Inman, J. M.; Kozhich, O. A.; Brownstein, M. J. A new strategy to amplify degraded RNA from small tissue samples for microarray studies. *Nucleic Acids Res* 2003, 31, e53.

CHAPTER 6

MICROBIAL DETOXIFYING ENZYMES INVOLVED IN BIODEGRADATION OF ORGANIC CHEMOPOLLUTANTS

PRADNYA PRALHAD KANEKAR and SEEMA SHREEPAD SARNAIK

CONTENTS

6.1   INTRODUCTION

Rapid industrialization has resulted in rising up of chemical industries for manufacturing a variety of chemicals to fulfill the demands of public. The high growth of chemical industries, both manufacturer and user industries lead to pollution of, air, water and soil environment. The industries responsible for environmental pollution include petrochemical industries, oil refineries, distileries and the units manufacturing chemicals like pesticides, explosives and detonators, dyes and paints, detergents and soaps and such other chemicals. These industries cause environmental pollution due to high volume of their effluents and the residual chemicals appearing in these effluents.

6.1.1   TOXICITY OF SOME ORGANIC CHEMOPOLLUTANTS

The chemicals like nitro compounds, organochlorine compounds including pesticides are known to be harmful to both aquatic and terrestrial life including human beings. The authors have reviewed the health hazards of some of these organic chemopollutants as detailed in Table 6.1 (Kanekar *et al.*, 2003, 2004; Sarnaik and Kanekar, 1999). Michalowicz and Duda (2007) reviewed toxicity of all types of phenols, such as chlorophenols, nitrophenols and aminophenols and described the toxicity as irritation to skin, necrosis, damage of vital organs like kidneys, liver and eyes. Deshmukh *et al.* (2009) reported the toxicity of chlorinated organic compounds as carcinogenic, mutagenic in nature and persistent in the environment requiring biodegradation of these toxic chemicals. The literature survey shows that all these organic compounds are toxic and hazardous in nature and tend to remain in the environment for a long time hence their detoxification is necessary.

TABLE 6.1   Toxicity of some organic chemopollutants.

Chemopollutants	Toxicity
Pesticides	Cause nausea, irritation of the skin and eyes, affect the nervous system, mimic hormones causing reproductive problems, some of them are carcinogenic in nature
Dyes	Carcinogenic and mutagenic in nature, cause headache, giddiness, irritation to mucous membrane of respiratory tract, slight bronchial irritation to fatal pulmonary edema, inflammation of skin resembling dermatitis

TABLE 6.1   *(Continued)*

Chemopollutants	Toxicity
Explosives	Carcinogenic in nature, cause irritation of the digestive tract, me-thaemoglobinaemia, severe jaundice, exhibit disturbed heart function, kidney troubles, dysfunction of the whole vascular system
Phenolic compounds	Corrosive in nature, cause irritation to skin, responsible for necrosis, affect central nervous system (CNS)

Environmental pollution particularly the pollution of water and soil due to toxic chemicals is quite a serious problem, which is to be solved with topmost priority. Some physical or chemical treatment methods are available for the purpose; however they are not economically/practically feasible ones. Hence, microbially assisted degradation/detoxification/mineralization of toxic chemical compounds seems to be a good solution for the present problem. Microorganisms play an important role in the degradation of man-made chemicals that are released in the environment. These microorganisms have ability to evolve enzyme systems required for degradation or detoxification of different chemicals like nitro compounds, pesticides especially the organochlorine pesticides, which lead to their application for bioremediation purposes. Different enzymes are involved in break down of the chemopollutants. However, involvement of oxidoreductases or also called as oxygenases and hydrolases is found to contribute to a major extent in biodegradation of chemopollutants.

6.1.2   GENERAL INFORMATION ABOUT ENZYMES

Enzymes are the biological entities that catalyze chemical reactions. Enzymes are the group of catalytic proteins that are produced by living cells and that initiate and enhance the chemical processes of life without having any alterations in them. These are not being destroyed during the chemical reaction. Almost all chemical reactions in a biological cell need enzymes in order to occur at rates sufficient for life. Enzymes are the proteins in nature, specialized to catalyze biological reactions. They lower down the energy of activation for a reaction, and increase the rate of reaction. These enzymes possess very diverse substrate specificity, reactivity and some other physicochemical and biological characteristics, which are useful in different industrial and medical applications. Enzymes, the natural biocatalysts are widely distributed in plants, animals, microorganisms and even in human beings. Enzymes could also function outside living cells. Microorganisms are endowed with the property of secreting several enzymes, which can be used for different purposes.

The use of enzymes either alone or as a part of living cell can be traced long back. Now-a-days, the enzymes are used very commonly and therefore many of them are commercially manufactured. Due to ease of cultivation and limited nutritional requirements, microbial enzymes are assuming commercial applications in different industries like detergent, food processing, pharmaceutical and others.

6.1.3 CLASSIFICATION OF ENZYMES

Based on the mechanisms of action of these enzymes, International Commission of Enzymes has arranged all the enzymes in six main classes namely oxidoreductases that carry out oxidation – reduction reactions, transferases, which transfer functional groups, hydrolases, which hydrolyze macromolecules, lyases, which help in addition of double bonds, isomerases carrying out isomerization reactions and ligases, which carry out formation of bonds with ATP cleavages. Among these enzymes, oxidoreductases and hydrolases play an important role in biodegradation/detoxification of the toxic chemical compounds.

6.2   DETOXIFYING ENZYMES

6.2.1 OXIDOREDUCTASES

Oxidoreductases are widely found in plants and animals along with higher occurrence in microorganisms. These enzymes catalyze the reactions involving exchange of electrons or redox equivalents between donor and acceptor molecules, in reactions involving electron transfer, hydrogen extraction, hydride transfer, oxygen insertion, or other key steps (Xu, 2005). Oxidoreductases can be classified according to their sequence or three-dimensional structure, which is very informative for the study of structure function relationship, enzyme evolution and functional genomics. Monooxygenases are the enzyme systems that catalyze the insertion of one atom of molecular oxygen into an organic substrate and dioxygenases are the enzyme systems that catalyze the insertion of both atoms of molecular oxygen into an organic substrate. These oxidoreductases can be further classified into different subclasses depending upon their action on the group of donors such as CH-OH group, CH-CH group, $CH-NH_2$ CH-NH group, sulfur group, heme group and others. Their classification is also based on their action on NADH or NADPH, other nitrogenous compounds, diphenols and related substances, iron-sulfur proteins and reduced flavodoxin. For application purpose, these oxidoreductases are classified according to their signature catalysis and coenzyme-dependence. Thiol oxidases have Fe, Cu or Cys+FAD as active center, heme peroxidases have heme as active center, other peroxidases have Cys or Se-Cys as active center, Flavin monooxygenases have FMN/FAD as active center. Cu containing oxygenases have Cu as active center (Xu, 2005). Cytochrome p-450 are highly diversified set of heme containing proteins, having unusual reduced carbon monoxide difference spectrum with absorbance at 450 nm. Primarily these enzymes carry out the hydroxylation of substances. In addition to this, these enzymes perform the reactions like N-oxidation, sulfoxidation, epoxidation, N-, S-, and O-dealkylation, peroxidation, deamination, desulfurization and dehalogenation (White and Coon, 1980).

6.2.2 HYDROLASES

Hydrolases are the enzymes, which catalyze the hydrolysis of a chemical bond. These enzymes belong to one of the six main classes of enzymes. They catalyze the hydrolytic cleavage of complex macromolecules like proteins, starch, fats and nucleic acids through cleavage of ester bonds by the addition of water. Systematic names of hydrolases are formed as substrate hydrolase. Proteases hydrolyze proteins and a nuclease cleaves nucleic acids. The enzymes like proteases, amylases, esterases form the major component of the hydrolytic enzymes or hydrolases. Proteases are the enzymes hydrolyzing the proteins into smaller peptides and amino acid units, likewise esterases are the enzymes attacking the ester bonds of lipids or oils and break down into corresponding carboxylic acids and alcohols. Amylases or amylolytic hydrolases are the enzyme systems causing cleavage of starch or other complex starchy materials into simple carbohydrate monomers. All these hydrolases are studied extensively in the literature for a variety of applications, however, their use in biodegradation/detoxification of toxic chemicals is to a limited extent.

Hydrolases can be further classified into various subclasses. Based on their action on chemical bonds, these enzymes can be classified into different 13 subclasses such as esterases acting on ester bonds, glycozylases acting on glycozyl compounds, peptidases acting on peptide bonds, and other esterases acting on ether bonds, carbon-nitrogen bonds other than peptides, carbon-carbon bonds, carbon-sulfur bonds, carbon-phosphorus bonds, sulfur-sulfur bonds, sulfur-nitrogen bonds and those acting on acid anhydrides. These enzymes also can be classified according to the reactions they catalyze, such as nucleases carrying out the hydrolysis of nucleic acids, urease that hydrolyze urea and amylase hydrolyze glycosidic bond in carbohydrates.

Esterases comprise the major subclass of the hydrolases. These enzymes form one group of hydrolase enzymes. These act on ester bonds of lipids and split them into corresponding acid and an alcohol in presence of water. The esterases are involved in degradation of various types of lipids and oils. Lipases and phosphotriesterases comprise major subclasses of esterases along with other esterases like acetylesterase, cholinesterase, phosphatase, pectinesterase are some of the examples of commonly occurring esterases. Phosphotriesterases are the enzyme systems catalyzing the hydrolysis of organophosphate triesters. Generally, these are the binuclear zinc enzymes, which catalyze the hydrolysis of hazardous organophosphate pesticides.

Degradation of organophosphorus pesticides by the action of microorganisms is generally through the hydrolysis of P-O alkyl and P-O aryl bonds. The hydrolase enzyme responsible for catalyzing this reaction is phosphotriesterase or phosphatase, which releases phosphates from organophosphorus compounds and another enzyme, esterase acts on C-N bond of organophosphorus compounds and releases methyl amine. Therefore, the esterases and phosphotriesterases are the enzymes studied in more detail since they are involved in biodegradation/detoxification of the toxic organophosphate pesticides.

6.3 ENZYMATIC DEGRADATION OF SOME OF THE ORGANIC POLLUTANTS

Microorganisms degrading xenobiotic chemicals elaborate enzyme systems such as oxidoreductases, dechlorinases/dehalogenases, oxygenase, phosphatase, esterase and hydrolases. The enzymes like cytochrome p-450 are special class of enzymes, which are widely distributed in biological systems like plants, animals and microorganisms. Discovery of several new prokaryotic and eukaryotic cytochrome p-450 in recent years showed their ubiquitous distribution in microorganisms. These enzymes play an important role in oxidative metabolism of variety of chemicals of agricultural and environmental concern.

Aerobic degradation of the aromatic and related compounds is monitored by the enzymes like Rieske nonheme iron oxygenases. Such type of Rieske-oxygenase enzymes help various aerobic microorganisms to use variety of toxic and recalcitrant aromatic compounds as their carbon and energy source. The oxidative degradation of some organic substrates or their intermediates by *Pseudomonas* occasionally involves the participation of oxygenases. Some of the oxygenases have a rather broad specificity, contributing to the nutrition-al versatility of the strains. Oxygenases acting on aliphatic compounds such as alkanes may be part of complex oxidative systems. Some of the enzymes involved in degradation of organic chemopollutants are summarized in Table 6.2. The literature survey thus shows that the enzymes like cytochrome p-450, phosphotriesterase, organophosphorus hydrolase, dehalogenase, laccase and nitro reductase play an important role in degradation of organic hazardous chemopollutants.

TABLE 6.2 Microbial enzymes involved in degradation of organic chemopollutants.

Organic chemopol-lutant	Enzyme involved	Microorganism	References
Endosulfan	Monooxygenase	*Arthrobacter* sp.	Weir *et al.*, 2006
Hexachlorocyclohex-ane	Dehalogenase	*Sphingomonas* sp.	Manickam *et al.*, 2008
Triazine pesticides	s-triazine hydro-lase	*Rhodococcus corallinus*	Mulbry, 1994
Herbicides, Atrazine	Cytochrome p-450	*Rhodococcus* sp.	Nagy, 1995a,b
Phenoxybutyrate herbicides	Cytochrome p-450	*Rhodococcus erythro-polis*	Strauber *et al.*, 2003
Dyes	Azoreductase	*Pseudomonas* sp.	Zimmermann *et al.*, 1982
Dyes	Peroxidase	*Geotrichum candidum*	Kim and Shoda, 1999
Textile dyes	Laccase	*Tremates hirsute*	Abdulla *et al.*, 2000
Dyes	Laccase	*Tramates modesta*	Nyahongo *et al.*, 2002
Dyes	Laccase	*Pleurotus pulmonarius*	Zilly *et al.*, 2002
Dyes	β 1-4, endoxyla-nase	*Bacillus* sp.	Mishra and Thakur, 2011
Nitro explosives	Type I nitroreduc-tase	Members of *Entero-bacteriaceae*	Kitts *et al.*, 2000

TABLE 6.2 *(Continued)*

Organic chemopollutant	Enzyme involved	Microorganism	References
Polychlorinated biphenyls	Biphenyl dioxygenase	*Burkholderia xenovorans*	Kumar *et al.*, 2011
Camphor	Cytochrome p-450	*Pseudomonas putida*	Sariaslani, 1991
Long chain fatty acids	Cytochrome p-450	*Bacillus megaterium*	Sariaslani, 1991
Cyclohexane	Cytochrome p-450	*Xanthomonas* sp.	Sariaslani, 1991
Alkanes	Cytochrome p-450	*Acinetobacter calcoaceticus*	Sariaslani, 1991
Monoterpene cineole-1	Cytochrome p-450	*Citrobacter braakii*	Hawkes *et al.*, 2002
4-phenyldiamine	Aryl amine-N acetyl transferase	*Bacillus cereus*	Mulyono *et al.*, 2007
Organophosphorus pesticides	Phosphotriesterases	*Pseudomonas diminuta*	Dumas *et al.*, 1989
Organophosphorus pesticides	Organophosphorus hydrolase	Recombinant *Escherichia coli*	Kaneva *et al.*, 1998
Organophosphorus pesticides	Organophosphorus hydrolase	*Nocardiodes simplex*	Mulbry, 2000
Organophosphorus pesticides	Organophosphorus hydrolase	*Moraxella* sp.	Shimazu *et al.*, 2001
Organophosphorus pesticides	Phosphotriesterases	*Pseudomonas aeruginosa, Clavibacter michiganense*	Das and Singh, 2006
Parathion	Parathion hydrolase	*Streptomyces lividans*	Rowland *et al.*, 1991

6.3.1 PESTICIDES

India is predominantly an agrarian country. The agriculture setup more or less controls the economy of the country. Majority of the population is dependent on agriculture related activities. To increase the production of food grains for fulfilling the demands of the population and to have minimum losses of agriculture produces, pesticides are used enormously. Many of these chemicals used as pesticides are synthetic in nature. The variety of pesticides is used during agricultural practices. Depending on the action, these pesticides can be broadly classified as insecticides, herbicides and fungicides. Depending on the chemical nature of the pesticides, these pesticides can be grouped as organophosphates, organochlorine, carbamate, triazine type of pesticides and others. The application of pesticides starts from presowing stage and can be applied to soil, for seed treatment and as foliar spray. The excessive and indiscriminate use of different types of pesticides contaminates the environment. The fate of released pesticides is dependent on the microbial flora in the vicinity since microorganisms are known to degrade various types of pesticides by secreting different types of enzyme systems as per requirement.

Organochlorine pesticides are highly toxic in nature; however, due to their efficacy they have been used for a long time. Hexachlorocyclohexane (HCH), endosulfan, dichlorodiphenyl- trichloroethane (DDT), dichlorophenoxyacetic acid (2,4-D) are the most frequently used pesticides. Although these have been banned for last few years, their residues are still detected in the environmental samples. Phillips *et al.* (2005) studied the microbial degradation of HCH, the enzymes involved in degradation of HCH and cloning, sequencing and characterization of the gene products in the selected bacterial strains. Enantioselective transformation of hexachlorocyclohexane (HCH) by dehydrochlorinases LinA1 and LinA2 from soil bacterium *Sphingomonas paucimobilis* B90A were reported by Suar *et al.* (2005). The dehalogenase enzyme system from *Sphiongobium indicum* B90A causing transformation of β- and δ-hexachlorocyclohexane was studied by Sharma *et al.* (2006). Manickam *et al.* (2008) isolated the HCH degrading *Sphingomonas* sp. using the dehalogenase assay and characterized the genes involved in degradation of γ-HCH. Presence of dehalogenases indicates the dechlorinating ability of the organism and in turn detoxification of the chloro compounds. The isolate *Sphingomonas* sp. reported by the authors harbored linABCDE genes. Camacho-Perez *et al.* (2012) reviewed the aerobic and anaerobic degradation of γ-hexachlorocyclohexane (HCH) and the genes and enzymes involved in the metabolic pathways of γ-HCH degradation. The authors have also reported the intermediate metabolites formed during degradation of HCH and involvement of phenol hydroxylases and catechol deoxygenases in the degradation of intermediate compounds.

Carbamate or dithiocarbamates can be used as either insecticides or fungicides. Due to their low cost and high efficiency, carbamate type of pesticides are used widely leading to the environmental pollution of soil and water. Involvement of enzymes like dehydrogenase (Nagy *et al.*, 1995a) and cytochrome p-450 system in degradation of herbicide EPTC (S-ethyldipropyl thio carbamate). Puranik (2000) carried out some work on mineralization of ethylenethiourea (ETU), a toxic metabolite of carbamate group of pesticide Mancozeb by the soil isolates *Arthrobacter sulfureus* MCM B-412, *Microbacterium lacticum* MCM B-413 and *Pseudomonas aeruginosa* MCM B-422 to carbon dioxide and ammonia through formation of intermediate metabolites namely ethylenediamine and ethyleneurea. The author also reported the involvement of enzymes like decarboxylase and deaminase. Cheesman *et al.* (2007) reviewed the biological degradation of carbamate type of pesticides and developed enzymatic bioremediation technique using formulations of detoxifying enzyme systems to clean up the environmental sites contaminated with carbamate type of pesticides. Karayilanoglu *et al.* (2008) reported the bacterial degradation of toxic carbamate type of pesticide aldicarb through hydrolysis by the enzyme esterase. The authors thought that the bacteria including *Stenotrophomonas maltophilia* could be useful for detoxifying aldicarb and in turn can be suitable for developing biotechnological methods for decontamination of sites polluted with aldicarb and similar type of pesticides.

Triazines are the six membered ring compounds with three nitrogen atoms in their structure and exhibit herbicidal activity. These are the most commonly used herbicides and have pre as well as post emergence applications. Because of their immense use, they are prevalent in the environment. Mulbry (1994) had isolated, purified and characterized an inducible s-triazine hydrolase from *Rhodococcus corallinus* NRRL B-15444R degrading several triazine compounds. Bacterial dehalogenases degrading triazines were studied by Villemur *et al.* (2005). Weir *et al.* (2006) isolated *Arthrobacter* sp. degrading endosulfan and endosulfan sulfate and also the genes encoding enzyme capable of degrading both isomers of endosulfan and endosulfan sulfate. The enzyme was found to be two component flavin-dependent monooxygenase. Involvement of cytochrome p-450 system in degradation of herbicide atrazine by *Rhodococcus* sp. is reported by some researchers (Nagy *et al.*, 1995b; Shao and Bekhi, 1996). Vaishampayan (2004) studied the complete degradation of atrazine, an organochlorine herbicide by *Arthrobacter* sp. strain MCM B-436 isolated from atrazine exposed soil and presence of trzN gene, as novel chlorohydrolase enzyme similar to that reported by Mulbry *et al.* (2002). The chlorohydrolase enzyme from *Arthrobacter* sp. MCM B-436 exhibited K_m

and V_{max} values as 23.8 µM and 1.11µmol of atrazine converted to hydroxy-atrazine per minute per mg of protein respectively. Hydrolases and dioxygenases by *Rhodoferax* sp. P230 and *Delftia* (*Comamonas*) *acidovorans* MC1 degrading phenoxypropionate and phenoxyacetate herbicides were reported by Muller *et al.* (2001). Phenylurea hydrolase (puhA) by *Arthrobacter globiformis* strain D47 capable of degrading phenyl urea herbicides was reported by Turnbull *et al.* (2001). Strauber *et al.* (2003) showed the cytochrome p-450 catalyzed cleavage of ether bond of phenoxybutyrate herbicides in *Rhodococcus erythropolis* K2–3. Carbaryl hydrolase, the enzyme degrading carbaryl was purified by Hashimoto *et al.* (2002).

Organophosphorus Hydrolase (OPH) is a bacterial enzyme, which has been shown to degrade a variety of pesticides belonging to organophosphate group. Kanekar *et al.* (2004) reviewed the biodegradation of organophosphorus pesticides (OPs). Production of OPH on the cell surface is highly host specific and OPH activity is dependent on growth conditions. Biodegradation of organophosphorus pesticides by surface-expressed organophosphorus hydrolase was studied by Richins *et al.* (1997). These enzymes were also studied by Mulbry (2000), Cho *et al.* (2002), Zhang *et al.* (2005), and Mee-Hie *et al.* (2006). Singh and Walker (2006) reviewed the microbial degradation of organophosphorus (OP) compounds with a thought that bioremediation can be a cheap and efficient method for decontamination of polluted sites. OPH or phosphotriesterase is involved in the initial step of degradation of OPs by bacteria. Phosphatase/phosphotriesterases are also reported to be involved in degradation of OPs like paraoxon, parathion, diazinon, methyl parathion, cyanophos, fensulfothion, monocrotophos, dimethoate, and coumaphos (Dumas *et al.*, 1989; Rabie, 1995; Liu *et al.*, 2001; Cai and Xun, 2002; Das and Singh, 2006). Bhadbhade *et al.* (2002a) have described biomineralization of insecticide Monocrotophos (MCP) to phosphates, ammonia and carbon dioxide brought about through the formation of intermediate compounds namely methylamine, volatile fatty acids like acetic acid or *n*-valeric acid alongwith one unidentified metabolite. Three bacterial cultures mineralizing MCP, namely *Arthrobacter atrocyaneus*, *Bacillus megaterium* and *Pseudomonas mendocina* were found to possess two enzymes namely phosphatase and esterase involved in the degradation of MCP by the organisms. Hassal (1990) described the role of special enzymes, carboxylamidases (esterases) in degradation of dimethoate and release of methylamine.

Liu *et al.* (2001) purified and characterized a dimethoate degrading enzyme from *A. niger*. While studying microbial degradation of dimethoate, attempts were made by Deshpande (2002) to identify the intermediates like dimethyldithiophosphate, acetic acid, and methylamine. Further, CO_2 ammonia, orthophosphates and H_2S were also detected as mineralization products. A pathway was proposed for degradation of dimethoate and possible role of enzymes like phosphatase and esterase involved. When, cell free preparations were used for studying enzyme systems, it was observed that extracellular enzymes were involved in dimethoate degradation (Deshpande, 2002).

6.3.2 DYES

Dyes are obtained as natural chemical compounds or may be from man made route. Natural dyes are mainly of vegetable origin and synthetic dyes are derivatives of aromatic hydrocarbons having coal tar as main source. Dyes are employed mainly in textile industry for printing and dying purpose alongwith their application for coloring a wide variety of materials like paper, toys, food items, cosmetics and others to improve their quality, as staining agents as well as sensitizers. Dyes are classified broadly as acid dyes, basic dyes, azo dyes, nitroso dyes and vat dyes. Dyes are known to be toxic to both aquatic as well as terrestrial life since some of them are carcinogenic and/or mutagenic in nature. These dyes enter environment at the site of production as well as application. Majority of the dyes are water soluble, hence they appear in the wastewaters generated during their production or application. Some of the microorganisms are known to detoxify or degrade some of the dyes by elaborating essential enzyme systems. Sarnaik and Kanekar (1999) have studied biodegradation of triphenyl methane dye, methyl violet.

Oxidoreductases like laccases and peroxidases have applications in textile industries mainly for their role in cotton fiber whitening, dye finishing and waste treatment. Laccase-catalyzed textile dye-bleaching is useful in dyed cotton fabric and being commercialized in recent years. Some researchers have worked on microbial especially fungal peroxidases and/or laccases used in degradation of dyes and in turn in treatment of dye industry effluents. Zimmermann et al. (1982) studied the properties of Orange II azoreductase, the enzyme system initiating azo-dye degradation by Pseudomonas sp. The enzyme systems used in degradation/decolorization of dyes like azo dyes, TPM dyes, heterocyclic dyes, polymeric dyes, Amido Black, Congo Red, Trypan Blue, Methyl Green, Remazol Brilliant Blue R, Methyl violet, Ethyl violet and Brilliant Cresyl Blue were reported by Olikka et al. (1998), Kim and Shoda (1999), Abdulla et al. (2000), Zilly et al. (2002) and Nyanhongo et al. (2002) using organisms like Pleurotus pulmonarius, Trametes modesta, Trametes hirsute, Trametes versicolor, Sclerotium rolfsii, Phanerochaete *chrysosporium* and *Geotrichum candidum*. Pereira *et al.* (2009) studied the decolorization of structurally different synthetic dyes using recombinant bacterial CotA-laccase from *Bacillus subtilis* at alkaline pH and in the absence of redox mediators. The authors have also studied the enzymatic biotransformation of the azo dye Sudan Orange G (SOG) in more detail. Saratale *et al.* (2010) carried out the studies on decolorization and biodegradation of a variety of reactive dyes including sulfonated reactive dye Green HE4BD using the consortium of *Proteus vulgaris* and *Micrococcus glutamicus*. The authors observed the involvement of oxidoreductase type of enzyme systems in the decolorization and degradation of reactive dyes. Mishra and Thakur (2011) isolated *Bacillus* sp. from sludge and sediment samples of pulp and paper mill, which was able to produce β-1,4 endoxylanase thermoalkali tolerant in nature and was able to decolorize variety of recalcitrant dyes. Thus this enzyme system has an application for bioremediation of dye containing wastewater over higher range of temperature and wide range of pH. Kolekar *et al.* (2012) developed aerobic granules from textile wastewater sludge to test their ability to degrade dye-reactive blue 59 (RB 59) and found that the enzymes azoreductase and cytochrome p-450 play an important role in dye degradation. Lade *et al.* (2012) developed a defined consortium – AP consisting of fungus *Aspergillus ochraceus* NCIM-1146 and a bacterium *Pseudomonas* sp. SUK1 and assessed its potential for enhanced decolorization and detoxification of azo dye Rubine GFL and textile effluent. The authors also studied the generation of laccase, veratryl alcohol oxidase, azo reductase and NADH-DCIP reductase in the consortium-AP which suggested the synergetic reactions of fungal and bacterial cultures for enhanced decolorization and detoxification of the dye and wastewater.

6.3.3 EXPLOSIVES

Explosives by definition are the chemical compounds that detonate at very high speed and release hot gases. Majority of explosives are organic nitro compounds. Kanekar *et al.* (2003) reviewed the biodegradation of nitroexplosives. Microorganisms are reported in the literature to degrade nitroaromatic compounds by aerobic as well as anaerobic mechanisms involving removal or productive metabolism of nitro groups. Aerobic mechanism of degradation involves presence of the enzymes monooxygenase and/or dioxygenase, which can add one/two oxygen atoms and eliminate the nitro group from a variety of nitroaromatic compounds. There is reduction of nitro compounds by the addition of a hydride ion to form a hydride-Meisenheimer complex, which subsequently rearomatizes with the elimination of nitrite or reduction of the nitro group to the corresponding hydroxylamine. Anaerobic mechanism involves reduction of nitro group via nitroso and hydroxylamino intermediates to the corresponding amines by anaerobic microorganisms (Spain, 1995).

Degradation of nitroexplosives by the enzyme systems like nitroreductases, manganese peroxidases and laccases are reported in the literature (Binks et al., 1995; Spain, 1995; Sheremata and Hawari, 2000; Rodgers and Bunce, 2001; Coleman et al., 2002). Under aerobic conditions, cytochrome p-450 enzymes are known to accept polynitro compounds as electron acceptors. Degradation of nitro explosives by the reductive pathway could be due to non-specific nitroreductase enzymes present in both aerobic and anaerobic organisms (Rodgers and Bunce, 2001). Since nitroaromatic compounds containing only nitro group substituents, are not direct substrates for lignin-degrading enzymes like lignin peroxidase and/or manganese peroxidase, these enzymes reduce the aromatic nitro group to an amine and then undergo reduction thus resulting in degradation of nitroaromatic compounds (Sheremata and Hawari, 2000).

Nitrate reductases are ubiquitous enzymes in diverse groups of microorganisms and their physiological role is to reduce nitrate to nitrite via a two-electron transfer. Enzymes such as nitroreductases (types I and II), hydrolases, hydrogenases are implicated in transformation of cyclic nitramines (Kitts et al., 2000; Hawari et al., 2001). Many of the facultative bacteria of Enterobacteriaceae family are reported to posses these enzyme systems including Providentia rettgeri (Dautpure, 2007). Coleman et al. (2002) studied the biodegradation of RDX by Rhodococcus strain DN22 which was found to be plasmid borne involving cytochrome p-450 enzyme system. Resorcinol 2,3-dioxygenase from Candida pulcherima degrading dinitrobenzene was purified and characterized by Dey (1991). Singh et al. (2012) in their review on microbial remediation of explosives mentioned about biodegradation and biotransformation pathways of some explosives. The authors have discussed about the metabolism of these explosives, isolation, characterization of the detoxifying enzymes and molecular basis of degradation of these toxic compounds. The authors think that this information will be useful for developing economically feasible methods for remediation of sites contaminated with explosives.

6.3.4 ORGANOCHLORINE COMPOUNDS

Dehalogenases are the enzymes detoxifying the halogenated organic compounds by cleaving their carbon-halogen bond. Chan et al. (2010) have described screening of microbial cultures for dehalogenases based on their sequence and activity. Since many microbial genomes harbor enzyme systems like esterases, phosphatases alongwith dehalogenases, the authors have emphasized on sequence and activity based screening of dehalogenases. Janssen *et al.* (2005) reviewed the bacterial degradation of xenobiotic compounds particularly organochlorine compounds with reference to evolution and distribution of novel enzyme systems involved therein. The authors have stated that dehalogenases having different catabolic mechanism can be classified in different protein superfamilies and can have diverse sequences. The authors have also mentioned that genetic adaptations and mutations in structural genes will improve the catabolic ability of bacteria towards halogenated organic compounds and will take part in making the environment free of these toxic and recalcitrant chemical entities. Bhatt *et al.* (2007) reviewed biodegradation of chlorinated compounds and the enzyme systems involved therein. Olaniran and Igbinosa (2011) reviewed the properties, distribution and microbial degradation processes of chlorophenols and other related derivatives of environmental concern. The authors have reported that chlorophenols are chlorinated aromatic compounds, commonly found in pesticide preparations as well as in industrial wastewaters. These are recalcitrant to biodegradation and consequently persistent in the environment. The key enzymes involved in their degradation include mono- and dioxygenases. The authors have postulated the enhanced degradation of these compounds by engineered enzymes.

Adebusoye and Miletto (2011) characterized the bacterial isolates from pristine and contaminated sites mineralizing 2,4-dichlorobenzoic acid. The strains exhibiting 2,4-dichlorobenzoic acid (2,4-diCBA) catabolism were mainly representatives of α-, β- and γ-*Proteobacteria*. Some of them exhibited presence of enzyme systems like catechol 1,2-dioxygenase types I and II and protocatechuate 3,4-dioxygenase. Pieper (2005) reviewed microbial degradation of polychlorinated biphenyls (PCBs) under aerobic conditions. Biphenyl 2,3-dioxygenases are the key enzymes involved in PCB degradation. Kumar *et al.* (2011) have reported the structural insight into the polychlorinated biphenyl (PCB) degrading ability of a biphenyl dioxygenases of *Burkholderia xenovorans* LB400. They have described the enzyme as a multicomponent Rieske-type oxygenase that catalyzes the dihydroxylation of biphenyl and many polychlorinated biphenyls (PCBs). The authors have also stated that these studies provide important insight about the expansion of substrate range of these Rieske-type oxygenases through mutations, which increase the flexibility and mobility of protein segments during the catalytic activity.

6.3.5 PHENOLIC COMPOUNDS

Phenolic compounds are natural as well as man made chemical entities and have various industrial applications like production of plastics, paints, resins, dyes and antioxidants. The presence of phenols in the environment is related with the production and decomposition of numerous pesticides and other xenobiotic chemicals. Wastewaters generated from petrochemical, textile and coal industries contain high concentration of phenolic compounds (Basha *et al.*, 2010). The handling and transportation activities of phenolic compounds also lead to the environmental pollution. Simple pure phenol is also a very toxic and corrosive chemical having high solubility in water. It is likely that the wastewaters generated in the phenol manufacturing, using industrial units contain higher concentrations of phenol in their wastewaters. Phenols include various types of compounds like chlorophenols, nitrophenols, hydroxyphenols, aminophenols, which are known to be more toxic than pure phenol.

Extensive research has been done on biodegradation of phenols with various aspects. Nair *et al.* (2008) have described various microorganisms responsible for degradation of phenols and related compounds like chlorophenols, nitrophenols belonging to the genera *Pseudomonas, Bacillus, Acinetobacter, Ralstonia, Nocardiodes, Alcaligenes, Sphingomonas, Achromobacter*, and fungi like *Trametes, Phanerochaete, Pleurotus, Trichosporon, Fusarium* and *Termitomyces*. The authors have also mentioned the involvement of enzymes like phenol oxidase, phenol hydroxylases, polyphenol oxidase, peroxidase, laccase, catechol dioxygenase and tyrosinase in the degradation of phenols. Microbial degradation of phenol has been reviewed by Durojaiye and Solomon (2008) wherein a variety of microorganisms degrading phenol, factors affecting their degradability, mechanisms and kinetics of degradation with the involvement of necessary enzymes are described. Basha *et al.* (2010) reviewed recent advances in the biodegradation of phenol. The authors have described various aerobic and anaerobic organisms involved in the degradation of phenol and the metabolic pathways of both aerobic and anaerobic degradation of phenol with the involvement of various enzymes like hydroxylase, monooxygenase, dioxygenase, dehydrogenase, isomerase and decarboxylase.

Udaysoorian and Prabu (2005) carried out studies on biodegradation of mono-dihydroxy phenols and methoxy phenols with the help of enzymes laccase and polyphenol oxidase of the lignolytic fungus *Trametes versicolor*. Nagamani *et al.* (2009) had isolated and characterized a soil bacterium *Xanthobacter flavus*, which was able to degrade phenol using the enzyme catechol 1,2 dioxygenase. Krupinski and Dlugonski (2011) studied the biodegradation of nonylphenols by some microbial cultures, which can use nonylphenols as sole source of carbon and energy. The authors also studied active participation of hydroxylases, cytochrome p-450 and lignin degrading enzymes, such as laccases, manganese peroxidases and lignin peroxidase. While describing metabolic pathways for the phenol degradation, Sridevi *et al.* (2012) have stated that aerobic degradation of phenol initiates by oxygenation into catechols as intermediates followed by the ring cleavage at either ortho or meta position. The ring fission is catalyzed by catechol dioxygenase. Duffner *et al.* (2000) carried out the studies on degradation of phenol at 65°C using the thermophilic strain of *Bacillus thermoglucosidasius* and observed the degradation of phenol through meta cleavage pathway. The authors also reported the cloning and sequence analysis of five genes involved in degradation of phenol.

Microbial or enzymatic remediation techniques can be developed for treatment of phenol containing industrial effluents. Microbial remediation of phenol containing wastewaters was described by Sarnaik and Kanekar (1995) and Kanekar *et al.* (1999). Bevilaqua *et al.* (2002) developed a process for removal of phenol using a biological method alongwith the enzymatic treatment using tyrosinase extracted from a mushroom *Agaricus bispora* for pretreatment of the effluent under study or polishing the biologically treated effluent. This finding suggests possible application of tyrosinase of microbial origin for treatment of phenol containing wastewaters. Ryan *et al.* (2007) were successful in carrying out bioremediation of phenolic wastewater by *Tramates versicolor*, which was dependent on the growth and age of organism and in turn the production of enzyme laccase.

6.3.6 OTHER XENOBIOTIC CHEMICALS

Sariaslani (1991) reviewed thoroughly the microbial cytochrome p-450 and their active role in degradation of a variety of xenobiotic compounds. The author has given detailed account of enzyme systems from different microorganisms, for example enzymes involved in degradation of camphor by *Pseudomonas putida* with structure of enzyme protein; *Bacillus megaterium* hydroxylating long chain fatty acids and their aldehydes and alcohols; *Xanthomonas* sp. performing NADPH-dependent hydroxylation of cyclohexane to cyclohexanol; *Acinetobacter calcoaceticus* carrying out oxidation of alkanes; several cytochrome p-450 of unspecified function from *Rhizobium japonicum* alongwith some enzyme systems from actinomycetes, yeasts and fungi. Hawkes *et al.* (2002) carried out isolation and characterization of cytochrome p-450$_{cin}$ (CYP176A) from *Citrobacter braakii* using monoterpene cineole-1 as sole source of carbon and energy. Parales and Ju have illustrated the degradation pathways for the compounds like toluene, benzene, xylene, stirene, biphenyl and naphthalene wherein involvement of dioxygenases is emphasized and the microorganisms used included species of *Pseudomonas*, *Rhodococcus*, *Burkholderia* and *Ralstonia*. Some reports are available on role of microbial oxygenases/cytochrome p-450 in degradation of chemopollutants. Roy *et al.* (2012) isolated a strain of *Sphingobium* sp. for degradation of phenantherene a toxic compound belonging to a group of polyhydroxy aromatic hydrocarbons (PAHs), which are persistent organic pollutants and observed the enzyme systems taking part in metabolism of phenantherene through meta-cleavage.

Mooney *et al.* (2006) reviewed different aspects of microbial degradation of the environmentally and industrially important and toxic compound, stirene with respect to the biochemistry and molecular genetics and the enzyme systems involved therein. Mulyono *et al.* (2007) reported presence of arylamin N-acetyltransferase (NAT) in a strain of *Bacillus cereus* converting 4-phenelenediamine to 4-aminoacetanilide, which is useful in detoxifying 4-phenylenediamine. Cavalea *et al.* (2007) thought that ancillary enzyme activity in microorganisms could facilitate the utilization of polycyclic aromatic hydrocarbons as sole source of carbon. The authors have detected the glutathione-S-transferase activity in *Sphingobium chlorophenolicum* grown on phenanthrene and rhodanese like protein in *Rhodococcus aetherovorans*, *R. opaceus* and *Mycobacterium smegmatis* grown on phenanthrene. Thus these organisms and their enzyme systems are useful in detoxifying polycyclic aromatic hydrocarbons. Wu *et al.* (2008) reported the role of fungal laccase in remediation of soil contaminated with polycyclic aromatic hydrocarbons (PAHs). The authors have stated about the direct application of free laccases in the remediation of soil polluted with PAHs like anthracene, benz(o)pyrene through oxidation. The authors observed that the free laccase could transform benz(o)pyrene to a significant extent within 14 days of incubation.

Studies on degradation of caprolactam, an industrially important compound, which is being used in the manufacture of synthetic fiber nylon-6 and appearing as major pollutant of the nylon-6 effluent were carried out by Kulkarni and Kanekar (1998a). The author reported the complete degradation of caprolactam by *Pseudomonas putida* to carbon dioxide through intermediate metabolites namely aminocaproic acid, amino acid, adipic acid, caproic acid, propionic acid and acetic acid and expected to follow the metabolic pathway described by Fukumura (1966) involving the enzymes like aminocaproic acid transaminase and adipic semialdehyde dehydrogenase.

6.4 MOLECULAR BASIS OF DETOXIFYING ENZYMES

Investigating the genetic basis of biodegradation of xenobiotic compounds has always been remained interesting. Many of the microorganisms capable of metabolizing synthetic organic compounds like pesticides, harbor large degradative plasmids. Pesticide degrading genes in microbes have been found to be located on plasmids, transposones, and/or on chromosomes. Extensive work has been done on hexachlorocyclohexane (HCH) or lindane degrading genes including their organization, diversity and distribution in *Sphingomonas paucimobilis* and *Sphingobium indicum* by Dogra *et al.* (2004), Lal *et al.* (2006) and Malhotra *et al.* (2007). Enantioselective transformation of hexachlorocyclohexane (HCH) by dehydrochlorinases LinA1 and LinA2 from soil bacterium *Sphingomonas paucimobilis* B90A were reported by Suar *et al.* (2004, 2005). Guha *et al.* (2000) reported the degradation of endosulfan and its utilization as a sole source of carbon and energy by plasmid harboring and cured strains of *Micrococcus* sp. Sutherland *et al.* (2002) had successfully cloned the gene sequence of enzyme system degrading Endosulfan and the translated product of gene (Esd) had up to 50% sequence homology with the unusual family of monooxygenase enzymes that use reduced flavins, provided by separate flavin reductase enzyme as cosubstrate.

Omotayo *et al.* (2011) carried out research on atrazine degradation by soil bacteria and involvement of trzND, atzBC genes. The authors have isolated bacteria using conventional technique for isolation as well as *in situ* enrichment technique using porous Bio-Sep beads fortified with atrazine. Vaishampayan *et al.* (2007) have illustrated the role of genes atzA, atzB and atzC involved in degradation of atrazine. Sequence analysis of atzB, atzC and atzD genes from *Arthrobacter* sp. exhibited high sequence similarity with atzB, atzC and atzD genes from *Pseudomonas* sp. strain ADP reported by Matrinez *et al.* (2001) and Piutti *et al.* (2003). Plasmids were detected in *Arthrobacter sulfureus* and *Microbacterium lacticum* degrading ethylene thiourea (ETU), a breakdown toxic product of fungicide Mancozeb (Puranik, 2000).

The enzyme phosphotriesterases (parathion hydrolase) was cloned from a *Flavobacterium* sp. into *Streptomyces lividans* by Rowland *et al.* (1991). Segers *et al.* (1994) reported plasmid controlled parathion degradation in *Brevundimonas diminuta*. Somara and Siddavattam (1995) described *Flavobacterium balustinum* harboring an indigenous plasmid of approximately 86 kb in size. The degradative enzyme parathion hydrolase was found to be encoded by this plasmid. Kumar *et al.* (1996) have reviewed microbial degradation of pesticides with special emphasis on the role of catabolic genes in relation to pesticide degradation and the application of recombinant DNA technology in development of 'superbug,' an organism, which can simultaneously degrade several xenobiotics. Possible involvement of plasmids in degradation of malathion and chlorpyriphos has been reported by Guha *et al*. (1997). Mulbry *et al*. (1986) performed the studies on identification of a plasmid-borne parathion hydrolase gene from *Flavobacterium* sp. by southern hybridization with opd from *Pseudomonas diminuta*. Organophosphorus hydrolase (OPH) gene adpB was studied by Mulbry (1998) from *Nocardia* sp. strain B-1 capable of degrading parathion. Chen and Mulchandani (1998) studied use of live biocatalysts for pesticide detoxification. They have discussed the use of genetically engineered *E. coli* with surface expressed organophosphorus hydrolase and suggested the ultimate creation of 'super biocatalyst' capable of degrading several pesticides rapidly and cost effectively. Similar studies on isolation of a methyl parathion-degrading *Pseudomonas* sp. that possesses DNA homologous to the opd gene from a *Flavobacterium* sp. were carried out by Chaudhari *et al*. (1988). Serdar *et al*. (1982) reported the involvement of plasmid in parathion hydrolysis by *Pseudomonas diminuta*. Kaneva *et al*. (1998) focused on the studies of the factors influencing parathion degradation by recombinant *Escherichia coli* with surface-expressed organophosphorus hydrolase.

Zhongli *et al*. (2001) identified *Plesiomonas* sp. capable of hydrolyzing methyl parathion to p-nitrophenol. A novel organophosphate hydrolase (OPH) gene designated as mpd was selected from its genomic library prepared by shotgun cloning. The nucleotide sequence of the mpd gene was determined. The gene could be effectively expressed in *E. coli*. Singh and Walker (2006) have mentioned that the OPH is encoded with the gene opd and is common in all OP degrading bacteria. The authors have also stated that the gene opd has been sequenced and cloned in various taxonomically diverse microorganisms and altered for better activity and stability. They have also mentioned about the progress carried out in isolation and characterization of genes with similar function but with different sequences. Studies carried out to determine the molecular basis of Monocrotophos (MCP) and Dimethoate degradation clearly indicated presence of plasmid taking active part in degradation of MCP by *Pseudomonas mendocina* MCM B-424 (Bhadbhade et. al. 2002b) and dimethoate by *Brevundimonas* sp. MCM B-427 (Deshpande *et al*., 2001) and the plasmids could be transferred to and expressed in *E. coli* Nova Blue. Kanekar *et al*. (2004) have reviewed the work on genetic basis of degradation of organophosphorus pesticides.

A soil isolate of *Moraxlla* sp. capable of growing on p-nitrophenol (PNP), was genetically engineered so as to get simultaneous degradation of OPs and PNP. The truncated ice nucleation protein (INPNC) anchor was used to target the OPH, onto the surface of *Moraxlla* sp. A shuttle vector, pPNCO33, coding for INPNC-OPH was constructed and translocation, surface display, and functionality of OPH were demonstrated in both *E. coli* as well as in *Moraxlla* sp. 70-fold more activity was observed in *Moraxlla* sp. as compared to that in *E. coli*. The resulting *Moraxlla* sp. degraded organophosphates and PNP rapidly within 10 h (Shimazu *et al.*, 2001). Hawkes *et al.* (2002) carried out studies on cloning of p-450$_{cin}$ gene (cinA) from *Citrobacter braakii* degrading monoterpene cineole-1. Sequencing revealed three open reading frames identified as NADPH-dependent flavodoxin/ferrodoxin reductase and a flavodoxin. This gene was successfully subcloned and expressed in *E. coli*. Kitagawa *et al.* (2004) identified and characterized PNP degrading gene cluster from *Rhodococcus opacus* SAO101. Tsirogianni *et al.* (2004) carried out studies on mass spectrometric mapping of the enzymes like catechol dioxygenases from *Pseudomonas* sp. which play an important role in degradation of phenol and were able to identify 10 different enzymes, which were able to degrade phenol. Nordin *et al.* (2005) studied the gene cluster (cphAI and cphAII genes) of *Arthrobacter chlorophenolicus* able to degrade 4-chlorophenol via hydroxyphenols by secreting the functional enzyme hydroxyquinol 1,2-dioxygenases. Movahedyan *et al.* (2009) developed a technique for detection of phenol degrading bacteria using polymerase chain reaction technique. For this purpose the authors used multicomponent phenol hydroxylase (LmPH) gene and gene coding the N fragment in *Pseudomonas putida* derived methyl phenol operon (DmpN gene) through PCR.

Wierckx *et al.* (2011) reported the importance of degradation of furfural, hydroxymethyl furfural furanic aldehydes – the toxic components of lignocellulose hydrolysate to nontoxic compounds. Eight HMF genes are present in the enzyme systems involved in these degradation pathways and the cluster of 5 genes is highly conserved in the microbes capable of degrading furfurals. Studies on genetic basis of degradation of caprolactam carried out by Kulkarni and Kanekar (1998b) revealed that *Pseudomonas putida* harbored a plasmid, probably a highly stable one and under adverse environmental conditions it may allow the organism to successfully degrade caprolactam appearing in the industrial effluent. Dey (1999) has reported degradation of dinitrobenzene by *Streptomyces aminophilus*. The nitroaryl reductase gene from *S. aminophilus* was cloned and expressed in *E. coli* and *S. lividans* TK64. Johri *et al.* (1999) have extensively reviewed characterization and regulation of catabolic genes providing an insight in the recent advances made on characterization and expression of catabolic genes that encode the degradation/detoxification of persistent and toxic xenobiotic compounds.

6.5   FUTURE PROSPECTS

Newer and newer chemical compounds are being discovered and exploited for modernization of life. Production of chemical compounds can not be stopped and hence the choice left is to live with these chemicals. To save life from hazardous chemical pollutants released in the environment, we have to use our wisdom to explore natural microbial resources for detoxifying enzymes. In comparison to mammalian enzymes, microbial cytochrome enzymes are studied to a limited extent. Presently ~500 enzyme products are investigated for about 50 different applications among which oxidoreductases contribute to a restricted extent. Oxidoreductases are the more preferred enzymes as compared to other enzymes like hydrolases, lyases or transferases, may be due to their higher stereoselectivity and ability to act on substances like hydrocarbons, which are considered as inert to other enzymes. The oxidoreductases have very low specific activity and need cofactors. These limitations could be overcome by applying modern techniques like gene cloning and genetic engineering of proteins.

These oxidoreductases can also be employed for biomineralization purposes in mining industries especially in recovery of gold from ores. Munnecke (1976) has postulated the enzymatic hydrolysis of organophosphate insecticides as a possible pesticide disposal method. Ryan et al. (2007) were successful in carrying out bioremediation of phenolic wastewater using the enzyme laccase from *Tramates versicolor* thus indicating the possible use of laccases in treatment of phenolic wastewaters. Likewise, microorganisms producing enzyme dehalogenase may be useful in removal of chlorine from the organochlorine compounds and degradation of these compounds. Development of microbial amperometric biosensor based on a carbon paste electrode containing genetically engineered cells expressing organophosphorus hydrolase (OPH) on the cell surface for the direct measurement of organophosphate nerve agents as described by Mulchandani et al. (2001) seems to be interesting for exploration. The success of directed evolution of OPH for improved hydrolysis of methyl parathion, can be easily extended in creating other OPH variants with improved activity against poorly degraded pesticides. The researchers can attempt to obtain such OPH variants.

Although some reports are available on organophosphate degrading (opd) genes, the picture is yet not very clear on mechanism of degradation of these pesticides. The use of the opd gene as a probe may accelerate progress toward understanding the complex interactions of soil microorganisms with organophosphates. Constructing recombinant strains for degradation of a number of organophosphorus pesticides and their metabolites would be a challenge. These would be a boon to soil bioremediation processes. The information regarding the degradation of nitroexplosives, their metabolic pathway and the enzyme systems involved therein would be useful to monitor the degradation of explosives particularly nitroexplosives in terms of fate of the metabolites, toxicity and mineralization for complete removal of the compound. Additionally, this enzymatic strategy would also give an idea about the most likely microbial species involved in natural attenuation of nitroexplosives. Nitroreductases can be used to develop biosensors for detection of nitroexplosives in soil. In addition to oxidoreductases and hydrolases reported so far, search for novel enzymes for degradation of new chemicals being discovered will continue.

6.6   CONCLUSION

The oxidoreductases possess very diverse substrate specificity, reactivity and some other physicochemical and biological characteristics, which are useful, in different industrial and medical applications. These are present ubiquitously in plants, animals and microorganisms. Microorganisms play an important role in the degradation of these man-made chemicals, which are released in the environment. These microorganisms are endowed with the property of secreting enzymes required for degradation or detoxification of toxic chemicals like nitro compounds, pesticides especially the organochlorine pesticides. In general the enzymes involved in degradation of chemopollutants include oxidases, hydrolases, dehalogenases, dealkylases and deaminases. The special type of enzymes like cytochrome p-450 are widely distributed in biological systems like plants, animals and microorganisms. They play an important role in oxidative metabolism of a variety of chemicals of agricultural and environmental concern, such as pesticides. In comparison to mammalian enzymes, microbial cytochrome enzymes are studied to a limited extent. There is a wide scope for researchers to explore novel applications of these oxidoreductases in medical field, as biosensors, and also in detoxifying the newer toxic chemicals manufactured as on demands. The knowledge about the three dimensional structures of the cytochrome p-450s will help in application of these in bioremediation and other purposes. Thus, ecofriendly technologies could be developed for bioremediation purposes using these enzyme systems.

KEYWORDS

- **Enzymatic degradation**
- **Ethylenethiourea**
- **Flavin monooxygenases**
- **Hexachlorocyclohexane**
- **Microbial detoxification**
- **Organic pollutants**
- **Organochlorine compounds**
- **Organophosphorus hydrolase**
- **Polychlorinated biphenyls**

REFERENCES

Abdulla, E.; Tzanov, T.; Costa, S.; Robra, K. H.; CavacoPaulo, A.; Gubitz, G. M. Decolorization and detoxification of textile dyes with a laccase from *Tramates hirsute. Appl Environ Microbiol* 2000, 66, 3357–3362.

Adebusoye, S. A.; Miletto, M. Characterization of multiple chlorobenzoic acid-degrading organisms from pristine and contaminated systems: mineralization of 2,4-dichlorobenzoic acid. *Biores Technol* 2011, 102(3), 3041–3048.

Basha, K. M.; Rajendran, A.; Thangavelu, V. Recent advances in the biodegradation of phenol: a review. *Asian J Exp Biol Sci* 2010, 1(2), 219–234.

Bevilaqua, J. V.; Cammarota, M. C.; Freire, D. M. G.; Santanna, G. L. Jr. Phenol removal through combined biological and enzymatic treatment. *Brazilian J Chem Eng* 2002, 19(2), 151–158.

Bhadbhade, B. J.; Sarnaik, S. S.; Kanekar, P. P. Biomineralization of an organophosphorus pesticide Monocrotophos by soil bacteria. *J Appl Microbiol* 2002a, 93, 224–234.

Bhadbhade, B. J.; Dhakephalkar, P. K.; Sarnaik, S. S.; Kanekar, P. P. Plasmid associated biodegradation of an organophosphorus pesticide Monocrotophos by *Pseudomonas mendocina. Biotechnol Lett* 2002b, 24, 647–650.

Bhatt, P.; Suresh Kumar, M.; Mudliar, S.; Chakrabarti, T. Biodegradation of chlorinated compounds: a review. *Crit Rev Environ Sci Technol* 2007, 37(2), 165–198.

Binks, P. R.; Nicklin, S.; Bruce, N. C. Degradation of hexahydro-1,3,5-trinitro-1,3,5-triazine (RDX) by *Stenotrophomonas maltophilia* PB1. *Appl Environ Microbiol* 1995, 61, 1318–1322.

Cai, M.; Xun, L. Organization and regulation of pentachlorophenol degrading genes in *Sphingomonas chlorophenolicum* ATCC 39723. *J Bacteriol* 2002, 184(17), 4672–4680.

Camacho-Pérez, B.; Ríos-Leal, E.; Rinderknecht-Seijas, N.; Poggi-Varaldo, H. M. Enzymes involved in the biodegradation of hexachlorocyclohexane: a mini review. *J Environ Mang* 2012, 95, S306-S318.

Cavalea, L.; Guerricri, N.; Colombo, M.; Pagani, S.; Andreoni, V. Enzymatic and genetic profiles in environmental strains grown on polycyclic aromatic hydrocarbons. *Antonie van Leeuwenhoek* 2007, 91(4), 315–325.

Chan, W. Y.; Wong, M.; Guthrie, J.; Savechenko, A. V.; Yakunin, A. F.; Pai, E. F.; Edwards, E. A. Sequence- and activity- based screening of microbial genomes for novel dehalogenases. *Microb Biotechnol* 2010, 3(1), 107–120.

Chaudhari, G. R.; Ali, A. N.; Wheeler, W. B. Isolation of a methyl parathion-degrading *Pseudomonas* sp. that possesses DNA homologous to the opd gene from a *Flavobacterium* sp. *Appl Environ Microbiol* 1988, 54(2), 288–293.

Cheesman, M. J.; Home, I.; Weir, K. M.; Pandey, G.; Williams, M. R.; Scott, C.; Russell, R. J.; Oakshott, J. G. Carbamate pesticides and their biological degradation: prospects for enzymatic bioremediation. In *Rational environmental management of agrochemicals: risk assessment, monitoring and remedial action*, American Chemical Society, USA, 2007; pp 288–305.

Chen, W.; Mulchandani, A. The use of live biocatalysts for pesticide detoxification. *Trends Biotechnol* 1998, 16(2), 71–76.

Cho, C. M.; Mulchandani, A.; Chen, W. Cell surface display of organophosphorus hydrolase for selective screening of improved hydrolysis of organophosphate nerve agents. *Appl Environ Microbiol* 2002, 68(4), 2026–2030.

Coleman, N. V.; Spain, J. C.; Duxbury, T. Evidence that RDX biodegradation by *Rhodococcus* strain DN22 is plasmid-borne and involves a cytochrome P-450. *J Appl Microbiol* 2002, 93, 463–472.

Das, S.; Singh, D. K. Purification and characterization of phosphotriesterases from *Pseudomonas aeruginosa* F10B and *Clavibacter michiganense* subsp. *insidiosum* SBL11. *Can J Microbiol* 2006, 52(2), 157–168.

Dautpure, P. S. Microbial degradation of nitroexplosive HMX (High Melting Explosive). Ph D. Thesis, University of Pune, Pune, India, 2007.

Deshpande, N. M.; Dhakephalkar, P. K.; Kanekar, P. P. Plasmid mediated dimethoate degradation in *Pseudomonas aeruginosa* MCM B-427. *Lett Appl Microbiol* 2001, 33, 275–279.

Deshpande, N. M. Biodegradation of Dimethoate – a carbamate group of organophosphorus insecticides. Ph. D. Thesis, University of Pune, Pune, India, 2002.

Deshmukh, N. S.; Lapsia, K. L.; Savant, D. V.; Chiplonkar, S. A.; Yeole, T. Y.; Dhakephalkar, P. K.; Ranade, D. R. Upflow anaerobic filter for the degradation of adsorbable organic halides (AOX) from bleach composite wastewater of pulp and paper industry. *Chemosphere* 2009, 75, 1179–1185.

Dey, S. Purification and properties of resorcinol 2,3-dioxygenase from *Candida pulcherima*. *Indian J Appl Pure Biol* 1991, 6, 107–111.

Dey, S. Cloning and expression of nitroaryl reductase gene from *Streptomyces aminophilus* to *E. coli* and *S. lividans* TK64. *Indian J Exp Biol* 1999, 37, 787–792.

Dogra, C.; Raina, V.; Pal, R.; Suar, M.; Lal, S.; Gartemann, K. H.; Holliger, C.; van der Meer, J. R.; Lal, R. Organization of lin genes and IS6100 among different strains of hexachlorocyclohexane degrading *Sphingomonas paucimobilis* strains: evidence of natural horizontal transfer. *Journal of Bacteriology* 2004, 186, 2225–2235.

Duffner, F. M.; Kirchner, U.; Bauer, M. P.; Muller, R. Phenol/cresol degradation by the thermophilic *Bacillus thermoglucosidasius* A7: cloning and sequence analysis of five genes involved in the pathway. *Gene* 2000, 256, 215–221.

Dumas, D. P.; Caldwelol, S. S. R.; Wild, J. R.; Raushel, F. M. Purification and properties of the phosphotriesterase from *Pseudomonas diminuta*. *J Biol Chem* 1989, 264(33), 19659–19665.

Durojaiye, A. O.; Solomon, B. O. Microbial degradation of phenols: a review. *Intl J Environ Polln* 2008, 32(1), 12–18.

Fukumura, T. Two bacterial enzymes hydrolyzing oligomers of 6-amino acaroic acid. *J Biochem* 1966, 59, 537–544.

Guha, A.; Kumari, B.; Bora, T. C.; Roy, M. K. Possible involvement of plasmids in degradation of malathion and chlorpyriphos by *Micrococcus* sp. *Folia Microbiol (Praha)* 1997, 42(6), 574–576.

Guha, A.; Kumari, B.; Bora, T. C.; Deka, P. C.; Roy, M. K. Bioremediation of endosulfan by *Micrococcus* sp. *Indian J. Environ Health* 2000, 42(1), 9–12.

Hashimoto, M.; Fukui, M.; Hayano, K.; Hayatsu, M. Nucleotide sequence and genetic structure of a novel carbaryl hydrolase gene (cehA) from *Rhizobium* sp. strain AC100. *Appl Environ Microbiol* 2002, 68(3), 1220–1227.

Hassal, A. K. Organophosphorus insecticides. In *The biochemistry and uses of pesticides*, ELBS Publication: London, 1990; pp 81–124.

Hawari, J.; Halasz, A.; Beaudet, S.; Paquet, L.; Ampleman, G.; Thiboutot, S. Biotransformation routes of octahydro- 1,3,5,7- tetranitro- 1,3,5,7- tetrazocine by municipal anaerobic sludge. *Environ Sci Technol* 2001, 35, 70–75.

Hawkes, D. B.; Adams, G. W.; Burlingame, A. L.; Ortiz de Montellano, P. R.; De Voss, J. J. Cytochrome p-450cin (CYP176A), isolation, expression and characterization. *J Biol Chem* 2002, 277(31), 27725–27732.

Janssen, D. B.; Dinkla, I. J. T.; Poelarends, G. J.; Terpstra, P. Bacterial degradation of xenobiotic compounds: evolution and distribution of novel enzyme activities. *Environ Microbiol* 2005, 7(12), 1868–1882.

Johri, A. K.; Dua, M.; Singh, A.; Sethunathan, N.; Legge, R. L. Characterization and regularization of catabolic genes. *Crit Rev Microbiol* 1999, 25(4), 245–273.

Kanekar, P. P.; Sarnaik, S. S.; Kelkar, A. S. Bioremediation of Phenol by alkalophilic bacteria isolated from alkaline Lake of Lonar, India. *J Appl Microbiol* 1999, 85, S128-S133.

Kanekar, P. P.; Dautpure, P. S.; Sarnaik, S. S. Biodegradation of nitroexplosives. *Indian J Exp Biol* 2003, 41, 991–1001.

Kanekar, P. P.; Bhadbhade, B. J.; Deshpande, N. M.; Sarnaik, S. S. Biodegradation of organophosphorus pesticides. *Proc Indian Natn Sci Acad* 2004, B70, 57–70.

Kaneva, I.; Mulchandani, A.; Chen, W. Factors influencing parathion degradation by recombinant *Escherichia coli* with surface-expressed organophosphorus hydrolase. *Biotechnol Prog* 1998, 14(2), 275–278.

Karayilanoglu, T.; Kenar, L.; Serdar, M.; Kose, S.; Aydin, A. Bacterial biodegradation of Aldicarb and determination of bacterium, which has the most biodegradative effect. *Turk J Biochem* 2008, 33(4), 209–214.

Kim, S. J.; Shoda, M. Purification and characterization of a novel peroxidase from *Geotrichum candidum* Dec-1 involved in decolorization of dyes. *Appl Environ Microbiol* 1999, 65, 1029–1035.

Kitagawa, W.; Kimura, N.; Kamagata, Y. A novel p-nitrophenol degradation gene cluster from a gram positive bacterium, *Rhodococcus opacus* SAO101. *J Bacteriol* 2004, 186(15), 4894–4902.

Kitts, C. L.; Green, C. E.; Otley, R. A.; Alvarez, M. A.; Unkefe, P. J. Type I nitroreductases in soil enterobacteria reduce TNT (2,4,6-trinitrotoluene) and RDX (hexahydro-1,3,5-trinitro-1,3,5-triazine). *Can J Microbiol* 2000, 46, 278–282.

Kolekar, Y. M.; Nemade, H. N.; Markad, V. L.; Adav, S. S.; Patole, M. S.; Kodam, K. M. Decolorization and biodegradation of azo dye, reactive blue 59 by aerobic granules. *Biores Technol* 2012, 104, 818–822.

Krupinski, M.; Dlugonski, J. Biodegradation of nonylphenols by some microorganisms. *Postepy Mikrobiol* 2011, 50(4), 313–319.

Kulkarni, R. S.; Kanekar, P. P. Bioremediation of ε-caprolactam from a nylon-6 waste water using *Pseudomonas aeruginosa* MCM B-407 isolated from nylon-6 industry activated sludge. *Curr Microbiol* 1998a, 37, 191–194.

Kulkarni, R. S.; Kanekar, P. P. Effect of some curing agents on phenotypic stability in *Pseudomonas putida* MCM B-408 degrading ε-caprolactam. *World J Microbiol Biotechnol* 1998b, 14, 255–257.

Kumar, P.; Mohammadi, M.; Viger, J. -F.; Barriault, D.; Gomez-Gil, L.; Eltis, L. D.; Bolin, J. T.; Sylvestre, M. Structural insight into the expanded PCB-degrading abilities of a biphenyl dioxygenase obtained by directed evolution. *J Mol Biol* 2011, 405(2), 531–547.

Kumar, S.; Mukerji, K. G.; Pal, R. Molecular aspects of pesticide degradation by microorganisms. *Crit Rev Microbiol* 1996, 22, 1–26.

Lade, H. S.; Waghmode, T. R.; Kadam, A. A.; Govindwar, S. P. Enhanced biodegradation and detoxification of disperse azo dye Rubine GFL and textile industry effluent by defined fungal-bacterial consortium. *Int Biodeter Biodegr* 2012, 72, 94–107.

Lal, R.; Dogra, C.; Malhotra, S.; Sharma, P.; Pal, R. Diversity, distribution and divergence of lin genes in hexachlorocyclohexane degrading sphingomonads. *Trends Biotechnol* 2006, 24, 121–130.

Liu, Y.; Chung, Y.; Xiong, Y. Purification and characterization of a dimethoate-degrading enzyme of *Aspergillus niger* ZHY 256, isolated from sewage. *Appl Environ Microbiol* 2001, 67, 3746–3749.

Malhotra, S.; Sharma, P.; Kumari, H.; Singh, A.; Lal, R. Localization of HCH catabolic genes (lin genes) in *Sphingobium indicum* B90A. *Indian J Microbiol* 2007, 47, 271–275.

Manickam, N.; Reddy, M. K.; Saini, H. S.; Shaker, R. Isolation of hexachlorocyclohexane- degrading *Sphingomonas* sp. by dehalogenase assay and characterization of genes involved in γ-HCH degradation. *J Appl Microbiol* 2008, 104, 952–960.

Matrinez, B.; Tomkins, J.; Wackett, L. P.; Wing, R.; Sadowsky, M. J. Complete nucleotide sequence and organization of the atrazine catabolic plasmid pADP-1 from *Pseudomonas* sp. strain ADP. *J Bacteriol* 2001, 183, 684–697.

Mee-Hie, C. C.; Mulchandani, A.; Chen, W. Functional analysis of organophosphorus hydrolase variants with high degradation activity towards organophosphate pesticides. *Protein Eng Des Sel* 2006, 19(3), 99–105.

Michalowicz, J.; Duda, W. Phenols – sources and toxicity. *Polish J Environ Stud* 2007, 16(3), 347–362.

Mishra, M.; Thakur, I. S. Purification, characterization and mass spectroscopic analysis of thermoalkalitolerant β-1,4 endoxylanase from *Bacillus* sp. and its potential for dye decolorization. *Int Biodet Biodegrad* 2011, 65(2), 301–308.

Mooney, A.; Ward, P. G.; O'Connor, K. E. Microbial degradation of stirene: biochemistry, molecular genetics, and perspectives for biotechnological applications. *Appl Microbiol Biotechnol* 2006, 72, 1–10.

Movahedyan, H.; Khorsandi, H.; Salehi, R.; Nikaeen, M. Detection of phenol degrading bacteria and *Pseudomonas putida* in activated sludge by polymerase chain reaction. *Iran J Environ Health Sci Eng* 2009, 6(2), 115–120.

Mulbry, W. W. Purification and characterization of an inducible s-triazine hydrolase from *Rhodococcus corallinus* NRRL B-15444R. *Appl Environ Microbiol* 1994, 60(2), 613–618.

Mulbry, W. Selective deletions involving the organophosphorus hydrolase gene adpB from *Nocardia* strain B-1. *Microbiol Res* 1998, 153(3), 213–217.

Mulbry, W. Characterization of a novel organophosphorus hydrolase from *Nocardiodes simplex* NRRL B-24074. *Microbiol Res* 2000, 154(4), 285–308.

Mulbry, W. W.; Karns JS.; Kearney PC.; Nelson JO.; McDaniel CS.; Wild JR. Identification of a plasmid-borne parathion hydrolase gene from *Flavobacterium* sp. by southern hybridization with opd from *Pseudomonas diminuta*. *Appl Environ Microbiol* 1986, 51(5), 926–930.

Mulbry, W. W.; Zhu H.; Nour SM.; Topp E. The triazine hydrolase gene trzn from *Nocardioides* sp. strain C190: cloning and construction of gene specific primers. *FEMS Microbiol Lett* 2002, 206, 75–79.

Mulchandani, P.; Chen, W.; Mulchandani, A.; Wang, J.; Chen, L. Amperometric microbial biosensor for direct determination of organophosphate pesticides using recombinant microorganism with surface expressed organophosphorus hydrolase. *Biosens Bioelectron* 2001, 1666(7–8), 433–437.

Muller, R. H.; Kleinsteuber, S.; Babel, W. Physiological and genetic characteristics of two bacterial strains using phenoxypropionate and phenoxyacetate herbicides. *Microbiol Res* 2001, 156(2), 121–131.

Mulyono.; Takenaka, S.; Sasano, Y.; Murakami, S.; Aoki, K. *Bacillus cereus* strain 10-L-2 produces two arylamine N-acetyltransferases that transform 4-phenylenediamine into 4-aminoacetanilide. *J Biosci Bioeng* 2007, 103(2), 147–154.

Munnecke, D. M. Enzymatic hydrolysis of organophosphate insecticides, a possible pesticide disposal method. *Appl Environ Microbiol* 1976, 32(1), 7–13.

Nagamani, A.; Soligalla, R.; Lowry M. Isolation and characterization of phenol degrading *Xanthobacter flavus*. *African J Biotech* 2009, 8(20), 5449–5453.

Nagy, I.; Compernolle, F.; Ghys, K.; Vanderleyden, J.; de Mot, R. A single cytochrome p450 system is involved in degradation of the herbicides EPTC (S-ethyldipropyl thio carbamate) and atrazine by *Rhodococcus* sp. strain N186/21. *Appl Environ Microbiol* 1995a, 61(5), 2056–2060.

Nagy, I.; Schoofs, G.; Compernolle, F.; Proost, P.; Vanderleyden, J.; de Mot, R. Degradation of the thiocarbamate herbicide EPTC (S-ethyldipropylcarbamothioate) and biosafening by *Rhodococcus* sp. strain N186/21 involve an inducible cytochrome p-450 system and aldehyde dehydrogenase. *J Bacteriol* 1995b, 177(3), 676–687.

Nair, C. I.; Jayachandran, K.; Shashidhar, S. Biodegradation of phenol. *African J Biotech* 2008, 7(25), 4951–4958.

Nordin, K.; Unell, M.; Jansson, J. K. Novel 4-chlorophenol degradation gene cluster and degradation route via hydroxyquinol in *Arthrobacter chlorophenolicus* A6. *Appl Environ Microbiol* 2005, 71(11), 6538–6544.

Nyanhongo, G. S.; Goves, J.; Gubitz, G.; Zvauya, R. Production of laccase by a newly isolated strain of *Trametes modesta*. *Biores Technol* 2002, 84, 259–263.

Olaniran, A. O.; Igbinosa, E. O. Chlorophenols and other related derivatives of environmental concern: properties, distribution and microbial degradation processes. *Chemosphere* 2011, 83(10), 1297–1306.

Ollikka, P.; Harjunpaa, T.; Palmu, K.; Mantsala, P.; Suominon, I. Oxidation of Crocein Orange G by lignin peroxidase isoenzymes kinetics and effect of H_2O_2. *Appl Biochem Biotechnol* 1998, 75, 307–321.

Omotayo, A. E.; Ilori, M. O.; Amund, O. O.; Ghosh, D.; Roy, K.; Radosevich, M. Establishment and characterization of atrazine degrading cultures from Nigerian agricultural soil using traditional and Bio-Sep bead enrichment techniques. *Appl Soil Ecol* 2011, 48(1), 63–70.

Phillips, T. M.; Seech, A. G.; Lee, H.; Trevors, J. T. Biodegradation of hexachlorocyclohexane (HCH) by microorganisms. *Biodegradation* 2005, 16, 363–392.

Pieper, D. H. Aerobic degradation of polychlorinated biphenyls. *Appl Microbiol Biotechnol* 2005, 67, 170–191.

Pereira, L.; Coelho, A. V.; Viegas, C. A.; Correia dos Santos, M. M.; Robalo, M. P.; Martins, L. O. Enzymatic biotransformation of the azo dye Sudan Orange G with bacterial CotA-laccase. *J Biotech* 2009, 139(1), 68–77.

Piutti, S.; Semon, E.; Landry, D.; Hartmann, A.; Dousset, S.; Lichtfouse, E.; Topp, E.; Soulas, G.; Martin-Laurent, F. Isolation and characterization of *Nocardioides* sp.12, an atrazine-degrading bacterial strain possessing the gene trzN from bulk- and maize rhizosphere soil. *FEMS Microbiol Lett* 2003, 221(1), 11–17.

Puranik, K. P. Microbial bioremediation of ethylenethiourea from mancozeb pesticide industrial effluent. Ph.D. Thesis, University of Pune, Pune, India, 2000.

Rabie, G. H. Biodegradation of the organophosphorus insecticide monocrotophos by *Penicillium corylophilum*. *Zagazig J Phar Sci* 1995, 4(2), 14–19.

Richins, R. D.; Kaneva I.; Mulchandani A.; Chen W. Biodegradation of organophosphorus pesticides by surface-expressed organophosphorus hydrolase. *Nat Biotech* 1997, 15(10), 984–987.

Rodgers, J. D.; Bunce, N. J. Treatment methods for the remediation of nitroaromatic explosives. *Water Res* 2001, 35, 2101–2111.

Rowland, S. S.; Speedie MK.; Pogell BM. Purification and characterization of a secreted recombinant phosphotriesterase (parathion hydrolase) from *Streptomyces lividans*. *Appl Environ Microbiol* 1991, 57(2), 440–444.

Roy, M.; Khara, P.; Dutta, T. K. Meta-Cleavage of hydroxynaphthanoic acids in the degradation of phenanthrene by *Sphingobium* sp. strain PNB. *Microbiol* 2012, 158 (3), 685–695.

Ryan, D.; Leukes, W.; Burton, S. Improving the bioremediation of phenolic wastewaters by *Trametes versicolor*. *Biores Technol* 2007, 98(3), 579–587.

Saratale, R. G.; Saratale, G. D.; Chang, J. S.; Govindwar, S. P. Decolorization and biodegradation of reactive dyes and dye wastewater by a developed bacterial consortium. *Biodegradation* 2010, 21, 999–1015.

Sarnaik, S.; Kanekar, P. Bioremediation of color of methyl violet and phenol from a dye-industry waste effluent using *Pseudomonas* spp. isolated from factory soil. *J Appl Bacteriol* 1995, 79, 459–469.

Sarnaik, S. S.; Kanekar, P. P. Biodegradation of methyl violet by *Pseudomonas mendocina* MCM B-402. *Appl Microbiol Biotechnol* 1999, 52, 251–254.

Sariaslani, F. S. Microbial cytochromes p-450 and xenobiotics metabolism. *Adv Appl Microbiol* 1991, 35, 133–178.

Segers, P.; Vancanneyt, M.; Pot, B.; Torck, U.; Hoste, B.; Dewettinck, D.; Falsen, E.; Kersters, K.; De Vos, P. Classification of *Pseudomonas diminuta* Leifson and Hugh 1954 and *Pseudomonas vesicularis* Busing, Doll and Freytag 1953 in *Brevundimonas* gen. nov. as *Brevundimonas diminuta* comb. nov. and *Brevundimonas vesicularis* comb. nov., respectively. *Int J Sys Bacteriol* 1994, 44, 499–510.

Serdar, C. M.; Gibson, D. T.; Munnecke, D. M.; Lancaster, J. H. Plasmid involvement in parathion hydrolysis by *Pseudomonas diminuta*. *Appl Environ Microbiol* 1982, 44, 246–249.

Shao, Z. Q.; Bekhi, R. Characterization of the expression of the thcB gene, coding for a pesticide degrading cytochrome p-450 in *Rhodococcus* strains. *Appl Environ Microbiol* 1996, 62(2), 403–407.

Sharma, P.; Raina, V.; Kumari, R.; Malhotra, S.; Dogra, C.; Kumari, H.; Kohler, H.-P. E.; Buser, H. R.; Holliger, C.; Lal, R. Haloalkane dehalogenase LinB is responsible for β- and δ-hexachlorocyclohexane transformation in *Sphingobium indicum* B90A. *Appl Environ Microbiol* 2006, 72, 5720–5727.

Sheremata, T. W.; Hawari, J. Mineralization of RDX by the white rot fungus *Phanerochaete chrysosporium* to carbon dioxide and nitrous oxide. *Environ Sci Technol* 2000, 34, 3384–3388.

Shimazu, M.; Mulchandani, A.; Chen, W. Simultaneous degradation of organophosphorus pesticides and p-nitrophenol by a genetically engineered *Moraxlla* sp. with surface-expressed organophosphorus hydrolase. *Biotech Bioeng* 2001, 76(4), 318–324.

Singh, B.; Kaur, J.; Singh, K. Microbial remediation of explosive waste. *Crit Rev Microbiol* 2012, 38(2), 152–167.

Singh, B. K.; Walker, A. Microbial degradation of organophosphorus compounds. *FEMS Microbiol Rev* 2006, 30(3), 428–471.

Somara, S.; Siddavattam, D. Plasmid mediated organophosphate pesticide degradation by *Flavobacterium balustinum*. *Biochem Mol Biol Int* 1995, 36(3), 627–631.

Spain, J. C. Biodegradation of nitroaromatic compounds. *Ann Rev Microbiol* 1995, 49, 523–555.

Strauber, H.; Muller, R. H.; Babel, W. Evidence of cytochrome p-450 catalyzed cleavage of the ether bond of phenoxybutyrate herbicides in *Rhodococcus erythropolis* K2–3. *Biodegradation* 2003, 14(1), 41–50.

Sridevi, V.; Chandana Lakshmi, M. V. V.; Manasa, M.; Sravani, M. Metabolic pathways for the biodegradation of phenol. *Int J Eng Sci Adv Tech* 2012, 2(3), 695–705.

Suar, M.; van der Meer, J. R.; Lawlor, K.; Holliger, C.; Lal, R. Dynamics of multi lin gene expression in *Sphingomonas paucimobilis* B90A in response to different hexachlorocyclohexane (HCH) isomers. *Appl Environ Microbiol* 2004, 70, 6650–6656.

Suar, M.; Hauser, A.; Poiger, T.; Buser, H. R.; Muller, M. D.; Dogra, C.; Raina, V.; Holliger, C.; van der Meer, J. R.; Lal, R.; Kohler, H. P. Enantioselective transformation of HCH by dehydrochlorinases LinA1 and LinA2 from soil bacterium *Sphingomonas paucimobilis* B90A. *Appl Environ Microbiol* 2005, 71, 8514–8518.

Sutherland, T. D.; Horne, I.; Russell, R. J.; Oakshott, J. G. Gene cloning and molecular characterization of a two-enzyme system catalyzing the oxidative detoxification of beta-ES. *Appl Environ Microbiol* 2002, 68, 6237–6245.

Tsirogianni, I.; Aivaliotis, M.; Karas, M.; Tsiotis, G. Mass spectrometric mapping of the enzymes involved in the phenol degradation of an indigenous soil pseudomonad. *Biochim Biophys Acta* 2004, 1700(1), 117–123.

Turnbull, G. A.; Ousley, M.; Walker, A.; Shaw, E.; Morgan, J. A. Degradation of substituted phenyl urea herbicides by *Arthrobacter globiformis* strain D47 and characterization of a plasmid associated hydrolase gene puhA. *Appl Environ Microbiol* 2001, 67(5), 2270–2275.

Udaysoorian, C.; Prabu, P. C. Biodegradation of phenols by lignolytic fungus *Trametes versicolor*. *J Biol Sci* 2005, 5(6), 824–827.

Vaishampayan, P. A. Bioremediation of a herbicide, atrazine from soil. Ph.D. Thesis, University of Pune, Pune, India, 2004.

Vaishampayan, P. A.; Kanekar, P. P.; Dhakephalkar, P. K. Isolation and characterization of *Arthrobacter* sp. strain MCM B-436, an atrazine degrading bacterium from rhizospheric soil. *Int Biodet Biodegrad* 2007, 60, 273–278.

Villemur, R.; Beaudet, R.; Lanthier, M.; Gauthier, A.; Boyer, A.; Thiboodeau, J.; Lepine, F.; Duguay, M.; Page-Belanger, R. Molecular analysis of *Desulfitobacterium frappieri* pcp-1 involved in reductive dehalogenation of pentachlorophenol. *Water Sci Technol* 2005, 52(1–2), 101–106.

White, R. E.; Coon, M. J. Oxygen activation by cytochrome p-450. *Ann Rev Biochem* 1980, 49, 315–356.

Weir, K. M.; Sutherland, T. D.; Horne, I.; Russell, R. J.; Oakeshott, J. G. A single monooxygenase, Ese is involved in metabolism of the organochlorides Endosulfan and endosulfate in an *Arthrobacter* sp. *Appl Environ Microbiol* 2006, 72(5), 3524–3530.

Wierckx, N.; Koopman, F.; Ruijssenaars, H. J.; de Winde, J. H. Microbial degradation of furanic compounds: biochemistry, genetics, and impact. *Appl Microbiol Biotechnol* 2011, 92, 1095–1105.

Wu, Y. C.; Teng, Y.; Li, Z. G.; Liao, X. W.; Luo, Y. M. Potential role of polycyclic aromatic hydrocarbons (PAHs) oxidation by fungal laccase in the remediation of an aged contaminated soil. *Soil Biol Biochem* 2008, 40(3), 789–796.

Xu, F. Applications of oxidoreductases: recent progress. *Industrial Biotechnology* 2005, 1(1), 38–50.

Zhang, R.; Cui, Z.; Jiang, J.; He, J.; Gu, X.; Li, S. Diversity of organophosphorus pesticide-degrading bacteria in a polluted soil and conservation of their organophosphorus hydrolase genes. *Can J Microbiol* 2005, 51(4), 3337–3343.

Zhongli, C.; Shunpeng, L.; Guoping, F. Isolation of methyl parathion-degrading strain M6 and cloning of the methyl parathion hydrolase gene. *Appl Environ Microbiol* 2001, 67(10), 4922–4925.

Zilly, A.; Souza, C. G.; Barbosa-Tessmann, I. P.; Peralta, R. M. Decolorization of industrial dyes by a Brazilian strain of *Pleurotus pulmonarius* producing laccase as the sole phenol oxidizing enzyme. *Folia Microbiol (Praha)* 2002, 47, 273–277.

Zimmermann, T.; Kulla, H. G.; Leizinger, T. Properties of purified Orange II azoreductase, the enzyme initiating azo dye degradation by *Pseudomonas* KF 46. *European J Biochem* 1982, 120, 197–203.

CHAPTER 7

AN INSIGHT INTO HORIZONTAL GENE TRANSFER TRIGGERING WIDESPREAD ANTIMICROBIAL RESISTANCE IN BACTERIA

ZAKIR HOSSAIN

CONTENTS

7.1   INTRODUCTION

Rapid distribution of antimicrobial resistance among bacterial populations currently poses a major therapeutic challenge worldwide. A lot of extensive research has been carried out to understand the potential factors attributing this resistance. Horizontal gene transfer (HGT), also called lateral gene transfer (LGT), via the process known as conjugation or mating has been established as the predominant reason for the dissemination of antibiotic resistance through bacterial species (OECD, 2010; Malik et al., 2008). This phenomenon refers to a unidirectional exchange (donor to recipient) of genetic material between closely related or phylogenetically distant organisms. Mobile genetic elements (MGEs) or mobilomes, such as plasmids or conjugative transposons of bacteria can be transferred from a donor to a recipient by conjugation, which has been described as intimate physical contact between bacterial populations due to the activity of specialized protein complex, mating pair formation (Mpf) system. Mpf serves as an apparatus for the formation of surface-exposed sex pili that establish the conjunction between mating cells, and functions along with coupling protein (CP) associated with processing of deoxyribo nucleic acid (DNA) for transport (OECD, 2010; Schroder and Lanka, 2005; Daugelavicius et al., 1997). Thus, horizontal transfer of plasmid-encoded antibiotic resistance genes from bacteria to bacteria is induced and more bacterial populations develop the same phenotype. Emergence and propagation of resistant bacterial mutants occur in the environment due to antibiotic selective pressure, which is very common in hospitals as a consequence of intensive use of antimicrobial agents to treat general and nosocomial infections (Rai et al., 2012; Dzidik and Bedekovic, 2003). These mutants may even carry multiple antibiotic resistance genes when patients are provided with prolonged treatment with various antimicrobials. Antibacterial drugs administered to treat infections are only partly metabolized by patients' body while large amounts of these compounds are discharged directly into hospital effluents or the environment (Malik et al., 2008). Therefore, prolonged exposure of a large bacterial community in the environment to such antimicrobial agents results in the emergence of resistance genes, which are then transported to other closely or distantly related species by means of lateral transfer. A wide variety of gram-negative as well as gram-positive bacteria has been found to harbor small to large molecular-weight plasmids, which are potentially associated with high antimicrobial resistance (Hossain et al., 2011; Grohmann et al., 2003). Conjugative transfer of multiple antibiotic resistance genes by plasmid-containing bacteria may occur at high frequency and efficiency (Carattoli, 2003). Increased antimicrobial resistance and subsequent evidence of horizontal transfer of resistance genes may pose more serious threat to global healthcare in years to come.

7.2 PLASMID BIOLOGY

Plasmids are considered to be a pool of extrachromosomal DNA shared among populations (Del Solar *et al.*, 1998). They are self-replicating genetic materials having a covalently closed, circular/linear and double-stranded DNA structure (Carattoli, 2003), and are commonly observed in bacteria, archaea, and yeasts. Plasmids were initially found in bacteria belonging to the family of enterobacteriaceae and then almost in every single identified strain. Their sizes may vary from smaller than one kilobase (kb) to several hundred kilobases. Despite their existence in supercoiled and often circular form by nature (Fig. 7.1), plasmids could be detected as linear or in open circle by alkaline lysis followed by agarose gel electrophoresis (Miljkovic-Selimovik *et al.*, 2007). Interestingly, these mobile elements are sometimes lost from the cell when the organism encounters adverse or nonselective environmental conditions. Plasmids have simple genetic configuration, and can be easily isolated from host as well as manipulated *in vitro* (Del Solar *et al.*, 1998). Therefore, researchers have the opportunity to exploit such exceptional features to use bacterial plasmids in multidisciplinary research in microbiology and biotechnology. Plasmids are considered to be very useful models in the study of the regulation of DNA replication, and are frequently used as vectors in genetic engineering (Wegrzyn and Wegrzyn, 2002).

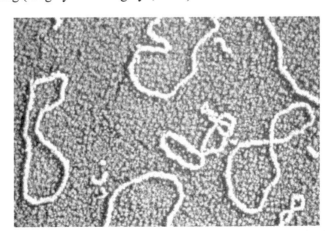

FIGURE 7.1   Electron micrograph of small bacterial plasmids (Bennett, 2008).

7.2.1 PLASMID REPLICATION

The analysis of plasmid replication and its control has contributed to some important breakthroughs in genetic research, including the discovery of anti-sense RNA, mechanisms of DNA replication, and control of gene expression. Although plasmids replicate autonomously by using their own machinery, they broadly depend on host enzymes for this process. The idea of such obvious plasmid-host interaction has led researchers working in environmental and in evolutionary fields to improved understanding in relevant studies. Replication of plasmids has mostly been described by using examples of gram-negative bacteria, including *Escherichia coli*. Three replication mechanisms, namely theta-type, strand displacement and rolling circle have so far been elucidated. Of these, theta-type replication has been suggested to be abundant among plasmids from gram-negative bacteria. However, this mechanism has also been sporadically found in gram-positive bacterial plasmids (Del Solar *et al.*, 1998). Plasmid replication is generally divided into three stages-initiation, elongation and termination. It recruits DNA helicase I, DNA gyrase, DNA polymerase III, endonuclease and ligase. It begins at a specific site called origin of vegetative replication (*oriV*) when catalyzed by one or a few plasmid-encoded initiation proteins that determine plasmid-specific DNA sequences. Plasmids that have two different origins of replication linked to the bidirectional transfer between two bacterial strains are called shuttle vectors. During replication, host-cell cytoplasmic enzyme DNA helicase I uncoils the DNA chain which is later transcribed from the unwinding structure into a negative, wound form by DNA gyrase (Miljkovic-Selimovik *et al.*, 2007). Gyrase is essential for adenosine triphosphate (ATP)-dependent negative supercoiling of DNA duplex (Reece and Maxwell, 1991). The synthesis of RNA primers by either host RNA polymerase (RNAP) or bacterial or plasmid primase is followed by DNA synthesis continuously on leading strand and discontinuously on lagging strand. DNA polymerase III is sought for the missing chain replacement by elongation of the synthesized DNA (Del Solar *et al.*, 1998).

7.2.2 REPLICATION CONTROL

Initiation and frequency of replication are controlled by the functioning of "Rep" proteins at *orivV*. The copy number of a plasmid in the cell as a result of replication directly depends on the frequency of initiation of a replication (Miljkovic-Selimovik *et al.*, 2007; Del Solar *et al.*, 1998). Replication control mechanism was first analyzed in plasmid R1 by the isolation of bacterial mutants, which showed an increased copy number. The determinants of this control were found to exist in the plasmid of those mutants itself, and negative regulators or inhibitors assembled at the initiation of replication were associated with this control. The copy number could be in the range of 25–50 copies/cell if the expression vector is derived from low-copy-number plasmids, such as pBR322, and 150–200 copies/cell if derived from high-copy-number plasmids, such as pUC which are derivatives of pBR322 (Lin-Chao et al., 1992; Lupski et al., 1986).

7.3 ROLE OF PLASMID IN ANTIMICROBIAL RESISTANCE

Bacteria acquire resistance to antibiotics by mechanisms of mutational change in their DNA or by acquisition of resistance-encoding genetic elements in order to survive a selective pressure. They have been found to be capable of compensating the cost of the development of resistance by subsequent secondary mutations to adjust the altered configuration in a manner, which makes the mutants as fit as the wild-type strains. Studies have shown that incorporation of an additional resistance into bacteria already protected against certain antibiotics due to possession of a chromosomal mutation or a self-transmissible resistance plasmid increases the fitness of the strains. Hence, this type of gene interactions termed as sign epistasis, where acquisition of further resistance has antagonistic effect on deleterious mutations, is likely to be beneficial to the bacteria (Silva et al., 2011).

Uptake of resistance determinants by plasmid is widely associated with transfer of the acquired resistance between bacteria of the same population or taxonomically distant species (Dzidic and Bedekovic, 2003). This distinctive attribute facilitates a large pathogenic bacterial community in the environment with adaptation to the lethal change in their surroundings. Many studies have suggested that bacterial plasmids often carry genes encoding antimicrobial resistance (Lunn et al., 2010; Morita et al., 2010). Plasmid-mediated resistance to quinolone and β-lactam antibiotics in gram-negative bacteria has been widely documented (Cattoir et al., 2007). Plasmid-encoded AmpC β-lactamases were first described in the late 1980s, and production of these enzymes has been reported in E. coli, Klebsiella spp., Proteus mirabilis, and Salmonella spp. (Black et al., 2005). Extended-spectrum beta-lactamase (ESBL) producing gram-negative bacilli causing nosocomial infections are one of the major concerns around the world (Rawat and Nair, 2010). β-lactamases are capable of hydrolyzing a range of antimicrobial agents that include penicillins, cephalosporins and monobactams. As ESBL is mainly encoded by the bacterial plasmid (Dzidic and Bedekovic, 2003), these bacteria have advantage of transferring the resistance gene to more pathogenic bacterial genera, particularly of high clinical importance. Although quinolone resistance is generally determined by bacterial chromosome through mutations (Martinez *et al.*, 1998), certain plasmids have been reported to possess a mutator effect on the frequency of mutations in *E. coli*, thereby conferring resistance to this group of antibiotics (Jacoby, 2005). Increasing evidences of cotransfer of quinolone-resistance determinant on ESBL-containing plasmids (Fig. 7.2) support the hypothesis that resistance to quinolone drugs in bacterial pathogens could be at least partly plasmid-induced (Mammeri *et al.*, 2005; Wang *et al.*, 2004). Emergence of such resistance is triggered by *qnr*, *qepA*, and *aac(6')-lb-cr* genes encoded in the plasmid. These genes control bacterial susceptibility to the drugs via different mechanisms. Protection of DNA gyrase from quinolones by *qnr* is a result of the synthesis of a 218-aa protein belonging to the pentapeptide repeat family (Tran and Jacoby, 2002), and reduced susceptibility to norfloxacin is triggered by the efflux action of *qepA* (Yamane *et al.*, 2007). The gene *aac(6')-lb-cr*, which is modified from aminoglycoside N-acetyltransferase, inactivates ciprofloxacin by N-acetylation at the amino nitrogen on its piperazinyl substituent (Robicsek *et al.*, 2006).

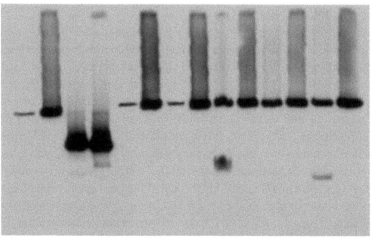

FIGURE 7.2   Analysis of plasmid DNAs from donor and transconjugant (recipient) strains of *Klebsiella pneumoniae* hybridized with *qnr* probe. Lanes showing designations with an ending letter T are plasmids from transconjugants carrying *qnr* gene in a manner, which is common among donors in other lanes (Wang *et al.*, 2004).

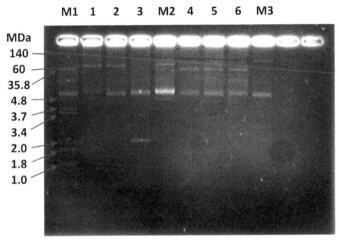

FIGURE 7.3 Plasmid profile of *Escherichia coli* analyzed by agarose gel electrophoresis. Lanes M1, M2 and M3 are molecular markers isolated and purified from different reference *E. coli* strains showing large plasmids of 35.8, 60 and 140 MDa, respectively; lanes 1–6 are plasmids of isolates showing similar plasmid patterns (Hossain *et al.*, 2011).

Unlike gram-negative bacteria, which usually have at least one large plasmid (Fig. 7.3) for multiple-drug resistance (Hossain *et al.*, 2011), some gram-positive bacteria have been reported to harbor smaller plasmids encoding resistance genes (Miljkovic-Selimovik *et al.*, 2007). There are evidences of the presence of resistance plasmids (R plasmids) in streptococci, which were first described in *Streptococcus fecalis* (reclassified as *Enterococcus fecalis*) and *Streptococcus mutans* (Schaberg and Zervos, 1986). The antibiotic resistance genes are commonly located in transposons (Tn) or insertion sequences (IS) or integron gene cassettes also known as jumping genes, which have ability to move from one DNA molecule to another by transposition or site-specific recombination (Bennett, 2008). Transposons capable of relocating themselves from one plasmid to another or from plasmid to chromosome are thought to be highly associated with widespread antimicrobial resistance among bacteria of different genera (Miljkovic-Selimovik *et al.*, 2007). Likewise, resistance plasmids can be constructed by the assembly of resistance-encoding gene cassettes captured by bacterial integrons via site-specific recombination system, instead of transposition. IS like elements called *ISCR* that have small cryptic sequences are also thought to transpose by rolling circle (RC) transposition, which may result in resistance-plasmid evolution (Bennett, 2008). The virulence plasmid of an enterohaemorrhagic *Escherichia coli* (EHEC) strain O26 EHEC has been found to carry a Tn*21*-derived complex antibiotic resistance gene locus encoding resistance to antibiotics and mercuric compounds. The Tn*21* derivative has been inserted within the conjugal transfer gene *traC*, which is an essential element of the plasmid conjugation system. This transposon is thought to have accumulated the resistance genes through In2 or other insertion events. Tn*21* and its derivatives exist in a range of plasmids, including those having a broad host range, and are well disseminated among the enterobacteriaceae. Distribution of antimicrobial resistance in a large number of clinically significant bacterial species is now attributed to the presence of resistance genes on such transposon derivatives. Tn*21* harbored by *Shigella flexineri* plasmid NR1 possesses genes *sul1* and *aadA1* encoding resistance to a range of sulfonamides and aminoglycosides, respectively. There is evidence that a Tn*21* derivative exists on plasmid pRMH760 of *Klebsiella pneumoniae*, and encodes the resistance genes *aphA1* (kanamycin and neomycin), *bla*$_{TEM}$ (ampicillin), *aadB* (aminogycoside), *sul1* (sulfonamide), and *dfrA10* (trimethoprim) (Venturini *et al.*, 2010). The IncF (incompatibility) plasmid pRSB107 isolated from activate-sludge bacteria of a waste-water treatment plant has demonstrated a Tn*21* derivative that encodes resistance to seven antibiotics (Szczepanowski *et al.*, 2005). Besides transposition of resistance determinants, significant genetic diversity of plasmids as a result of nonsynonymous nucleotide substitution (mutation) may also contribute to plasmid-mediated drug resistance (Zhao *et al.*, 2010).

Antibiotic resistance by conjugational transfer has been detected in the bacterium *Salmonella enterica* serovar Typhimurium isolated in Hong Kong. These isolates have shown resistance to streptomycin, tetracycline, chloramphenicol, and sulphonamide, and possessed plasmid of incompatible group H1 discovered in many strains of *S. enterica* serovar Typhi resistant to chloramphenicol (Ling and Chau, 1987). In another study, a *S. typhi* strain resistant to cephalosporin was isolated and characterized. This isolate harbored a transferrable plasmid encoding the CTX-M-15 extended-spectrum-β-lactamase (Morita *et al.*, 2010). A 56-megadalton (MDa) plasmid was found in strains of gentamycin-resistant *K. pneumoniae* associated with outbreaks reported in Peter Brigham Hospital in Boston, USA during 1975–1976. This epidemic plasmid was spread out on strains of *E. coli*, and eventually cycled among strains of *Serratia marcescens*. This incidence was suggested to be due to the conjugational transfer of the plasmid onto underlying strains distributed among the flora of hospitalized patients (O'Brien *et al.*, 1980). In an investigation into antibiotic multiresistance in aquatic and clinical isolates of EHEC serotype O157, plasmid-mediated resistance has been demonstrated by curing of plasmids followed by antimicrobial susceptibility testing. Mutant (cured) strains previously found resistant showed susceptibility to the same panel of antibiotics due to loss of plasmid (Chigor *et al.*, 2010). Another study has elicited a significant association between multiple antibiotic resistance and middle order transmissible plasmids having molecular weights ranging from 44 to 77 MDa in a large number of *E. coli* and *K. pneumoniae* strains isolated from patients with urinary tract infections (UTI) (Lina *et al.*, 2007). Large plasmid-mediated ESBL gene bla_{SHV-12} and AmpC β-lactamase gene bla_{CMY-2} have recently been detected in canine *Enterobacter* isolates (Sidjabat *et al.*, 2007). Such global scenario of antimicrobial resistance among bacterial populations reinforces the fact that plasmids have strong association with acquisition and dissemination of multiple resistance genes in the environment.

7.4 MECHANISMS OF BACTERIAL CONJUGATION

Conjugation is known to be a replicative process that helps donate a copy of plasmid to a recipient bacterial cell while leaving the other copy with the donor (Wilkins, 1995). This system was first described by using the intestinal *E. coli* strain K-12 in 1947 by Tatum and Lederberg, who detected the presence of recombinant types of this bacterium in mixed cultures showing characteristics of prototrophs, and confirmed that the machinery involved in this recombination was independent of typical transformation (Tatum and Lederberg, 1947). Since then, the idea of conjugational exchange of genetic elements among bacteria has received broad attention around the world. Pathogenic bacteria have ability to transfer resistance-encoded genetic elements to other species by conjugation. Studies have demonstrated that the transfer of antibiotic resistance genes can take place in the intestine between a variety of different gram-positive or gram-negative bacilli (Dzidic and Bedekovic, 2003). Therefore, presence of at least one conjugative or transferrable plasmid encoding the acquired resistance is essential for a successful LGT. The nature of surface receptor on the potential recipient cell is also essential for gene transfer and determination of host range of plasmids (Bennett, 2008). Many bacterial plasmids are conjugative, and encode genes coding for the transfer of resistance determinants between bacterial cells. However, nonconjugative resistance plasmids, which do not carry such vital genes enabling cells to couple, can also be mobilized by coresiding conjugative plasmids. Therefore, mobilizable plasmids encode functions associated with the transfer of their own DNA. Conjugative plasmids are relatively larger ($\geq$30 kb) than mobilizable ones, which are merely less than 10 kb in size (Bennett, 2008).

Since conjugation systems in gram-positive bacteria have not been studied in great detail, most conjugation mechanisms have been described by using gram-negative models.

7.4.1   GRAM-NEGATIVE BACTERIA AS A PARADIGM FOR CONJUGATION

Conjugation in gram-negative bacteria is considered to be a rolling-circle replication (RCR) process, which depends on the bacterial type IV secretion system (T4SS) pathway (Lawley et al., 2003; Llosa et al., 2002). The mechanism recruits plasmid-encoded genes, which are linked to development of a complex extracellular filament called sex pilus or conjugation tube that functions like a grappling hook to join donor and recipient cells (Miljkovic-Selimovik et al., 2007). Interestingly, sex pili do not function as apparatus through which DNA is transferred although they have appeared to function as a channel for this transfer in a study on E. coli conjugation (Shu et al., 2008). These pili are predominantly involved in cell-cell contact that triggers the conjugational exchange. The transfer is mediated through DNA transfer pore that forms due to envelope-envelope contact via the pili, which are retracted upon the formation of the pore enabling a connection between cytoplasmic compartments of the two conjugative cells (Wilkins, 1995; Grossman and Silverman, 1989). In gram-negative bacteria, sex pili are essential for a close physical contact between a donor cell and a recipient cell. Two types of pili encoded by F- and P-plasmids have been studied widely (Fig. 7.4). Of these, F-pili are relatively longer and more flexible than P-pili (Lawley et al., 2003). Conjugation is initiated by the development of a sex pilus, and begins at the cis-acting site, the origin of transfer (oriT). It additionally involves a number of trans-acting functions essential for mating pair formation (Mpf), initiation and operation of DNA transfer, and conjugation control (Lawley et al., 2003; Furste et al., 1989; Grossman and Silverman, 1989). Pili formation is followed by segregation of two DNA chains of which one is transferred into the recipient. Conjugational synthesis of complete plasmid DNA occurs in both cells, and plasmid replicons retain their circular arrangement (Miljkovic-Selimovik et al., 2007). One of the best studied systems is IncP transfer (tra) system of the broad-host-range promiscuous plasmid RP4 belonging to incompatibility (Inc) group. IncP transfer system comprises two regions, Tra1 and Tra2 with 30 transfer functions. Two protein complexes, namely relaxome and Mpf complex (also T4SS) are connected via interaction with TraG-like coupling protein (CP), and are thought to play a key role in trafficking of the donor

DNA strand (Grohmann *et al.*, 2003). Mpf is a plasmid-encoded multiprotein complex located in the cell membranes, and is responsible for an efficient transport of DNA from a donor cell to a recipient cell (Daugelavicius *et al.*, 1997). Protein transport is also thought to be mediated by Mpf. Since gram-negative bacteria possess an outer membrane and an inner cytoplasmic membrane well-separated by the periplasm, the Mpf system is assumed to form a connection (channel) between these membranes for an efficient transport of the genetic materials (Grohmann *et al.*, 2003). Furthermore, DNA transfer and replication (Dtr) proteins alone or in combination with relaxomes apparently increase cell permeability in an Mpf complex-induced conjugational pathway (Daugelavicius *et al.*, 1997). The high-precision nucleoprotein complex relaxome (Grahn *et al.*, 2000), which has been reported to be important for the superhelical structure of DNA (Furste *et al.*, 1989), acts as an intermediate in the initiation reaction. Proteins of relaxomes are encoded by *tra* gene in the *oriT* (Grahn *et al.*, 2000). One of the most important proteins called DNA relaxase is involved in inducing the conjugative transfer by catalyzing the cleavage of a specific phosphodiester bond in the *nic* site within *oriT* in a site- and strand-specific manner (Grohmann *et al.*, 2003; Grahn *et al.*, 2000; Furste *et al.*, 1989). Subsequent strand displacement takes place via the rolling-circle mechanism. The covalent DNA relaxase adduct recruited in the transfer reaction has been suggested to be obligatory for the recircularization of the cleaved plasmid after the end of transfer (Grohmann *et al.*, 2003).

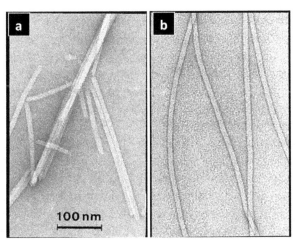

FIGURE 7.4   Electron micrographs of purified P- (a) and F-pili (b) prepared by negative staining (Eisenbrandt *et al.*, 1999).

7.4.2   CONJUGATIVE TRANSFER IN GRAM-POSITIVE BACTERIA

Conjugation in gram-positive pathogens has been reported in a number of studies. These bacteria often become resistant to antimicrobial agents by the acquisition of R plasmids (Schaberg and Zervos, 1986). Many gram-positive bacteria including streptococci and enterococci possess broad-host-range R plasmids (Clewell, 1990). In contrast, resistance-encoding plasmids are thought to be of narrow host range in the genus *Staphylococcus* (Grohmann *et al.*, 2003). In an epidemiological investigation of clinical staphylococci isolates associated with outbreaks among neonates, conjugative transfer of self-transmissible gentamicin-resistance plasmids has been observed between *Staphylococcus epidermidis* and *S. aureus* (Archer and Johnston, 1983). However, conjugation mechanism in gram-positive bacteria is complex, and has not been understood clearly yet. These bacteria do not depend on sex pili for a successful cell-cell coupling unlike gram-negative species. The fact that conjugative plasmids in gram-positive bacteria are generally smaller than those in gram-negative bacteria is perhaps an answer to why this coupling mechanism is rather different in these organisms (Bennett, 2008), that is, due to their small size, these plasmids do not code for proteins associated with the pili formation. Another reason for this exception could be the different cell-wall structure of gram-positive bacteria. This implies that outer membrane proteins, which are associated with the synthesis of pilin subunits leading to their assembly into mature sex pili (Beard and Connolly, 1975), may have less interaction with conjugative plasmids encoding the pili (Eisenbrandt *et al.*, 1999). Gram-positive bacteria also recruit in conjugation the T4SS, which is encoded by virulence (*vir*) gene operons *virB* and *virD* (Bauer *et al.*, 2011; Llosa and de la Cruz, 2005; Christie, 2004). Gene products of *virB* are the Mpf complex discussed previously in case of gram-negative bacterial conjugation, and are involved in the elaboration of membrane-spanning structure for DNA substrate (T-DNA) transfer. The *virD* operon codes for Dtr proteins associated with processing of T-DNA for transfer (Christie, 2004). DNA relaxases are linked to binding and cleaving of single-stranded plasmid DNA, are central to the initiation of the transfer reaction. The mobilization protein MobM and the IncQ-type TraA encoded by broad-host-range plasmid pIP501 are relaxases originating from gram-positive plasmids (Kopec *et al.*, 2005). Conjugation activity has been investigated in a 65-kb pLS20 plasmid isolated from *Bacillus subtilis*, which can also mobilize nonconjugative plasmids. It has been evident that conjugation machinery assembles at a single cell pole along the lateral cell membrane exclusively during extended stationary phase and during lag phase, and disperses during the active cell cycle. Conserved proteins *virB1*, *virB4*, *virD2* (relaxase), and *virD4* (coupling protein) that localize at the cell pole at different phases of cell cycle seem to play an important role in the conjugation system. Also, frequent accumulation of the third adenosine triphosphatase (ATPase) motif protein *virB11* at a single pole may be highly associated with the system. Therefore, the transfer of DNA possibly takes place at the cell pole of the bacterium (Bauer *et al.*, 2011). Conjugation in gram-positive bacteria may uniquely occur via secretion of sex pheromones by the recipient cell as found in enterococci (Clewell, 1990). These molecules allow clumping of the donor and the recipient, thereby enabling the exchange of nucleic acid (Tenover, 2006; Grohmann *et al.*, 2003). Pheromone-inducible conjugation has been widely studied in the enterococcal tetracycline-reisistance plasmid pCF10 (Dunny *et al.*, 1995). Pheromones are heat-stable and protease sensitive, and have specificities for donors harboring conjugative plasmids. The donor strain binds the pheromone with high affinity, and synthesizes an aggregation substance (AS) or adhesin that promotes the formation of mating aggregates by attachment to the recipient through a chromosomally encoded complementary receptor, the binding substance (BS) encoded by *prgZ* gene (Grohmann *et al.*, 2003; Dunny *et al.*, 1995; Kreft *et al.*, 1992). The cell surface component AS is a 150-kilodalton (kDa) protein encoded by *prgB* gene of pCF10 (Dunny *et al.*, 1995; Bensing and Dunny, 1993). Aggregation of the two cells facilitates formation of a mating channel through which the donor transfers the plasmid into the recipient strain. There seems to lie a plasmid-encoded function that has capacity to restrict the new donor cell from responding to its own pheromone. This function can be termed as surface exclusion (SE) or entry exclusion, and the protein regulating this feature is encoded by the structural gene *prgA*. Surface protein expression is controlled by two regulatory circuits, positive-control and negative-control, and are responsible for upstream and downstream regulations of expression, respectively (Dunny *et al.*, 1995). Non-pheromone-induced small plasmid transfer system has also been detected in some gram-positive bacteria. It has been described as a specific protease-sensitive unidirectional coaggregation between strains of *Bacillus thuringiensis* subsp. *israelensis*. Aggregation pathway in this bacterium is established through the interactions of two different phenotypes, Agr+ and Agr-, which induce the plasmid transfer from the Agr+ cell to the Agr- cell (Andrup *et al.*, 1993).

7.5   TOOLS FOR STUDYING RESISTANCE TRANSFER

Detection and characterization of antimicrobial resistance transfer and its subtle mechanisms throughout the kingdom of microbial pathogens involve implementation of available modern research tools. However, further techniques need to be developed in the field of microbiology and molecular biotechnology to obtain more promising data relating to horizontal exchange of resistance plasmids among bacterial species or genera. Various methods have been used widely to screen for diverse events involved in the conjugational pathway. Molecular typing is an important tool for comparison of genetic fingerprint patterns of a pool of resistant strains, and for locating resistance determinants on plasmids associated with horizontal mode of dissemination. In a study on florfenicol resistance in avian *E. coli*, random amplification of polymorphic DNA (RAPD) has been performed to determine the presence of a clonal strain in chicken isolates. RAPD analysis has revealed distinctive patterns of *E. coli* suggesting that the specific resistance was independent of any clonal spread (Fig. 7.5). DNA-DNA hybridization with *flo*-gene-specific probe generated by polymerase chain reaction (PCR) has shown location of this gene on different sized *Xba*I fragments originating from different sized high-molecular-weight plasmids. The results of this experiment have indicated a lateral nature of distribution of the resistance gene (Fig. 7.5) among most avian *E. coli* strains (Keyes *et al.*, 2000).

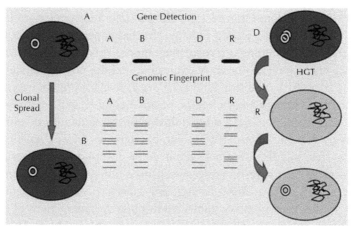

FIGURE 7.5   Molecular typing of bacteria to elucidate underlying mechanism of resistance transfer. Detection of identical resistance determinants on hybridized plasmids from genetically diverse bacteria is indicative of nonclonal (horizontal) spread of resistance; HGT = Horizontal Gene Transfer, D = Donor, R = Recipient.

Different molecular approaches have been applied in a detailed investigation of molecular diversity and evolutionary relationships of Tn*1546*-like elements in vancomycin-resistant enterococci (VRE) isolated from humans and animals. Pulse field gel electrophoresis (PFGE) has been used to plot the genetic relationship between strains. Restriction fragment length polymorphism (RFLP) analysis has been performed by restriction digestion of chromosomal DNA with *Hae*III and *Xba*I followed by hybridization with internal Tn*1546* PCR fragments. Overlapping DNA fragments have been amplified with different Tn*1546* primers along with an IS*1216* primer, and sequenced to determine mutations in the *vanA* and *vanY* genes. Divergent RFLP patterns of Tn*1546* types have suggested polymorphisms in these transposons related to point mutation, insertions and deletions. Therefore, these data conclude that DNA polymorphism among Tn*1546* variants is a successful tool for identifying the routes of transmission of vancomycin resistance genes (Willems *et al.*, 1999). Detection of identical VanA transposon types among genetically different enterococci in Denmark and the United Kingdom unambiguously indicates that the observed vancomycin resistance spread among the strains due to horizontal gene transfer (Woodford *et al.*, 1998; Jensen *et al.*, 1998; Aarestrup *et al.*, 1996).

7.6   CONCLUSION

Significant breakthroughs in researches have surely improved our understanding of microbial behavior over the years. It is now possible to propose that antimicrobial resistance is a colossal genetic event among pathogenic bacteria, which survive in the environment through molecular diversity and subtle interactions with neighboring cells. Horizontal spread of antibiotic resistance genes is, indeed, a major concern for health professionals and researchers as it greatly limits the option for treatment of serious infections. Although there are established opinions around the globe about controlling the extensive use of antimicrobials to facilitate the pathogens with much less selective pressure, it is dubious whether or not the resistance could be abolished by this strategy. As bacterial populations are likely to be capable of sharing a common genetic pool by recruiting gene transfer machinery, emergence of antibiotic resistance throughout the diversity may sustain irrevocably. These pathogens may additionally benefit from the underlying mechanism of sign epistasis and maintain less detrimental genetic recombination. Future challenge would be the development of new and more powerful research tools with the help of nanoparticles to study complex cellular systems of pathogenic bacteria associated with the spread of antimicrobial resistance in diverse niches. These devices will also be revolutionary in molecule by molecule characterization of bacterial protein functions such as those of type VI secretion system, which could be a promising drug target.

KEYWORDS

- Antimicrobial resistance
- Bacterial conjugation
- DNA relaxase
- Extended-spectrum beta-lactamase
- Genetic recombination
- Horizontal gene transfer
- Mating pair formation
- Mobilomes
- Plasmid replication
- Pulse field gel electrophoresis
- Resistance transfer
- Vancomycin-resistant enterococci

REFERENCES

Aarestrup, F. M.; Ahrens, P.; Madsen, M.; Pallesen, LV.; Poulsen, RL.;Westh, H. Glycopeptide susceptibility among Danish *Enterococcus faecium* and *Enterococcus fecalis* isolates of animal and human origin and PCR identification of genes within the VanA cluster. *Antimicrob Agents Chemother* 1996, 40, 1938.

Andrup, L.; Damgaard, J.; Wassermann, K. Mobilization of small plasmids in *Bacillus thuringiensis* subsp. *israelensis* is accompanied by specific aggregation. *J Bacteriol* 1993, 175, 6530.

Archer, G. L.; Johnston, J. L. Self-transmissible plasmids in staphylococci that encode resistance to aminoglycosides. *Antimicrob Agents Chemother* 1983, 24, 70–77.

Bauer, T.; Rosch, T.; Itaya, M.; Graumann, P. L. Localization pattern of conjugation machinery in a gram-positive bacterium. *J Bacteriol* 2011, 193, 6244.

Beard, J. P.; Connolly, J. C. Detection of a protein, similar to the sex pilus subunit, in the outer membrane of *Escherichia coli* cells carrying a derepressed F-like R factor. *J Bacteriol* 1975, 122, 59–65.

Bennett, P. M. Plasmid encoded antibiotic resistance: acquisition and transfer of antibiotic resistance genes in bacteria. *Br J Pharmacol* 2008, 153, S347-S357.

Bensing, B. A.; Dunny, G. M. Cloning and molecular analysis of genes affecting expression of binding substance, the recipient-encoded receptor(s) mediating mating aggregate formation in *Enterococcus fecalis. J Bacteriol* 1993, 175, 7421–7429.

Black, J. A.; Moland, E. S.; Thomson, K. S. AmpC disk test for detection of plasmid-mediated AmpC β-lactamases in *Enterobacteriaceae* lacking chromosomal AmpC β-lactamases. *J Clin Microbiol* 2005, 43, 3110–3113.

Carattoli, A. Plasmid-mediated antimicrobial resistance in *Salmonella enterica. Curr Issues Mol Biol* 2003, 5, 113–122.

Cattoir, V.; Poirel, L.; Rotimi, V.; Soussy, C.; Nordmann, P. Multiplex PCR for detection of plasmid-mediated quinolone resistance *qnr* genes in ESBL-producing enterobacterial isolates. *J Antimicrob Chemother* 2007, 60, 394–397.

Chigor, V. N.; Umoh, V. J.; Smith, S. I.; Igbinosa, E. O.; Okoh, A. I. Multidrug resistance and plasmid patterns of *Escherichia coli* O157 and other *E. coli* isolated from diarrheal stools and surface waters from some selected sources in Zaria, Nigeria. *Int J Environ Res Public Heath* 2010, 7, 3831–3841.

Christie, P. J. Type IV secretion: the *Agrobacterium* VirB/D4 and related conjugation systems. *Biochim Biophys Acta* 2004, 1694, 219–234.

Clewell, D. B. Movable genetic elements and antibiotic resistance in enterococci. *Eur J Clin Microbiol Infect Dis* 1990, 9, 90–102.

Daugelavicius, R.; Bamford, J. K. H.; Grahn, A. M.; Lanka, E.; Bamford, D. H. The IncP plasmid-encoded cell envelope-associated DNA transfer complex increases cell permeability. *J Bacteriol* 1997, 179, 5195–5202.

Del Solar, G.; Giraldo, R.; Ruiz-Echevarria, M. J.; Espinosa, M.; Diaz-Orejas, R. Replication and control of circular bacterial plasmids. *Microbiol Mol Biol Rev* 1998, 62, 434–464.

Dunny, G. M.; Leonard, B. A. B.; Hedberg, P. J. Pheromone-inducible conjugation in *Enterococcus fecalis*: interbacterial and host-parasite chemical communication. *J Bacteriol* 1995, 177, 871–876.

Dzidic, S.; Bedekovic, V. Horizontal gene transfer-emerging multidrug resistance in hospital bacteria. *Acta Phamacol Sin* 2003, 24, 519–526.

Eisenbrandt, R.; Kalkum, M.; Lai, E.; Lurz, R.; Kado, C. I.; Lanka, E. Conjugative pili of IncP plasmids, and the Ti plasmid T pilus are composed of cyclic subunits. *J Biol Chem* 1999, 274, 22548–22555.

Furste, J. P.; Pansegrau, W.; Ziegelin, G.; Kroger, M.; Lanka, E. Conjugative transfer of promiscuous IncP plasmids: interaction of plasmid-encoded products with the transfer origin. *Proc Natl Acad Sci* 1989, 86, 1771–1775.

Grahn, A. M.; Haase, J.; Bamford, D.H.; Lanka, E. Components of the RP4 conjugative transfer apparatus form an envelope structure bridging inner and outer membranes of donor cells: Implications for related macromolecule transport systems. *J Bacteriol* 2000, 182, 1564–1574.

Grohmann, E.; Muth, G.; Espinosa, M. Conjugative plasmid transfer in gram-positive bacteria. *Microbiol Mol Biol Rev* 2003, 67, 277–301.

Grossman, T. H.; Silverman, P.M. Structure and function of conjugative pili: inducible synthesis of functional F pili by *Escherichia coli* K-12 containing a *lac-tra* operon fusion. *J Bacteriol* 1989, 171, 650–56.

Hossain, Z.; Sultana, P.; Deb, S.; Ahmed, M. M. Multidrug resistance in large-plasmid-associated presumptive enterohaemorrhagic *Escherichia coli* isolated from contaminated lake water. *Bangladesh J Microbiol* 2011, 28, 33–40.

Jacoby, G. A. Mechanisms of resistance to quinolones. *Clin Infec Dis* 2005, 41, S120-S126.

Jensen, L. B.; Ahrens, P.; Dons, L.; Jones, R. N.; Hammerum, A. M.; Aarestrup, F. M. Molecular analysis of Tn*1546* in *Enterococcus faecium* isolated from animals and humans. *J Clin Microbiol* 2005, 36, 437–442.

Keyes, K.; Hudson, C.; Maurer, J. J.; Thayer, S.; White, D. G.; Lee, M. D. Detection of florfenicol resistance genes in *Escherichia coli* isolated from sick chickens. *Antimicrob Agents Chemother* 2000, 44, 421–424.

Kopec, J.; Bergmann, A.; Fritz, G.; Grohmann, E.; Keller, W. TraA and its N-terminal relaxase domain of the gram-positive plasmid pIP501 show specific *oriT* binding and behave as dimers in solution. *Biochem J* 2005, 387, 401–409.

Kreft, B.; Marre, R.; Schramm, U.; Wirth, R. Aggregation substance of *Enterococcus fecalis* mediates adhesion to cultured renal tubular cells. *Infect Immun* 1992, 60, 25–30.

Lawley, T. D.; Klimke, W. A.; Gubbins, M. J.; Frost, L. S. F factor conjugation is a true type IV secretion system. *FEMS Microbiol Lett* 2003, 224, 1–15.

Lin-Chao, S.; Chen, W. T.; Wong, T. T. High copy number of the pUC plasmid results from a Rom/Rop-suppressible point mutation in RNA II. *Mol Microbiol* 1992, 6, 3385–3393.

Lina, T. T.; Rahman, S. R.; Gomes, D. J. Multiple-antibiotic resistance mediated by plasmids and integrons in uropathogenic *Escherichia coli* and *Klebsiella pneumoniae*. *Bangladesh J Microbiol* 2007, 24, 19–23.

Ling, J.; Chau, P. Y. Incidence of plasmids in multiply resistant salmonella isolates from diarrheal patients in Hong Kong from 1973–82. *Epidem Inf* 1987, 99, 307–321.

Llosa, M.; Gomis-Ruth, F. X.; Coll, M.; de la Cruz Fd, F. Bacterial conjugation: a two-step mechanism for DNA transport. *Mol Microbiol* 2002, 45, 1–8.

Llosa, M.; de la Cruz Fd, F. Bacterial conjugation: a potential tool for genomic engineering. *Res Microbiol* 2005, 156, 1–6.

Lunn, A. D.; Fabrega, A.; Sanchez-Cespedes, J.; Vila, J. Prevalence of mechanisms decreasing quinolone-susceptibility among *Salmonella* spp. clinical isolates. *Int Microbiol* 2010, 13, 15–20.

Lupski, J. R.; Projan, S. J.; Ozaki, L. S.; Godson, G. N. A temperature dependent pBR322 copy number mutant resulting from a Tn5 position effect. *Proc Natl Acad Sci USA* 1986, 83, 7381–7385.

Malik, A.; Selik, E.; Bohn, C.; Bockelmann, U.; Knobel, K.; Grohmann, E. Detection of conjugative plasmids and antibiotic resistance genes in anthropogenic soils from Germany and India. *FEMS Microbiol Lett* 2008, 279, 207–216.

Mammeri, H.; Van De Loo, M.; Poirel, L.; Martinez-Martinez, L.; Nordmann P. Emergence of plasmid-mediated quinolone resistance in *Escherichia coli* in Europe. *Antimicrob Agents Chemother* 2005, 49, 71–76.

Martinez, J. L.; Alonso, A.; Gomes-Gomes, J. M.; Baquero, F. Quinolone resistance by mutations in chromosomal gyrase genes. Just the tip of the iceberg? *J Antimicrob Chemother* 1998, 42, 683–688.

Miljkovic-Selimovic, B.; Babic, T.; Kocic, B.; Stojanovic, P.; Ristic, L.; Dinic, M. Bacterial plasmids. *Acta Med Medianae* 2007, 46, 61–65.

Morita, M.; Takai, N.; Terajima, J.; Watanabe, H.; Kurokawa, M.; Sagara, H.; Ohnishi, K.; Izumiya, H. Plasmid-mediated resistance to cephalosporins in *Salmonella enterica* serovar Typhi. *Antimicrob Agents Chemother* 2010, 54, 3991–3992.

O'Brien, T. F.; Ross, D. G.; Guzman, M. A.; Medeiros, A. A.; Hedges R. W.; Botstein D. Dissemination of an antibiotic resistance plasmid in hospital patient flora. *Antimicrob Agents Chemother* 1980, 17, 537–543.

OECD. Safety assessment of transgenic organisms: OECD consensus documents. *OECD Publishing* 2010, 4, DOI: http://dx.doi.org/10.1787/9789264096158-en.

Rai, M. K.; Deshmukh, S. D.; Ingle, A. P.; Gade, A. K. Silver nanoparticles: the powerful nanoweapon against multidrug-resistant bacteria. *J Appl Microbiol* 2012, 112, 841–852.

Rawat, D.; Nair, D. Extended-spectrum β-lactamases in Gram negative bacteria. *J Glob Infect Dis* 2010, 2, 263–274.

Reece, R. J.; Maxwell, A. DNA gyrase: structure and function. *Crit Rev Biochem Mol Biol* 1991, 26, 335–375.

Robicsek, A.; Strahilevitz, J.; Jacoby, G. A.; Macielag, M.; Abbanat, D.; Park, C. H.; Bush, K.; Hooper, D. C. Fluoroquinolone-modifying enzyme: a new adaptation of a common aminoglycoside acetyltransferase. *Nat Med* 2006, 12, 83–88.

Schaberg, D. R.; Zervos, M. J. Intergeneric and interspecies gene exchange in gram-positive cocci. *Antimicrob Agents Chemother* 1986, 30, 817–822.

Schroder, G.; Lanka, E. The mating pair formation system of conjugative plasmids – a versatile secretion machinery for transfer of proteins and DNA. *Plasmid* 2005, 54, 1–25.

Shu, A. C.; Wu, C. C.; Chen, Y. Y.; Peng, H. L.; Chang, H. Y.; Yew, T. R. Evidence of DNA transfer through F-pilus channels during *Escherichia coli* conjugation. *Langmuir* 2008, 24, 6796–6802.

Sidjabat, H. E.; Hanson, N. D.; Smith-Moland, E.; Bell, J. M.; Gibson, J. S.; Filipich, L. J.; Trott, D. J. Identification of plasmid-mediated extended spectrum and AmpC *β*-lactamases in *Enterobacter* spp. isolated from dogs. *J Med Microbiol* 2007, 56, 426–434.

Silva, R. F.; Mendonca, S. C. M.; Carvalho, L. M.; Reis, A. M.; Gordo, I.; Trindade, S.; Dionisio, F. Pervasive sign epistasis between conjugative plasmids and drug-resistance chromosomal mutations. *PLoS Genet* 2011, 7(7), e1002181.

Szczepanowski, R.; Braun, S.; Riedel, V.; Schneiker, S.; Krahn, I.; Puhler, A.; Schluter, A. The 120592 bp IncF plasmid pRSB107 isolated from a sewage-treatment plant encodes nine different antibiotic-resistance determinants, two iron-acquisition systems and other putative virulence-associated functions. *Microbiology* 2005, 151, 1095–1111.

Tatum, E. L.; Lederberg, J. Gene recombination in the bacterium *Escherichia coli*. *J Bacteriol* 1947, 53, 673–684.

Tenover, F. C. Mechanisms of antimicrobial resistance in bacteria. *Am J Med* 2006, 119, S3-S10.

Tran, J. H.; Jacoby, G. A. Mechanism of plasmid-mediated quinolone resistance. *Proc Natl Acad Sci USA* 2002, 99, 5638–5642.

Venturini, C.; Beatson S. A.; Djordjevic, S. P.; Walker, M. J. Multiple antibiotic resistance gene recruitment onto the enterohemorrhagic *Escherichia coli* virulence plasmid. *FASEB J* 2010, 24, 1160–1166.

Wang, M.; Sahm, D. F.; Jacoby, G. A.; Hooper, D. C. Emerging plasmid-mediated quinolone resistance associated with the *qnr* gene in *Klebsiella pneumoniae* clinical isolates in the United States. *Antimicrob Agents Chemother* 2004, 48, 1295–1299.

Wegrzyn, G.; Wegrzyn, A. Stress responses and replication of plasmids in bacterial cells. *Microb Cell Fact* 2002, 1, 2.

Wilkins, B. M. Gene transfer by bacterial conjugation: diversity of systems and functional specifications. In *Population genetics of bacteria*, Baumberg, S., Young, J. P. W., Eds.; Cambridge University Press: Cambridge, 1995; pp 59–88.

Willems, R. J. L.; Top, J.; Van den Braak, N.; Van Belkum, A.; Mevius, D. J.; Hendriks, G.; Van Santen-Verheuvel, M.; Van Embden, JDA. Molecular diversity and evolutionary relationships of Tn*1546*-like elements in enterococci from humans and animals. *Antimicrob Agents Chemother* 1999, 43, 483–491.

Woodford, N.; Adebiyi, A. A.; Palepou, M. I.; Cookson, B. D. Diversity of VanA glycopeptide resistance elements in enterococci from humans and nonhuman sources. *Antimicrob Agents Chemother* 1998, 42, 502–508.

Yamane, K.; Wachino, J.; Suzuki, S.; Kimura, K.; Shibata, N.; Kato, H.; Shibayama, K.; Konda, T.; Arakawa Y. New plasmid-mediated fluoroquinolone efflux pump, QepA, found in an *Escherichia coli* clinical isolate. *Antimicrob Agents Chemother* 2007, 51, 3354–3360.

Zhao, F.; Bai, J.; Wu, J.; Liu, J.; Zhao, M.; Xia, S.; Wang, S.; Yao, X.; Yi, H.; Lin, M.; Gao, S.; Zhao, T.; Xu, Z.; Niu, Y.; Bao, Q. Sequencing and genetic variation of multidrug resistance plasmids in *Klebsiella pneumoniae*. *PLoS One* 2010, 5(4), e10141.

CHAPTER 8

PRODUCTION AND BIOTECHNOLOGICAL APPLICATIONS OF RECOMBINANT PROTEINS BY METHYLOTROPHIC YEAST: PAST, PRESENT AND FUTURE PERSPECTIVES

MARCOS LÓPEZ-PÉREZ and RINA MARHA GONZÁLEZ-CERVANTES

CONTENTS

8.1   INTRODUCTION

Since the late twentieth and early 20-first century, the needs of a competitive pharmaceutical and food industry have contributed to the growth of the discipline of biotechnology. This growth has been particularly important in molecular biology and genetic engineering, which has been favored by the great market potential of these technologies in the development of new drugs, and the farming industry. Traditionally, organisms were chosen and used for their ability to produce a compound of interest. Currently, the techniques of genetic engineering allow us to use an organism whose physiology is suitable for use in the laboratory and in the synthesis of recombinant proteins, which in its wild (natural) state, would not. There are a variety of organisms employed for this purpose, having used bacteria such as *Escherichia coli*, filamentous fungi such as *Aspergillus niger*, higher plants such as *Arabidopsis thaliana* and higher mammals such as the cow (*Bos taurus*). However, the object of analysis in this chapter is to review the focus of the yeast organisms as producers of recombinant proteins, specifically we refer to the methylotrophic yeasts. The yeasts are suitable for this type of procedure for several reasons, for example, the physiology of growth is similar to the bacterial. Additionally, it refers to the processes of manipulation of genetic material, which are also very similar to the techniques used with bacteria, which in turn makes it easy to use in laboratory conditions. Other desirable characteristics of these organisms are that, unlike bacteria, they possess mechanisms to synthesize eukaryotic proteins, such as posttranscriptional modifications, proteolytic processing, protein folding mechanisms and glycozylation (Eckart, 1996). In addition, yeast cultures have high yields of biomass and in terms of money and effort costs are low (Creeg, 1999). Finally, another interesting feature of these microorganisms is that it has been found that yeast is free of endotoxins, an advantage for its applications in pharmaceuticals and agriculture (Glick and Pasternak, 1998). However, the yeasts also have some disadvantages, the most relevant are related to some complex procedures such as transcriptional modification, polyhydroxylation and amidation, as well as mechanisms of phosphorylation (Creeg, 1995). Although some studies indicate that these disadvantages can be corrected, which ensure the use of these organisms.

8.2   GENERAL ASPECTS OF THE APPLICATIONS OF YEASTS IN BIOTECHNOLOGY

The yeasts have traditionally been used in fermentation processes since ancient times. Throughout the history of civilizations, *Saccharomyces cerevisiae* has been used for the production of alcoholic beverages. Currently, modern biotechnology, is applying these microorganisms for the production of chemicals, pharmaceuticals and proteins. *S. cerevisiae* is the body of this group more widely used in these processes. The variety of processes in which this microorganism has been used is substantial, having obtained proteins such as insulin (Wang *et al.*, 2001), secondary metabolites (Yan *et al.*, 2005) or fuels such as ethanol (Ameh *et al.*, 1989). However, today other "nonconventional" yeast species such as *Hansenuela polymorpha*, *Pichia pastoris* or *Kluyveromyces lactis* are being used. In addition, recent research gives in a good perspective for the future use of yeasts in biotechnology such as Veen and Lang (2004), which describes the great potential for the production of lipids of high value. This new aspect of research, is based on the fact that it is a microorganism where the metabolic routes are well characterized, and where the investigations may focus on the *novo* synthesis of three major families of lipids, sterols, steroid hormones and polyunsaturated fatty acids such as Omega 3 and Omega 6, which of course are not synthesized naturally in yeast. This new approach, along with the enormous advances in the molecular biology of yeast relating to the production of recombinant proteins and secondary metabolites, offers an overview of the versatility of these microorganisms for the biotechnology industry.

8.3   THE SECRETION OF PROTEINS FROM YEAST

The production of recombinant proteins is a multimillion-dollar business. Sales in the biopharmaceutical industry reached $87 billion in 2008, with expectations of increments in the 2014 (Goodman, 2009). In 2009, approvals for the biopharmaceutical industry marketing for more than 151 products were released in USA and European Union from all of these products, approximately 50% were proteins produced by two organisms, *Escherichia coli* 30% and *S. cerevisiae* 20%, respectively. Expression of recombinant proteins using bacteria has always had the disadvantage of using microorganisms, which are genetically unstable, since the genetic information is maintained in integrative plasmids, which remain in the cytoplasm and can eventually be lost. In addition, the bacteria are not the most appropriate bodies in the production of eukaryotic proteins, due to the lack of mechanisms of posttranslational modification. Also, there are other difficulties intrinsic to the processes of production of recombinant proteins in eukaryotes using bacteria as an expression system, such as the occasional formation of inclusion bodies (IB), that have been observed in cultures where the production of proteins is very high (Baneyx and Mujacic, 2004; Carrió and Villaverde, 2002). In the search for an adequate system of expression, one of the main issues to be taken into account refers to what the characteristics of an efficient secretion system are. One way to assess the efficiency is by focusing on analyzing the amount of heterologous protein with respect to the amount of total protein in the culture medium. This methodology is considered very appropriate because of the complexity involved in the processes of protein secretion, where signals are involved in molecular recognition and transport phenomena between different cell compartments. Consequently, it seems appropriate to establish a comparative model with other proteins produced by the organism used for heterologous expression with the purpose of evaluating the efficiency in the secretion process. Schekman and co-workers in 1985, worked to isolate thermo-sensitive mutants, which accumulated intermediaries in the path of proteins secretion in different cell organelles. These mutants were divided into 23 groups of complementarity, depending on the characteristics of each of these groups, the secretion process was elucidated in the following steps: (a) transport from the endoplasmic reticulum, (b) assembly of secretory vesicles to the golgi apparatus and (c) release of secretory vesicles. The analysis of the organelles of these mutants has contributed to the establishment of a pathway of cell protein secretion in yeast, continue that once the RNA messengers are translated they continue on the path: ER (endoplasmatic reticulum) → Golgi apparatus → secretory vesicles → cell surface. Figure 8.1 shows the complexity of protein expression system in yeast, with the most important phases of the process.

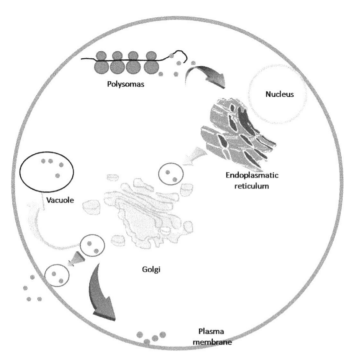

FIGURE 8.1 Scheme of yeast intracellular protein trafficking. The key stages in protein secretion in yeast: (i) Proteins destined for secretion are initially synthesized on ER-associated polysomes; (ii) These proteins are then discharged into the lumen of the ER; (iii) In the ER, proteolytic cleavage of a signal peptide and chaperone-assisted folding takes place along with glycozylation; (iv) Subsequently, proteins are directed from the ER by vesicles, which combine to the cis-Golgi apparatus, which is arranged in parallel arrays or stacks (COPII vesicles (Spang, 2004); (v) In the Golgi, further variation of carbohydrate side-chains on the proteins take place; (vi) Vesicles resultant from budding of the late Golgi then carry the proteins to the actively growing regions (bud region) of the cell (different vesicles are used in plasma membrane components and secreted enzymes); (vii) Fusion of secretory vesicles with the plasma membrane then delivers proteins to the periplasm.

8.3.1   TRANSLOCATION FROM THE CYTOPLASM TO ER

The ER is the organelle that couples protein biosynthesis to membrane trans-location, allowing segregation of secretory and membrane proteins from cytosolic or nuclear proteins. In yeast, secreted and membrane proteins are marked for translocation by a signal sequence, which is usually cleaved early in the biosynthesis of the proteins. The importance of signal peptides was shown in 1999 when Günter Blobel received the Nobel Prize in physiology or medicine for his discovery that "proteins have intrinsic signals that govern their transport and localization in the cell" (Blobel and Dobberstein, 1975). Signal sequences are normally N-terminal, consisting of around 20 amino acid residues, and are characterized by a central string of hydrophobic amino acids flanked by clusters of charged residues, and a residue with a small side chain precedes the cleavage site (Heijne, 1984). Once a signal sequence has emerged from a ribosome, that ribosome is targeted to the ER membrane, where the nascent chain is translocated across the membrane. This targeting is accomplished primarily by a cytoplasmic ribonucleoprotein complex, known as the signal recognition particle (SRP) (Walter *et al.*, 1981). Due to the intrinsic hydrophobicity of signal peptides, they themselves may interact with the lipid bilayer and thus aid the translocation process at this stage. A second route for protein to ingress into the ER is posttranslational and independent of SRP. In this route, which is more widely used by yeast than by higher eukaryotes, a complete polypeptide chain is made in the cytoplasm and held in an incomplete folded state by molecular chaperones such as cytosolic hsp70. Interestingly, the signal peptides derived from proteins in yeast work just as well or better than the peptide signals used for the production of heterologous proteins, although exceptions have been reported (Hiramatsu *et al.*, 1991). After numerous studies, Hofmarm and Schultz (1991) identified a peptide signal that worked for the production of recombinant proteins (yeast α-galactosidase gene). They were able to express human plasminogen efficiently. However, not all proteins were of the same level of efficiency in the secretion, so when this happened it has become necessary to seek the most appropriate signal peptide. Once these sequences are identified, the next step consists in making mutagenesis specifically in order to improve the efficiency in the process of secretion. Listed are some of more frequently signal peptides sequences used in the production of heterologous proteins (Table 8.1).

TABLE 8.1 Signal peptides used to express heterologous protein in yeast.

Signal peptide	Sequence	References
α-Galactosidase	MFAFYFLTACISLKGVFGVSPSYN-GLGL	Liljestrom, 1985
Acid phosphatase	MFKSVVYSILAASLANAGTIPLG-KLAD	Arima, 1983
Carboxypeptidase ψ	MKAFTSLLCGLGLSTTLAKAIS-LQRPL	Bird, 1987

TABLE 8.1   *(Continued)*

Signal peptide	Sequence	References
28 kDa killertoxin	MKIYHIFSVCYLITLCAAATTA-REEFF	Stark, 1986
Invertase	MLLQAFLFLLAGFAAKISASMT-NETSDRP	Taussig, 1983
Mating factor α–1	M R F P S I F T A V L F A A S S A - LAAPVNTTTEDE	Waters, 1988
PEP4	M F S L K A L L P L A L L L V S A N - QVAAKVHKAKIYKH	Ammerer, 1986
Mating factor α-2	MKFISTFLTFILAAVSVTASSDEDI-AQVP	Singh, 1983
BAR1	MSAINHLCLKLILASFAIINTI-TALTNDGTGHLE	MacKay, 1988
Kl toxin	MTKPTQVLVRSVSILFFITLLHLV-VALNDVAGPAET	Skipper, 1984
Glucoamylase	MVGLKNPYTHTMQRPFLLAYLV-LSLLFNSALGFPTALVPRGS	Yamashita, 1985
Killer plasmid 0RF2	MNIFYIFLFLLSFVQGLEHTHRRG-SL	Stark, 1987

The above problems reveal the high stringency of *S. cerevisiae* in the recognition of strange signal peptides by the internalization machinery of the cell at the ER level. For that reason, secretion vectors have been made by DNA recombinant techniques in which the gene sequence for the mature foreign protein is combined in frame to a yeast DNA coding for a signal sequence that replaces that of the protein to be expressed. Therefore, at present, the problem of proteins not delivered effectively into the ER may be solved by screening many different secretion leader sequences for positive ER delivery activity. Signal sequences of α-factor (Brake *et al.*, 1984; Green *et al.*, 1986), killer toxin (Tokunaga *et al.*, 1987) and invertase (Smith *et al.*, 1985; Nishizawa *et al.*, 1987) are employed.

8.3.2   TRANSLOCATION FROM ER TO GOLGI

Many of the natural proteins secreted from yeast are glycoproteins. Their progress from the ER to the Golgi can be easily monitored by the degree of carbohydrate addition; core glycozylation occurs in the ER and outer chains are added in the Golgi. The mechanisms of glycozylation along the departed from Golgi apparatus have been described by numerous studies (Newmana *et al.*, 1990a; Hicke *et al.*, 1990; Kaiser *et al.*, 1990; Newmana *et al.*, 1990b).

8.3.3   GOLGI TRANSPORT TO EXTRACELLULAR COMPARTMENTS

A vast majority of the heterologous proteins from yeast, which are success-fully secreted involve the use of a signal peptide also known as *prepro-leader factor*. One of the processes that are of great relevance that happens in this cellular compartment is referred to proteases that act on specific fusion pro-teins such as Kex 2 (Redding *et al.*, 1991). The processing of a protein from the pro-α-leader indicates that the protein has probably been transported to the Golgi. There are studies that indicate that some proteins have difficulties to pass through the Golgi Apparatus, due to phenomena of hyperglycosilation that eventually, they can produce.

8.3.4   FUSION OF THE VESICLES WITH THE MEMBRANE OF S. CEREVISIAE

In addition, we shall cite the importance of the cytoskeleton in the processes of exportation of proteins. In this regard we must include elements of the cytoskeleton as the actin and tubulin microfilaments that are involved in the establishment of the directionality of the processes of secretion. In reference to the processes of secretion that has as a result the secretion of extracellular proteins, we must emphasize that the vesicles containing these proteins that are ultimately transported to the plasma membrane poured its contents to the outside through the specific interaction with receptors located on the outer membrane. On the other hand, not all proteins are sorted in the above manner; a pathway independent of the ER-Golgi vesicle route has been identified in yeast.

8.4   HETEROLOGOUS PROTEIN PRODUCTION IN METHYLOTROPHIC YEAST

The facultative methylotrophic yeast species have been described in four genus, *Candida, Pichia, Hansenula* and *Turolopsis* (Egli *et al.*, 1980; Gleeson and Sudvery, 1988). Koichi Ogata at the end of the 1960s discovered that there were yeast species able to use methanol as a source of carbon and energy. Subsequently, the Phillips Petroleum Company developed the culture media and protocols for *P. pastoris* cultivation with methanol as a carbon source, in a fed-batch system in order to obtain high cell densities (more than 130 g L^{-1} wet weight) (Wegner, 1990). In subsequent years, the researchers in La Jolla (Salk Institute Biotechnology/Industrial Associates), designed a system for *P. pastoris*, for the production of heterologous proteins, using the isolation of the promoter gene for alcohol oxidase and the development of the first vector and strains constructed for this purpose, in addition to the relevant protocols for its correct genetic manipulation. However, despite of the fact that the yeast is not of the Saccharomyces type, *P. pastoris* is actually the methylotrophic yeast more frequently used, there are a large number of studies where researchers have used different methylotrofic yeasts as expression systems for the production of recombinant proteins, such as *Hansenula polymorpha, Pichia methanolica,* and *Candida boidinii* (Table 8.2).

TABLE 8.2   Non conventional strains of yeast use in the production of heterologous protein.

Strain, number, phenotype and *genotype*		References
Pichia	Y-11430 *wild-type*	Sreekrishna *et al.*, 1996
pastoris	GS115 *his4*/Mut⁺His⁻	Sreekrishna *et al.*, 1996
	KM71 *vaox1::SARG4 his4 arg4*/MutˢHis⁻	Cregg *et al.*, 1988
	MC 100-3 *aox1Δ::SARGaox2ΔPhis4His4arg4/* Mut⁻His⁻	Cregg *et al.*, 1989 White *et al.*, 1995
	SMD1168 Δ*pep4: :URA3 his4 ura3*/Mut⁺His⁻	Cregg *et al.*, 1985
	GS200 *arg4 his4*/Prot⁻	Cregg *et al.*, 1998
	JC254 *ura3*	Cereghino *et al.*, 1999
	JC304 *ade1 his4*	Cereghino *et al.*, 1999
	JC300 *ade1 arg4 his4*	Gleeson *et al.*, 1998
	SMD1165 *prb1 his4*	Gleeson *et al.*, 1998
	SMD1163 *pep4 prb1 his4*	

TABLE 8.2   *(Continued)*

Hansenula	Y-7560 *wild-type*	Levine *et al.*, 1973
polymorpha	NCYC 495 *leul-1*/Leu⁻	Gleeson *et al.*, 1988
	LR9 *ura3*/Ura⁻	Roggenkamp *et al.*, 1986
	A16 *leu2*	Gleeson *et al.*, 1988
	RB10	Fellinger *et al.*, 1991
	uDLB11/ Leu-, Ura-, Pep4–	Kang, *et al.*, 2005
	NAG1996/ Yni1-, Leu–	Brito *et al.*, 1996
	DL10/ Leu-, Ura-	Kang *et al.*, 2005
Pichia	IAM 12901 *wild-type*	Raymond *et al.*, 1998
methanolica	PMAD11 *ade2-11*/Ade⁻	Chang *et al.*, 2008
	PMAD16 *ade2-11 pep4Δ prb1Δ*/Ade⁻Aug⁻	Hong *et al.*, 2006
	PMAD12 *ade2-11 aug1Δ*/Ade⁻pep4Prbl⁻Prot⁻	Raymond, 1998
	PMAD16/pMETαB/HSA *ade2-11 pep4Δ prb1Δ ADE2HAS*/Ade⁺ Mut⁺	
Candida	AOU-1 *wild-type*	Sakai *et al.*, 1991
boidinii	TK62 *ura3*/Ura⁻	Sakai *et al.*, 1991
	2201	Komeda *et al.*, 2002

8.4.1 REGULATION OF METHANOL-INDUCIBLE GENE EXPRESSION IN METHYLOTROPHIC YEAST

With the purpose of understanding the overall process for production of heterologous proteins in conditions of induction with the presence of methanol, we will first describe the general biochemical pathway of the oxidation of methanol in these yeasts. All methylotrophic yeasts use a common methanol-using pathway. In order to express the genes that encode the enzymes involved in the degradation of methanol, it is necessary to add methanol in the culture medium as the sole carbon source (Egli *et al.*, 1980; Veenhuis *et al.*, 1983). The first step is catalyzed by the alcohol oxidase enzyme (AOX), where the methanol is oxidized to formaldehyde, and hydrogen peroxide. Formaldehyde is a central intermediate situated at the branch point between assimilation and dissimilation pathways (Fig. 8.2). Within the peroxisome and through the action of an enzyme catalase, peroxide is degraded to oxygen and water. Part of formaldehyde generated by the AOX is oxidized to CO_2 by the action in enzyme dehydrogenases. The rest of the formaldehyde is used to generate bio-mass in the various cellular constituents. This route, begins with the synthesis of xylulose 5-monophosphate, reaction catalyzed by the enzyme peroxisome dihydroxyacetone synthase (DHAS). Whose final product is glyceraldehyde 3-phosphate and dihydroxyacetone. They leave the peroxisome and penetrate a cyclic path whose purpose is to regenerate the xylulose 5-monophosphate. When this route has completed three cycles is a molecule of GAP that may be derived from to another path of cellular metabolism to form biomass.

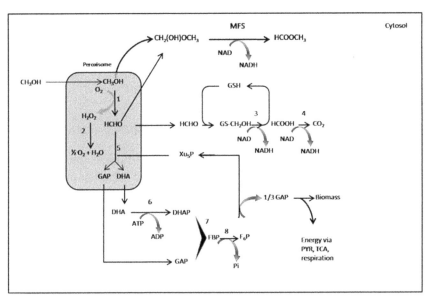

FIGURE 8.2 Methanol pathway in *P. pastoris.* 1, alcohol oxidase; 2, catalase; 3, formaldehyde dehydrogenase; 4, formate dehydrogenase; 5, dihydroxyacetone synthase; 6, dihydroxyacetone kinase; 7, fructose 1,6-biphosphate aldolase; 8, fructose 1,6-bisphosphatase. MFS: methylformate synthase; DHA: dihydroxyacetone; DHAP: dihydroxyacetone phosphate; GAP: glyceraldehyde 3-phosphate; Xu5P: xylulose 5-phosphate; Pi: phosphate; GSH: glutathione; F_6P: fructose 6-phosphate.

Methanol-induced gene expression occurs via two activation steps: (1) glucose derepression (gene activation independent of methanol) and (2) methanol-specific induction. For example, when grown on glycerol as a sole carbon source, in *C. boidinii* and *H. polymorpha* exhibited ~10% and 80% of the maximum methanol-induced *AOD* (alcohol oxidase) expression level, respectively. Glucose-limited chemostat culture experiments also showed that the levels of *AOD* in *H. polymorpha* progressively increased as the dilution decreased, whereas the derepression of *AOD* was lower in *C. boidinii* than in *H. polymorpha* (Egli *et al.*, 1980). Consequently, the extent and type of derepression differs among the methylotrophic yeast species.

8.4.2 FACTORS INFLUENCING SECRETION EFFICIENCY IN METHYLOTROPHIC YEAST

The processes of secretion of recombinant proteins in methylotrophic yeast may be affected by two large groups of physiological and molecular factors. The fundamental approach to modify the physiological factors refers to the conditions of fermentation. On the other hand the alterations that can be performed on the molecular factors, mainly refers to changes in the DNA. Below is an overview of the factors that can affect the processes of secretion of recombinant proteins in methylotrophic yeast.

8.4.2.1   UNTRANSLATED REGIONS (UTR) MRNA 5' AND 3'

The nucleotide sequence, the secondary structure and the length of the UTR sequence, can affect more or less in the process of production of a recombinant protein. There are a number of studies where it is reported that the suitability of this sequence helped a notable increase in the levels of recombinant protein expression (Demolder *et al.*, 1992; Sreekrishna *et al.*, 1993).

8.4.2.2   CODON BIS USE

There are a number of studies that show that the codons bias use (triplets that yeast use specifically to codify amino acid in the translation) does not have a clear effect in the production of recombinant proteins yeast (Kurland, 1991). However, there is a differential effect depending on the kind of yeast. For example, in *P. pastoris* there are reports where the codon use has had a clear effect on the overexpression of recombinant proteins (Bin Fang *et al.*, 2007). This type of effect has also been reported in other methylotrophic yeast such as *P. methanolica*, where there was more lactoferrin heterologous through the codon bias (Wang *et al.*, 2007). Another aspect that can affect the amount of recombinant protein is determined inherently by the nucleotide sequence of the heterologous gene. Within ORF regions rich in A+T it can happened an early termination during the process of translation. This has been reported for production recombinant gp120 protein of HIV 1 in *P. pastoris* (Scorer et at.,

1993). Consequently, it is a good idea to make changes in the nucleotide sequence of the structural gene introducing the specific codons of the species to use in heterologous expression, additionally, changes also need to be considered in order to get rid of potential instabilities in the primary and secondary structure of mRNA.

8.4.2.3   GLYCOSILATION

A factor that is of great relevance to the conditions of the efficiency in the processes of secretion of protein in the yeast is glycozylation. One of the main drawbacks observed at the production of recombinant proteins on this factor, has been reported in *S. cerevisiae*, where events occur in hyperglycosilation, adding chains of more than 40 monomers of mannose. Consequently enzymatic activity is being adversely affected. However, an advantage in methylotrophic yeast such as *P. pastoris* and *H. polymorpha* infrequent hyperglycosilation processes have been observed where chains of mannose, ranging from 8 to 14 residues, must be added (Buckholz *et al.*, 1991; Gellissen *et al.*, 1997).

8.4.2.4   SECRETION SIGNALS

In order to get an efficient product recovery and facilitate the purification step in heterologous protein production process in yeast, the use of an extracellular signal peptide is crucial. Signal peptide sequences are highly conserved in several groups from bacteria to mammals and reports are found where a native signal peptide was used to secrete recombinant proteins from heterologous hosts (Roggenkamp *et al.*, 1981; Rothstein *et al.*, 1987). Romanos *et al.* (1992) made a recombinant construction of a yeast signal peptide sequence in frame with a heterologous protein sequence with good results for protein production.

8.4.2.5   HOST STRAIN AND VECTOR DESIGN

Cloning vectors used for the construction of recombinant methylotrophic yeast are mainly of the integrative kind. The integrative process can be random or through homologous recombination. This last mechanism requires the presence of a sequence for a specific loci. In *P. pastoris* and *P. methanolilca* target sequences are *AOX1* and *AOX2* genes and *HIS4* locus (Creeg, 1999; Raymond, 1998). In *H. polymorpha* this integration can be located in two places: *MOX/TRP3* locus (Agaphonov *et al.*, 1995), or in the autonomous replication sequence (ARS) locus (Sohn *et al.*, 1996). Usually, hybrid plasmids from bacteria and yeast cloning vectors are constructed for this procedure, with a bacterial replication origin *Ori c* and antibiotic resistance genes, which are useful for bacterial transformant selection, on the other hand sequences derived from yeast are complementation genes such as *HIS4* for *P. pastoris* (Creeg, 1999), *ADE2* in *P. methanolica* (Raymond, 1998), *LEU2* and *URA3* in *H. polymorpha* (Agaphonov *et al.*, 1994) and *URA3* in *C. boidinii* (Sakai *et al.*, 1996a).

8.5   SOME CONSIDERATIONS ABOUT MORE RELEVANT CHARACTERISTICS IN METHYLOTROPHIC YEAST

8.5.1   *HANSENULA POLYMORPHA*

H. polymorpha is a methyltrophic yeast like *P. pastoris* and is able to grow in the presence of methanol as a sole carbon and energy source. One of the features described in the literature regarding *H. polymorpha* and production of recombinant proteins is the use of integrative vectors and plasmids with autonomous replication. The stable integration within the genome of *H. polymorpha* has been reported using the cassette of expression pur3510 developed from the plasmid Pte13 of *S. cerevisiae* (Veale, 1992). On the other hand, generations of recombinant *H. polymorpha* strains that have a CBS4732 background, typically employ vectors that are mitotically stable and integrated into the genome of the host (Gellisen and Hollenberg, 1997). These plasmids may integrate into the host DNA over a number of generations, resulting in strains with as many as 100 integrated plasmids present in tandem repeats (Gellissen, 2000; Kang *et al.*, 2002). For the expression of proteins both homologous and heterologous inducible form, they fundamentally use the promoters *MOX* and *FDM*, both derived from the metabolic pathway for degradation of methanol. These promoters regulate the expression of the ORF that need the presence of methanol as the sole carbon source, using these promoters along with the use of strains that can integrate large number of copies of the vector, we can get a large amount of protein. An example of this combination has been reported with phytase, secreted phytase product levels of up to 13.5 gL^{-1} have been obtained, this extremely high productivity was elicited by the FMD promoter applying conditions of glucose starvation (Mayer *et al.*, 1999). For the constitutive production of proteins other promoters have been described, examples of constitutive promoters are ACT (Kang *et al.*, 2001), GAP (Heo *et al.*, 2003), and PMA1 (Cox *et al.*, 2000). *H. polymorpha* has some characteristics in regard to the secretion mechanism of recombinant proteins, which has no precedent in other organisms used for the same purpose, so that one can manipulate the secretion path to direct the protein to different organelles. Signal sequences used in the production of recombinant proteins therefore, are different in function of the route of secretion. Targeting signals have now been identified for various cell compartments of *H. polymorpha* (mitochondrion, peroxisome, endoplasmic reticulum, secretory pathway and vacuole). This route of secretion has been used when the protein or the metabolite that occurs have toxic effects, directing the protein to the culture medium, or to the cell surface. One of the most widely used signal sequences derived from the PHO1 gene that encodes an acid phosphatase (Phongdara *et al.*, 1998), another signal sequence that is widely used is derived from the GAM1 gene of *S. occidentalis* (Weydemann *et al.*, 1995; Van Djik *et al.*, 2000), the signal sequence most frequently used is MFα 1 prepro signal peptide from *S. cerevisiae* (Brake *et al.*, 1984; Gellissen, 2000). Some of the recombinant proteins produced in *H. polymorpha*, are shown in the Table 8.3.

TABLE 8.3   Heterologous protein produced by methylotrophic yeast.

Host organism	Protein	Source	References
H. polymorpha	β–lactamase	*Escherichia coli*	Janowicz *et al.*, 1998
	Phytase	*A. niger*	Mayer *et al.*, 1999
	Glucose oxidase	*A.niger*	Hodgkins *et al.*, 1993
	Invertase	*P. anomala*	Rodriguez *et al.*, 1996
P. methanolica	Lignin peroxidase	*P. chrisosporeum*	Wang *et al.*, 2004
	Huridin	*Hirudo medicinalis*	Weydemann *et al.*, 1995
	α-Galactosidase	*Cyamopsis tetragonoloba*	Fellinger *et al.*, 1991
	Xilanase	*Neocallimastix frontalis*	Chang *et al.*, 2008
	Lactoferrin	*Sus scrofa*	Shan *et al.*, 2007
	Polymerases	*Hepatitis B virus*	Choi *et al.*, 2002
	Laccase	*Trametes versicolor*	Guo *et al.*, 2006
C. boidinii	Glucoamylase	*Rhizopus oryzae*	Sakai *et al.*, 1996a
	Adenylate kynase	*S. cerevisiae*	Sakai *et al.*, 1995

TABLE 8.3 *(Continued)*

Host organism	Protein	Source	References
P. pastoris	Insulin	*Homo sapiens*	Wang *et al.*, 2001
	Malaria vaccine antigen	*Plasmodium falciparum*	Brady *et al.*, 2001
	Antithrombin III	*H. sapiens*	Mochizuki *et al.*, 2001
	Tannase	*A. oryzae*	Stapleton *et al.*, 2004
	Laccase	*T. versicolor*	López *et al.*, 2010
	Heavy-chain fragment of botulinum neutoxin	*Clostridium botulinum*	Potter *et al.*, 1998

8.5.2 PICHIA METHANOLICA

P. methanolica is a homotalic and haploid yeast, which currently is widely used in the production of recombinant proteins. This species was originally studied under the name of *Pichia pinus* MH4 and was subsequently appointed as *P. methanolica*. Auxotrophic mutants of *P. methanolica* have been used to develop DNA transformation procedures (Faber *et al.*, 1994). In the metabolic pathway for methanol degradation in *P. methanolica* there are some differences with the previously described for *P. pastoris* in Fig. 8.2. As in *P. pastoris*, two genes are responsible for the degradation of methanol to acetaldehyde, *AUG1* and *AUG2* (Raymond *et al.*, 1998), *AUG1* is responsible for most of the alcohol oxidase activity of the cell, therefore, the promoter of this gene has been cloned and inserted in cloning vectors to regulate the synthesis of heterologous proteins in presence of methanol (Raymond *et al.*, 1998). Another similarity with *P. pastoris* is that like *AOX1* and *AOX2*, the sequence of *AUG1* and *AUG2* have a high percentage of similarity. The loss of gene *AUG1*, entails most of the alcohol oxidase activity, resulting in the phenotype Mut^s, strains with an aug1Δ genotype (methione auxotroph mutant) grow slowly on methanol. Slow growth of methanol allows isolation of Mut^s strains (aug1) (Raymond *et al.*, 1998). Another characteristic phenotype is the Mut$^+$ (Methanol utilization plus), which keeps the same ability to metabolize methanol than the wild phenotype. Post-translational modifications that occur in *P. methanolica* are not well characterized, however, efficient secretion of heterologous proteins have been shown (Raymond *et al.*, 1998). There are other factors that have an effect on the activity of the proteins produced, such as glycozylation, mentioned above. One disadvantage is that *P. methanolica* in production of recombinant proteins from mammal origin is due to the different patterns of glycozylation that occur in both types of organisms. To mitigate this adverse effect, the strategy is based on an analysis of the amino acid sequence in the recombinant protein to identify potential points of glycozylation, thus allowing the design of mutagenesis a strategy most appropriate in eliminating these potential sites of glycozylation in the complementary DNA sequence. Another strategy used in *P. methanolica* to avoid hyperglycosilation is to treat the heterologous proteins with endoglycosidase H or peptide N-glycosidase F. Some of the recombinant proteins produced in *H. polymorpha*, are shown in the Table 8.3.

8.5.3 CANDIDA BOIDINII

C. boidinii is also a methyltrophic yeast, which is currently being used for the elucidation of the metabolic pathway for methanol oxidation, some specific particularities of this species have been found having an impact on the recombinant protein production process. In *C. boidinii*, *ODA* is the enzyme responsible for the first oxidation of methanol (alcohol oxidase). The detailed analysis of the specific degradation routes for methanol, allows us to infer, which promoters could be the most suitable for the production of heterologous proteins. An example describing in this strategy, which determined that *C. boidinii* FLD1-disrupted strain (fld1Δ) mutant phenotypes are unable to grow in the presence of methanol (Lee *et al.*, 2002). They concluded that the promoter of this gene could be a good candidate to regulate the processes of protein production. However, there are other promoters that have already been used for the same purpose such as *AOD1* (Sakai *et al.*, 1991; Sakai *et al.*, 1997). Using these promoters in *H. polymorpha* has been achieved where 20% of the total soluble proteins present in the culture medium happened to be the protein of interest (Gellissen *et al.*, 1991). Compared to other methylotrophic yeast, *C. boidinii* has been much less used in the production of recombinant proteins (Sakai *et al.*, 1995, 1996b), some of the recombinant proteins produced by *C. boidinii* are shown in Table 8.4.

TABLE 8.4 Promoters used for heterologous protein production in *P. pastoris.*

Promoter	Regulation	References
AOX1	Inducible methanol	Tschopp *et al.*, 1987a
FLD1	Inducible methanol, methilamine	Shen *et al.*, 1998
GAP8	Constitutive	Waterham *et al.*, 1997
PEX8	Inducible methanol, oleate	Johnson *et al.*, 1999
YPT1	Constitutive	Sears *et al.*, 1998
TEF1	Constitutive	Jungoh *et al.*, 2007

8.6 PICHIA PASTORIS EXPRESSION SYSTEM

P. pastoris is currently a methyltrophic yeast widely used in the production of heterologous proteins, mainly because of factors such as the simplicity of the techniques required for handling this organism, the capacity to produce a large amount of recombinant proteins, some features of its cell physiology that allows a suitable posttranslational process in the production of proteins from higher eukaryotes and finally the commercial availability of specific kits for the construction of recombinant strains that produce heterologous proteins.

8.6.1 METHANOL METABOLISM IN P. PASTORIS

The biochemical pathway described for methanol oxidation in *P. pastoris* is shown (Fig. 8.2) for all the methylotrophic yeasts, however, it should be emphasized that there are some differences in specific enzymes in the *P. pastoris* methanol pathway. The most important expression system for the production of foreign proteins is *AOX*. The alcohol oxidase enzyme (AOX) catalyzes the first step in the methanol utilization pathway, the methanol oxidation to formaldehyde and hydrogen peroxide. Another very important feature of the regulation refers to the fact that two of the methanol pathway enzymes, AOX and DHAS, are present at high levels in cells grown on methanol, but are not detectable in cells grown on most other carbon sources (e.g., glucose, glycerol or ethanol). In cells that were fed with methanol at growth-limiting rates in fermentions cultures, AOX levels are highly induced, constituting >30% of total soluble proteins (Couderc *et al.*, 1980; Roggenkamp *et al.*, 1984).

8.6.1.1 AOX1 PROMOTER

There are two genes for the alcohol oxidase activity, *AOX1* and *AOX2* in *P. pastoris*. The gene responsible for most of the alcohol oxidase activity is *AOX1* (Tschopp *et al.*, 1987a; Cregg *et al.*, 1989). Being its promoter the more widely used in *P. pastoris* (Cereghino *et al.*, 2001), a variety of promoters has been reported in several studies in *P. pastoris* (Table 8.4).

8.6.1.2 AOX2 PROMOTER

Despite the fact that most of the alcohol oxidase activity is derived from the gene *AOX1*, there have been reports describing an increase in the level of expression with *AOX2* promoter (Mochizuki, 2001). In particular, in this study an antifoaming agent was used in order to prevent oxygen limitation and human serum albumin gene expression was increased using AOX2 promoter showing that this promoter can optimize gene expression.

8.6.1.3 FLD1 PROMOTER

In addition to the alcohol oxidase promoter gene other promoters of interest have been isolated related to methanol oxidation, such as the promoter gene for formaldehyde dehydrogenase (FLD) (Veenhuis *et al.*, 1983). The activity of this gene has been reported to be associated with resistance to toxicity created by the presence of formaldehyde as a result of methylamine metabolism (Shen *et al.*, 1998). The *FLD1* gene is inducible by both methanol and methylamine. The Fig. 8.3 shows the step in which the methylamine, may enter methanol degradation pathway. Using this promoter has the advantage of using methylamine for the induction of foreign protein expression and glucose or glycerol can be used as the carbon source instead of methanol. Methanol itself can be used as the sole carbon source and also for induction (Reilander and Weiss, 1998).

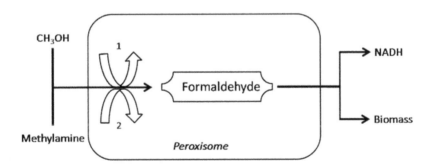

FIGURE 8.3   Synthesis of formaldehyde in methanol pathway in *P. pastoris*. 1: alcohol oxidase; 2: amine oxidase.

8.6.2 METHANOL UTILIZATION PHENOTYPES

Specific *P. pastoris* phenotypes are characterized by their ability to grow with methanol as a sole carbon source. This is supported by the fact that the strain with mutations in *AOX1* are better producers of foreign proteins that the wild phenotype (Mut⁺, Methanol utilization positive) (Tschopp *et al.*, 1987b; Chiruvolu *et al.*, 1997). An important feature of these strains is that they do not require large quantities of methanol during the fermentation. The most important strains of this are designed *Mut* (Methanol utilization). Particularly in the production of heterologous proteins the strain KM71, is widely used where the gene *AOX1* has been partially deleted and replaced with the *ARG4* gene of *S. cerevisiae* (Cregg *et al.*, 1988; Cregg, 1999). This strain is also known as Mut^s (Methanol utilization slow), the ability to use the methanol in this strain depends on the activity of the *AOX2* gene, which has a lower level of expression than the wildtype phenotype. When the strains presents deletions in both *AOX* genes are designed as Mut⁻ (Methanol utilization negative), these strains do not have growth capacity in methanol. However, the production of recombinant proteins is much more regulated since all of these strains retain the ability to induce expression at high levels from the *AOX1* promoter (Chiruvolu *et al.*, 1997).

8.6.3 CONSTRUCTION OF EXPRESSION VECTOR

Protein recombinant expression systems using *P. pastoris* as the host organism can be organized in three steps: (1) insertion of heterologous gene in a specific cloning vector specific for *P. pastoris*, (2) transformation of a *P. pastoris* strain, and (3) efficiency of the transformed strain in biomass production and gene. Vectors with efficient heterologous gene expression systems in *P. pastoris* have been described by Cereghino and Cregg (2000) and Koutz *et al.* (1989). Most studies with *P. pastoris* require the presence of methanol as the only source of carbon for the expression and production of recombinant proteins (Egli, 1980). The Fig. 8.4 shows some of the components, which are required in a transformation vector to clone heterologous protein in *P. pastoris*.

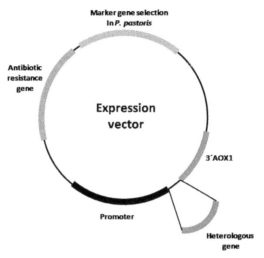

FIGURE 8.4   General structure of a *P. pastoris* expression vector.

8.6.4   EXPRESSION OF HETEROLOGOUS PROTEIN IN P. PASTORIS

For the expression of recombinant proteins in *P. pastoris* there is usually a two-phase system. The first phase is aimed to increase biomass in order to reach high cell densities, where glycerol is used as a source of energy, and the second stage is present once the glycerol is consumed and methanol is added. Production of foreign proteins through this methodology requires two prerequisites: (1) no traces of glycerol and glucose or other sources of carbon left in the culture medium, and (2) the presence of methanol in the culture medium leads to expression conditions. In the case of the existence of any other source of carbon enhancement with methanol that has been reported is a gradient of suppression of the expression of the AOX in function of the concentration of the other source of carbon. Some heterologous proteins expressed in *P. pastoris* are shown in the Table 8.3.

8.6.5   P. PASTORIS AND SOLID STATE FERMENTATION

The traditional fermentation process of the production of recombinant proteins is submerged fermentation. However, in this system *P. pastoris* demands plenty of oxygen, being sometimes necessary to inject pure oxygen in the process. This fact clearly affects the cost of the fermentation process when the culture reaches high cell densities (Yinliang *et al.*, 1997), (100 g L^{-1}). Biomass production in solid state fermentation eliminates problems of oxygen transfer, and has already been employed in the production of recombinant proteins in filamentous fungi such as *Aspergillus niger* (Téllez *et al.*, 2006), showing very encouraging results. Currently there is only a report that describes the production of a recombinant protein using *P. pastoris* on solid support (López *et al.*, 2010). Results of this work inferred that the production processes of recombinant proteins using *P. pastoris*, can be optimized not only from the analysis of biochemical pathways and genetic engineering of the elements involved in the regulation of the expression, but from the point of view of the conditions of the physiology of growth that directly affect the oxygen availability.

8.6.6 *RECENT PATENTS IN P. PASTORIS EXPRESSION SYSTEM*

The number of publications and commercial kits relating to the production of recombinant proteins using *P. pastoris* reflects a growing interest in this microorganism as an alternative to the production of bio-pharmaceuticals and other proteins in *E. coli* and mammalian cell lines. Some of the recent patents focused on optimizing the processes of secretion, are related to what is called glycoengineering, which has to do with the genetic modification of the potential glycozylation sites whose purpose is to optimize the activity of the heterologous proteins produced. Some patents related to this feature are prevention of high mannose glycan structures by inactivation of *OCH1*, combined with the heterologous expression of a Δ-1,2-mannosidase (Choi *et al.*, 2003), and production of glycoproteins with complex human galactosylated N-glycans (Bobrowicz *et al.*, 2004; Jacobs *et al.*, 2009). Another important area where patents are being developed referred to regulatory sequences for the production of heterologous proteins. In this sense, some of the patents developed in recent years are focused to avoid some drawbacks of the traditional system based in the *AOX1* promoter, which requires methanol as inductor to start the synthesis, a compound, which is dangerous, flammable, and highly volatile. An alternative is the utilization of *MOX* promoter from *H. polymorpha* (Sudbery *et al.*, 1988). Another example of this perspective has led to the design of systems for production of recombinant proteins using constitutive promoters of primary metabolism such as the glyceraldehyde 3 phosphate (PGAP) promoter gene, which has been used to obtain high yields without the need to induce the expression with methanol, these patents are based on the work of Waterham *et al.* (1997).

8.7 FUTURE PERSPECTIVES

The methylotrophic yeasts have become a cost effective alternative in the market for the production of recombinant proteins, competing with bacteria, filamentous fungi and mammalian cell lines. One line of research focuses on a less expensive production through process optimization. The two main processes that can be optimized for future research refer to the production of biomass, and the production of heterologous proteins. With respect to the production of biomass, it can explore different carbon sources that contribute to keep the pH in some values more or less close to neutral as it has been reported that there is an inhibitory effect on the growth at very acidic pH values. Another aspect that can explore refers to the modification of the conditions of fermentation, particularly solid state fermentation could reduce the costs of biomass production by decreasing steps in the process such as injection of pure oxygen. The other great process that can be optimized refers to genetic manipulation. Such as, the search for new promoters, the use of other signal sequences, and deeper understanding in the physiology of growth of these yeasts in the presence of methanol, which in turn will contribute to the design of new vectors that maximize the amount of recombinant protein produced per gram of substrate. Finally, we must emphasize that the release of the methylotrophic yeast genome sequence offers a novel tool for the further engineering development.

KEYWORDS

- *Candida boidinii*
- **Formaldehyde dehydrogenase**
- **Gene expression**
- *Hansenula polymorpha*
- **Heterologus proteins**
- **Methanol metabolism**
- **Methanol utilization**
- **Methylotrophic yeast**
- *Pichia methanolica*
- **Recombinant proteins**
- **Signal recognition particle**
- **Solid state fermentation**

REFERENCES

Agaphonov, M. O.; Poznyakovski, A. I.; Bogdanova, A. I.; Ter-Avanesyan, M. D. Isolation and characterization of the *LEU2* gene of *Hansenula polymorpha*. *Yeast* 1994, 10, 509–513.

Agaphonov, M. O.; Bevurov, M. Y.; Ter-Avanesyan, M. D.; Smirnov, V. N. A disruption-displacement approach for the targeted integration of foreign genes in *Hansenula polymorpha*. *Yeast* 1995, 11, 1241–1247.

Ameh, J. B.; Okagbue, R. N.; Ahman, A. A. Isolation and characterization of local yeast strains for ethanol production. *Nigerian J Technol Res* 1989, 1, 47–52.

Ammerer, G.; Hunter, C. P.; Rothman, J. H.; Saari, G. C.; Valls, L. A.; Stevens, T. H. PEP4 gene of *Shaccaromyces cerevisiae* encodes proteinase K, a vacuolar enzyme required for processing of vacuolar precursors. *Mol Cell Biol* 1986, 6, 2490–2499.

Arima, K.; Oshima, T.; Kubota, I.; Nakamura, N.; Mizunaga, T.; Tohe, A. The nucleotide sequence of the yeast PHO5 gene: a putative precursor of repressible acid phosphatase contains a signal peptide. *Nucleic Acids Res* 1983, 11, 1657–1672.

Baneyx, F.; Mujacic, M. Recombinant protein folding and misfolding in *Escherichia coli*. *Nat Biotechnol* 2004, 22, 1399–1408.

Bin-Fang, B. L.; Guang-Yuan, H. E. Synonymous codon usage bias and overexpression of a synthetic gene encoding interferon α2b in yeast. *ViroSin* 2007, 22(3), 226–232.

Bird, P.; Gething, M. J.; Sambrook, J. Translocation in yeast and mammalian cells: not all signal sequences are functionally equivalent. *J Cell Biol* 1987, 105, 2905–2914.

Bobrowicz, P.; Davidson, R. C.; Li, H.; Potgieter, T. I.; Nett, J. H.; Hamilton, S. R.; Stadheim, T. A.; Miele, R. G.; Bobrowicz, B.; Mitchell, T.; Rausch, S.; Renfer, E.; Wildt, S. Engineering of an artificial glycozylation pathway blocked in core oligosaccharide assembly in the yeast *Pichia pastoris*: production of complex humanized glycoproteins with terminal galactose. *Glycobiology* 2004, 14, 757–766.

Blobel, G.; Dobberstein, B. Characterization of molecules involved in protein translocation using a specific antibody. *J Cell Biol* 1975, 67, 835–851.

Brady, C. P.; Shimp, R. L.; Miles, A. P.; Whitmore, M.; Stowers, A. W. High-level production and purification of P30P2 MSP119, an important vaccine antigen for malaria, expressed in the methylotrophic yeast *Pichia pastoris. Protein Expr Purif* 2001, 23, 468–475.

Brake, A. J.; Merryweather, J. P.; Coit, D. G.; Heberlein, U. A.; Masiarz, F. R.; Mullenback, G. T.; Urdea, M. S.; Valenzuela, P.; Barr, P. J. a-Factor-directed synthesis and secretion of mature foreign proteins in *Saccharomyces cerevisiae. Proc Natl Acad Sci USA* 1984, 81, 4642–4646.

Brito, N.; Pérez, M. D.; González, C.; Siverio, J. The genes YNI1 and YNR1 encoding nitrite reductase and nitrate reductase, respectively, in the yeast *Hansenula polymorpha*, are clustered and coordinately regulated. *Biochem J* 1996, 317, 89–95.

Buckholz, R. G.; Gleeson, M. A. Yeast systems for the commercial production of heterologous proteins. *Biotechnology* 1991, 9, 1067–1072.

Carrio, M.; Villaverde, A. Construction and deconstruction of bacterial inclusion bodies. *J Biotechnol* 2002, 96, 3–12.

Cereghino, J. L.; Cregg, J. M. Heterologous protein expression in the methylotrophic yeast *Pichia pastoris*. *FEMS Microbiol Rev* 2000, 24, 45–66.

Cereghino, G. P. L.; Cereghino, J. L.; Sunga, A. J. New selectable marker/auxotrophic host strain combinations for molecular genetic manipulation of *Pichia pastoris*. *Gene* 2001, 263(1–2), 159–169.

Cereghino, G. P. L.; Lim, M.; Johnson, M. A.; Cereghino, J. L.; Sunga, A. J.; Raghavan, D.; Gleeson, M.; Cregg, J. M. New selectable marker/auxotrophic host strain combinations for molecular genetic manipulation of *Pichia pastoris*. *Gene* 2001, 24(263), 159–69.

Chang, T.; Ching-Tsan, H. Overexpression of the *Neocallimastix frontalis* xylanase gene in the methylotrophic yeasts *Pichia pastoris* and *Pichia methanolica*. *Enzyme Microb Tech* 2008, 42, 459–465.

Chiruvolu, V.; Cregg, J. M.; Meagher, M. M. Recombinant protein production in an alcohol oxidase-defective strain of *Pichia pastoris* in fed-batch fermentations. *Enzyme Microb Tech* 1997, 21, 277–283.

Choi, J.; Kim, E. E.; Park, Y. I.; Han, Y. S. Expression of the active human and duck hepatitis B virus polymerases in heterologous system of *Pichia methanolica*. *Antivir Res* 2002, 55, 279–290.

Choi, B. K.; Bobrowicz, P.; Davidson, R. C. Use of combinatorial genetic libraries to humanize N-linked glycozylation in the yeast *Pichia pastoris*. *Proc Natl Acad Sci USA* 2003, 100, 5022–5027.

Couderc, R.; Baratti, J. Oxidation of methanol by the yeast *Pichia pastoris*: purification and properties of alcohol oxidase. *Agric Biol Chem* 1980, 44, 2279–2289.

Cox, H.; Mead, D.; Sudbery, P.; Eland, M.; Evans, L. Constitutive expression of recombinant proteins in the methylotrophic yeast *Hansenula polymorpha* using the PMA1 promoter. *Yeast* 2000, 16, 1191–1203.

Cregg, J. M.; Barringer, K. J.; Hessler, A. Y.; Madden, K. R. *Pichia pastoris* as a host system for transformations. *Mol Cell Biol* 1985, 5, 3376–3385.

Cregg, J. M.; Madden, R. K. Development of the methylotrophic yeast, *Pichia pastoris* as a host for the production of foreign proteins. *Dev Ind Microbiol* 1988, 29, 33–41.

Cregg, J. M.; Madden, K. R.; Barringer, K. J.; Thill, G. P.; Stillman, C. A. Functional characterization of the two alcohol oxidase genes from the yeast *Pichia pastoris*. *Mol Cell Biol* 1989, 9, 1316–1323.

Cregg, J. M.; Shen, S.; Johnson, M.; Waterham, H. R. Classical genetic manipulation. *Methods Mol Biol* 1998, 103, 17–26.

Cregg, J. M. Expression in the methylotrophic yeast *Pichia pastoris*. In *Gene expression systems: using nature for the art of expression*, Fernandez, J. M., Hoeffler, J. P., Ed.; Academic Press: San Diego, 1999; pp 157–191.

Demolder, J.; Fiers, W.; Contreras, R. Efficient synthesis of secreted murine interleukin-2 by *Saccharomyces cerevisiae*. Influence of 3¢ untranslated regions and codon usage. *Gene* 1992, 111, 207–213.

Eckart, M. R.; Bussineau, C. M. Quality and authenticity of heterologous proteins synthesized in yeast. *Curr Opin Biotechnol* 1996, 7, 525–530.

Egli, T.; Van Dijken, J. P.; Veenhuis, M.; Harder, W.; Fietcher, A. Methanol metabolism in yeast; regulation of the synthesis of catabolic enzymes. *Arch Microbiol* 1980, 124, 115–124.

Faber, K. N.; Haima, P.; Harder, W.; Veenhuis, M.; Ab, G. Highly efficient electrotransformation of the yeast *Hansenula polymorpha. Curr Genet* 1994, 25, 305–310.

Fellinger, A. J.; Verbakel, J. M. A.; Veale, R. A.; Sudbery, P. E.; Bom, I. J.; Overbeeke, N.; Verrips, C. T. Expression of the α-galactosidase from *Cyamopsis tetragonoloba* (Guar) by *Hansenula polymorpha. Yeast* 1991, 7, 463–473.

Heijne, G. How signal sequences maintain cleavage specificity. *J Mol Biol* 1984, 173, 243–251.

Gellissen, G. Z. A.; Janowicz, A.; Merckelbach, M.; Piontek, P.; Keup, U.; Weydemann, C. P. Heterologous gene expression in *Hansenula polymorpha*: efficient secretion of glucoamylase. *Bio Technology* 1991, 9, 291–295.

Gellissen, G.; Hollenberg, C. P. Applications of yeast in gene expression studies: a comparison of *Saccharomyces cerevisiae, Hansenula polymorpha* and *Kluyveromyces lactis*, a review. *Gene* 1997, 190, 87–97.

Gellissen, G. Heterologous protein production in methylotrophic yeasts. *Appl Microbiol. Biotechnol.* 2000, 54, 741–750.

Gleeson, M. A.; Sudbery, P. F. Genetic analysis in the methylotrophic yeast *Hansenula polymorpha. Yeast* 1988, 4, 293–303.

Glesson, M. A.; Sudvery, P. F. The methylotrophic yeast. *Yeast* 1988, 4, 1–15.

Gleeson, M. A.; White, C. E.; Meininger, D. P.; Komives, E. A. Generation of protease-deficient strains and their use in heterologous protein expression. *Methods Mol Biol* 1998, 103, 81–94.

Glick, B. R.; Pasternak, J. J. *Molecular biotechnology: principles and applications of recombinant DNA*, 2nd Ed.; American Society for Microbiology: Washington, D.C., 1998.

Goodman, S. Market watch: sales of biologics to show robust growth through to 2013. *Nat Rev Drug Discov* 2009, 8(11), 837–837.

Green, R.; Schaber, M. D.; Shieds, D.; Kramer, R. Secretion of somatostatin by *Saccharomyces cerevisiae*. Correct processing of an a-factor-somatostatin hybrid. *J Biol Chem* 1986, 262, 7558–7565.

Guo, M.; Lu, F.; Du, L.; Pu, J.; Bai, D. Optimization of the expression of a laccase gene from Trametes versicolor in *Pichia methanolica. Appl Microbiol Biotechnol* 2006, 71, 848–852.

Heo, J. H.; Hong, W. K.; Cho, E. Y.; Kim, M. W.; Kim, J. Y.; Kim, C. H.; Rhee, S. K.; Kang, H. A. Properties of the *Hansenula polymorpha*-derived constitutive GAP promoter, assessed using an HSA reporter gene. *FEMS Yeast Res.* 2003, 4, 175–184.

Hicke, L.; Shekman.; R. Molecular machinery required for protein transport from the endoplasmic reticulum to the golgi complex. *BioEssays* 1990, 12, 253–258.

Hiramatsu, R.; Horinouchis, S.; Beppu, T. Isolation and characterization of human prourokinase and its mutants accumulated within the yeast secretory pathway. *Gene* 1991, 99, 235–241.

Hodgkins, M.; Mead, D.; Balance, D. J.; Goodey, A.; Sudbery, P. Expression of the glucose oxidase gene from *Aspergillus niger* in *Hansenula polymorpha* and its use as a reporter gene to isolate regulatory mutations. *Yeast* 1993, 9, 625–35.

Hofmarm, S. Mutations of the alpha-galactosidase signal peptide, which greatly enhanced secretion of heterologous protein by yeast. *Gene* 1991, 15, 101(1), 105–111.

Hong, L. Z.; Chong, X.; Yang, W.; Xue, Q. Y.; Zhi, M. L. Optimization of the expression of a laccase gene from *Trametes versicolor* in *Pichia methanolica. Appl Microbiol Biotechnol* 2006, 71(2), 848–852.

Jacobs, P. P.; Geyzens, S.; Vervecken, W.; Contreras, R.; Calleweart, N. Engineering complex-type N-glycozylation in *Pichia pastoris* using GlycoSwitch technology. *Nat Protocols* 2009, 4, 58–70.

Janowicz, Z. A.; MerckKelbach, A.; Eckart, M.; Weydemann, U.; Roggenkamp, R.; Comberbach, M.; Hollenberg, C. P. Expression system based on the methylotrophic yeast *Hansenula polymorpha*. *Yeast* 1998, 20, 145–155.

Johnson, M. A.; Waterham, H. R.; Ksheminska, G. P.; Fayura, L. R.; Cereghino, J. L.; Stasyk, O. V. Positive selection of novel peroxisome biogenesis-defective mutants of the yeast *Pichia pastoris*. *Genetics* 1999, 151, 1379–1391.

Jungoh, A.; Hong, J.; Lee, H.; Park, M.; Lee, E.; Kim, C.; Choi, E.; Jung, J.; Lee. H. Translation elongation factor 1-α gene from *Pichia pastoris*: molecular cloning, sequence, and use of its promoter. *Appl Microbiol Biotechnol* 2007, 74(3), 601–608.

Kaiser, C. A.; Schekman, R. Distinct sets of *SEC* genes govern transport vesicle formation and fusion early in the secretory pathway. *Cell* 1990, 61, 723–733.

Kang, H. A.; Hong, W. K.; Sohn, J. H.; Choi, E. S.; Rhee, S. K. Molecular characterization of the actin-encoding gene and the use of its promoter for a dominant selection system in the methylotrophic yeast *Hansenula polymorpha*. *Appl Microbiol Biotechnol* 2001, 55, 734–741.

Kang, H. A.; Sohn, J. H.; Agaphonov, M. O.; Choi, E. S.; Ter-Avanesyan, M. D.; Rhee, S. K. *Hansenula polymorpha, biology and applications*, Wiley VCH: Weinheim, 2002; pp 124–146.

Kang, H. A.; Gellissen, G. *Hansenula polymorpha*. In *Production of recombinant proteins novel microbial and eukaryotic expression systems*, Gellissen, G., Ed.; Wiley-VCH: Weinheim, 2005; pp 111–142.

Komeda, T.; Sakai, Y.; Kato, N.; Kondo, K. Construction of protease-deficient *Candida boidinii* strains useful for recombinant protein production: cloning and disruption of proteinase A gene (*PEP4*) and proteinase B gene. *Biosci Biotechnol Biochem* 2002, 66(3), 628–631.

Koutz, P.; Davis, G. R.; Stillman, C. Structural comparison of the *Pichia pastoris* alcohol oxidase genes. *Yeast* 1989, 5, 167–177.

Kurland, C. G. Codon bias and gene expression. *FEBS Lett* 1991, 285, 165–169.

Lee, B.; Yurimoto, H.; Sakai, Y.; Kato, N. Physiological role of the glutathione-dependent form-aldehyde dehydrogenase in the methylotrophic yeast *Candida boidinii*. *Microbiol* 2002, 148, 2697–2704.

Liljestrom, P. L. The nucleotide sequence of the yeast MEL1 gene coding for alpha-galactosi-dase in *Escherichia coli* K-12. *Nucleic Acids Res.* 1985, 13, 7257–7268.

López, M.; Loera, O.; Guerrero-Olazarán, M.; Viader-Salvadó, J. M.; Gallegos-López, J. A.; Fernández, F. J.; Favela-Torres, E.; Viniegra-González, G. Cell growth and *Trametes versicolor* laccase production in transformed *Pichia pastoris* cultured by solid-state or submerged fermentations. *J Chem Technol Biotechnol* 2010, 85, 435–440.

MacKay, V. L.; Welch, S. K.; Insley, M. Y.; Manney, T. R.; Holly, J.; Saari, G. C.; Parker, M. L. The *Saccharomyces cerevisiae* BAR1 gene encodes an exported protein with homology to pepsin. *Proc Natl Acad Sci USA* 1988, 85, 55–59.

Mayer, A. F.; Hellmuth, K.; Schlieker, H.; Lopez-Ulibarri, R.; Oertel, S.; Dahlems, U. An expression system matures: a highly efficient and cost-effective process for phytase production by recombinant strains of *Hansenula polymorpha. Biotechnol Bioeng* 1999, 63(3), 373–81.

Mochizuki, S.; Hamato, N.; Hirose, M.; Miyano, K.; Ohtni, W.; Kameyama, S.; Kuwae, S.; Tokuyama, T.; Ohi, H. Expression and characterization of recombinant human antithrombin III in *Pichia pastoris. Protein Expr Purif* 2001, 23, 55–65.

Newmana, P.; Shimj Ferro-Novick, S. *BET1, BOS1,* and *SEC22* are members of a group of interacting yeast genes required for transport from the endoplasmic reticulum to the golgi complex. *Mol Cell Biol* 1990a, 10, 3405–3414.

Newmana, P.; Ferro-Novick, S. Defining components required for transport from the ER to golgi complex in Yeast. *BioEssays* 1990b, 12, 485–491.

Nishizawa, M.; Ozawa, F.; Hishinuma, F. Construction of an expression and secretion vector for the yeast *Saccharomyces cerevisiae. Agric Biol Chem* 1987, 51, 515–521.

Ogata, K.; Nishikawa, H.; Ohsugi, M. A yeast capable of using methanol. *Agric Biol Chem* 1969, 33, 1519–1520.

Phongdara, A.; Merckelbach, A.; Keup, P.; Gellissen, G.; Hollenberg, C. P. Cloning and characterization of the gene encoding a repressible acid phosphatase (PHO1) from the methylotrophic yeast *Hansenula polymorpha. Appl Microbiol Biotechnol* 1998, 50, 77–84.

Potter, K. J.; Bevins, M. A.; Vassilieva, E. V.; Chiruvolu, V. R.; Smith, T.; Smith, L. A.; Meagher, M. M. Production and purification of the heavy-chain fragment C of botulinum neurotoxin, serotype B, expressed in the methylotrophic yeast *Pichia pastoris. Protein Expr Purif* 1998, 13, 357–365.

Raymond, C. T.; Bukowski, S.; Holderman, A.; Ching, E.; Vanaja, M. Development of the methylotrophic yeast *Pichia methanolica* for the expression of the 65 kilodalton isoform of human glutamate decarboxylase. *Yeast* 1998, 14, 11–23.

Redding, K.; Holcomb, C.; Fuller, R. S. Immunolocalization of Kex2 protease identifies a putative late golgi compartment in the yeast *Saccharomyces cerevisiae. J Cell Biol* 1991, 113, 527–538.

Reilander, H.; Weiss, H. M. Production of G-protein-coupled receptors in yeast. *Curr Opin Biotechnol* 1998, 9, 510–517.

Rodriguez, J.; Perez, J. A.; Ruiz, T.; Rodriguez, L. Characterization of the invertase from *Pichia anomala. Biochem Journal* 1995, 306, 235–239.

Roggenkamp, R.; Kustermann-Kun, B.; Hollenberg, C. P. Expression and processing of bacterial b-lactamase in the yeast *Saccharomyces cerevisiae. Proc Natl Acad Sci USA* 1981, 78, 4466–4470.

Roggenkamp, R.; Janowicz, Z.; Stanikowski, B.; Hollenberg, C. P. Biosynthesis and regulation of the peroxisomal methanol oxidase from the methylotrophic yeast *Hansenula polymorpha. Mol Gen Genet* 1984, 194, 489–493.

Roggenkamp, R. O.; Hansen, H.; Eckart, M.; Janowicz, Z. A.; Hollenberg, C. P. Transformation of the methylotrophic yeast *Hansenula polymorpha* by autonomous replication and integration vectors. *Mol Gen Genet* 1986, 202, 302–308.

Romanos, M. A.; Scorer, C. A.; Clare, J. F. Foreign gene expression in yeast: a review. *Yeast* 1992, 8, 423–488.

Rothstein, S. J.; Lahners, K. N.; Lazarus, C. M.; Baulcombe, D. C.; Gatenby A. A. Synthesis and secretion of wheat α-amylase in *Saccharomyces cerevisiae*. *Gene* 1987, 55, 353–356.

Sakai, Y.; kazarimoto, T.; Tani, Y. Transformation system for asporogenous methylotrophic yeast, *Candida boidinii*: cloning for the Orotidina 5′phosphate descarboxilase gene (URA3) isolation of uracil auxotrophic mutants, and use of the mutant for integrative transformation. *J Bacteriol* 1991, 173, 7458–7463.

Sakai, Y.; Rogi, T.; Takeuchi, R.; Kato, N.; Tani, Y. Expression of Saccharomyces adenylate kinase gene in *Candida boidinii* under the regulation of its alcohol oxidase promoter. *Appl Microbiol Biotechnol* 1995, 42, 860–864.

Sakai, Y.; Akiyama, M.; Kondoh, H.; Shibano, Y.; Kato, N. High-level secretion of fungal glucoamylase using the *Candida boidinii* gene expression system. *Biochim Biophys Acta* 1996a, 1308, 81–87.

Sakai, Y.; Saigannji, A.; Yurimoto, H.; Takabe, K.; Saiki, H.; Kato, N. The absence of Pmp47, a putative yeast peroxisomal transporter, causes a defect in transport and folding of a specific matrix enzyme. *J Cell Biol* 1996b, 134, 37–51.

Sakai, Y.; Murdanoto, A. P.; Konishi, T.; Iwamatsu, A.; Kato, N. Regulation of the formate dehydrogenase gene, FDH1, in the methylotrophic yeast *Candida boidinii* and growth characteristics of an FDH1-disrupted strain on methanol, methylamine, and choline. *J Bacteriol* 1997, 179, 4480–4485.

Schekman, R. Protein localization and membrane traffic in yeast. *Annu. Rev Cell Biol* 1985, 1, 115–143.

Scorer, C. A.; Buckholz, R. G.; Clare, J. J.; Romanos, M. A. The intracellular production and secretion of HIV-1 envelope protein in the methylotrophic yeast *Pichia pastoris*. *Gene* 1993, 136, 111–119.

Sears, I. B.; O'Connor, J.; Rossanese, O. W.; Glick, B. S. A versatile set of vectors for constitutive and regulated gene expression in *Pichia pastoris*. *Yeast* 1998, 14, 783–790.

Shan, T.; Wang, Y.; Liu, J.; Xu, Z. Effect of dietary lacto-ferrin on the immune functions and serum iron level of weanling piglets. *J. Anim. Sci.* 2007, 85, 2140–2146.

Shen, S.; Sulter, G.; Jeffries, T. W.; Cregg, J. M. A strong nitrogen source regulated promoter for controlled expression of foreign genes in the yeast *Pichia pastoris*. *Gene* 1998, 216, 93–102.

Singh, A.; Chen, E. Y.; Lugovoy, J. M.; Chang, C. N.; Hitzeman R. A.; Seeburg, P. H. *Saccharomyces cerevisiae* contains two discrete genes coding for the alpha-factor pheromone. *Nucleic Acids Res.* 1983, 11, 4049–4063.

Skipper, N.; Thomas, D. Y.; Lau, P. C. K. Cloning and sequencing of the preprotoxin-coding region of the yeast M1 double-stranded RNA. *EMBO J* 1984, 3, 107–111.

Smith, R. A.; Duncan, M. J.; Moir, D. T. Heterologous protein secretion from yeast. *Science* 1985, 229, 1219–1224.

Sohn, J. H.; Choi, E. S.; Kim, C. H.; Agaphonov, M. O.; Ter-Avanesyan, M. D.; Rhee, J. S.; Rhee, S. K. A novel autonomously replicating sequence (ARS) for multiple integration in the yeast *Hansenula polymorpha* DL-1. *J Bacteriol* 1996, 178, 4420–4428.

Spang, A. Vesicle transport: a close collaboration of Rabs and effectors. *Curr Biol* 2004, 14, 33–34.

Sreekrishna, K. Strategies for optimizing protein expression and secretion in the methyotrophic yeast *Pichia pastoris*. In *Industrial microorganisms, basic and applied molecular genetics*, Baltz, R. H., Hegeman, G. D., Skatrud, P. L., Eds.; American Society of Microbiology: Washington DC, 1993; pp 119–126.

Sreekrishna, K.; Kropp, K. *Pichia pastoris*. In *Nonconventional yeasts in biotechnology*, Wolf K., Ed.; Springer-Verlag: Heidelberg, 1996; pp 203–252.

Stapleton, P. C.; O'Brien, M. M.; O'Callaghan, J.; Dobson, A. D. Molecular cloning of the cellobiose dehydrogenase gene from *Tramates versicolor* and expression in *Pichia pastoris*. *Enz Microb Technol* 2004, 34, 55–63.

Stark, M. J.; Boyd, A. The killer toxin of *kluyveromyces lactis*: characterization of subunits and identifications of the genes, which encoded then. *EMBO J* 1986, 5, 1995–2002.

Stark, M. J.; Mileham, A. J.; Romanos, M. A.; Boyd, A. Resolution of sequence discrepancies in the ORF1 region of the *Kluyveromyces lactis* plasmid K1. *Nucleic Acids Res* 1988, 16(2), 771.

Stephans, J. C.; Powers, M. A.; Gyenes, A.; Van Nest, G. A.; Miller, E. T.; Higgins, K. W.; Luciw, P. A. Antigenicity and immunogenicity of domains of the human immunodeficiency virus (HIV) envelope polypeptide expressed in the yeast *Saccharomyces cerevisiae*. *Vaccine* 1987, 5(2), 90–101.

Sudbery, P. E.; Gleeson, M. A.; Veale, R. A.; Ledeboer, A. M.; Zoetmulder, M. C. *Hansenula polymorpha* as a novel yeast system for the expression of heterologous genes. *Biochem Soc Trans* 1988, 16, 1081–1083.

Taussig, R.; Carlson, M. Nucleotide sequence of the yeast SUC2 gene for invertase. *Nucleic Acids Res* 1983, 11, 1943–1954.

Téllez, A.; Arana, A.; González, A. E.; Viniegra, G.; Loera, O. Expression of a heterologous laccase by *Aspergillus niger* cultured by solid-state and submerged fermentations. *Enzyme Microbiol Technol* 2006, 38, 665–669.

Tokunaga, M.; Wada, N.; Hishinuma, F. A novel yeast secretion vector using secretion signal of killer toxin encoded on the yeast linear DNA plasmid PGKLl. *Biochem Biphys Res Commun* 1987, 144, 613–619.

Tschopp, J. F.; Brust, P. F.; Cregg, J. M.; Stillman, C. A.; Gingeras, T. R. Expression of the LacZ gene from two methanol-regulated promoters in *Pichia pastoris*. *Nucleic Acids Res.* 1987a, 15, 3859–3876.

Tschopp, J. F.; Sverlow, G.; Kosson, R.; Craig, W.; Grinna, L. High level secretion of glycozylated invertase in the methylotrophic yeast, *Pichia pastoris*. *Biotechnology* 1987b, 5, 1305–1308.

TZ Shan, Y. Z.; Wang, G. F.; Liu, H. Q. Expression of recombinant porcine lactoferrin N-lobe in *Pichia methanolica* and its antibacterial activity. *J Anim Feed Sci* 2007, 16, 283–292.

Van Djik, R.; Faber, K. N.; Jakw, K.; Veenhuis, M.; Klei, I. J. The methylotrophic yeast *Hansenula polymorpha*: a versatile cell factory. *Enzyme Microbiol Technol* 2000, 26, 793–800.

Veale, R. A.; Giuseppiu, M. L. F.; Van Eigk, H. M. J.; Sudbery, P. E.; Verrips, C. T. Development of a strain of *Hansenula polymorpha* for the efficient expression of Guar α-galactosidase. *Yeast* 1992, 8, 361–372.

Veen, M.; Lang, C. Production of lipid compounds in the yeast *Saccharomyces cerevisiae*. *Appl Microbiol Biotechnol*. 2004, 63, 635–646.

Veenhuis, M.; Van Dijken, J. P.; Harder, W. The significance of peroxisomes in the metabolism of one-carbon compounds in yeast. *Adv Microb Physiol* 1983, 24, 1–82.

Walter, P.; Ibrahimi, I.; Blobel, G. Translocation of proteins across the endoplasmic reticulum. I. Signal recognition protein (SRP) binds to *in-vitro*-assembled polysomes synthesizing secretory protein. *J Cell Biol* 1981, 91, 545–550.

Wang, Y.; Liang, Z. H.; Zhang, Y. S.; Yao, S. Y.; Xu, Y. G.; Tang, Y. H.; Zhu, S. Q.; Cui, D. F.; Feng, Y. M. Human insulin from a precursor overexpressed in the methylotrophic yeast *Pichia pastoris* and a simple procedure for purifying the expression product. *Biotechnol Bioeng* 2001, 73, 74–79.

Wang, H.; Lu, F.; Sun, Y.; Du, L. Heterologous expression of lignin peroxidase of *Phanerochaete chrysosporium* in *Pichia methanolica*. *Biotechnol Lett* 2004, 26, 1569–1573.

Wang, H.; Zhao, X.; Lu, F. Heterologous expression of bovine lactoferricin in *Pichia methanolica*. *Biochemistry (Moscow)* 2007, 72, 640–643.

Waters, M. G.; Evans, E. A.; Blobel, G. Preproalpha-factor has a cleavable signal sequence. *J Biol Chem* 1988, 263, 6209–6214.

Waterham, H. R.; Digan, M. E.; Koutz, P. J.; Lair, S. V.; Cregg, J. M. Isolation of the *Pichia pastoris* glyceraldehyde-3-phosphate dehydrogenase gene and regulation and use of its promoter. *Gene* 1997, 186, 37–44.

Wegner, G. Emerging applications of the methylotrophic yeasts. *FEMS Microbiol Rev* 1990, 7, 279–283.

Weydemann, U.; Keup, P.; Piontek, M.; Strasser, A. W.; Schweden, J.; Gellissen, G.; Janowicz, Z. A. High-level secretion of hirudin by *Hansenula polymorpha* authentic processing of three different preprohirudins. *Appl Microbiol Biotechnol* 1995, 44, 377–385.

White, C. E.; Hunter, M. J.; Meiniger, D. P.; White, L. R.; Komives, E. A. Large scale expression, purification and characterization of small fragment of trombomudilin, The role of sixth domain and methionine 338. *Protein Eng* 1995, 8, 1177–1187.

Yamashita, I.; Suzuki, K.; Fukui, S. Nucleotide sequence of the extracellular glucoamylase gene STA1 in the yeast *Saccharomyces diastaticus*. *J Bacteriol* 1985, 161, 567–573.

Yan, A.; Kohli, A.; Koffas, M. A. G. Biosynthesis of natural flavanones in *Saccharomyces cerevisiae*. *Appl Environ Microbiol* 2005, 71, 5610–5613.

Yinliang, C.; Cino, J.; Hart, G.; Freedman, D.; White, C.; Komives, E. A. High protein expression in fermentation of recombinant *Pichia pastoris* by a fed-batch process. *Process Biochem* 1997, 32(2), 107–111.

CHAPTER 9

MECHANISM OF TOLERANCE AND ENGINEERING OF YEAST STRAIN RESISTANT TO NOVEL FERMENTATION INHIBITORS OF BIOETHANOL PRODUCTION

LAHIRU NIROSHAN JAYAKODY, NOBUYUKI HAYASHI, and HIROSHI KITAGAKI

CONTENTS

9.1 INTRODUCTION

High uncertainty and fragile nature of the fossil fuel-based global economy and the environment issues of global warming have highlighted the importance of biomass-based biofuel production for both sustainable energy economy and climate change mitigation. Among the biofuels, bioethanol has the most promising attraction and importance as a next generation liquid biofuel for automobile. Its industrialization has been proved to be practical in several countries and present annual global bioethanol production has reached 86.9 billion liters. However, the main drawback of the current production system is that it abundantly consumes starch and sugar as raw materials, and competes with arable land and food production, finally leading to intense food scarcity in the world (Ogg, 2009; Bastianoni and Marchettini, 1996). Therefore, production of bioethanol from cellulose, which does not compete with food stock, is desired.

Annual cellulose production in the world is accounted about 50 billion tones, it made up with glucose monomer units linked with β 1–4 glycoside linkage. Cellobiose is basic repetitive unit and molecular weight of cellulose depends on degree of polymerization, basically its range from 300,000–500,000 (Bobleter, 1994). Degradation of β 1–4 linked D-glucopyranose contains cellulose into fermentable sugar is a key step in ethanol production from cellulose. Several pretreatment methods are available to degrade celluloses into simple sugar, including acid treatment, steaming explosion, ammonium freeze explosion, wet oxidation and enzyme degradation. Primarily acid or alkaline catalysts are widely used for the pretreatment process (Kumar et al., 2009), but there are several drawbacks in these methods. One major disadvantage of acid hydrolysis is that it leads to generation of massive amount of various kind of waste in the process, acid recovery facilities has to be run with higher cost, and corrosion problems are prominent within the facilities (Yu et al., 2007). On the other hand the enzymatic hydrolysis was recognized as promising hydrolyzing technique. However, it is not commercialized due to very high cost factor. Compared to the former pretreatment methods (Kumar et al., 2009), hot-compressed water treatment is a novel and promising pretreatment method to recover sugars for production of bioethanol from cellulose and hemicelluloses (Nakata et al., 2006; Kumagai et al., 2004; Adschiri et al., 1993; Bonn et al., 1983).

Hot-compressed water is defined as water in subcritical or a super-critical stage or the temperature above 150°C with various pressures. It has reactant, solvent and catalytic capacities. Depending on the temperature and pressure, it supports ionic or free radical reaction. It breaks down celluloses into various compound basically through pyrolytic cleavage, swelling and dissolution of the glycosidic bond in the cellulose. In general, when cellulose is treated with hot-compressed water, a high concentration of H^+ ions attack the lone electron pairs on the oxygen atoms of acetal C-O bonds, which increases the positive charge of the 1st carbon of glucose and enables the nucleophilic attack of OH- ions to the 1st carbon atom, resulting in the degradation of the acetal bond of cellulose and the generation of a single glucose unit (Yu et al., 2007; Lu et al., 2009). The degradation of cellulose with hot-compressed water at the temperature range from 230–400°C mainly yielded glucose, fructose, erythrose, dihydroxyacetone, pyruvaldehyde, cellobiose, cellotriose, cellopentaose and cellohexaose. Treatment of cellulose with hot-compressed water has several advantages. No hazardous wastes are produced in the process, the reaction rate is quite fast, and economically feasible with mass scale production (Kumar et al., 2009). The drawback of this method is that the resultant solution has an inhibitory effect on ethanol fermentation by yeast cells. Because the glucose yielded in hydrolysis process of cellulose is further decomposed to form furfural, 5-hydroxymethylfurfural and methylglyoxal. So far, it has been believed that those substances are responsible for the inhibition of ethanol fermentation (Klinke, 2004). During the treatment of celluloses with pressurized hot water, glycolaldehyde is produced (Fig. 9.1) at the concentration of 1 mM to 22 mM through retro-aldol condensation of glucose, fructose and mannose (Katsunobu and Shiro, 2002; Lu et al., 2009). For the first time in the world, we identified the significant of glycolaldehyde in the ethanol fermentation (Jayakody et al., 2011).

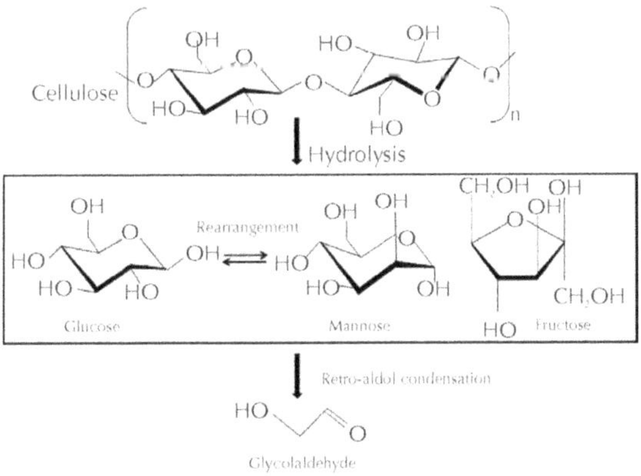

FIGURE 9.1 Chemical degradation pathway of glycolaldehyde production in hot compressed water treatment of cellulose.

Glycolaldehyde is an α-hydroxyaldehyde with a hydroxyl bond next to the aldehyde bond, which discriminates this molecule from other general aldehydes, and there have been very few studies of glycolaldehyde in the field of fermentation. Generally, aldehydes are characterized by their polarized π-electron clouds surrounding the carbonyl bonds and low pKa of α-carbon because of inductive effect. However, since α-hydroxyaldehyde bears a hydroxyl bond in the vicinity of their carbonyl bond, it is considered to form Schiff's base with amino bases of proteins followed by Amadori rearrangement, conversion to aldoamine, regeneration of carbonyl base and cross-linking of proteins (Acharya and Manning, 1983) or followed by formation of carboxymethyllysine and related advanced glycation end products (Takeuchi et al., 2000; Glomb and Monnier, 1995). Glycolaldehyde is peculiar in α-hydroxyaldehydes because it has only two carbons, thus generating a drastic molecular characteristic such as the 2109-fold increased activity of Maillard reaction relative to glucose (Hayashi and Namiki, 1986).

There has been little information about detoxification mechanism of glycolaldehyde or the metabolic fate of glycolaldehyde in a reductive environment. Hence, in this chapter, we present the overview of glycoladehyde as the primary fermentation inhibitor in hot-compressed water-treated cellulose hydrolysate and strategy for engineering a yeast strain with improved tolerance to the hot-compressed water-treated cellulose by attenuating the toxicity of the glycolaldehyde present in the hot compressed water-treated cellulose.

9.2   GLYCOLALDEHYDE AS THE KEY FERMENTATION INHIBITOR

9.2.1   INHIBITION OF YEAST GROWTH BY GLYCOLALDEHYDE

Glycolaldehyde significantly inhibits cell growth of yeast even at a concentration as low as 0.01 mM (Jayakody et al., 2011). The IC_{50} value of glycolaldehyde on Saccharomyces cerevisiae is about 10 mM. Hence, the concentration of glycolaldehyde contained in the actual pressurized hot water-treated cellulose hydrolysate is high enough to inhibit yeast growth (Jayakody et al., 2011; Jayakody et al., 2012). Furthermore, growth analysis indicated that glycolaldehyde affects both cell growth rate and lag phase of cell growth. The lag time of untreated cells was 12 h, while that of cells treated with 1 mM was 15 h, indicating that glycolaldehyde lengthens the lag time of growth. The growth rate of yeast cells treated with 1 mM glycolaldehyde was 0.0728 h^{-1}, while that of untreated yeast cells was 0.0829 h⁻¹, indicating that glycolaldehyde slows the growth rate (Table 9.1).

TABLE 9.1 Effect of glycolaldehyde on maximal specific growth rate and cell number of yeast *Saccharomyces cerevisiae*.

Concentration of glycol-aldehyde (mM)	m_{max}[a] (h^{-1})	Cell number[b] (per mL)
0.00	0.205	3.76×10^7
0.01	**0.171**	3.56×10^7
0.50	**0.177**	3.40×10^7
1.00	**0.156**	3.07×10^7
5.00	**0.129**	1.60×10^7
10.00	**0.122**	6.20×10^6
20.00	**0.050**	1.40×10^6

[a] Represents maximal specific growth rate at logarithmic growth phase around OD_{600} of 1.0;
[b] Cell number represents the number of cells per ml incubated for 48 h; bold letters indicate values that were significantly ($p < 0.01$) lower than the nontreated cells as judged by unpaired one-tailed Student's t-test from independent triplicate experiments.

Furfutal and 5-HMF have been recognized as the major fermentation inhibitors in ethanol fermentation. Moreover, further degradation of tautomerization intermediate of glucose and fructose yields furfural and 5-HMF in hot-compressed water-treated cellulose at the concentration of 10–21 mM and 8–24 mM, respectively (Klinke *et al.*, 2004). Therefore, it is highly important to compare the individual and combinational inhibitory effects of furfural and 5-HMF with glycolaldehyde. Growth analysis at the concentration of 5 mM of glycolaldehyde, furfural and 5-HMF revealed that the specific growth rate and the cell dry weight of glycolaldehyde-treated cells are smaller than those of 5-HMF- and furfural-treated cells (Jayakody *et al.*, 2012). In addition, the combinational effect of these inhibitors at the concentrations contained in the actual hot-compressed water-treated cellulose (Lu *et al.*, 2009) is further inhibit the yeast growth (Fig. 9.2), indicating that glycolaldehyde has a statistically significant combinational inhibitory effect when added with 5-HMF and furfural. These results establish that glycolaldehyde has a greater inhibitory effect than 5-HMF and furfural, and it exhibits a combinational effect with them in the hot-compressed water treated hydrolysate.

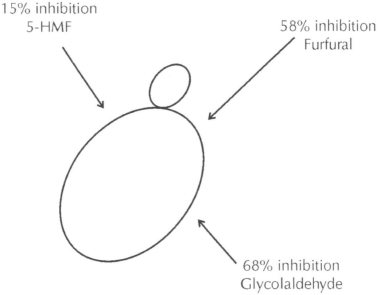

FIGURE 9.2   Inhibitory effects of glycolaldehyde, 5-HMF and furfural. The inhibition figures indicate inhibition ratios of OD_{600} of BY4743 + pRS426 incubated in media containing 2.3 mM glycolaldehyde, 3.3 mM furfural, 3.5 mM 5-HMF.

9.2.2   THE EFFECT OF GLYCOLALDEHYDE ON ETHANOL PRODUCTION

Glycoaldehyde not only inhibits yeast growth but also greatly reduces the ethanol production. The effect of glycolaldehyde on ethanol production by yeast depends on the glycolaldehyde concentration. Moreover, glycolaldehyde higher than 1 mM significantly decreases ethanol production (Jayakody et al., 2011). The detailed analysis at various timepoints during the ethanol fermentation reveals that glycolaldehyde reduces concentrations of ethanol production as well as the glucose consumption (Jayakody et al., 2011). After 48 h of fermentation at the concentration of 2 mM glycolaldehyde ethanol concentration is around 1.66±0.13 % (v/v) and glucose consumption is 4.86±0.072% (v/v), where the untreated cells ethanol concentration is 3.42±0.050 % (v/v) and glucose consumption is around 9.2±0.079 % (v/v). Furthermore, ethanol yield (g produced ethanol/g consumed glucose) of untreated cells was 0.320±0.0039, while that of 2 mM glycolaldehyde-treated cells was 0.254±0.017. The etha-nol yield of cells treated with 2 mM glycolaldehyde was significantly lower than that of untreated cells ($p = 0.040$). In conclusion, glycolaldehyde signifi-cantly decreases ethanol production, glucose consumption and ethanol yield by yeast. Furthermore, the concentration required for fermentation inhibition by glycolaldehyde is as low as the concentrations required for inhibition of yeast fermentation by furfural, 5-hydroxy methyl furfural, indicating that gly-colaldehyde is the important inhibitory substance during bioethanol produc-tion from pressurized hot water-treated cellulose.

9.2.3   GENOME-WIDE ANALYSIS OF GLYCOLALDEHYDE TOXICITY

Genome-wide screening of genes is powerful biotechnology technique in order to obtain insight into the molecular mechanism of inhibition of yeast ethanol fermentation by glycolaldehyde. Therefore, we adopted a genome-wide analysis of glycolaldehyde toxicity towards yeast cells. As a result, 170 genes were identified as genes required for glycolaldehyde tolerance (the mu-tants, which are more than 10% sensitive relative to the wildtype and whose sensitivity is significant at $p < 0.05$) by screening the complete mutant col-lection of *Saccharomyces cerevisiae* BY4743 comprizing of 4848 homozy-gous diploid deletion strains with 0.01 mM glycolealdehyde (Jayakody *et al.*, 2011). The categories involved in glycolaldehyde resistance were extracted from the list of genes. The list of sensitive mutants was submitted to the GO yeast databases on the FunSpec web-based clustering tool (Robinson *et al.*, 2002). The categories that were statistically most unlikely to occur by chance are those involved with mitochondrial respiratory chain complex IV (*COX9*, *COX6* and *COX5B*), ubiquitin ligase complex (*SLX8* (0.69 ± 0.0017 fold), *BUL2* (0.72 ±0.038 fold) and YNL311c), polysome (*PBP1*, *CTK1* and *SSB2*), elongator holozyme complex (*ELP2* and *IKI3*) in GO cellular component, re-sponse to acid (*BCK1*, *MID2* and *RLM1*), golgi to vacuole transport (*VPS54*, *VPS45*, *APS3* and *APL2*) and mitochondrial electron transport, cytochrome *c* to oxygen (*COX9*, *COX6* and *COX5B*) in GO Biological Process and chroma-tin DNA binding (*GON7* and *RED1*) and phospholipase activity (*PLB1* and *YOR022C*) in GO molecular function (Table 9.2).

TABLE 9.2   Functional categories that are over represented in the sensitive mutants ($p <$ 0.01).

GO functional categories	Gene functions
GO cellular component	Mitochondrial respiratory chain complex IV, Ubiquitin ligase complex, Polysome, Elongator holoenzyme complex
GO biological process	Response to acid, Golgi to vacuole transport, Mitochondrial electron transport, cytochrome c to oxygen
GO molecular function	Chromatin DNA binding, Phospholipase activity

Gene functions were identified by addressing the GO database with the FunSpec statistical evaluation program. Probability of the functional set occurring as a chance event is shown.

This study shows that mutants defective in ubiquitin ligase complex and polysomes were significantly sensitive to glycolaldehyde. Glycolaldehyde has been reported to cross-link proteins through its electrophilic attack towards the lone electron-pair of the nitrogen atom of amino groups of proteins (Glob and Monnier, 1995) and thiolate anion of cysteine of proteins (Hayashi and Namiki, 1986). Moreover, glycation has been reported to decrease total cellular proteasome activity in human fibroblast and keratinocytes, which is suggested to have a critical role in aging and diabetes. The observed sensitivity of mutants involved in these systems suggests that glycolaldehyde attacks proteins that are being translated from mRNA and hinder proper folding of proteins, which is alleviated by ubiquitin ligase complex.

There are several studies that report mutants sensitive to other aldehydes such as formaldehyde, 5-HMF, furfural, acetoaldehyde and vanillin (Grey *et al.*, 1996; Gorsich *et al.*, 2005; Matsufuji *et al.*, 2008; Endo *et al.*, 2008). Maybe because of the molecular characteristics of glycolaldehyde, these mutants do not significantly overlap with mutants detected in glycolaldehyde screen. For example, mutants sensitive to acetaldehyde and furfural include the pentose phosphate pathway mutants. However, in glycolaldehyde screen, no pentose phosphate pathway mutant except *prs3* was identified. Mutants sensitive to vanillin included those involved in chromatin remodeling and vesicle transport, but these categories were not overrepresented in glycolaldehyde screen. The ubiquitin ligase complex mutants were included in glycolaldehyde sensitivity test and in the sensitivity test against fufural but not in the sensitivity test against acetaldehyde or vanillin (Endo *et al.*, 2008; Matsufuji *et al.*, 2008). Moreover, mutants defective in oxidative defense such as *sod1* and *yap1* (Gorsich *et al.*, 2005), or stress defense such as *msn2* and *bcy1* (Gorsich *et al.*, 2005) were included in neither of the screens. These comparisons lead to the conclusion that glycolaldehyde has a unique function inside the cells because of its peculiar molecular feature.

9.3 ENGINEERING A YEAST STRAIN RESISTANT TO GLYCOLALDEHYDE

9.3.1 *MECHANISM OF IN SITU DETOXIFICATION OF GLYCOLALDEHYDE*

Based on the results of genome wide analysis and the molecular structure and function of glycoaladehyde, it was suggested that the plus charge of the α-carbon of the glycolaldehyde molecule plays a key role in the inhibition of yeast, because electrophilic attack of the plus charge of carbonyl carbon of glycolaldehyde to negatively charged molecules inside cells is the main cause of the toxicity. Hence, the reduction of plus charge of the carbonyl carbon of glycolaldehyde molecule by NADH was implemented as the principle strategy to develop a resistant strain (Fig 9.3). Although not detected in the functional categories of glycolaldehyde resistant genes in GO-based analysis, a mutant defective in aldehyde dehydrogenases such as *adh1* (0.72±0.0037 fold) was obtained as sensitive genes in the glycolaldehyde screen (Jayakody *et al.*, 2011). This result suggests that these dehydrogenases function to confer glycolaldehyde tolerance, and that glycolaldehyde functions as an aldehyde within cells and the enzymes that reduce the glycolaldehyde to ethylene glycol is effective to mitigate the damage. This result is consistent with the previous study that reported the role of aldehyde dehydrogenase Adh6 against 5-hydroxymethylfurfural (Petersson *et al.*, 2006). Moreover, ethylene glycol was not toxic to yeast cells when it was administered with the same concentration as glycolaldehyde (Jayakody *et al.*, 2012). Since *adh1* was sensitive to glycolaldehyde and Adh1 is capable of reducing short-chain aldehydes such as acetaldehyde and formaldehyde by using NADH as a cofactor (Leskovac *et al.*, 2002; Grey *et al.*, 1996), it was selected for biochemical reduction of glycolaldehyde into ethyleneglycol (Fig. 9.3) (Jayakody *et al.*, 2012). This hypothesis was verified by constructing *ADH1* overexpression strain.

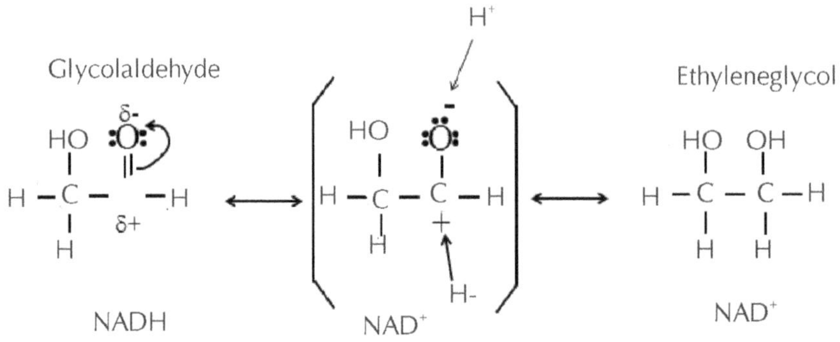

FIGURE 9.3   NADH-dependent glycolaldehyde reduction into ethylene glycol.

9.3.2   CONVERSION OF GLYCOLALDEHYDE INTO ETHYLENE GLYCOL BY ADH1

It turned out that, in the presence of glycolaldehyde, the strain harboring *ADH1*-overexpressing plasmid shows remarkably improved the growth in the presence of glycolaldehyde as compared to a strain harboring an empty vector (Jayakody *et al.*, 2012). Consistent with our hypothesis, *ADH1*-overexpressing strain significantly decreased the concentrations of extracellular glycolaldehyde and increased the extracellular concentration of ethylene glycol, which is the reduced form of glycolaldehyde (Jayakody *et al.*, 2012). The conversion ratio of glycolaldehyde to ethylene glycol is 39±2.9% when the control strain is used; this ratio increases to 89±3.1% in the case of the *ADH1*-overexpressing strain. These results indicate that *ADH1*-overexpressing strain is highly capable of converting glycolaldehyde to ethylene glycol. Moreover, the developed *ADH1*-overexpressing strain exhibits an improved fermentation profile in a glycolaldehyde-containing medium. The strain harboring *ADH1*-overexpressing plasmid significantly increased glucose consumption, ethanol production, and acetic acid production and markedly decreased the glycerol production, relative to the strain harboring an empty vector in the presence of glycoladehyde (Fig. 9.4b).

9.3.3   IMPROVEMENT OF FERMENTATION PROFILE IN HOT-COMPRESSED WATER-TREATED CELLULOSE

On the basis of the above-mentioned results, we hypothesized that this strain would exhibit an improved tolerance to hot-compressed water-treated cellulose.

To verify this hypothesis, hot-compressed water-treated cellulose hyroly-sate (cellulose treated at 280°C, 5 MPa for 1 min) was inoculated with the *ADH1*-overexpressing strain and the control strain as the model substance. The glucose concentration of the hot-compressed water-treated cellulose was 24 ± 1.1 g/L ($n = 3$), which was sufficient to perform ethanol fermentation. The glycolaldehyde concentration in the hot-compressed water-treated cellulose hydrolysate was 8.9 ± 0.34 mM ($n = 3$), which was sufficient for exhibiting an inhibitory effect on fermentation. In media containing the hot-compressed water-treated cellulose, the *ADH1*-overexpressing strain exhibited a statisti-cally significantly improved growth, ethanol production, glucose consump-tion, glycolaldehyde consumption, and ethylene glycol production as com-pared to the control strain (Jayakody *et al.*, 2012). The ratio of conversion of glycolaldehyde to ethylene glycol was $72\pm1.7\%$ in the case of the *ADH1*-overexpressing strain, while it was $33\pm0.85\%$ in the case of the control strain (Jayakody *et al.*, 2012). These results clearly support our hypothesis that the *ADH1*-overexpressing strain exhibits an improved fermentation profile in a medium containing the hot-compressed water-treated cellulose by reducing glycolaldehyde to ethylene glycol.

9.3.4   DETOXIFICATION MECHANISM OF GLYCOLALDEHYDE

The Adh1-mediated detoxification of glycolaldehyde elucidated in this study has many similarities with the detoxification mechanisms of short aldehydes (carbons less than 2). For example, the overexpression of *ADH1* confers re-sistance to formaldehyde (Grey *et al.*, 1996), and Adh1 reduces acetaldehyde (Leskovac *et al.*, 2002). However, native Adh1 does not have a reducing abil-ity towards 5-HMF or furfural (Almeida *et al.*, 2008). Therefore, Adh1 is con-sidered to have a reducing activity towards short aldehydes (carbons less than 2) such as acetaldehyde, formaldehyde, and glycolaldehyde by using NADH as a cofactor and it does not reduce 5-HMF or furfural. The only exception is the mutated Adh1 in TBM3000, which has an ability to reduce 5-HMF or fur-fural by using NADH as a cofactor (Laaden *et al.*, 2008). Hence, the reduction of 5-HMF to furan-2,5-dimethanol or furfural to furan methanol seems to have a different mechanism from that of glycolaldehyde. Enzymes, which reduce 5-HMF and furfural mostly use NADPH as a cofactor, and detoxification of these inhibitors requires genes involved in the pentose phosphate pathway or the biosynthesis of aromatic amino acids (Heer *et al.*, 2009; Liu *et al.*, 2008; Petersson *et al.*, 2006). For example, cinnamaldehyde dehydrogenases Adh6 and Adh7 reduce 5-HMF and furfural by using NADPH as a cofactor (Almeida *et al.*, 2008; Liu *et al.*, 2008; Petersson *et al.*, 2006). Also, the overexpression of *ZWF1*, which encodes glucose-6-phosphate dehydrogenase in the pentose phosphate pathway, confers resistance to furfural, although it does not directly reduce furfural (Gorsich *et al.*, 2006). Furthermore, the YGL157W gene product also reduced furfural and 5-HMF by using NADPH as a cofactor (Liu and Moon, 2009). Moreover, YKL071W, which encodes an uncharacterized ORF and was significantly overexpressed in 25 mM and 36 mM furfural, exhibited a striking similarity to the Sniffer carbonyl reductases using NADPH as a cofactor (Heer *et al.*, 2009).

9.3.5 METABOLIC SHIFT CAUSED BY GLYCOLALDEHYDE

The analysis of the fermentation products revealed that the cells are adjusting their metabolic carbon fluxes in order to adapt to the change in the carbon flux and redox balance caused by the glycolaldehyde treatment and the overexpression of *ADH1* (Fig. 9.4a, b) (Jayakody *et al.*, 2012). Glycerol production by the reduction of dihydroxyacetone phosphate is reported to use NADH and competes with the production of ethanol by the reduction of acetaldehyde (Cordier *et al.*, 2007). Therefore, the decrease in the glycerol production in glycolaldehyde-treated cells and *ADH1*-overexpressing strain is explained by the competition of the reaction from dihydroxyacetone phosphate to glycerol-3-phosphate with the reaction of glycolaldehyde to ethylene glycol for the reductive potential of NADH. This result is also consistent with the decrease in glycerol by the mutated *ADH1*-overexpressing strain (Almeida *et al.*, 2009). In contrast, it has been reported that acetic acid is produced by the oxidation of acetaldehyde by using mainly $NADP^+$ and partially NAD^+ as cofactors (Saint-Prix *et al.*, 2004; Wang *et al.*, 1998). The increase in acetic acid in glycolaldehyde-treated cells can be explained by the accumulated acetaldehyde because of the competition between acetaldehyde and glycoaldehyde for Adh1 and NADH. The increase of acetic acid in the glycolaldehyde-treated *ADH1*-overexpressing strain can be explained by the increase in NAD^+ because of the increased reaction of Adh1-catalyzed reduction of acetaldehyde and glycolaldehyde coupled with oxidation of NADH. This hypothesis is further supported by several other reports, which observed an increase of acetic acid in cells overexpressing H_2O-forming NADH oxidase (Heux *et al.*, 2006) or NADH-dependent *GPD1* (Remize *et al.*, 1999; Michnick *et al.*, 1997).

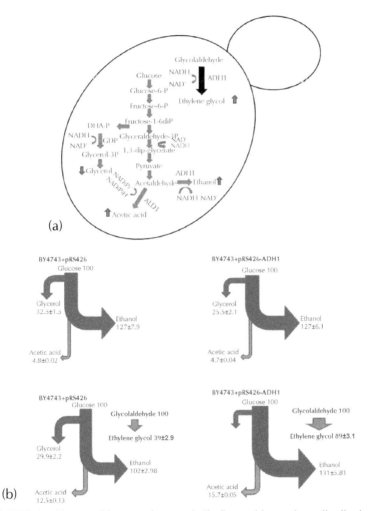

FIGURE 9.4 Change of fermentation metabolic flux and its products distribution due to glycolaldehyde reduction. (a) The effect of glycolaldehyde reduction on glycolysis pathway due to NADH demand of *ADH1* expression strain. Red color arrows are indicated the changes of product level. (b) Fermentation products distribution of 100 mol of consumed glucose with or without glycolaldehyde. Dark grey color numbers indicate statistically significant value as compared to the control strain ($p < 0.01$, $n = 3$).

9.3.6 RELATION OF GLYCOLALDEHYDE TO XYLOSE FERMENTATION

The material used for biethanol production is biomass, derived from photobiosynthesis by plants. The key components of biomass are cellulose, hemicel-

lulose and lignin. For example, woody plants, which are typical biomass materials, consist of approximately 40–50% cellulose, 20–30% hemicellulose,
and 20–28% lignin along with other substances (Sun and Cheng, 2002). As
hemicellulose contains pentoses such as xylose and arabinose, the fermentation of pentose is a critical process for the efficient production of bioethanol
from hemicellulose. However, *S. cerevisiae* does not ferment pentoses. For
example, xylose fermentation requires xylose reductase, which reduces xylose to xylitol by using NADPH as a cofactor and xylitol dehydrogenase,
which converts xylitol to xylulose by using NAD^+ as a cofactor (Matsushika
et al., 2008). Alternatively, xylulose can be produced from xylose by xylose
isomerase (Walfridsson *et al.*, 1996). Xylulose can be metabolized by *S. cerevisiae* after being phosphorylated to form xylulose 5-phosphate, and being incorporated into the pentose phosphate pathway. Glycolaldehyde is generated
from xylose by the treatment of hemicellulose with hot-compressed water
through retroaldol condensation (Yu *et al.*, 2008). The main constraint of this
pathway is occurred due to different cofactor involvement in conversion process of D-xylose into D-xylulose. It leads to NAD^+ shortage in xylitol dehydrogenase reaction and facilitates xylitol excretion and subsequently reduce
xylose assimilation (Jeffries *et al.*, 2004). However, it has been shown that,
the increase of intracellular pool of NAD^+ enhanced the carbon conversion by
reducing xylitol excretion (Bruinenberg *et al.*, 1983). Recently, Almeida *et al.*
(2009) demonstrated that overexpression of mutated *ADH1-S110P-Y295C* in
engineered xylose fermenting yeast leads high carbon conversion and low xylitol excretion in anaerobic fermentation of xylose/glucose mix media in the
presence of HMF. The NADH dependent Adh1 in the *ADH1-S110P-Y295C*
strain has ability to converts HMF into 2,5-bis-hydroxy methyl furan. It generates NAD^+ pool in the cell and facilitate xylitol dehydrogenase activity. The
overexpression of mutated *ADH1* has changed $NADH/NAD^+$ fluxes reactions
significantly in engineered xylose-fermenting yeast. It leads to increase detoxification of HMF, carbon conversion and utilization of xylose (Almeida *et
al.*, 2009). Our obtained results of glycolaldehyde reduction to ethylene glycol
by overexpression *ADH1* strain has also followed the more or less similar
$NADH/NAD^+$ fluxes reactions and it can generate NAD^+ pool inside the cell
and subsequently improved xylose assimilation in xylose fermenting engineered yeast. Since 5-HMF and furfural have been reported to function as external electron acceptors for NADH and to increase xylose assimilation during xylose fermentation (Almeida *et al.*, 2009; Wahlbom and Hahn-Hägerdal,
2002), the glycolaldehyde derived from hemicellulose might function as an

external electron acceptor of NADH and improve xylose fermentation, which makes an important topic for future studies (Fig. 9.5).

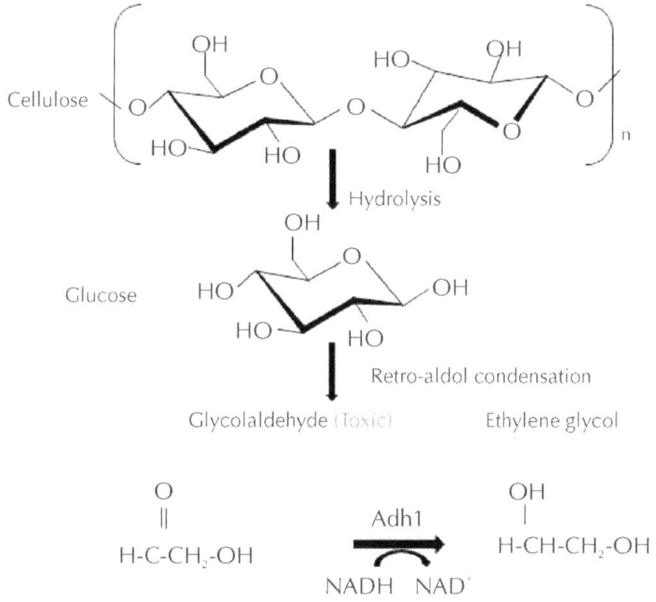

FIGURE 9.5   Detoxification mechanism of glycolaldehyde.

9.4   CONCLUSION

In the preceding researches on the ethanol fermentation of cellulose and hemi-cellulose after a hot-compressed water treatment, only 5-HMF and furfural have been focused upon as fermentation inhibitors. Glycolaldehyde was not considered in these studies, although it was generated in the hydrolysate and exhibited inhibitory effects. However, recent findings highlighted glycolalde-hyde as the key toxic compound in bioethanol fermentation. Therefore, toxic-ity of glycolaldehyde and its detoxification mechanism is of high significance in this field. Furthermore, the strategy of reducing glycolaldehyde to ethylene glycol proposed in this chapter is a promising strategy to decrease the toxicity of hot-compressed water-treated cellulose hydrolysate. This novel informa-tion will certainly be valuable to develop biocatalyst for sustainable cellulosic ethanol production system with hot-compressed water treatment.

KEYWORDS

- **Bioethanol production**
- **Ethylene glycol**
- **Fermentation inhibitor**
- **Glycolaldehyde**
- **Hot-compressed water-treated cellulose**
- *In situ* **detoxification**
- **Metabolic shift**
- **Xylose fermentation**
- **Yeast strain**

REFERENCES

Adschiri, T.; Hirose, S.; Malaluan, R.; Arai, K. Noncatalytic conversion of cellulose in super-critical and subcritical water. *J Chem Eng Jpn* 1993, 26, 676–680.

Almeida, J. R.; Röder, A.; Modig, T.; Laadan, B.; Lidén, G.; Gorwa-Grauslund, M. F. NADH- vs NADPH-coupled reduction of 5-hydroxymethyl furfural (HMF) and its implications on product distribution in *Saccharomyces cerevisiae*. *Appl Microbiol Biotechnol* 2008, 78, 939–945.

Almeida, J. R.; Bertilsson, M.; Hahn-Hägerdal, B.; Lidén, G.; Gorwa-Grauslund, M. F. Carbon fluxes of xylose-consuming *Saccharomyces cerevisiae* strains are affected differently by NADH and NADPH usage in HMF reduction. *Appl Microbiol Biotechnol* 2009, 84, 751–761.

Bastianoni, S.; Marchettini, N. Ethanol production from biomass: analysis of process efficiency and sustainability. *Biomass Bioenerg* 1996, 11, 411–418.

Bobleter, O. Hydrothermal degradation of polymers derived from plants. *Prog Polym Sci* 1994, 19, 797–841.

Bonn, G.; Concin, R.; Bobleter, O. Hydrothermolysis – a new process for the utilization of biomass. *Wood Sci Technol* 1983, 17, 195–202.

Bruinenberg, P. M.; Bot, P. H. M.; van Dijken, J. P.; Scheffers, W. A. The role of redox balances in the anaerobic fermentation of xylose by yeasts. *Appl Microbiol Biotechnol* 1983, 18, 287–292.

Cordier, H.; Mendes, F.; Vasconcelos, I.; François, J. M. A metabolic and genomic study of engineered *Saccharomyces cerevisiae* strains for high glycerol production. *Metab Eng* 2007, 9, 364–378.

Endo, A.; Nakamura, T.; Ando, A.; Tokuyasu, K.; Shima, J. Genome-wide screening of the genes required for tolerance to vanillin, which is a potential inhibitor of bioethanol fermentation, in *Saccharomyces cerevisiae*. *Biotechnol Biofuels* 2008, 1, 3.

Glomb, M. A.; Monnier, V. M. Mechanism of protein modification by glyoxal and glycolaldehyde, reactive intermediates of the Maillard reaction. *J Biol Chem* 1995, 270, 10017–10026.

Gorsich, S. W.; Dien, B. S.; Nichols, N. N.; Slininger, P. J.; Liu, Z. L.; Skory, C. D. Tolerance to furfural-induced stress is associated with pentose phosphate pathway genes *ZWF1*, *GND1*, *RPE1*, and *TKL1* in *Saccharomyces cerevisiae*. *Appl Microbiol Biotechnol* 2005, 1, 339–349.

Grey, M.; Schmidt, M.; Brendel, M. Overexpression of *ADH1* confers hyper-resistance to form-aldehyde in *Saccharomyces cerevisiae*. *Curr Genet* 1996, 9, 437–440.

Hayashi, T.; Namiki, M. Role of sugar fragmentation in an early stage browning of amino-carbonyl reaction of sugar with amino acid. *Agr Biol Chem* 1986, 50, 1965–1970.

Heer, D.; Heine, D.; Sauer, U. Resistance of *Saccharomyces cerevisiae* to high concentrations of furfural is based on NADPH-dependent reduction by at least two oxireductases. *Appl Environ Microbiol* 2009, 75, 7631–7638.

Heux, S.; Cachon, R.; Dequin, S. Cofactor engineering in *Saccharomyces cerevisiae*: expression of a H_2O-forming NADH oxidase and impact on redox metabolism. *Metab Eng* 2006, 8, 303–314.

Jayakody, L. N.; Hayashi, N.; Kitagaki, H. Identification of glycolaldehyde as the key inhibitor of bioethanol fermentation by yeast and genome-wide analysis of its toxicity. *Biotechnol Lett* 2011, 33, 285–292.

Jayakody, L. N.; Horie, K.; Hayashi, N.; Kitagaki, H. Improvement of *Saccharamyces cerevisiae* to hot-compressed water treated cellulose by expression of *ADH1*. *Appl Microbiol Biotechnol* 2012, 94, 273–283.

Jeffries, T. W.; Jin, Y. S. Metabolic engineering for improved fermentation of pentoses by yeast. *Appl Microbiol Biotechnol* 2004, 63, 495–509.

Katsunobu, E.; Shiro, S. A comparative study on chemical conversion of cellulose between the batch-type and flow-type systems in supercritical water. *Cellulose* 2002, 9, 301–311.

Klinke, H. B.; Thomsen, A. B.; Ahring, B. K. Inhibition of ethanol-producing yeast and bacteria by degradation products produced during pretreatment of biomass. *Appl Microbiol Biotechnol* 2004, 66, 10–26.

Kumagai, S.; Hayashi, N.; Sasaki, T.; Nakada, M.; Shibata, M. Fractionation and saccharification of cellulose and hemicellulose in rice hull by hot-compressed-water treatment with two-step heating. *J Jpn Inst Energy* 2004, 83, 776–781.

Kumar, P.; Diane, M. B.; Michael, J. D.; Stroeve, P. Methods for pretreatment of lignocellu-losic biomass for efficient hydrolysis and biofuel production. *Ind Eng Chem Res* 2009, 48, 3713–3729.

Laaden, B.; Almeiida, R. M.; Radsrom, P.; Hagerdal, B. H.; Grauslund, M. G. Identification of an NADH-dependent 5-hydroxy methyl furfural reducing alcohol dehydrogenase in *Saccharomyces cerevisiae*. *Yeast* 2008, 25, 191–198.

Leskovac, V.; Trivić, S.; Pericin, D. The three zinc-containing alcohol dehydrogenases from baker's yeast. *Saccharomyces cerevisiae*. *FEMS Yeast Res* 2002, 2, 481–494.

Liu, Z. L.; Moon, J. A novel NADPH-dependent aldehyde reductase gene from *Saccharomyces cerevisiae* NRRL Y-12632 involved in the detoxification of aldehyde inhibitors derived from lignocellulosic biomass conversion. *Gene* 2009, 446, 1–10.

Lu, X.; Yamauchi, K.; Phaiiboonsilpa, N.; Saka, S. Two-step hydrolysis of Japanese beech as treated by semiflow hot-compressed water. *J Wood Sci* 2009, 55, 367–375.

Matsufuji, Y.; Fujimura, S.; Ito T.; Nishizawa, M.; Miyaji, T.; Nakagawa, J.; Ohyama, T.; To-
mizuka, N.; Nakagawa, T. Acetaldehyde tolerance in *Saccharomyces cerevisiae* involves the
pentose phosphate pathway and oleic acid biosynthesis. *Yeast* 2008, 25, 825–833.

Matsushika, A.; Watanabe, S.; Kodaki, T.; Makino, K.; Sawayama, S. Bioethanol production
from xylose by recombinant *Saccharomyces cerevisiae* expressing xylose reductase, NADP+-
dependent xylitol dehydrogenase, and xylulokinase. *J Biosci Bioeng* 2008, 105, 296–299.

Michnick, S.; Roustan, J. L.; Remize, F.; Barre, P.; Dequin, S. Modulation of glycerol and ethanol yields during alcoholic fermentation in *Saccharomyces cerevisiae* strains overexpressed or
disrupted for *GPD1* encoding glycerol 3-phosphate dehydrogenase. *Yeast* 1997, 13, 783–793.

Nakata, T.; Miyafuji, H.; Saka, S. Bioethanol from cellulose with supercritical water treatment
followed by enzymatic hydrolysis. *Appl Biochem Biotechnol* 2006, 129, 476–485.

Ogg, C. W. Avoiding more biofuel surprises: the fuel, food and forest trade-offs. *JDAE* 2009,
1, 12–17.

Petersson, A.; Almeida, J. R.; Modig, T.; Karhumaa, K.; Hahn-Hägerdal, B.; Gorwa-Grauslund,
M. F.; Lidén, G. 5-Hydroxymethyl furfural reducing enzyme encoded by the *Saccharomyces
cerevisiae ADH6* gene conveys HMF tolerance. *Yeast* 2006, 23, 455–464.

Remize, F.; Roustan, J. L.; Sablayrolles, J. M.; Barre, P.; Dequin, S. Glycerol overproduction by
engineered *Saccharomyces cerevisiae* wine yeast strains leads to substantial changes in by-
product formation and to a stimulation of fermentation rate in stationary phase. *Appl Environ
Microbiol* 1999, 65, 143–149.

Robinson, M. D.; Grigull, J.; Nohammad, N.; Hughes, T. R. FunSpec: a web-based cluster interpreter for yeast. *BMC Bioinformatics* 2002, 3, 35.

Saint-Prix, F.; Bönquist, L.; Dequin, S. Functional analysis of the ALD gene family of *Saccharomyces cerevisiae* during anaerobic growth on glucose: the NADP+-dependent Ald6p and
Ald5p isoforms play a major role in acetate formation. *Microbiology* 2004, 150, 2209–2220.

Sun, Y.; Cheng, J. Hydrolysis of lignocellulosic materials for ethanol production: a review. *Bioresour Technol* 2002, 83(1), 1–11.

Wahlbom, C. F.; Hahn-Hägerdal, B. Furfural, 5-hydroxymethyl furfural, and acetoin act as external electron acceptors during anaerobic fermentation of xylose in recombinant *Saccharomyces cerevisiae*. *Biotechnol Bioeng* 2002, 78, 172–178.

Walfridsson, M.; Bao, X.; Anderlund, M.; Lilius, G.; Bülow, L.; Hahn-Hägerdal, B. Ethanolic
fermentation of xylose with *Saccharomyces cerevisiae* harboring the *Thermus thermophilus
xylA* gene, which expresses an active xylose (glucose) isomerase. *Appl Environ Microbiol*
1996, 62, 4648–4651.

Wang, X.; Mann, C. J.; Bai, Y.; Ni, L.; Weiner, H. Molecular cloning, characterization, and potential roles of cytosolic and mitochondrial aldehyde dehydrogenases in ethanol metabolism
in *Saccharomyces cerevisiae*. *J Bacteriol* 1998, 180, 822–830.

Yu, Y.; Lou, X.; Wu, H. Some recent advances in hydrolysis of biomass in hot compressed water
and its comparison with other hydrolysis methods. *Energ Fuel* 2007, 22, 46–60.

CHAPTER 10

BIOTECHNOLOGICAL POTENTIAL OF ALKALIPHILIC MICROORGANISMS

PRADNYA PRALHAD KANEKAR, AMARAJA ABHAY JOSHI, SNEHAL OMKAR KULKARNI, SUCHITRA BABURAO BORGAVE, SEEMA SHREEPAD SARNAIK, SMITA SHRIKANT NILEGAONKAR, ANITA SATISH KELKAR, and REBECCA SANDEEP THOMBRE

CONTENTS

10.1   INTRODUCTION

Microorganisms play an important role in development of biotechnological processes like production of bioactive molecules like enzymes, biomaterials like exopolysaccharides, biodegradable plastic, secondary metabolites like antimicrobial compounds and environmental biotechnological processes like bioremediation and bioaugmentation. Many 'moderate' microorganisms are known to contribute to industrial and environmental biotechnology. Last four decades have witnessed recognition of unusual microorganisms, extremophiles, which live in extreme environments of temperature, pH and concentration of salt, sugar and atmospheric pressure. They follow different metabolic pathways and therefore are looked upon as a new source of biotechnological products. In this chapter, biotechnological potential of alkaliphilic microorganisms is envisaged. In general, the microorganisms widely occur in normal environment such as neutral pH, temperature of 30 to 37°C. However, on earth, some unusual environments like extreme acidic or alkaline pH and very low or high temperature are created due to geological and other events. It was thought that such extreme environments may not support life (Horikoshi, 1991). Microorganisms that get adapted to such extreme conditions are found to grow in such unusual environments and are termed as extremophiles such as alkaliphiles, halophiles, thermophiles, acidophiles indicating the type of habitat that they live in (Horikoshi, 1991).

10.2   ALKALIPHILES

Alkaliphiles are interesting group of extremophilic microorganisms that thrive at pH of 9.0 and above. Since many microorganisms show more than one pH optimum for growth in response to different growth conditions such as nutrients and temperature, it is difficult to define and differentiate between alkaliphilic and alkalitolerant microorganisms. Even then, alkaliphiles are defined as the microorganisms that grow optimally at pH 9 and above, mostly in the range of 10 to 12 and do not grow or grow only slowly at near neutral pH (Horikoshi, 1999).

10.2.1   HISTORY OF ALKALIPHILES

In Japan, since ancient times, alkaliphilic microorganisms have been in use in reduction of a basic dye indigo from indigo leaves at very high alkaline pH. This process was called as indigo fermentation. Horikoshi and Akiba (1982) however rediscovered these alkaliphilic microorganisms. Many alkaliphilic microorganisms were isolated and explored for alkaline enzymes especially alkaline protease (Horikoshi 1971a, b). Since then, alkaliphilic microorganisms have been studied for their taxonomy, physiological and ecological aspects, their enzymes, genetics and molecular biology.

10.2.2 HABITATS OF ALKALIPHILES

Alkaliphiles have been isolated from various natural and manmade environments. Alkalinity in the terrestrial and aquatic environments may be generated by a number of factors, either naturally occurring or brought about by man's activities. Alkaline environments are created by commercial processes like cement manufacturing, production of indigo dye, mining operations, paper and pulp production and food processing effluents. Soda lakes and soda deserts are the most stable and worldwide naturally occurring alkaline environments, which are in remote places and therefore not easily approachable, prohibiting their detailed and systematic study (Jones *et al.*, 1998). The best-studied soda lakes are those of East African Rift valley (Grant *et al.*, 1990; Jones *et al.*, 1994). Microbiological studies of central Asian soda lakes have also been well documented (Issatchencko, 1951; Zhilina and Zavarzin, 1994; Joshi *et al.*, 2005, 2007, 2008).

10.3 ALKALIPHILIC MICROORGANISMS AS A SOURCE OF BIOMOLECULES

Alkaliphilic microbes possess potential applications in various fields of biotechnology and hence are being globally studied. As mentioned earlier, alkaliphilic microorganisms have been investigated in last 40 years for production of biomaterials, secondary metabolites, enzymes and development of bioremediation processes.

10.3.1 EXOPOLYSACCHARIDE (EPS)

Microorganisms synthesize polysaccharides, which are water-soluble polymers. They may be ionic or non ionic in nature. These polysaccharides are formed of very regular, branched or unbranched repeating units interconnected by glycosidic linkages. Microbial EPS vary considerably in their composition resulting in different chemical and physical properties. Some are neutral macromolecules, but the majorities are polyanionic due to the presence of either uronic acids or ketal-linked pyruvate. Polyanionic status is attributed to inorganic residues, such as phosphate or rarely sulfate (Sutherland, 1990). A very few EPS may even be polycationic. The primary conformation of the EPS depends upon composition and structure of the polysaccharides. In some of these polymers, rigidity is considerably conferred by the backbone composition of sequences of 1, 4-β- or 1, 3-β-linkages, for example in cellulosic backbone of xanthan from *Xanthomonas campestris*. Other linkages in polysaccharides may yield more flexible structures like 1, 2-α- or 1, 6-α-linkages found in many dextrans. The presence or absence of acyl substituents such as *O*-acetyl or *O*-succinyl esters or pyruvate ketals influences the transition in solution from random coil to ordered helical aggregates (Sutherland, 1997). The polysaccharides are essentially very long, thin molecular chains with molecular mass of the order of $0.5–2.0 \times 10^6$ Da.

In the recent years, a major emphasis has been laid on search for novel microbial polysaccharides having many interesting physicochemical and rheological properties. Functional consortia are formed by microorganisms through biosynthesis of extracellular polymeric substances or exopolysaccharide (EPS), which is attached to solid surfaces or released in the form of slime. Bacteria attach to surfaces with the help of EPS thus, improving nutrient acquisition. EPS may be having a role in protection from environmental stresses and host defenses (Iqbal *et al.*, 2002). Production of EPS by microbes is being studied since long because of its many interesting physicochemical and rheological properties with novel functionality. Microbial polysaccharides find applications in the food, pharmaceutical, petroleum and other industries (Sutherland 1990, 1998; Desai and Banat, 1997; Tombs and Harding, 1998). They are also found to remove heavy metals from industrial effluents and potable water (Norberg and Perrsson, 1984). Some of the applications of EPS are listed in Table 10.1. Microbial polysaccharides have different properties than that of traditional polymers of plant or algal origin.

TABLE 10.1   Established applications of microbial EPS.

Types of EPS	Applications
ß–D-Glucans	Antitumour agents
Hyalouronic acid	Eye and joint surgery
Bacterial cellulose	Wound dressings
Curdlan, Pullulan, Scleroglucan	Oligosaccharide preparation
Xanthan	Food, thixotropic paints, paper coating
Dextran	Food
Gellan	Gelling agents in food

Extreme environments become a rich source of microorganisms, which synthesize novel molecules (Arias *et al.*, 2003). It has been hypothesized that EPS protects microorganisms from stress in extreme habitats (Nicholaus *et al.*, 1999a, b). EPS forms a substantial component of the extracellular polymers surrounding most microbial cells in extreme environments. The extremophiles adapt themselves to compensate for the deleterious effects of extreme conditions such as high temperature, salt, low pH or temperature, high radiation. Biosynthesis of EPS is one of the most common adaptation strategies. The unusual metabolic pathways found in some extremophiles make them as potential producers of EPSs with novel and unique characteristics and functional activities under extreme conditions. Some of the alkaliphilic and halophilic organisms producing EPS are listed in Table 10.2. Though the knowledge on the structural and rheological properties of EPS from extremophiles is limited, it indicates a variety in properties, not observed in traditional polymers.

TABLE 10.2 Exopolysaccharide (EPS) producing alkaliphilic bacteria with potential industrial applications.

Organisms	Application	References
Alkaliphilic *Bacillus* spp.	Not reported	Carsaro *et al.* (1999)
Halomonas alkaliantartica	Not reported	Poli *et al.* (2007)
Haloalkaliphilic *Bacillus* sp.	Coating material for drugs	Ganesh kumar *et al.* (2004)
Halomonas alkaliphila	Not reported	Romano *et al.* (2006)
Vagococcus carniphilus	Flocculation	Joshi and Kanekar (2011)

There are a few reports on production of exopolysaccharides by alkaliphilic bacteria. EPS produced by alkaliphilic *Bacillus* spp. isolated from Lake Natron has been characterized (Carsaro *et al.*, 1999) and found to have D-galactopyranuronic acid (GalpA), 2,4-diacetamido-2,4,6-trideoxy-D-glucopyranose (QuipNAc4NAc), 2-acetamido-2-deoxy D-mannopyranuronic acid (ManpNAcA) and one uncommon unit of D-galactopyranuronic acid with the carboxyl group amide-linked to glycine [GalpA(Gly)]. Haloalkaliphilic *Bacillus* isolated from soil samples of heavily polluted tidal mudflats of the Korean Yellow Sea produced an exopolysaccharide with very good flocculating activity (Ganesh Kumar *et al.*, 2004). Production of EPS by *Halomonas alkaliphila* isolated from salt pool Campania, Italy has been reported (Romano *et al.*, 2006). EPS producing *Oceanobacillus oncorhynchi* subsp. *incaldanensis* has been studied by Romano *et al.* (2006). EPS-producing *Vagococcus* sp. W31 was isolated and characterized from a sewage sample collected from Little Moon River in Beijing (Gao *et al.*, 2006). The organism showed good flocculating activity for Kaolin clay thus having a potential application in wastewater treatment. EPS producing *Halomonas alkaliantarctica*, isolated from cape Russell in Antarctica has been reported (Poli *et al.*, 2007). *Vagococcus carniphilus*, isolated from sediment sample of the alkaline Lonar lake, India produced exopolysaccharide with good flocculating activity and showed 75% similarity with dextran (Joshi and Kanekar, 2011). Production of biosurfactant from an alkaliphilic bacterium *Cronobacter sakazakii* was studied (Jain *et al.*, 2012). It was found to contain total sugars (73.3%), reducing sugars (1.464%), protein (11.9%), uronic acid (15.98%) and sulfate (6.015%). It showed low viscosity with pseudoplastic rheological behavior and exhibited significant emulsification activity with oils and thus used for bioremediation of oil and hydrocarbons. The EPS production by alkaliphilic bacteria exhibited great industrial potential because of their unique physicochemical characteristics, pH stability and structural variations. There is still hope for the newer polysaccharides with novel applications in future by exploring more and more alkaliphiles.

10.3.2   POLYHYDROXYALKANOATE

Polyhydroxyalkanoates (PHAs) are polyesters of hydroxyacids (HAs). PHAs identified so far are primarily linear, head to tail polyesters composed of 3-hydroxy fatty acid monomers. PHAs are structurally simple macromolecules synthesized by certain bacteria. They are accumulated as discrete granules to the levels as high as 90% of the cell dry weight and are generally believed to play a role as a sink of carbon and reducing equivalents. Native PHA granules were isolated from cell extracts of *Bacillus megaterium* (Merrick and Doudoroff, 1964), *Azotobacter beijerinckii* (Ritchie and Dawes, 1969) and *Zooglea ramigera* (Barnard and Sanders, 1988, 1989).

10.3.2.1   DIVERSITY OF PHA

Of all PHAs, PHB is the most extensively characterized polymer, since it was the first to be discovered in 1926 by Lemoigne at the Institute of Pasteur. The diversity of bacterial polyhydroxyalkanoates has changed dramatically. Until 1970s, 3HB was considered as the only constituent of PHAs. In 1980s, PHAs having other monomers besides 3HB were shown to be accumulated by many bacteria with addition of certain precursors in the production medium. Widely studied polyhydroxyalkanoates are PHB, PHV and copolymer of PHB-*co*-PHV. Today more than 150 different monomers of PHAs are synthesized by different microorganisms.

10.3.2.2   APPLICATIONS OF PHA

PHAs have wide range of applications because of their novel features. They are natural thermoplastic polyesters thus becoming suitable replacements for petrochemical polymers currently in use for packaging and coating application. To begin with, the focus was on molding applications particularly for consumer packaging items such as bottles, cosmetic containers, pens and diaper back sheets. PHAs also have been processed into fibers, which then were used to construct materials such as non-woven fabrics. They have also been described as hot melt adhesives. In agriculture, they may find applications in encapsulation of seeds and encapsulation of fertilizers for slow release. PHAs have numerous biomedical applications. The main advantage in the biomedical field is that a PHA can be introduced in human body and is not required to be removed again. Since PHA is a product of cell metabolism, it has an ideal biocompatibility (Zinn *et al.*, 2001). In the pure form or as a composite, PHAs are used as sutures, swabs, repair patches, orthopedic pins, adhesion barriers, stents, nerve guides and bone marrow scaffolds (Verlinden *et al.*, 2007). PHAs are used particularly as the osteosynthetic materials in the stimulation of bone growth, in bone plates, surgical sutures and blood vessel replacements. PHAs are used for the slow delivery of drugs, hormones and medicines.

10.3.2.3   BIODEGRADABILITY OF PHA

Besides the typical polymeric properties of PHA, an important characteristic of PHA is their biodegradability. Microorganisms in nature are able to degrade PHAs using PHA hydrolases and PHA depolymerases. Microorganisms colonize on the surface of the polymer and secrete enzymes, which degrade PHB or PHB-*co*-PHV into HB and HB and HV units, respectively. These monomers are then used up by the cell as a carbon source for the biomass growth. The rate of polymer biodegradation depends on the variety of factors such as surface area, pH, temperature, moisture and pressure. The end products of PHA degradation in aerobic environment are CO_2 and water while methane is produced in anaerobic condition (Ojumu *et al.*, 2004).

10.3.2.4   PRODUCTION OF POLYHYDROXYALKANOATE USING DIFFERENT MICROORGANISMS

Among the candidates for biodegradable plastics, PHAs have attracted much attention because of their material properties similar to conventional plastics and complete biodegradability. For past two decades, the whole research was focused on the preparation of biodegradable polymer. There exist a number of reports on production of PHAs (Anderson and Dawes, 1990; Mergaert *et al.*, 1992a; Lee, 1996; Madison and Huisman, 1999; Lee *et al.*, 1999; Ojumu *et al.*, 2004; Khanna and Srivastava, 2005; Verlinden *et al.*, 2007). The fact that PHA can be produced from renewable resources, it does not lead to the depletion of finite resources and its good processibility make PHA suitable for applications in several areas as a partial substitute for non biodegradable synthetic polymers. Over 250 different bacteria including Gram-negative and Gram-positive species phylogenetically representing both eubacteria and archaea have been reported to accumulate various PHAs (Steinbuchel, 1991). Some of the common organisms producing PHA are *Cupriavidus necator*, *Azotobacter vinelandi*, *Alcaligenes lactus*, *Pseudomonas* sp. and *Bacillus* sp.

10.3.2.5   PRODUCTION OF PHA BY EXTREMOPHILIC BACTERIA

Most of the microorganisms reported so far for production of PHA are isolated from moderate environments. Extremophilic microorganisms such as halophiles and alkaliphiles are only recently being explored for production of PHA. The first report of extremophilic organism producing PHB was of halophilic archaebacterium, *Haloferax mediterranei* (Lillo and Rodriguez-Valera, 1990). This archon was further studied by Chen *et al.* (2006) to produce PHB-*co*-PHV from enzymatic extruded starch. Quillaguaman *et al.* (2005) have shown the potential of *Halomonas boliviensis* as a source of PHB from simple substrates such as butyric acid, glucose and sucrose.

10.3.2.6   PRODUCTION OF PHA BY ALKALIPHILIC BACTERIA

There are only a couple of reports of alkaliphilic bacteria producing PHA. Joshi and Kanekar *et al.* (2011) have investigated alkaliphilic strain of *Bacillus cereus* isolated from alkaline Soda Lake of Lonar, India for production of a copolymer PHB-*co*-PHV using glucose as a simple and single carbon source. Kulkarni *et al.* (2010, 2011) have reported production of PHA by moderately haloalkalitolerant strain of *Halomonas campisalis* MCM B-1027 isolated from alkaline soda lake of Lonar, India. The polymer produced is found to be a copolymer of PHB-*co*-PHV from a single carbon source as maltose, biodegradable in nature and having tensile strength comparable to the reported organisms. 10–100 µ thick films could be drawn from the polymer by solvent casting technique and leak proof bags prepared from the polymer film could be used as packaging material for food commodity items. Many more alkaliphilic bacteria should be explored for this ecofriendly biodegradable plastic.

10.3.3 ANTIMICROBIALS FROM ALKALIPHILES

Microbial natural products become the source of most of the antibiotics today. Microorganisms have evolved to produce antimicrobials in response to external stimuli and stresses such as other microbial competitors. The scarcity of new antibiotics currently under development in the pharmaceutical industry is alarming. Still, microbial natural products remain the most promising source of novel antibiotics. However, new approaches are required to improve the efficiency of the discovery process. Microbes from extreme environments have attracted considerable attention in recent years. It is difficult to believe that extremophiles well produce antimicrobials as they are likely to have few competitors in extreme environment. Microorganisms produce a large number of chemicals for no obvious role. However, any large collection of chemicals will contain molecules having high biological activity. Indeed, the ability to produce and retain a rich chemical diversity surely enhances the production of the very rare compounds. Many microorganisms produce chemicals with potent antibiotic properties, but only some of them are suitable for commercial production. In general, the extremophilic bacteria follow different metabolic pathways thus leading to formation of a variety of secondary metabolites like antibiotics, growth promoters and enzyme inhibitors. In recent years, extremophiles have been identified as a treasure of novel bioproducts including antimicrobials (Da Costa *et al.*, 1988; Horikoshi, 1995). Little information is available on production of antimicrobial compounds from alkaliphilic bacteria. This information emphasizes the value screening of novel natural products and derived products as lead compounds from alkaliphiles. The alkaliphilic microorganisms represent a relatively unexplored resource, and may provide valuable antimicrobials (Horikoshi, 1999). The discovery of new antimicrobial drugs is greatly influenced by the need to develop new agents active against organisms resistant to earlier generations of drugs (Hancock and Strohl, 2001).

Some alkaliphilic bacteria are reported for the production of antimicrobial compounds as described in Table 10.3. Since the discovery of alkaliphiles, many Japanese pharmaceutical companies have tried using alkaline media to isolate new microorganisms having ability to produce new antibiotics (Sato *et al.*, 1983; Tsujibo *et al.*, 1988). Reports of antibiotic production under alkaline conditions are very less, which could be due to the instability exhibited by antibiotics under higher pH (Horikoshi, 1999). A novel alkaliphilic *Streptomyces* strain AK409 produced pyrocoll, an antibiotic, antiparasitic and antitumor compound (Dieter *et al.*, 2003). An alkaliphilic *Streptomyces sannanensis* strain RJT-1 secreted a potent antibiotic, which showed activity against Gram positive bacteria (Vasavada *et al.*, 2006). Lawton and co-workers, in 2008, first reported the production of a two-peptide lantibiotic (lanthionine-containing peptide antibiotic) haloduracin by an alkaliphilic *Bacillus halodurans* C-125. Haloduracin was active against a wide range of Gram positive bacteria. Alkaliphilic and salt tolerant *Streptomyces tanashiensis* strain A2D isolated from soil of phoomdi in Loktak Lake of Manipur, India produced bioactive metabolite, which showed activity against *Candida albicans* and *Fusarium moniliforme* (Singh *et al.*, 2009).

TABLE 10.3   Alkaliphilic bacteria producing antimicrobial compounds.

Alkaliphilic bacteria	Antimicrobial compound produced	Inhibitory activity against	Reference
Paecilomyces lilacinus	Peptide antibiotic No. 1907-VIII	*Bacillus subtilis, Staphylococcus aureus, Pseudomonas aeruginosa, Escherichia coli, Candida albicans, Aspergillus niger*	Sato *et al.* (1983)
Nocardiopsis dassonvillei OPC-15	Antibiotics I and II	*Proteus mirabilis, Bacillus subtilis*	Tsujibo *et al.* (1988)
Corynebacterium sp. YUA 25	*N*-2-methylbutanoyl tyramine	Aldose reductase inhibitor	Bahn *et al.* (1998)
Streptomyces sp. AK409	Pyrocoll	*Arthrobacter* strains, Filamentous fungi, pathogenic protozoa, human tumor cell lines	Dieter *et al.* (2003)

TABLE 10.3 *(Continued)*

Alkaliphilic bacteria	Antimicrobial compound produced	Inhibitory activity against	Reference
Streptomyces sannanensis RJT-1	Antibiotic	*Staphylococcus aureus, Bacillus cereus, Bacillus megaterium, Bacillus subtilis*	Vasavada *et al.* (2006)
Bacillus halodurans C-125	Haloduracin	*Lactobacillus, Listeria, Streptococcus, Enterococcus, Bacillus, Pediococcus*	Lawton *et al.* (2008)
Streptomyces tanashiensis A2D	Bioactive metabolite	*Candida albicans, Fusarium moniliforme*	Singh *et al.* (2009)
Streptomyces sp.	Antibacterial substance	*Bacillus subtilis, Staphylococcus aureus, Escherichia coli, Proteus vulgaris, Salmonella typhi*, human lung carcinoma A549 cell line	Kharat *et al.* (2009)
Synechocystis aquatilis	Antimicrobial compound	*Staphylococcus aureus, Proteus vulgaris, Pseudomonas aeruginosa, Bacillus subtilis, Escherichia coli*	Deshmukh and Puranik, 2010
Nocardiopsis sp. YIM DT266	Naphthospironone A	*Bacillus subtilis, Staphylococcus aureus, Escherichia coli, Aspergillus niger* and HeLa, L929, AGZY cells	Ding *et al.* (2010)
Streptomyces aburaviensis Kut-8	Antibiotic	*Staphylococcus aureus, Bacillus cereus, Bacillus megaterium, Bacillus subtilis*	Thumar *et al.* (2010)

TABLE 10.3   *(Continued)*

Alkaliphilic bacteria	Antimicrobial compound produced	Inhibitory activity against	Reference
Bacillus halodurans B20	Antibacterial compound	*Staphylococcus aureus, Enterococcus faecium, Enterococcus faecalis, Streptococcus* sp.	Danesh, 2011
Marinilactibacillus psychrotolerance ALK 9, *Facklamia tabacinasalis* ALK	Antimicrobial compound	*Listeria* sp.	Roth *et al.* (2011)
Planococcus sp., *Oceanobacillus* sp., *Alcanivorax* sp.	Antibacterial substance	*Bacillus subtilis, Pseudomonas aeruginosa, Staphylococcus aureus, Escherichia coli, Proteus vulgaris*	Deshmukh *et al.* (2011)
Bacillus asahii, B. gibsonii, B. pseudofirmus, B. pumilus, B. flexus, B. lehensis	Antibacterial substance	*Escherichia coli, Klebsiella pneumoniae, Proteus vulgaris, Enterococcus aerogenes, Staphylococcus aureus*	Tambekar and Dhundale, 2012

Alkaliphilic microorganisms isolated from Lonar Lake, India have been reported for production of antimicrobial compounds (Kharat *et al.*, 2009; Deshmukh and Puranik, 2010; Deshmukh *et al.*, 2011; Tambekar and Dhundale, 2012). Recently, the production of Naphthospironone A (1) from alkaliphilic *Nocardiopsis* sp. (YIM DT266) was reported by Ding *et al.* (2010). This metabolite showed cytotoxic and antibiotic activity. A halotolerant alkaliphile *Streptomyces aburaviensis* Kut-8, isolated from saline desert of Kutch, Western India is reported to secret antibiotic active against Gram-positive bacteria (Thumar *et al.*, 2010). *Bacillus halodurans* B20 isolated from samples collected in eastern Africa produced antibacterial compounds (Danesh, 2011). Facultative anaerobic halophilic and alkaliphilic bacteria *Marinilactibacillus psychrotolerance* ALK 9 and *Facklamia tabacinasalis* ALK 1 isolated from a natural smear ecosystem inhibit Listeria growth in early ripening stages (Roth *et al.*, 2011). Borgave *et al.* (2012) studied alkaliphilic bacteria isolated from

Lonar Lake for their antimicrobial activity against clinical pathogens (*Salmonella typhi*, *Pseudomonas aeruginosa*, *Escherichia coli*, *Klebsiella pneumoniae* and *Staphylococcus aureus*) and phytopathogenic fungi (*Fusarium oxysporum*, *Fusarium moniliforme*, *Aspergillus parasiticus*, *Rhizoctonia solani* and *Colletotrichum gloeosporioides*). Out of 78 alkaliphilic bacteria, 25 strains showed either antibacterial or antifungal activity. The culture supernatants of 25 isolates were tested using agar well diffusion method out of which, moderately alkaliphilic halotolerant *Halomonas campisalis*, *Planococcus maritimus* and *Paenibacillus* sp. L55 stood out inhibitor of nine of the 10 indicator microorganisms. An extensive exploratory screening of alkaliphiles for antibiotics is very important for newer antimicrobial compounds.

10.3.4 PROTEASES FROM ALKALIPHILIC BACTERIA

Microbes live in an environment where the nutrients are mainly macromolecular in nature. These nutrients are not utilizable by the microbes unless cleaved into smaller molecules that they can absorb. The cleavage of macromolecular nutrients into smaller molecules is accomplished by the enzymes secreted by the microbes themselves. For the cleavage of proteins or polypeptides in the environment, microbes produce extracellular and intracellular proteases. Proteases have applications in detergent, food, pharmaceutical and leather tanning (Taylor and Richardson, 1979). Currently, only 1–2% of the microorganisms on the Earth have been commercially exploited and among these there are only a few examples of extremophiles (Gomes and Steiner, 2004). Fabulous work is going on globally on proteases secreted by alkaliphilic bacteria. Most of the work is on the genus *Bacillus*. Alkaliphilic bacteria have been described previously for production of proteases (Horikoshi 1971 a, b; Manachini *et al.*, 1988; Takami *et al.*, 1989, 1990; Chaphalkar and Dey, 1994; Kwon *et al.*, 1994). Most of the proteases are with stability in alkaline pH and temperature stability thus indicating their potential in detergent industry.

10.3.4.1 DETERGENT PROTEASE

Detergents are chemical compounds used for cleaning purposes. However, stains of blood, soil being proteinaceous in nature are not easily removed by detergent chemicals. Proteases have important role in enhancing cleaning performance of detergents. Since the pH of detergents is highly alkaline, alkali stable proteases produced by alkaliphilic bacteria bear importance in detergent industries. Alkaliphilic bacteria *Arthrobacter ramosus* and *Bacillus alcalophilus* isolated from sediment samples of the alkaline Lonar Lake, India produced protease using soyacake as a sole source of carbon and nitrogen. The enzyme was thermo stable upto 65°C, stable at pH 7–12 and also active in the presence of commercial detergent. This enzyme removed bloodstains from cotton fabric indicating its potential use in detergent formulations (Kanekar *et al.*, 2002). Alkaline protease of *Bacillus pumilus* exhibited optimum pH 11.5 and temperature 50–60°C (Kumar, 2002). The crude serine alkaline protease of *Bacillus* strains, isolated under extreme alkaline conditions from Izmir, Turkey, was bleach stable, with an optimum temperature of 60°C and a pH of 11, indicating its potential applications in the detergent industry (Genckal and Tari, 2006). An oxidant and SDS-stable alkaline protease secreted by *Bacillus clausii* I-52 was found to be highly compatible and stable to the commercial detergent components and their preparations. Analysis of wash performance using EMPA test fabrics revealed that protease removed protein stains in the presence of commercial detergents as well as surfactants efficiently (Joo *et al.*, 2006). An extracellular alkaline protease from a novel haloalkaliphilic bacterium *Bacillus pseudofirmus* (Ve1) was purified to the homogeneity and found to have a molecular weight of 30–32 kDa. The study on this enzyme assumes significance in the light of dual extremities of pH and stability to salt and moderate temperature (Patel *et al.*, 2006). High stability of AprB towards surfactants and oxidizing agents, an optimal pH of 10.0, and an optimal temperature of 60°C suggest that this high alkaline protease bears potential industrial applications (Deng *et al.*, 2010). An alkaline protease, 31.0 kDa from *B. cereus*, showed maximum activities at pH 10, 50°C and only Cu^{2+} ions enhanced the relative enzyme activity up to 112%. The application of alkaline protease for the removal of bloodstains from cotton cloth indicates its potential use in detergent formulations. The protease showed excellent stability in the presence of locally available detergents and retained about 60% of its activity with most of them even after 3 h of incubation at temperature of 50°C (Abou-Elela *et al.*, 2011). Alkaline protease of *Nesterenkonia* sp. AL-20 was active at optimum temperature 74°C and pH 10.0 with high stability against H_2O_2 and sequestering agents (Bakhtiar *et al.*, 2003).

10.3.4.2   DEHAIRING PROTEASE

Leather industry is one of the large-scale industries, which use a number of chemicals for different steps in leather making. Dehairing of animal hide is conventionally done using chemicals leading to environmental pollution. Enzymatic dehairing is a clean process and hence being adapted by many tanneries.

Dehairing proteases are commercially available yet research is continued for better activity. Apart from detergent industry the crude enzyme preparation of *Bacillus cereus* VITSN04 has shown effective dehairing of goat skins, which is an important step in leather processing (Sundararajan *et al.*, 2011). Thermostable alkaline protease produced from *Bacillus* sp. JB 99 exhibited keratinolytic and dehairing activity to buffalo and goat hide without damaging the collagen layer, thus making it a potential candidate for application in leather industry to avoid pollution problem caused by the use of chemicals in the industry (Shrinivas and Naik, 2011). The proteolytic activity of *Bacillus* sp. CA15 has a wide range of pH, at low temperature and compatability to surfactants, indicating that the organism could be a potential source of alkaline protease for use in additive in detergent formulation or in the leather industry (Uyar *et al.*, 2011). Patil and Chaudhari (2009) showed that purified protease of alkaliphilic *Pseudomonas aeruginosa* MTCC 7926 was resistant to surfactants, solvents, metals, and bleaching agents. The enzyme could be used in detergent formulations, dehairing of animal skin, processing of X-ray film and possibly nonaqueous enzymatic synthesis of peptides.

10.3.4.3   RECOVERY OF SILVER FROM USED X-RAY FILMS

X-ray films are prepared by using gelatin, silver nitrate and sheets of plastic. Hence, used or waste X-ray film becomes a source of precious metal silver after treatment with gelatinase. An obligate alkaliphilic *Bacillus sphaericus* strain, isolated from alkaline soils in the Himalaya, produced an extracellular protease, which was stable in presence of laundry detergents comparable to that of commercial proteases. The gelatin layer in 25 g of used X-ray films was efficiently hydrolyzed within 12 min at 50°C, pH 11.0 and 25 U protease ml^{-1} indicating efficiency in silver recovery from used X-ray films (Singh *et al.*, 1999). The processing capacity for the recovery of silver from used X-ray film was increased by employing a thermostable alkaline protease (Protease B18) obtained from a thermophilic alkaliphile (Fujiwara *et al.*, 1991).

10.3.4.4   KERATINOLYTIC PROTEASE

Keratinaceous wastes like bird feathers, animal hair are traditionally hydrolyzed using mineral acids to obtain amino acids. The chemical treatment leads to acidic waste with high nitrogen content. Keratinolytic bacteria and fungi have been extensively studied. The enzyme Keratinase is used for hydrolysis of bird feathers to get a mixture of amino acids. Keratinase of *Nocardiopsis* sp. was stable from 70–75°C and pH 11.0–11.5 (Mitsuiki *et al.*, 2002). The protease of *B. pseudofirmus* has potential for the enzymatic and/or microbiological hydrolysis of feather to be used as animal feed supplement (Gessesse *et al.*, 2003). The enzyme produced from immobilized cells of thermoalkaliphilic *Bacillus halodurans* JB 99 efficiently degraded chicken feathers in the presence of a reducing agent suggesting its application in the management of keratin-rich waste and obtaining value-added products in the poultry industry (Shrinivas *et al.*, 2012).

10.3.4.5   THERMOSTABLE PROTEASE

The thermostable alkaline protease from alkaliphilic *Bacillus pumilus* MK6–5 has application in ultrafiltration membrane cleaning due to its activity in broad pH and temperature ranges, and tolerance to detergents, unlike the mesophilic proteases, which face these limitations (Kumar, 2002). The crude alkaline protease produced by *Bacillus halodurans* (gb10172612) isolated from Wadi El-Natrun, an Egyptian soda lake, showed reasonable activity at temperature range of 65 to 75°C with maximum activity at 70°C and had a relatively wide pH range of activity between pH 8 to 11, with maximum enzyme activity at pH 10 in 50 mM Tris-HCl buffer at 70°C, indicating the thermo-alkaline nature of the proteases (Ibrahim, 2007). *Serratia liquiefaciens* collected from different alkaline environments around Orissa state, India indicated the potential use of these microorganisms as biotechnological tools for various industrial activities especially in production of protease (Smita *et al.*, 2012).

10.3.4.6   RECOMBINANT PROTEASE

To obtain a new serine protease from alkaliphilic *Bacillus* sp. NKS-21, shotgun cloning was carried out. The cloned intracellular protease was expressed in *Escherichia coli* purified and characterized. The enzyme showed stability under alkaline condition at pH 10 and tolerance to surfactants (Yamagata *et al.*, 1995). The demand of alkaline proteases in detergents, leather industry, etc. continues and hence more and more alkaliphiles should be explored for proteases.

10.3.5   CYCLODEXTRIN GLYCOSYL TRANSFERASE

Cyclodextrin glycosyl transferase (CGTase; EC 2.4.1.19) is an enzyme that converts starch into cyclodextrins (CDs), which are closed-ring structures having six or more glucose units joined by means of α-1,4 glucosidic bonds. CGTases are classified in the α-amylase family and are known to catalyze four different transferase reactions: cyclization, coupling, disproportionation, and hydrolysis (Wind *et al.*, 1995). Three major types of cyclodextrins are produced by CGTases depending on number of glucose units, α-CD, β-CD and γ-CD (Li *et al.*, 2007). In 1969, Corn Products International started producing CD's using CGTase from *Bacillus macerans* (Horikoshi, 1999). The structural and genetic studies on CGTase were reported in the late 1980s. Kaneko *et al.* (1988) elucidated the nucleotide sequence and studied the cloning of genes coding for CGTase. Maekelae *et al.* (1988) purified and studied the enzyme in the same year. Kimura *et al.* (1989) proposed an additional polypeptide in the -COOH region of CGTase. Nakamura *et al.* (1992) have studied functional relationships between Cyclodextrin glycosyl transferase from an alkaliphilic *Bacillus* and alpha-amylases and have reported that catalytic mechanisms of both enzymes are similar. Nakamura *et al.* (1993) investigated the active residues in the substrate binding sites of CGTase and identified them as three histidine residues. Haga *et al.* (1994) crystallized CGTase from alkaliphilic *Bacillus* and performed some preliminary experiments on X ray diffraction studies.

10.3.5.1   APPLICATIONS OF CYCLODEXTRINS

The first Cyclodextrin-containing pharmaceutical product was marketed in Japan in 1980's. Later Cyclodextrin-containing products appeared on the European and American market in 1997 (Szejtli, 2004). New Cyclodextrin-based technologies are constantly being developed and thus even 100 years after their discovery, cyclodextrins are still regarded to have novel unexplored potential. Cyclodextrins have a variety of applications in food, pharmaceutical and cosmetic industry. Cyclodextrins are used as drug carriers and tableting vehicles. They are used to reduce the bitter or irritant taste and bad smell of drugs. Cyclodextrins can improve the stability of active pharmaceutical ingredients and increase the shelf life of drugs (Szejtli, 2004). They can improve the cord strength of polyester fibers used for reinforcement of rubbers (Szejtli, 2004). It is observed that there is a distinct increase in the rate of formation of penicillin-G from phenyl acetic acid and 6-aminopenicillanic acid when both the substrates are in a γ-Cyclodextrin complexed form. In environmental biotechnology, Cyclodextrin is used to remove organic pollutants and heavy metals from soil, water, and atmosphere. Fava *et al.* (1998) found that γ-Cyclodextrin has the potential in the bioremediation of chronically polychlorinated biphenyl-contaminated soils. Cyclodextrins are declared to be "Generally Recognized as Safe" (GRAS) and have no adverse effects on the absorption of certain nutrients (Munro *et al.*, 2004). It has been proposed that they can be used in all kinds of food and nutraceutical applications as a food ingredient and additive. They can also stabilize emulsions of fats and oils. This property is useful for the preparation of bread spreads, dairy ice creams and breads (Munro *et al.*, 2004).

10.3.5.2 ALKALIPHILES AS CANDIDATES FOR PRODUCTION OF CGTASE

It is evident from the above-mentioned applications that CGTase enzyme is industrially a very important enzyme because of its application in production of CD's. However, there are two problems associated with CGTase production, namely yield of cyclodextrin is low (hence cost is high) and toxic solvents like trichloroethylene and bromobenzene are used to precipitate CD's due to low conversion rate (Horikoshi, 1999). CGTase from alkaliphilic *Bacillus* sp. overcame all these problems and led to mass production of α-CD, β-CD and γ-CD (Horikoshi, 1999). Since then it is known that alkaliphiles are candidate organisms for CGTase production. Thatai *et al.* (1999) have described the enzyme from *Paenibacillus pabuli*, *P. graminis* and *Thermoanaerobicum*. Antranikian *et al.* (2009) have isolated and identified gene for CGTase production from extremophile *Anaerobranca gottschalkii*, an extreme thermo alkaliphile. Thombre and Kanekar (2011) have reported CGTase production from *Exiguobacterium aurantiacum*, *Paenibacillus* sp. L55, Lake Bogoria isolate 25 B, *Bacillus firmus*, *Bacillus licheniformis* and *Bacillus fusiformis* isolated from Lonar Lake. On going research now emphasizes on screening novel extremophiles for CGTase production and exploration of novel applications of cyclodextrins due to their unique structure.

10.4   ALKALIPHILES IN ENVIRONMENTAL BIOTECHNOLOGY

Environmental pollution is caused by industrial waste waters containing toxic chemical compounds. Some of the wastewaters are alkaline in nature for example phenol bearing wastewaters, wastewaters generated in chlor-alkali industries, potato processing units, units manufacturing and using indigo dye and electroplating units. These wastewaters may contain either organic pollutants like phenols, chlorophenols, pesticides or inorganic pollutants like cyanides, thiocyanates, heavy metals like chromium and copper. It is convenient to use alkaliphilic microorganisms for treatment of these wastewaters. Some reports are available in the literature on bioremediation of alkaline wastewaters using alkaliphilic microorganisms.

10.5 BIOREMEDIATION OF CHEMOPOLLUTANTS USING ALKALIPHILIC BACTERIA

10.5.1 PHENOLIC COMPOUNDS

Phenols are used in a number of chemical processes as raw material for example in production of methyl violet. Due to its high solubility, phenol appears in the wastewater generated in the phenol producing and using industries. Kanekar *et al.* (1999) carried out some work on degradation of phenol by alkaliphilic bacteria. Since some of the phenol-bearing industrial wastewaters are alkaline in nature, it was thought worthwhile to explore the use of alkaliphilic bacteria like species of *Pseudomonas*, *Arthrobacter*, *Bacillus*, *Micrococcus*, *Citrobacter*, etc. isolated from Lonar Lake, India for the removal of phenol. Muller *et al.* (1998) isolated several Gram-negative bacterial strains from concrete debris of a demolished herbicide production plant, exhibiting ability to degrade 2,4-dichlorophenol (DCP), 4-chloro-2-methylphenol (MCP) and 4-chlorophenol (4-CP), and 2-chlorophenol (2-CP) by the strains of Gram negative bacteria closely related to *Ochrobactrum anthropi*. The strains displayed alkaliphilic nature with optimum DCP/MCP degradation at pH values around 8.5–9.5.

10.5.2 PESTICIDES

2,4-dichlorophenoxyacetic acid (2,4-D) is mainly used as a herbicide to kill broad leaf weeds in sugarcane, sorghum and grass. 2,4-D is readily broken down by microbes in soil and aquatic environments. Few reports are available on degradation of 2,4-D by alkaliphilic bacteria. Maltseva *et al.* (1996) isolated three 2,4-D degrading halophilic and alkaliphilic bacteria from the highly saline and alkaline Alkali Lake site in south-western Oregon contaminated with 2,4-D. The strain I-18, was the most efficient strain, degrading 2,4-D at the concentration of ~3,000 mg L^{-1} in 3 d, and could grow optimally on 2,4-D at pH 8.4–9.4 and at sodium ion concentration of 0.6–1.0 M. Kiesel *et al.* (2007) reported the potential of alkaliphilic bacteria towards chloroaromatic compounds appearing in many environmental systems as xenobiotics.

10.5.3   BENZOATES AND PYRENES

Benzoates and pyrenes are industrially important compounds having wide applications and appear in industrial effluents. Obligate alkaliphilic strains of *Bacillus krulwichiae* AM31D[T] and AM11D, which use benzoate and *m*-hydroxybenzoate, were isolated by Yumoto *et al.* (2003) from soil obtained from Tsukuba, Ibaraki, Japan, growing at pH 8–10, but not at neutral pH. Oiea *et al.* (2007) demonstrated the degradation of benzoate and salicylate by *Halomonas campisalis*, an alkaliphile and moderate halophile. Alkaliphilic *Mycobacterium* sp. strain MHP −1 isolated from soil was reported by Habea *et al.* (2004) to use pyrene as a sole source of carbon and energy. Around 50% pyrene was degraded by the isolate MHP-1 at pH 9, within 7 d when pyrene was present at the concentration of 0.1% (w/v). 4,5-phenanthrenedioic acid, 4-phenanthroic acid and phthalic acid were identified as intermediate metabolites during degradation of pyrene by alkaliphilic *Mycobacterium* sp.

10.5.4   OILS, HYDROCARBONS AND ALKANES

Oil refineries and petroleum industries generate waste containing hydrocarbons and alkanes, which contaminate soil, fresh water bodies and sea water due to leakage of oil from ships. Extensive studies have been carried out on bioremediation of oil-contaminated sites. A novel, facultatively psychrophilic alkaliphilic bacterium *Dietzia psychralcaliphila* sp. nov., growing on a chemically defined medium containing n-alkanes as the sole carbon source was isolated from a drain of a fish product-processing plant (Yumoto *et al.*, 2002). Sugimori *et al.* (2000) isolated an alkaliphilic *Dietzia* sp., strain GS-1, which degraded disodium terephthalate (DT), which degraded 19.3 mM of DT in 168 h at pH 10. Al-Awadhi *et al.* (2007) isolated alkaliphilic and halophilic oil using bacteria from the intertidal zone of the Arabian Gulf coast. The alkaliphilic oil-using bacteria belonged to the genera *Marinobacter*, *Micrococcus*, *Dietzia*, *Bacillus*, *Oceanobacillus*, and *Citricoccus*. Most of these isolates were able to use n-alkanes as sole sources of carbon and energy. The two agarolytic bacteria, viz. *Halomonas aquamarina* and *Alteromonas macleodii*, were isolated by Sorkhoh *et al.* (2010) from fouling material adhering to a navigation vehicle half-sunken in the coastal water of the Arabian Gulf and found to be slightly alkaliphilic (optimum pH 8) in nature. These two organisms were able to use hydrocarbons and oils, indicating their suitability as biological systems for bioremediation of oily marine ecosystems. Jain *et al.* (2012) studied the production of biosurfactant from an alkaliphilic bacterium *Cronobacter sakazakii,* which can be a potential candidate for bioremediation of oil and hydrocarbons. Diluted caustic (NaOH) solutions are often used in the petrochemical industry for the removal of acidic compounds, which leads to the formation of a waste product referred to as sulfidic spent caustic. De Graaff *et al.* (2011) reported the biological treatment of refinery-spent caustics under halo-alkaline conditions.

10.5.5 CHLOR-ALKALI WASTEWATER

Kulshreshtha *et al.* (2010) reported ability of *Exiguobacterium* sp., which is a facultative alkaliphilic bacterium to lower down the pH of highly alkaline wastewater from a chlor-alkali unit, which has initial pH of 12.0 to 7.5 after treatment within 2.0 h. Use of *Exiguobacterium* sp. for neutralization of highly alkaline wastewater generated in chlor-alkali unit, without addition of any external carbon source indicated potential use of the microbial culture as an alternative to the conventional acid neutralization method for treatment of alkaline wastewater. Jain *et al.* (2011) reported the biological neutralization of chlor-alkali industrial effluent by an alkaliphilic bacterium, isolated from the Gujarat coast, which was identified as *Enterococcus faecium* strain R-5. The isolate has ability to bring down the pH of wastewater from 12.0 to 7.0 within 3 h in the presence of carbon and nitrogen sources.

10.5.6 INORGANIC CHEMICALS WASTEWATER

Rhodococcus erythropolis, a bacterium possessing a vast catabolic potential, was adapted to grow at 4–37°C, pH 3–11 and in the presence of up to 7.5% sodium chloride and 1% copper sulfate (Carla and de Carvalho, 2012). This bacterium could be used for treatment of industrial wastewaters containing sodium chloride and copper sulfate.

10.5.7 CYANIDES

Presence of inorganic chemicals like thiocyanates, cyanides or sulfides also seem to be troublesome to living entities, therefore their removal from the wastewaters become imperative. Sorokin *et al.* (2001) have isolated alkaliphilic bacteria using thiocyanates (CNS-) at pH 10 from highly alkaline soda lake sediments and soda soils. Cyanide containing wastewaters are generated as a consequence of a large number of industrial activities such as gold mining, steel and aluminum manufacturing, electroplating and nitrile pesticides used in agriculture. Luque-Almagro (2011a, 2011b) used an alkaliphilic autochthonous bacterium *Pseudomonas pseudoalcaligenes* CECTS344 for treatment of cyanide containing wastewaters. The authors have also reported that the same alkaliphilic bacterium *Pseudomonas pseudoalcaligenes* CECT5344 was able to use cyanide as the sole source of nitrogen.

10.5.8 METALS

Presence of metal ions particularly the heavy metal ions in the wastewaters proves to be problematic since they are toxic to living entities. Microbial metal reduction has the potential for immobilizing toxic metals and radionuclides in diverse environments, however, the bacterial reduction of metals under extremely alkaline conditions has been demonstrated recently. Roh *et al.* (2007) carried out the studies on metal reduction and mineral formation using an alkaliphilic bacterium, *Alkaliphilus metalliredigens* (QYMF), isolated from a leachate-pond having pH 9.0–10.0 and containing high levels of salt. Ibrahim *et al.* (2011a) isolated a strain of *Amphibacillus* sp. $KSUCr_3$ from hypersaline soda lake, which has ability to reduce Cr (VI) – under alkaline conditions. In the other studies Ibrahim *et al.* (2011b) isolated Cr (VI) resistant alkaliphilic bacteria from sediment and water samples collected from Wadi Natrun hypersaline Soda lakes located in northern Egypt, which can tolerate Cr (VI) concentration up to 2.94 g/L in alkaline medium. The alkaliphilic bacteria thus have potential to detoxify various organic pollutants as well as inorganic pollutants and thus can be used for bioremediation of alkaline wastewaters generated during production of various industrial chemicals and their applications.

10.6 CONCLUSION AND FUTURE PERSPECTIVES

Newer and newer products are being developed as a consequence of demands by human being in health, agriculture and environment sectors. The requirement of industrial enzymes like proteases, amylases and lipases continues. Biodegradable plastic is a need of the day due to environmental pollution caused by synthetic plastics. Development of antibiotic resistance in infectious microorganisms has led to search for new antimicrobial compounds. Since manufacture of chemical compounds required for modern life cannot be stopped, environmental pollution through industrial wastes will continue in future. Alkaliphilic microorganisms seem to be promising candidates for biotechnological innovations to solve present day's problems, to fulfill demands of human beings, animals and clean the environment. Young researchers may search for more and newer alkaliphilic microorganisms and explore them for betterment of human life, sustainable agriculture and clean environment. The alkaliphilic microorganisms are being globally explored for novel biomolecules. The authors have made moderate efforts to gather knowledge on biotechnological processes developed using alkaliphilic microorganisms. Production of exopolysaccharides, biodegradable plastics, antimicrobial compounds, proteases, CGTase is only examples of employing alkaliphiles. The alkaliphiles can be looked upon as a biological resource of many more biomolecules.

KEYWORDS

- **Alkaliphilic bacteria**
- **Benzoates**
- **Biomolecules**
- **Chemopollutants**
- **Cyanide**
- **Cyclodextrin glycosyl transferase**
- **Environmental biotechnology**
- **Exopolysaccharide**
- **Extremophilic bacteria**
- **Polyhydroxyalkanoate**
- **Proteases**
- **Pyrenes**

REFERENCES

Abou-Elela, G. M.; Ibrahim, H. A. H.; Hassan, S. W.; Abd-Elnaby, H.; El-Toukhy, N. M. K. Alkaline protease production by alkaliphilic marine bacteria isolated from Marsa-Matrouh (Egypt) with special emphasis on *Bacillus cereus* purified protease. *Afr J Biotechnol* 2011, 10(22), 4631–4642.

Al-Awadhi, H.; Sulaiman, R. H. D.; Mahmoud, H. M.; Radwan, S. S. Alkaliphilic and halophilic hydrocarbon-using bacteria from kuwaiti coasts of the arabian gulf. *Appl Microbiol Biotechnol* 2007, 77(1), 183–186.

Anderson, A. J.; Dawes, E. A. Occurrence, metabolism, metabolic role and industrial use of bacterial polyhydroxyalkanoates. *Microbiol Rev* 1990, 54, 450–472.

Antranikian, G.; Ruepp, A.; Gordon, P. M. K.; Ballschmiter, M.; Zibat, A.; Stark, M.; Sensen, C. W.; Frishman, D.; Liebl, W.; Klenk, H-P. Rapid access to genes of biotechnologically useful enzymes by partial genome sequencing: the thermoalkaliphile *Anaerobranca gottschalkii*. *J Mol Microbiol Biotechnol* 2009, 16, 81–90.

Arias, S.; Moral, A. D.; Ferrer, M. R.; Tallon, R.; Quesada, E.; Bejar, V. Mauran, a exopolysaccharide produced by the halophilic bacterium *Halomonas maura*, with a novel composition and interesting properties for biotechnology. *Extremophiles* 2003, 7, 319–326.

Bahn, Y. S.; Park, J. M.; Bai, D. H.; Takase, S.; Yu, J. H. YUA001, a novel aldose reductase inhibitor isolated from alkaliphilic *Corynebacterium* sp. YUA25. I. Taxonomy, fermentation, isolation and characterization. *J Antibiotics* 1998, 51, 902–907.

Bakhtiar, S.; Andersson, M. M.; Gessesse, A.; Mattiasson, B.; Hatti-Kaul, R. Stability characteristics of a calcium-independent alkaline protease from *Nesterenkonia* sp. *Enzy Microb Technol* 2003,32, 525–531.

Barnard, G. N.; Sanders, J. K. M. Observation of mobile poly(β-hydroxybutyrate) in the storage granules of *Methylobacterium* AM1 by *in vivo* ^{13}C- NMR spectroscopy. *FEBS Lett* 1988, 231, 16–18.

Barnard, G. N.; Sanders, J. K. M. The poly β-hydroxybutyrate granule *in vivo*. A new insight based on NMR spectroscopy of whole cells. *J Biol Chem* 1989, 264, 3286–3291.

Borgave, S. B.; Joshi, A. A.; Kelkar, A. S.; Kanekar, P. P. Screening of alkaliphilic, haloalkaliphilic bacteria and alkalithermophilic actinomycetes isolated form alkaline soda lake of Lonar, India for antimicrobial activity. *Int J Pharma Biosci* 2012, 3(4) B, 258–274.

Carla, C. C. R.; de Carvalho, P. A. Adaptation of *Rhodococcus erythropolis* cells for growth and bioremediation under extreme conditions. *Res Microbiol* 2012, 163(2), 125–136.

Carsaro, M. M.; Grant, W. D.; Grant, S.; Marciano, C. E.; Parrilli, M. Structure determination of an exopolysaccharide from an alkaliphilic bacterium closely related to *Bacillus* spp. *Eur J Biochem* 1999, 264, 554–561.

Chaphalkar, S.; Dey, S. Some aspects of production of extracellular protease for *Streptomyces diastaticus*. *J Micro Biotechnol* 1994, 9(2), 85–100.

Chen, C. W.; Don, T. M.; Yen, H. F. Enzymatic extruded starch as a carbon source for the production of poly(3-hydroxybutyrate-*co*-3-hydroxyvalerate) by *Haloferax mediterranei*. *Pro Biochem* 2006, 41(11), 2289–2296.

Da Costa, M. S.; Duarte, J. C.; Williams, R. A. D. *Microbiology of extreme environments and its potential for biotechnology*, FEMS Symposium 49, Elsevier Applied Science: London, 1998.

Danesh, A. Production, purification and characterization of antibacterial biomolecules from an alkaliphilic *Bacillus*. Ph.D. Thesis, Lund University, Sweden, 2011.

de Graaff, M.; Bijmans, M. F. M.; Abbas, B.; Euverink, G. J. W.; Muyzer, G.; Janssen, A. J. H. Biological treatment of refinery spent caustics under halo-alkaline conditions. *Biores Technol* 2011, 102(15), 7257–7264.

Deng, A.; Wu, J.; Zhang, Y.; Zhang, G.; Wen, T. Purification and characterization of a surfactant-stable high-alkaline protease from *Bacillus* sp. B001. *Biores Technol* 2010, 101(18), 7100–7106.

Desai, J. D.; Banat, I. M. Microbial production of surfactants and their commercial potential. *Microbiol Mol Biol Rev* 1997, 61, 47–64.

Deshmukh, D. V.; Puranik, P. R. Application of Plackett-Burman design to evaluate media components affecting antibacterial activity of alkaliphilic cyanobacteria isolated from Lonar Lake. *Turk J Biochem* 2010, 35(2), 114–120.

Deshmukh, K. B.; Pathak, A. P.; Karuppayil, M. S. Bacterial diversity of Lonar Soda Lake of India. *Ind J Microbiol* 2011, 51(1), 107–111.

Dieter, A.; Hamm, A.; Fiedler, H. P.; Goodfellow, M.; Mueller, W.E.; Brun, R.; Beil, W.; Bringmann, G. Pyrocoll, an antibiotic, antiparasitic and antitumor compound produced by a novel alkaliphilic *Streptomyces* strain. *J Antibiot* 2003, 56(7), 639–646.

Ding, Z. G.; Li, M. G.; Zhao, J. Y.; Ren, J.; Huang, R.; Xie, M. J.; Cui, X. L.; Zhu, H. J.; Wen, M. L. Naphthospironone A: an un precedented and highly functionalized polycyclic metabolite from an alkaline mine waste extremophile. *Chemistry* 2010, 16(13), 3902–3905.

Fava, F.; Gioia, D.; Marchetti, L. Cyclodextrin effects on the *ex-situ* bioremediation of a chronically polychlorobiphenyl-contaminated soil. *Biotechnol Bioeng* 1998, 58, 345–355.

Fujiwara, N.; Kazuhiko, Y.; Akihiko, M. Utilization of a thermostable alkaline protease from an alkalophilic thermophile for the recovery of silver from used X-ray film. *J Ferment Bioeng* 1991, 72(4), 306–308.

Ganesh Kumar, C.; Joo, H.; Choi, J.; Koo, Y.; Chang, C. Purification and characterization of an extra cellular polysaccharide from haloalkaliphilic *Bacillus* sp. I-450. *Enzyme Microb Technol* 2004, 34, 673–681.

Gao, J.; Bao, H. I.; Xin, M.; Liu, Y.; Li, Q.; Zhang, Y. Characterization of a bioflocculant from a newly isolated *Vagococcus* sp. W31*. *J Zhejiang Univ Sci B* 2006, 7(3), 186–192.

Genckal, H.; Tari, C. Alkaline protease production from alkalophilic *Bacillus* sp. isolated from natural habitats. *Enz Microbial Technol* 2006, 39(2), 703–710.

Gessesse, A.; Hatti-Kaul, R.; Gashe, B. A.; Mattiasson, B. Novel alkaline proteases from alkaliphilic bacteria grown on chicken feather. *Enzy Microb Technol* 2003, 32, 519–524.

Gomes, J.; Steiner, W. The biocatalytic potential of extremophiles and extremozymes. *Food Technol Biotechnol* 2004, 42(4), 223–235.

Grant, W. D.; Mwatha, W. E.; Jones, B. E. Alkaliphiles: ecology, diversity and applications. *FEMS Microbiol Rev* 1990, 75, 255–270.

Habe, H.; Kanemitsu, M.; Nomura, M.; Takemura, T.; Iwata, K.; Nojiri, H.; Yamane, H.; Omori, T. Isolation and characterization of an alkaliphilic bacterium using pyrene as a carbon source. *J Biosci Bioeng* 2004, 98(4), 306–308.

Haga, K.; Harata, K.; Nakamura, A.; Yamane, K. Crystallization and preliminary X-Ray studies of Cyclodextrin glucanotransferase from alkaliphilic *Bacillus* sp. 1101. *J Mol Biol* 1994, 237, 163–164.

Hancock, E. W.; Strohl, W. R. Antimicrobials in the twenty-first century. *Curr Opin Microbiol* 2001, 4, 491–545.

Horikoshi, K. Production of alkaline enzymes by alkalophilic microorganisms. Part I. Alkaline protease produced by *Bacillus* No.221. *Agric Biol Chem* 1971a, 36, 1407–1414.

Horikoshi, K. Production of alkaline enzymes by alkalophilic microorganisms. Part II. Alkaline amylase produced by *Bacillus* No. A-40–2. *Agric Biol Chem* 1971b, 35, 1783–1791.

Horikoshi, K. General view of alkaliphiles and thermophiles. In *Super bugs: microorganisms in extreme environments*, Horikoshi, K., Grant, W. D., Eds.; Springer Verlag: Berlin, 1991; 3–13.

Horikoshi, K. Discovering novel bacteria with an eye to biotechnological applications. *Curr Opin Biotechnol* 1995, 6, 292–297.

Horikoshi, K. Alkaliphiles: some applications of their products for biotechnology. *Microbiol Mol Biol Rev* 1999, 63(4), 735–750.

Horikoshi, K.; Akiba, T. Alkalophilic *Microorganisms: A New Microbial World*, Springer-Verlag: Tokyo, 1982.

Ibrahim, A. S. S.; El-Shayeb, N. M. A.; Mabrouk, S. S. Isolation and identification of alkaline protease producing alkaliphilic bacteria from an Egyptian Soda Lake. *J Appl Sci Res* 2007, 3(11), 1363–1368.

Ibrahim, A. S. S.; El-Tayeb, M. A.; Elbadawi, Y. B.; Al-Salamah, A. A. Isolation and characterization of novel potent Cr(VI) reducing alkaliphilic *Amphibacillus* sp KSUCr3 from hypersaline soda lakes. *Electronic J Biotechnol* 2011a, 14(4), DOI: 10.2225/vol14-issue4.

Ibrahim, A. S. S.; Elbadawi, Y. B.; Al-Salamah, A. A. Bioreduction of Cr (VI) by potent novel chromate resistant alkaliphilic *Bacillus* sp. strain KSUCr$_5$ isolated from hypersaline Soda lakes. *Afr J Biotechnol* 2011b, 10(37), 7207–7218.

Iqbal, A.; Bhatti, N.; Nosheen, S.; Jamil, A.; Malik, M. Histochemical and physicochemical study of bacterial exopolysaccharides. *Biotechnology* 2002, 1, 28–33.

Issatchencko, B. L. *Chlorous, sulfate and soda lakes of the Kulunda Steppe and the biogenic process in them, Selected works*, Vol. 2, Academia Naukova: Leningrad, 1951, pp143–162.

Jain, R. M.; Mody, K. H.; Keshri, J.; Jha, B. Biological neutralization of chlor-alkali industry wastewater. *Mar Pollut Bull.* 2011, 62(11), 2377–2383.

Jain, R. M.; Mody, K.; Mishra, A.; Jha, B. Isolation and structural characterization of biosurfactant produced by an alkaliphilic bacterium *Cronobacter sakazakii* isolated from oil contaminated wastewater. *Carbohyd Polym* 2012, 87, 2320–2326.

Jones, B. E.; Grant, W. D.; Collins, N. C.; Mwatha, W. E. Alkaliphiles: diversity and identification. In *Bacterial diversity and systematics*, Priest, F. G., Eds.; Plenum Press: New York, 1994, 195–230.

Jones, B. E.; Grant, W. D.; Duckworth, A. W.; Owenson, G. G. Microbial diversity of soda lakes. *Extremophiles* 1998, 2, 191–200.

Joo, H. S.; Chung-Soon, C. Production of an oxidant and SDS-stable alkaline protease from an alkaliphilic *Bacillus clausii* I-52 by submerged fermentation: feasibility as a laundry detergent additive. *Enz Microb Technol* 2006, 38(1–2), 176–183.

Joshi, A. A.; Kanekar, P. P. Production of exopolysaccharide by *Vagococcus carniphilus* MCM B-1018 isolated from alkaline Lonar Lake, India. *Ann Microbiol* 2011, 61, 733–740.

Joshi, A. A.; Kanekar, P. P.; Kelkar, A. S.; Shouche, Y. S.; Wani, A. A.; Borgave, S. B.; Sarnaik, S. S. Cultivable bacterial diversity of alkaline Lonar lake, India. *Microbial Ecol* 2008, 55(2), 163–172.

Joshi, A. A.; Kanekar, P. P.; Sarnaik, S. S.; Kelkar, A. S.; Shouche, Y. S.; Wani, A. Moderately halophilic, alkalitolerant *Halomonas campisalis* MCM B-365 from Lonar Lake, India. *J Basic Microbiol* 2007, 47, 213–221.

Joshi, A. A.; Kanekar, P. P.; Sarnaik, S. S.; Kelkar, A. S. Bacterial diversity of Lonar lake ecosystem. In *Biodiversity of Lonar crater*, Banmeru, P. K., Banmeru, S. K., Mishra, V. R., Eds.; Anamaya publishers: New Delhi, 2005, 71–75.

Kanekar, P. P.; Nilegaonkar, S. S.; Sarnaik, S. S.; Kelkar, A. S. Optimization of protease activity of alkaliphilic bacteria isolated from an alkaline lake in India. *Biores Technol* 2002, 85, 87–93.

Kanekar, P. P.; Sarnaik, S. S.; Kelkar, A. S. Bioremediation of phenol by alkaliphilic bacteria isolated alkaline Lake of Lonar, India. *J Appl Microbiol* 1999, 85, 128S-133S

Kaneko, T.; Hamamoto, T.; Horikoshi, K. Molecular cloning and nucleotide sequence of the cyclomaltodextrin glucanotransferase gene from the alkalophilic *Bacillus* sp. strain no. 38–2. *J Gen Microbiol* 1988, 134, 97–105.

Khanna, S.; Srivastava, A. Recent advances in microbial polyhydroxyalkanoates. *Pro Biochem* 2005, 40, 607–619.

Kharat, K. R.; Kharat, A.; Hardikar, B. P. Antimicrobial and cytotoxic activity of *Streptomyces* sp. from lonar lake. *Afr J Biotechnol* 2009, 8(23), 6645–6648.

Kiesel, B.; Mueller, R. H.; Kleinsteuber, R. Adaptative potential of alkaliphilic bacteria towards chloroaromatic substrates assessed by a gfp-tagged 2,4-D degradation plasmid. *Eng Life Sci* 2007, 7(4), 361–372.

Kimura, K.; Kataoka, S.; Nakamura, A.; Takano, T.; Kobayashi, S.; Yamane, K. Functions of the COOH-terminal region of Cyclodextrin glucanotransferase of alkalophilic *Bacillus* sp. #1011: relation to catalyzing activity and pH stability. *Biochem Biophys Res Commun* 1989, 161, 1273–1279.

Kulkarni, S. O.; Kanekar, P. P.; Jog, J. P.; Patil, P. A.; Nilegaonkar, S. S.; Sarnaik, S. S.; Kshirsagar, P. R. Characterization of copolymer poly(hydroxybutyrate-cohydroxyvalerate) (PHB-*co*-PHU) produced by *Halomonas campisalis* MCMB-1027, its biodegradability and potential application. *Biores Technol* 2011, 102(11), 6625–6628.

Kulkarni, S. O.; Kanekar, P. P.; Nilegaonkar, S. S.; Sarnaik, S. S.; Jog, J. P. Production and characterization of biodegradable poly(hydroxybutyrate-cohydroxyvalerate) (PHB-co-PHV) copolymer by moderately haloalkalitolerant *Halomonas campisalis* MCM B-1027 isolated from Lonar Lake, India. *Biores Technol* 2010, 101(24), 9765–9771.

Kulshreshtha, N. M.; Kumar, A.; Dhall, P.; Gupta, S.; Bisht, G.; Pasha, S.; Singh, V. P.; Kumar, R. Neutralization of alkaline industrial wastewaters using *Exiguobacterium* sp. *Int Biodeter Biodegr* 2010, 64(3), 191–196.

Kumar, C. G. Purification and characterization of a thermostable alkaline protease from alkalophilic *Bacillus pumilus*. *Lett Appl Microbiol* 2002, 34(1), 13–17.

Kwon, Y. T.; Jin, O. K.; Sun, Y. M.; Hyune, H. L.; Hyune, M. R. Extracellular alkaline proteases from alkalophilic *Vibrio metschnikovii* strain RH530. *Biotechnol Lett* 1994, 16(4), 413–418.

Lawton, E. M.; Cotter, P. D.; Hill, C.; Ross, R. P. Identification of a novel two peptide Lantibiotic, Haloduracin, produced by the alkaliphile *Bacillus halodurans* C-125. *FEMS Microbiol Lett* 2008, 267, 64–71.

Lee, S. Y. Plastic bacteria? Progress and prospects for polyhydroxyalkanoates production in bacteria. *Trends Biotechnol* 1996, 14(11), 431–438.

Lee, S. Y.; Choi, J.; Wong, H. H. Recent advances in polyhydroxyalkanoate production by bacterial fermentation: mini-review. *Int J Biol Macromol* 1999, 25:31–36.

Li, Y.; Wiliana, T.; Tam, K. C. Synthesis of amorphous calcium phosphate using various types of cyclodextrins. *Materials Research Bulletin* 2007, 42(5), 820–827.

Lillo, J. G.; Rodriguez-Valera, F. Effects of culture conditions on Poly(β-hydroxybutyric acid) production by *Haloferax mediterranei*. *Appl Environ Microbiol* 1990, 56, 2517–2521.

Luque-Almagro, V. M.; Blasco, R.; Martinez-Luque, M.; Moreno-Vivian, C.; Castillo, F.; Roldan, M. D. Bacterial cyanide degradation is under review: *Pseudomonas pseudoalcaligenes* CECT5344, a case of an alkaliphilic cyanotroph. *Biochem Soc Trans* 2011a, 39, 269–274.

Luque-Almagro, V. M.; Merchan, F.; Blasco, R.; Igeno, M. I.; Martinez-Luque, M.; Moreno-Vivian, C.; Castillo, F.; Roldan, M. D. Cyanide degradation by *Pseudomonas pseudoalcaligenes* CECT5344 involves a malate: quinone oxidoreductase and an associated cyanide-insensitive electron transfer chain. *Microbiol* 2011b, 157, 739–746.

Madison, L. L.; Huisman, G. W. Metabolic engineering of Poly(3-Hydroxyalkanoates): from DNA to plastic. *Microbiol Mol Biol Rev* 1999, 63, 21–53.

Maekelae, M.; Mattsson, P.; Schinina, M. E.; Korpela, A. Purification and properties of cyclomaltodextrin glucanotransferase from an alkalophilic *Bacillus*. *Biotechnol Appl Biochem* 1988, 10, 414–427.

Maltseva, O.; McGowan, C.; Fulthorpe, R.; Oriel, P. Degradation of 2,4-dichloro phenoxyacetic acid by haloalkaliphilic bacteria. *Microbiol* 1996, 142, 1115–1122.

Manachini, P. L.; Fortina, M. G.; Parini, C. Alkaline protease produced by *Bacillus thermoruber* – a new species of *Bacillus*. *Appl Microbiol Biotechnol* 1988, 28, 409–413.

Martinez-Checa, F.; Toledo, F. L.; Vilchez, R.; Quesada, E.; Calvo, C. Yield, production, chemical composition and functional properties of emulsifier H28 synthesized by *Halomonas eurihalina* strain H-28 in media containing various hydrocarbons. *Appl Microbiol Biotechnol* 2002, 58(3), 358–363.

Mergaert, J.; Anderson, C.; Wouters, A.; Swings, J.; Kersters, K. Degradation of poly(hydroxyalkanoates). *FEMS Microbiol Rev* 1992a, 103, 317–322.

Merrick, J. M.; Doudoroff, M. Depolymerization of poly β-hydroxybutyrate by an intracellular enzyme system. *J Bacteriol* 1964, 88, 60–71.

Mitsuiki, S.; Sakai, M. Y.; Moriyama, M.; Goto, K.; Furukawa, T. Purification and some properties of a keratinolytic enzyme from an alkaliphilic *Nocardiopsis* sp. TOA-1. *Biosci Biotechnol Biochem* 2002, 66, 164–167.

Muller, R. H.; Jorks, S.; Kleinsteuber, S.; Babel, W. Degradation of various chlorophenols under alkaline conditions by Gram-negative bacteria closely related to *Ochrobactrum anthropi*. *J Basic Microbiol* 1998, 38(4), 269–281.

Munro, I. C.; Newberne, P. M.; Young, V. R.; Bär, A. Safety assessment of γ-cyclodextrin. *Regul Toxicol Pharm* 2004, 39, S3-S13.

Nakamura, A.; Haga, K.; Ogawa, S.; Kuwano, K.; Kimura, K.; Yamane, K. Functional relationships between Cyclodextrin glucanotransferase from an alkaliphilic *Bacillus* and alpha- amylases. Site-directed mutagenesis of the conserved two Asp and one Glu residues. *FEBS Letts* 1992, 296, 37–40.

Nakamura, A.; Haga, K.; Yamane, K. Three histidine residues in the active center of Cyclodextrin glucanotransferase from alkaliphilic *Bacillus* sp. 1011: effects of replacement on pH dependence and transition – state stabilization. *Biochem* 1993, 32, 6624–6631.

Nicholaus, B.; Lama, L.; Esposito, E.; Manca, M. C.; Importa, R.; Bellitti, M. R.; Duckworth, A. W.; Grant, W. D.; Gambacorta, A. *Haloarcula* spp. able to biosynthesize exo-endopolymers. *J Ind Microbiol Biotechnol* 1999a, 23, 489–496.

Nicholaus, B.; Lama, L.; Manca, M. C.; Gambacorta, A. Extremophiles: polysaccharides and enzymes degrading polysaccharides. *Recent Res Dev Biotechnol Bioeng* 1999b, 2, 37–64.

Norberg, A. B.; Persson, H. Accumulation of heavy metal ions by *Zoogloea rarigera*. *Biotechnol Bioeng* 1984, 26, 239–246.

Oie, C. S. I.; Albaugh, C. E.; Peyton, B. M.. Benzoate and salicylate degradation by *Halomonas campisalis*, an alkaliphilic and moderately halophilic microorganism. *Water Res* 2007, 41(6), 1235–1242.

Ojumu, T. V.; Yu, J.; Solomon, B. O. Production of polyhydroxyalkanoates, a bacterial biodegradable polymer. *Afr J Biotechnol* 2004, 3, 18–24.

Patel, R. K.; Mital, S.; Dodia, R.; Joshi, H.; Singh, S. P. Purification and characterization of alkaline protease from a newly isolated haloalkaliphilic *Bacillus* sp. *Proc Biochem* 2006, 41(9), 2002–2009.

Patil, U.; Chaudhari, A. Purification and characterization of solvent-tolerant, thermostable, alkaline metalloprotease from alkalophilic *Pseudomonas aeruginosa* MTCC 7926. *J Chem Technol Biotechnol* 2009, 84(9), 1255–1262.

Poli, A.; Esposito, E.; Orlando, P.; Lama, L.; Giordano, A.; de Appolonia, F.; Nicolaus, B.; Gambacorta, A. *Halomonas alkaliantarctica* sp. nov., isolated from saline lake Cape Russell in Antarctica, an alkalophilic moderately halophilic, exopolysaccharide-producing bacterium. *Syst Appl Microbiol* 2007, 30, 31–38.

Quillaguaman, J.; Hashim, S.; Bento, F.; Mattiason, B.; Hatti-Kaul, R. Poly(β hydroxybutyrate) production by a moderate halophile, *Halomonas boliviensis* LC1 using starch hydrolysate as a substrate. *J Appl Microbiol* 2005, 99(1), 151–157.

Ritchie, G. A. F.; Dawes, E. A. The noninvolvement of acyl-carrier protein in poly β-hydroxybutyrate synthesis in *Azotobacter beijerinckii*. *Biochem J* 1969, 112, 803–805.

Roh, Y.; Chon, C. M.; Moon, J. W. Metal reduction and biomineralization by an alkaliphilic metal-reducing bacterium, *Alkaliphilus metalliredigens* (QYMF). *Geosci J* 2007, 11(4), 415–423.

Romano, I.; Lama, L.; Nicolaus, B.; Poli, A.; Gambacorta, A.; Giordano, A. *Oceanobacillus oncorhynchi* subsp. *incaldanensis* subsp. nov, an alkalitolerant halophile Isolated from an algal mat collected from a sulfurous spring in Campania (Italy), and amended description of *Oceanobacillus oncorhynchi*. *J Syst Evol Microbiol* 2006, 56, 805–810.

Romano, I.; Lama, L.; Nicolaus, B.; Poli, A.; Gambacorta, A.; Giordano, A. *Halomonas alkaliphila* sp. nov. a novel halotolerant alkaliphilic bacterium isolated from a salt pool in Campania (Italy). *J Gen Appl Microbiol* 2006, 52, 339–348.

Roth, E.; Schwenninger, S. M.; Eugster-Meier, E.; Lacroix, C. Facultative anaerobic halophilic and alkaliphilic bacteria isolated from a natural smear ecosystem inhibits *Listeria* growth in early ripening stages. *Int J Food Microbiol* 2011, 147(1), 26–32.

Sato, M.; Beppu, T.; Arima, K. Studies on antibiotics produced at high alkaline pH. *Agric Biol Chem* 1983, 47(9), 2019–2027.

Shrinivas, D.; Kumar, R.; Naik, G. R. Enhanced production of alkaline thermostable keratino-lytic protease from calcium alginate immobilized cells of thermoalkalophilic *Bacillus halodurans* JB 99 exhibiting dehairing activity. *J Ind Microbiol Biotechnol* 2012, 39(1), 93–98.

Shrinivas, D.; Naik, G. R. Characterization of alkaline thermostable keratinolytic protease from thermoalkalophilic *Bacillus halodurans* JB 99 exhibiting dehairing activity. *Int Biodeter Biodegr* 2011, 65(1), 29–35.

Singh, J.; Vohra, R. M.; Sahoo, D. K.. Alkaline protease from a new obligate alkalophilic isolate of *Bacillus sphaericus. Biotechnol Lett* 1999, 21(10), 921–924.

Singh, L. S.; Mazumder, S.; Bora, T. C. Optimization of process parameters for growth and bioactive metabolite produced by a salt tolerant and alkaliphilic actinomycetes, *Streptomyces tanashiensis* strain A2D. *J Med Mycol* 2009, 19(4), 225–233.

Smita, G. S.; Ray, P.; Mohapatra, S. Quantification and optimization of bacterial isolates for production of alkaline protease. *Asian J Exp Biol Sci* 2012, 3(1), 181–186.

Sorkhoh, N. A.; Al-Awadhi, H.; Al-Mailem, D. M.; Kansour, M. K.; Khanafer, M.; Radwan, S. S. Agarolytic bacteria with hydrocarbon-utilization potential in fouling material from the Arabian Gulf coast. *Int Biodeter Biodegr* 2010, 64(7), 554–559.

Sorokin, D. Y.; Tourova, T. P.; Lyzenko, A. M.; Kuenen, J. G. Microbial thiocyanate utilization under highly alkaline conditions. *Appl Environ Microbiol* 2001, 67(2), 528–538.

Steinbuchel, A. Polyhydroxyalkanoic acids. In *Biomaterials: novel materials from biological sources*, Byrom, D., Ed.; Stockton: New York, 1991, 124–213.

Sugimori, D.; Dake, T.; Nakamura, S. Microbial degradation of disodium terephthalate by alka-liphilic *Dietzia* sp. strain GS-1. *Biosci Biotech Biochem* 2000, 64, 2709–2711.

Sundararajan, S.; Chandrababu, N. K.; Shanthi, C. Alkaline protease from *Bacillus cereus* VITSN04: Potential application as a dehairing agent. *J Biosci Bioeng* 2011, 111(2), 128–133.

Sutherland, I. Biotechnology of microbial exopolysaccharides. In *Cambridge Studies in Biotechnology*, Vol. 9; Baddiley, J., Higgins, N. H., Potter, W. G., Eds.; Cambridge University Press: Cambridge, 1990.

Sutherland, I. W. Microbial exopolysaccharides – structural subtleties and their consequences. *Pure Appl Chem* 1997, 69, 1911–1917.

Sutherland, I. W. Novel and established applications of microbial polysaccharides. *Trends Biotechnol* 1998, 16, 41–46.

Szejtli, J. Past, present, and future of Cyclodextrin research. *Pure Appl Chem* 2004, 76, 1825–1845.

Takami, H.; Akiba, T.; Horikoshi, A. Production of extremely thermostable alkaline protease from *Bacillus* sp. No. AH-101. *Appl Microbiol Biotechnol* 1989, 30, 120–124.

Takami, H.; Akiba, T.; Horikoshi, A. Characterization of an alkaline protease from *Bacillus* sp. No. AH-101. *Appl Microbiol Biotechnol* 1990, 33, 519–523.

Tambekar, D.; Dhundale, V. Isolation and characterization of antibacterial substance produced from Lonar Lake. *Int J Res Rev Pharma Appl Sci* 2012, 2(1), 41–54.

Taylor, M. J.; Richardson, T. Applications of microbial enzymes in food systems and biotechnology. *Adv Appl Microbiol* 1979, 25, 7–35.

Thatai, A.; Kumar, M.; Mukherjee, M. Single step purification process for cyclodextrin glucano-transferase from a *Bacillus* sp. isolated from soil. *Preparative Biochem Biotechnol* 1999, 29, 35–47.

Thombre, R. S.; Kanekar, P. P. Studies on cyclodextrin glycozyl transferase (CGTase) producing alkaliphilic bacteria from Lonar Lake. *Journal of Microbial and Biochemical Technology* 2011, 10(S1), DOI: 10.4172/1948–5948.1000001.

Thumar, J. T.; Dhulia, K.; Singh, S. P. Isolation of partial purification of an antimicrobial agent from halotolerant alkaliphilic *Streptomyces aburaviensis* strain Kut-8. *World J Microbiol Biotechnol* 2010, 26, 2081–2087.

Tombs, M.; Harding, S. E. *An introduction to polysaccharide biotechnology*, Taylor and Francis: London, 1998.

Tsujibo, H.; Sato, T.; Inui, M.; Yamamoto, H.; Inamorai, Y. Intracellular accumulation of phenazine antibiotics produced by an alkaliphilic actinomycetes. I. Taxonomy, isolation and identification of the phenazine antibiotics. *Agric Biol Chem* 1988, 52(2), 301–306.

Uyar, F.; Porsuk, I.; Kizil, G.; Ince, A.; Yilmaz, E. Optimal conditions for production of extracellular protease from newly isolated *Bacillus cereus* strain CA15. *Eurasia J Biosci* 2011, 5, 1–9.

Vasavada, S. H.; Thumar, J. T.; Singh, S. P. Secretion of a potent antibiotic by salt-tolerant and alkaliphilic actinomycetes *Streptomyces sannanensis* strain RJT-1. *Curr Sci* 2006, 91(10), 1393–1397.

Verlinden, R. A. J.; Hill, D. J.; Kenward, M. A.; Williams, C. D.; Radecka, I. Bacterial synthesis of biodegradable polyhydroxyalkanoates. *J Appl Microbiol* 2007, 102, 1437–1449.

Wind, R.; Liebl, W.; Buitlaar, R.; Penninga, D.; Spreinat, A.; Dijkhuizen, L.; Bahl, H. Cyclodextrin formation by thermo stable α-Amylase of *Thermoanaerobacterium thermosulfurigenes* EM1 and reclassification of the enzyme as cyclodextrin glycozyl transferase. *Appl Env Microbiol* 1995, 61, 1257–1265.

Yamagata, Y.; Ichishima, E. A new alkaline serine protease from alkalophilic *Bacillus* sp.: cloning, sequencing, and characterization of an intracellular protease. *Curr Microbiol* 1995, 30(6), 357–366.

Yumoto, A.; Nakamura, H.; Iwata, K.; Kojima, K.; Kusumoto, Y.; Nodasaka, H.; Matsuyama, A. *Dietzia psychralcaliphila* sp. nov., a novel facultatively psychrophilic alkaliphile that grows on hydrocarbons. *Int J Syst Evol Microbiol* 2002, 52, 85–90.

Yumoto, S.; Yamaga, Y.; Sogabe, Y.; Nodasaka, H.; Matsuyama, K.; Nakajima, A., Suemori, M. *Bacillus krulwichiae* sp. nov., a halotolerant obligate alkaliphile that uses benzoate and m-hydroxybenzoate. *Int J Syst Evol Microbiol* 2003, 53, 1531–1536.

Zhilina, T. N.; Zavarzin, G. A. Alkaliphilic anaerobic community at pH 10. *Curr Microbiol* 1994, 29, 109–112.

Zinn, M.; Withholt, B.; Egli, T. Occurrence, Synthesis and medical application of bacterial polyhydroxyalkanoate. *Advances in Drug Delivery Reviews* 2001, 53, 5–21.

CHAPTER 11

DESIGN AND TAILORING OF POLYHYDROXYALKANOATE-BASED BIOMATERIALS CONTAINING 4-HYDROXYBUTYRATE MONOMER

SEVAKUMARAN VIGNESWARI, KESAVEN BHUBALAN, and ABDULLAH AL-ASHRAF AMIRUL

CONTENTS

11.1 INTRODUCTION

Polyhydroxyalkanoate (PHA), a naturally occurring microbial biopolymer had been studied in detail in terms of its production, properties and applications. This biopolymer is completely biodegradable and some PHA is known for its biocompatibility. PHA and some of its blends and composites have been tailored as suitable biomaterials and are extensively studied for various pharmaceutical and medical applications (Williams and Martin, 2002; Frier, 2006). These PHA polymers have been tested in tissue engineering applications as surgical sutures, bone plates, implants, gauzes, osteosynthetic materials and also as matrix material assisting slow release of drugs and hormones (Zinn *et al.*, 2001; Williams and Martin, 2002; Sudesh, 2004). Most of the PHA-based biomaterials studied are constituents of this polymer containing 3-hydroxybutyrate (3HB) and/or 4-hydroxybutyrate (4HB) units. These monomers were identified as normal constituent of human blood (Wiggam *et al.*, 1997). Fresh human blood has been found to contain 0.17–1.51 mg/L of 4-hydroxybutyric acid (Sudesh and Doi, 2000). Hence, much effort has been devoted to produce P(4HB) homopolymer and P(3HB-*co*-4HB) copolymers with improved biomaterial properties. TephaFLEX® is an example of the latest biomaterial derived using these polymers for the fabrication of medical devices by Tepha Inc. (Tepha Medical Devices). P(4HB) and P(3HB-*co*-4HB) copolymers have gained much interest for a wide range of medical and pharmaceutical applications (Chee *et al.*, 2008; Yang *et al.*, 2002). This chapter will discuss the P(3HB-*co*-4HB) copolymer as well as its blends and composites as a biocompatible material in the medical and pharmaceutical fields. The fabrication of P(3HB-*co*-4HB) copolymer and its blends or composites with varying monomer compositions and properties will be reviewed. The application of resulting biomaterials in the medical field will be summarized.

11.2 PHA: AN OVERVIEW

PHA is comprised of various hydroxycarboxylic acids and it is accumulated as energy and carbon storage material in some bacteria (Anderson and Dawes, 1990; Lee, 1996). Since its discovery in 1926, much research had been devoted in PHA research by scientists worldwide. Many Gram positive and Gram negative bacteria have been identified to synthesize PHA with various constituents in the presence of excess carbon source and limiting-nutrient conditions (such as shortage of nitrogen, phosphorus, magnesium or sulfur) (Anderson and Dawes, 1990; Doi, 1990; Lee, 1996). The assimilated carbon sources are biochemically processed into hydroxyalkanoates units, polymerized and stored in the form of water insoluble inclusions (granules) in the cell cytoplasm (Potter and Steinbüchel, 2005; Rehm and Steinbüchel, 1999).

PHA granules can be stained specifically with Sudan black or light fluorescent stains such as Nile blue and Nile red (Gorenflo *et al.*, 1999; Spiekermann *et al.*, 1999). PHA granules are observed as light-refracting granules under phase contrast light microscope, whereas, under transmission electron microscope (TEM), PHA granules are displayed as electron transparent, discrete, spherical particles with clear boundaries (Fig. 11.1). To date, more than 150 different constituents of PHAs have been identified as either homopolymers or copolymers (Steinbüchel and Lütke-Eversloh, 2003). The properties of PHA resemble some of the properties of commodity plastics such as polypropylene or low-density polyethylene. Most PHAs are thermoplastics, which can be tailored into stiff packing materials or highly elastic elastomers (Anderson and Dawes, 1990; Sudesh and Iwata, 2008). The chemical and physical property of PHA polymers is influenced by the functionalized groups in the side chain of monomers. These groups includes halogen, carboxyl, hydroxyl, epoxyl and phenoxy (Kessler *et al.*, 2001; Kim and Lenz, 2001).

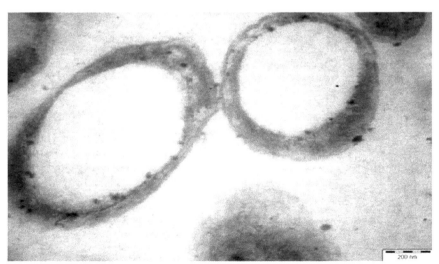

FIGURE 11.1   TEM showing PHA accumulation in the cytoplasm of *Cupriavidus* sp. USMAA1020.

This polymeric material is produced by bacteria using carbon sources such as simple sugars or complex plant oils (Loo and Sudesh, 2007; Tsuge, 2002).

The key enzyme in PHA biosynthesis is the PHA synthase (PhaC) (Sudesh *et al.*, 2000). Generally, the type of PHA synthesized is structurally related to the carbon source fed during fermentation and the substrate specificity of PhaC (Steinbüchel and Lütke-Eversloh, 2003; Taguchi and Doi, 2004). The building block of PHA is (R)-3-hydroxyalkanoic acid monomer unit (Fig. 11.2). PHA can be categorized into three main classes, which are short-chain-length PHA (SCL$_{PHA}$) with monomers consisting of 3 to 5 carbon atoms, medium-chain-length PHA (MCL$_{PHA}$), with monomers consisting of 6 to 14 carbon atoms and there is also a hybrid polymer with a combination of short-chain-length and medium-chain-length PHA (SCL-MCL$_{PHA}$).

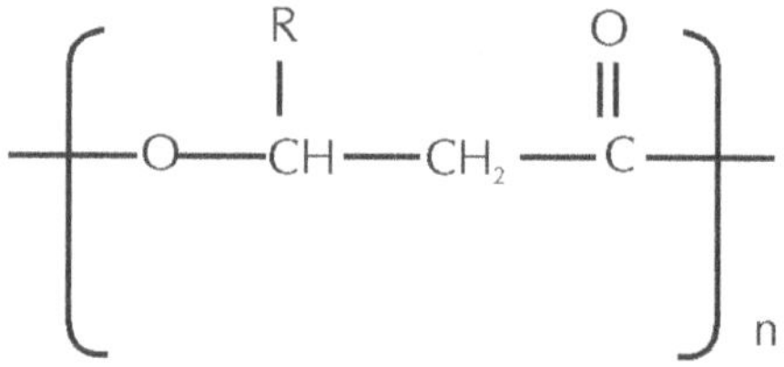

FIGURE 11.2 Chemical structure of PHA (R refers to side group while n refers to the number of repeating units).

The most common naturally occurring microbial PHA is poly(3-hydroxy-butyrate) [P(3HB)] (Doi, 1990). Besides P(3HB), other most commonly investigated SCL$_{PHA}$ includes homopolymer, copolymer or terpolymer consisting of 3-hydroxyvalerate (3HV) and 4-hydroxybutyrate (4HB) monomers. These SCL$_{PHA}$ is mostly associated with bacteria from the genus *Cupriavidus*, *Alcaligenes* and *Escherichia coli* transformants (Li *et al.*, 2007). On the other hand, MCL$_{PHA}$ and SCL-MCL$_{PHA}$ are commonly associated with bacteria from the genera *Pseudomonas* and *Aeromonas* (Sun *et al.*, 2007; Chen *et al.*, 2001).

The process of PHA production varies according to the type bacteria. PHA production for growth associated PHA producers is achieved by one-stage cultivation method. This mode of cultivation induces PHA accumulation together with growth of cells. On the other hand, two-stage cultivation is preferred for nongrowth associated PHA producers. In this mode of cultivation, cell growth phase is carried out in a separate nutrient enriched medium and PHA accumulation phase, which is initiated by transferring the cells into a nitrogen-free mineral salts medium. In large-scale or industrial scale production systems, fed-batch cultivation method is normally carried-out (Chen *et al.*, 2001; Kahar *et al.*, 2004). PHA accumulation in bacteria can be controlled by varying the ratio of carbon to nitrogen (C/N). The preferred C/N ratio falls in the range of 20–50 (Amirul *et al.*, 2008b; Lee *et al.*, 2008).

PHA is degraded internally by intracellular PHA depolymerases of PHA accumulating bacteria or extracellular PHA depolymerases of PHA degrading microorganisms. Intracellular degradation occurs in carbon limiting conditions and the accumulated PHA granules are hydrolyzed by the bacteria as carbon and energy source (Madison and Huisman, 1999). For example, P(3HB) is broken down to 3-hydroxybutyric acid by the PHA depolymerase and oligomer hydrolase (Kobayashi *et al.*, 2005). The 3-hydroxybutyric acid is further oxidized to acetyl-CoA which is then used for cell regeneration. In the natural environment, extracellular depolymerases are secreted by microorganisms to hydrolyze PHA polymer. The soluble fraction after hydrolysis normally consists of mixtures of oligomers or monomers. These products are then taken up for cell metabolism via absorption through the microorganism's cell wall (Doi, 1990). It has been reported that the rate of biodegradation is influenced by a number of factors such as environmental conditions, native microbial population and the physical properties of the polymers itself (Jendrossek *et al.*, 1996; Abou-Zeid *et al.*, 2001; Khanna and Srivastava, 2005). PHA copolymers degrade at a faster rate compared to P(3HB) homopolymer due to low crystallinity and porous surface (Mergaert *et al.*, 1993; Wang *et al.*, 2004; Sridewi *et al.*, 2006).

PHA is known for its biocompatibility, thus, making it a suitable biomaterial (Zinn *et al.*, 2001; Williams and Martin, 2002; Bhubalan *et al.*, 2011). The biodegradation of various PHA based biomaterials have been investigated *in vivo* and the results of these findings reports no formation of toxic compounds in the organisms (Zinn *et al.*, 2001; Williams and Martin, 2002). The main breakdown products of PHA are 3-hydroxyacids and they are naturally found in animals. It was reported that 3-hydroxybutyric acid and 4-hydroxybutyric acid are normal constituent of human blood (Adams *et al.*, 1987; Wiggam *et al.*, 1997). PHA polymer containing 4HB was found to exhibit relatively higher *in vivo* degradation rate compared to other PHAs and it can be controlled by varying the 4HB monomer composition (Saito *et al.*, 1996). Hence, PHA copolymer containing 4HB; P(3HB-*co*-4HB) copolymer has gained interest in a wide range of medical applications (Yang *et al.*, 2002; Chee *et al.*, 2008). Besides being degraded by depolymerases, 4HB can be degraded by eukaryotic lipases and esterases as well (Mukai *et al.*, 1994; Saito *et al.*, 1996). This further aids the biodegradation of biomaterials possessing 4HB monomers in living systems.

PHAs exhibit a wide variety of mechanical properties from hard crystalline to elastomeric materials. The physical property of PHA polymers utterly depends on the type of monomer incorporated into the polymer chains. PHA polymers could be tailored by controlling the type and composition of monomer incorporated. P(3HB) homopolymer is a highly crystalline, brittle and stiff material (Doi 1990). The M_w of P(3HB) produced from wild-type bacteria is usually in the range of 1×10^4–3×10^6 Da with a polydispersity of around two (Doi, 1990). In contrast, poly(4-hydroxybutyrate) P(4HB) homopolymer could be considered as a strong elastic material since its tensile strength and elongation at break is 104 MPa and 1000%, respectively (Saito and Doi, 1994). It possesses almost similar characteristics as ultrahigh molecular weight polyethylene.

The introduction of secondary monomers is the common strategy to produce better and more processable polymers. It was observed that the incorporation of other monomer in the 3HB polymer chain significantly improved the physical and thermal properties of resulting copolymers and changes were found to be very much dependent on the ratio of the monomer fractions (Matsusaki et al., 2000). It was proposed that SCL_{PHA} resembles conventional plastics, whereas MCL_{PHA} is regarded as elastomers and rubbers (Suriyamongkol et al., 2007). Besides the incorporation of secondary monomers, blends of PHA polymers with other biodegradable materials have also been investigated. PHA blends have been generated with components such as poly(ethylene glycols) (Foster et al., 2007), poly(vinyl alcohol) (Yoshie et al., 1995), poly(lactic acid) (Blümm and Owen, 1995; Koyama and Doi, 1997) and other natural products such as rubber (Bhatt et al., 2008) or bamboo fibers (Singh et al., 2008). PHA polymers could be designed and developed to suit specific needs by varying its physical properties.

Successful commercialization of PHA via large-scale microbial fermentation technology is evident as this bioplastic is currently available under various trademarks like Biomer®, Mirel™, Biogreen®, Biocycle® and Biopol® (Sudesh and Iwata, 2008). Renewable carbon feedstocks such as plant sugars and oils are used in industrial-scale PHA production. Some PHA have been identified for low-value high-volume commodity applications, while others for specific high-value low-volume applications in medical or pharmaceutical fields (Philip et al., 2007). Tepha Inc. has been evaluating the biocompatibility of poly(4-hydroxybutyrate) [P(4HB)] polymer produced by genetically engineered E. coli (Martin and Williams, 2003). The P(4HB) produced and marketed by Tepha is known to meet the standards set by the FDA for endotoxin levels. The biocompatible nature of PHAs allows it to be used for various medical applications. PHA polymers of P(3HB), P(4HB), poly(3-hydroxybutyrate-*co*-4-hydroxybutyrate) [P(3HB-*co*-4HB)], poly(3-hydroxy-butyrate-*co*-3-hydroxyvalerate) [P(3HB-*co*-3HV)], poly(3-hydroxybutyrate-*co*-3-hydroxyhexanoate) [P(3HB-*co*-3HHx)] and poly(3-hydroxyoctanoate) [P(3HO)] are of particular interest in tissue engineering applications, as surgical sutures, bone plates, implants, gauzes, osteosynthetic materials and also as matrix material assisting slow release of drugs and hormones (Zinn *et al.*, 2001; Williams and Martin, 2002; Sudesh, 2004).

11.3 P(3HB-*CO*-4HB): BIOSYNTHESIS AND PROPERTIES

P(3HB-*co*-4HB) (Fig. 11.3) copolymer production had been studied in detail and well characterized. Incorporation of 4HB monomer is achieved by adding precursor substrates such as 4-hydroxybutyric acid, γ-butyrolactone, and ω-alkanediols (Amirul *et al.*, 2008a, b; Amirul *et al.*, 2009; Vigneswari *et al.*, 2009; Vigneswari *et al.*, 2010; Rahayu *et al.*, 2008). Wild-type and recombinants of *C. necator* and *Delftia acidovorans* are the preferred strains investigated for P(3HB-*co*-4HB) production using various carbon precursors, but other strains have also been evaluated. Genetically engineered *E. coli* can produce this copolymer from unrelated carbon source such as glucose (Valentin and Dennis, 1997; Li *et al.*, 2010).

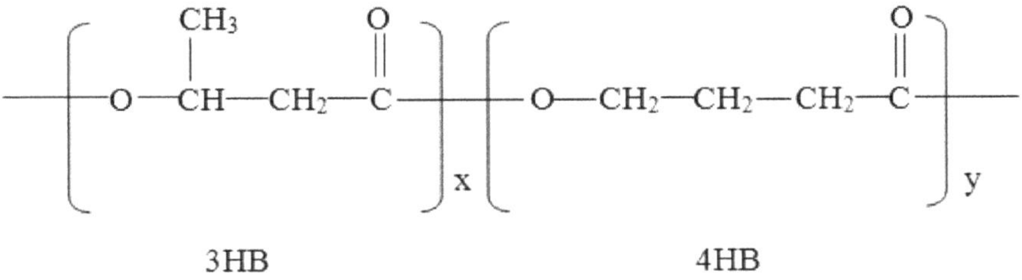

3HB   4HB

FIGURE 11.3   Chemical structure of P(3HB-*co*-4HB) (*x* and *y* refers to number of repeating units).

Biosynthesis of P(3HB-*co*-4HB) can be carried-out in shake-flasks or in bioreactors by employing batch and fed-batch systems. In shake-flasks cultivation, two-stage cultivation is usually preferred as it results in higher 4HB molar fraction (Amirul *et al.*, 2008a; Lee *et al.*, 2004; Sudesh *et al.*, 1999). Sugars such as fructose, sucrose, glucose as well as acetic acid and butyric acid are normally used for the generation of 3HB monomer (Lee *et al.*, 2004).

On the other hand, 4HB is generated using 4HB precursors. Factors affecting the biosynthesis of this copolymer include feeding of single or mixtures of carbon sources, C/N ratio, pH, inoculum concentration and aeration (Valentin and Dennis, 1997; Li *et al.*, 2004, 2010; Amirul *et al.*, 2008a).

Formation of P(3HB-*co*-4HB) copolymer involves different metabolic pathways with reference to the type of carbon source used (Fig. 11.4). The 4HB precursors will have to be converted into 4-hydroxybutyryl-CoA before being polymerized into 4HB monomer by the PhaC. Substrates such as γ-butyrolactone and ω-alkanediols are first converted into 4-hydroxybutyric acid and then transferase or thiokinase catalyzes the conversion of 4-hydroxy-butyric acid into 4-hydroxybutyryl-CoA. On the other hand, γ-butyrolactone is cleaved to 4-hydroxybutyric acid by the reaction of esterases or lactonases. As for ω-alkanediols such as 1,4-butanediol, 1,6-hexanediol, and 1,8-octanediol, these substrates are oxidized via enzymatic reactions including β-oxidation to form 4-hydroxybutyric acid before being converted to 4-hydroxybutyryl-CoA. Catabolism of 4-hydroxybutyric acid also leads to the formation of 3HB intermediate, 3-hydroxybutyryl-CoA. Generation of 3HB monomer leads to accumulation of P(3HB-*co*-4HB) copolymer.

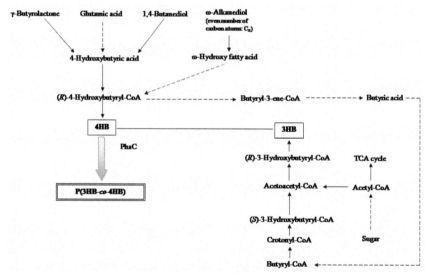

FIGURE 11.4 Proposed metabolic pathways involved in biosynthesis of P(3HB-*co*-4HB) from different carbon precursors (Braunegg *et al.*, 1998; Doi, 1990; Steinbüchel and Lütke-Eversloh, 2003).

The polymeric properties of P(3HB-*co*-4HB) copolymer are well defined and it is known to retain even though being produced by different bacteria. P(3HB-*co*-4HB) copolymers possess lower melting temperature (T_m) compared to P(3HB) homopolymer. The T_m ranges between 50 to 178°C with increasing 4HB molar fraction from 0 to 100 mol% (Saito and Doi, 1994; Mitomo *et al.*, 2001; Vigneswari *et al.*, 2009). On the other hand, glass transition temperature (T_g) is found to be in the range of 4°C to 48°C. The average molecular weight (M_n) of P(3HB-*co*-4HB) copolymers is reported to be in the range of 10^4 to 10^6 Da (Saito and Doi, 1994; Valentin and Dennis, 1997; Mitomo *et al.*, 2001; Amirul *et al.*, 2008b). Several studies report a decrease in M_n with increasing 4HB molar fraction (Kang *et al.*, 1995; Mitomo *et al.*, 2001; Amirul *et al.*, 2008a; Vigneswari *et al.*, 2009). Copolymer with high M_n (1.8×10^6 Da) was previously synthesized by recombinant *E. coli* using glucose as sole carbon source, but with low 4HB molar fraction (Valentin and Dennis, 1997). The P(3HB-*co*-4HB) copolymers may be in random or block formation. Amirul *et al.* described that P(3HB-*co*-4HB) with 4HB monomer composition in the range of 32 to 51 mol% were almost random copolymers as they exhibited *D* values between 1.3 to 1.8 (Amirul *et al.*, 2008a). On the other hand, P(3HB-*co*-23 mol% 4HB) was typically random with a D value of 1 (Vigneswari *et al.*, 2009). P(3HB-*co*-4HB) copolymers with high 4HB monomer composition are found to exhibit elastomeric property. The elongation to break values of P(3HB-*co*-4HB) copolymers are in the range of 5 to 1320% with increasing 4HB monomer composition (Saito and Doi, 1994). P(3HB-*co*-82 mol% 4HB) was found to be a very flexible polymer with an elongation to break of up to 1320% (Saito and Doi, 1994). On the other hand, the tensile strength of copolymers with 4HB composition of 0 to 64 mol% declined from 43 to 17 MPa but an increasing trend (17 to 104 MPa) was observed with copolymers with 64 to 100 mol% (Saito and Doi, 1994). The elastomeric properties of P(3HB-*co*-4HB) copolymer enables it to be fabricated for various applications demanding strong yet flexible materials.

11.4 BIOCOMPATIBILITY OF P(3HB-*CO*-4HB): MEDICAL AND PHARMACEUTICAL APPLICATIONS

Biocompatibility is defined as, the state of not having any toxic or injurious effects on biological systems. On the other hand, biomaterial is regarded as a material intended to interface with the biological systems to treat, augment, evaluate or replace any tissue, organ or any function in the body (Williams *et al.*, 1999). Biomaterials have to be easily processable, sterilizable, and capable of controlled stability or degradation in response to biological conditions. The biocompatibility of a medical implant can be influenced by a number of factors such as the toxicity of the material used, form and design of the implant, technique of inserting the device and *in situ* dynamics or movement of the implant based on the surrounding matrix (Williams *et al.*, 1999). Biomaterials are being used in increasingly diverse and complex situations, with applications now involving tissue engineering, invasive sensors, drug delivery and gene transfection systems, the medically oriented nanotechnologies and biotechnology (Sudesh *et al.*, 2000). PHA namely P(4HB) and P(3HB-*co*-4HB) have been attracting much attention with regards to its unique properties as the natural metabolite, which enables it to be processed as biodegradable biomaterials. This is perhaps not entirely surprizing since hydrolysis of P(3HB-*co*-4HB) yields 4HB, a natural human metabolite present in blood. It has in fact recently approved the administration of very large doses of this metabolite in the medical field in view of the relatively slow rate of conversion of P4HB to 4HB. P(4HB) exhibited high tolerance *in vivo* and is considered as an excellent biomaterial based on the strict regulations by Food and Drug Administration (FDA) and careful observations by Tepha, Inc. (Martin and Williams, 2003). Based on observation, L929 mouse fibroblast and ovine vascular cells were seeded onto P(4HB) films showed successful amount of cell proliferation and adhesion (Sodian *et al.*, 2000). P(3HB-*co*-4HB) copolymers have shown good surface characteristics and promote cell attachment, proliferation and differentiation. P(3HB-*co*-4HB) copolymers films with 4HB molar fractions in the range of 23 to 75 mol% were found to support L929 cell growth and proliferation was demonstrated on all films tested (Vigneswari *et al.*, 2009). In a study, P(3HB-*co*-4HB) copolymers with 5, 24 and 38 mol% of 4HB were evaluated for dermal as well as orthopedic support using L929 and human dermal fibroblasts (Chanprateep *et al.*, 2010). The L929 and human dermal fibroblast cell densities grown on P(3HB-*co*-4HB) were better compared to the positive control, poly(lactic-coglycolic acid) (PLGA) films. Nevertheless, no significant morphological changes were observed. The number of human dermal fibroblasts increased proportionally with 4HB monomer composition. In addition, Ee and co-workers demonstrated the cytotoxicity of P(3HB-*co*-4HB) films with V79 and L929 cells using MTT assay, acridine orange/propidium iodide staining and alkaline comet assay (Ee *et al.*, 2009). In a following study, the alkaline comet assay revealed no genotoxic effects as well as mutagenic and clastogenic effects from post sterilized P(3HB-*co*-4HB) films (Ee *et al.*, 2009). Ames and *umu* tests (*S. typhimurium* strains) and micronucleus assay (V79 fibroblasts) were carried out. From the results obtained, it was indicated that leachables of poststerilized P(3HB-*co*-4HB) caused no mutagenic and clastogenic effects. Chee *et al.*, investigated the influence of P(3HB-*co*-4HB) composition ratio and drug loading level (*Mitragyna speciosa* crude extracts) on the biocompatibility of P(3HB-*co*-4HB) (Chee *et al.*, 2008). Positive results were obtained by observing good cell viability and this was comparable with PLGA (positive control).

It is known that different cells prefer different surfaces for attachment. The morphology and cell's capacity for proliferation and differentiation are influenced by the quality of cellular adhesion on materials. P(3HB-*co*-4HB) films are known to possess different surface morphologies depending on the 4HB content (Chee *et al.*, 2008). Besides, surface modification of P(3HB-*co*-4HB) films has been carried out to improve cell adhesion and promote better cell growth. Various techniques including alkaline hydrolysis, ammonia plasma treatment and lipase treatment were demonstrated. Besides these, immobilization of biomolecules such as collagen is also known to improve cell proliferation. In a recent study, blends of depyrogenated P(3HB-*co*-4HB) copolymer with vitamin E and/or collagen were produced and investigated *in vitro* (Rao *et al.*, 2010). It was presumed that addition of either vitamin E or collagen will improve the biological performance of P(3HB-*co*-4HB). The *in vitro* assay investigated the cytocompatibility and proinflammatory response using NIH 3T3 fibroblast and murine J774 A-1 monocyte or macrophage cells. The flow cytometric analysis confirmed that combination of depyrogenation and blending strategies of P(3HB-*co*-4HB) with vitamin E and collagen significantly reduced apoptotic and necrotic cells. Hence, it was suggested that this type of blend extends the range of biomaterials suitable for tissue engineering.

In a separate study, nanofiber scaffold and films of P(3HB-*co*-4HB) copolymer prepared by phase separation process was investigated to mimic natural extracellular matrix (Xu *et al.*, 2010). Neural stem cells derived from Sprague-Dawley rats were seeded onto both P(3HB-*co*-4HB) nanofiber scaffold and film and incubated for a period of 10 days. At the end of the experiment, higher cell viability was observed with cells grown on P(3HB-*co*-4HB) film compared to PLA and P(3HB) films. Nevertheless, P(3HB-*co*-4HB) nanofiber scaffold significantly improved cell viability compared to film. The continuous fibrous network of the nanofiber scaffold, which results in highly interconnected porous structure (average diameters of pores = 50–500 nm) enabled the neurites to penetrate into the interior of the scaffold matrices and establish close connection between cells as well as absorb nutrients, receive signals and discard wastes. The nanofiber scaffold was suggested to be useful for repairing central nervous system injury.

Composite of PGA and P(4HB) evaluated *in vivo* as heart valves and pulmonary artery patches showed positive results (Shinoka *et al.*, 1995; Hoerstrup *et al.*, 2000; Stock *et al.*, 2000; Opitz *et al.*, 2004). PGA/P(4HB) heart valves or pulmonary artery patches were reported to be tested in sheep (Hoerstrup *et al.*, 2000; Stock *et al.*, 2000). Besides this, various polysaccharides such as hyaluronic acid, collagen, chitosan, pectin or alginic acid was individually immobilized into P(3HB) and P(4HB) matrices to promote formation of porous surfaces. In fact, electrospun nanofibers of P(3HB-*co*-4HB) and other PHA constituent were evaluated as scaffolds *in vivo* (Ying *et al.*, 2008). Ying and co-workers prepared P(3HB-*co*-7 mol% 4HB) and P(3HB-*co*-97 mol% 4HB) nanofibers using 1,1,1,3,3,3-hexafluoro-2-propanol (HFIP) as the solvent (Ying *et al.*, 2008). By average, the width of fibers for scaffold with 7 mol% 4HB was 190 nm, while scaffold with 97 mol% 4HB consisted of fibers with 220 nm. It was reported that the width of nanofibers increased proportionally with the molecular weight of polymer subjected for electrospinning. Nevertheless, after sterilization process, the matrix of P(3HB-*co*-7 mol% 4HB) remained unchanged, whereas, P(3HB-*co*-97 mol% 4HB) became less porous. Tissue tolerances and bioabsorption behavior of the electrospun P(3HB-*co*-4HB) scaffolds were tested *in vivo* by subcutaneous implantation in rats. Periodical histological observation of the implanted electrospun nanofibers showed that the scaffolds of P(3HB-*co*-7 mol% 4HB) and P(3HB-*co*-97 mol% 4HB) elicited mild tissue response. Thin connective tissue surrounding P(3HB-*co*-97 mol% 4HB) was observed and after 12 weeks of implantation, no fibrous encapsulation was found around the degraded scaffold. A considerable drop in the number of inflammatory cells projected minimal inflammatory response. On the other hand, the bioabsorption and degradation rate of P(3HB-*co*-97 mol% 4HB) was much higher compared to all the other PHA scaffolds tested. It was found that the mechanical properties of P(3HB-*co*-4HB) nanofiber scaffolds were comparable to those of human skin, suggesting that the scaffolds could provide sufficient biomechanical support when used for tissue engineering applications. Electrospun nano-fiber of P(3HB-*co*-4HB) nanofiber membranes were also prepared via electrospinning from a combination of solvent such as chloroform and dimethylformamide (DMF) (Yang and Cai, 2011; Vigneswari *et al.*, 2012). P(3HB-*co*-4HB) with various surface architecture were prepared using different techniques such as enzymatic degradation and salt-leaching technique to create a porous surface for biomedical application (Vigneswari *et al.*, 2012). Recently, P(3HB-*co*-70%4HB) copolymer was used to develop nanobiocomposites in combination with nanoclay reinforcement. Pronounced improvement in the optical transparency, mechanical and thermal properties were achieved through reinforcement of 5 wt% Claytone into P(3HB-*co*-70%4HB) having the lowest molecular weight as compared to the other polymers. P(3HB-*co*-70%4HB)/5 wt% Claytone composite also exhibited enhancement in the antimicrobial performance, which increased with the clay concentrations. P(3HB-*co*-70%4HB), a biocompatible and biodegradable nanocomposite, which had demonstrated salient features with comparable good performance is believed to create new prospects with special incidence in regenerative medicine and as environmentally friendly materials (green nanocomposites) (Wang *et al.*, 2012). A novel biocompatible P(3HB-*co*-4HB) blend with the natural polymer collagen was fabricated by facile blending technique with an addition of vitamin E and evaluated with cell lines J774-11 monocyte/macrophage and NIH 3T3 fibroblast cell lines as potential engineered tissue (Rao *et al.*, 2011).

11.5 P(3HB-*CO*-4HB) AS BLENDS AND COMPOSITES

Biobased P(3HB-*co*-4HB) containing 4.0 mol% 4-hydroxybutyrate (4HB) was melt-mixed with short glass fibers (SGF) via a co-rotating twin-screw extruder. The compositing conditions, average glass fiber length and distribution, thermal, crystallization, and mechanical properties of the P3HB/4HB/SGF composites were investigated. It was found that the tensile strength, tensile modulus, and impact strength of the composites improved 6.6 times comparatively of the P(3HB-*co*-4HB) copolymer (Yu *et al.*, 2012). Poly(butylene succinate) [PBS]/P(3HB-*co*-4HB) blends were prepared by mechanically melting mixing. The crystallinity, rheological behavior and mechanical properties of the PBS/P(3HB-*co*-4HB) were studied respectively by differential scanning calorimeter (DSC), rheometer and universal material testing machine. The crystallization temperature of PBS increase with the P(3HB-*co*-4HB) increasing. With the increase of the content of P(3HB-*co*-4HB), the storage modulus (G') and loss modulus of blends increased significantly. Hence, the addition of P(3HB-*co*-4HB) could improve the properties of PBS (Tang *et al.*, 2012). Biodegradable polymer blends based on biosourced polymers, namely polylactide (PLA) blend with P(3HB-*co*-4HB), were prepared by melt compounding. The blend was an immiscible system with the P(3HB-*co*-4HB) domains evenly dispersed in the PLA matrix. However, the T_g of P(3HB-*co*-4HB) component in the blends decreased compared with neat P(3HB-*co*-4HB), which might be attributed to that the presence of the phase interface between PLA and P(3HB-*co*-4HB) resulting in enhanced chain mobility near interface. The addition of P(3HB-*co*-4HB) enhanced the cold crystallization of PLA in the blends due to the nucleation enhancement of PLA caused by the enhanced chain mobility near the phase interface between PLA and P(3HB-co-4HB) in the immiscible blends. The elongation to break was increased significantly, indicating that the inherent brittlement of PLA was improved by adding P(3HB-co-4HB). The interesting aspect was that the biodegradability of PLA is significantly enhanced after blends preparation (Han et al., 2012). P(3HB-co-4HB) and cobalt-aluminum layered double hydroxide (LDH) were prepared via melt intercalation. The thermal stability, thermal combustion and thermo-mechanical properties for these bio-nanocomposites were systematically investigated. The microscale combustion calorimetry showed that the heat release capacity (HRC) of the bio-nanocomposites is an important parameter of the fire hazard, which is significantly reduced with the addition of LDH. The storage modulus of the bio-nanocomposites with small amount of LDH is remarkably enhanced measured by dynamic mechanical analysis (Zhang et al., 2012). Composite membrane of porous P(3HB-co-4HB)/calcium metaphosphate (CMP) was fabricated by simple blending to enhance the properties of P(3HB-co-4HB) (Wu et al., 2012). A series of alternating block polyurethanes (abbreviated as PU3/4HB-alt-PEG) and random block polyurethanes (abbreviated as PU3/4HB-ran-PEG) based on biodegradable P(3HB-co-4HB) and poly(ethylene glycol) (PEG) with similar chemical compositions were synthesized using 1, 6-hexamethylene diisocyanate (HDI) as coupling agent. The PU3/4HB-alt-PEG exhibited good hemocompatibility based on the platelet adhesion study as the composite possessed hydrophilic surface and evident microstructure surface. The cell culture assay demonstrated that fibroblasts and rat glial cells were more favorable for attachment on PU3/4HB-alt-PEG films. It is said alternating block polyurethanes provides a way to control the exact structure of the biomaterials and tailor better properties for biomedical requirements (Li et al., 2012; Qiu et al., 2013).

11.6   CONCLUSION AND FUTURE OUTLOOK

Tailoring surface properties of biocompatible and degradable polymer scaffolds is the key to progress in various tissue engineering strategies. The expansion of the modern medical sciences demands extensive research and development of biocompatible and biodegradable biomaterials. With the increasing global concern on environmental issues such as waste management and depletion of finite raw materials, utilization of biodegradable materials in biomedical research is very much welcomed. In the past 40 years, an increasing number of biodegradable polymers have been examined. However, materials with optimal properties deemed as potential biomaterial are still not available for many clinical applications. PHA especially with the 4HB monomer has showed excellent biocompatibility *in vitro* and *in vivo* and has gained much recognition over the last decade. Therefore, it has been extensively investigated for various pharmaceutical and medical applications. Fine-tuning the physicochemical surface characteristics of this biodegradable PHA material allows in modulating its application in tissue engineering. In order to ensure constant supply of this material, various approaches are currently being considered for the production of high yields of polymer through efficient fermentation processes.

KEYWORDS

- **4-hydroxybutyrate**
- **Biocompatibility**
- **Biomaterials**
- **Biosynthesis of poly(3-hydroxybutyrate-*co*-4-hydroxybutyrate)**
- **Composites**
- **Medical and pharmaceutical applications**
- **Polyhydroxyalkanoate**

REFERENCES

Abou-Zeid, D. M.; Müller, R. J.; Deckwer, W. D. Degradation of natural and synthetic polyesters under anaerobic conditions. *J Biotechnol* 2001, 86(2), 113–126.

Adams, J. H.; Irving, G.; Koeslag, J. H. Adrenergic blockade restores glucose's antiketogenic activity after exercise in carbohydrate-depleted athletes. *J Physiology* 1987, 386, 439–454.

Amirul, A. A.; Yahya, A. R. M.; Sudesh, K.; Azizan, M. N. M.; Majid, M. I. A. Isolation of poly(3-hydroxybutyrate-*co*-4-hydroxybutyrate) producer from Malaysian environment using γ-butyrolactone as carbon source. *World J Microb Biot* 2009, 25(7), 1199–1206.

Amirul, A. A.; Syairah, S. N.; Yahya, A. R. M.; Azizan, M. N. M.; Majid, M. I. A. Synthesis of biodegradable polyesters by Gram negative bacterium isolated from Malaysian environment. *World J Microb Biot* 2008a, 24(8): 1327–1332.

Amirul, A. A.; Yahya, A. R. M.; Sudesh, K.; Azizan, M. N. M.; Majid, M. I. A. Biosynthesis of poly(3-hydroxybutyrate-*co*-4-hydroxybutyrate) copolymer by *Cupriavidus* sp. USMAA1020 isolated from lake kulim, Malaysia. *Bioresource Technol* 2008b, 99(11), 4903–4909.

Anderson, A. J.; Dawes, E. A. Occurrence, metabolism, metabolic role, and industrial uses of bacterial polyhydroxyalkanoates. *Microbiol Reviews* 1990, 54(4), 450–472.

Bhatt, R.; Shah, D.; Patel, K. C.; Trivedi, U. PHA-rubber blends: synthesis, characterization and biodegradation. *Bioresource Technol* 2008, 99(11), 4615–4620.

Bhubalan, K.; Lee, W. H.; Sudesh, K. Polyhydroxyalkanoate. In: *Biodegradable polymers in clinical use and clinical development biopolymers*, Domb, A. J., Kumar, N., Arza, A., Eds.; WILEY-VCH: New Jersey, 2011; 249–315.

Blümm, E.; Owen, A. J. Miscibility, crystallization and melting of poly(3-hydroxybutyrate)/ poly(L-lactide) blends. *Polymer* 1995, 36, 4077–4081.

Chanprateep, S.; Buasri, K.; Muangwong, A.; Utiswannakul, P. Biosynthesis and biocompatibility of biodegradable poly(3-hydroxybutyrate-*co*-4-hydroxybutyrate). *Polym Degrad Stabil* 2010, 95(10), 2003–2012.

Chee, J. W.; Amirul, A. A.; Muhammad, T. S.; Majid, M. I. A.; Mansor, S. M. The influence of copolymer ratio and drug loading level on the biocompatibility of P(3HB-*co*-4HB) synthesized by *Cupriavidus* sp. (USMAA2–4). *Biochem Eng J* 38(3):314–318.

Chen, G.; Zhang, G.; Park, S.; Lee, S. Industrial scale production of poly(3-hydroxybutyrate-*co*-3-hydroxyhexanoate). *Appl Microbiol Biotechnol* 2001, 57(1), 50–55.

Chen, G. Q.; Wu, Q. The application of polyhydroxyalkanoates as tissue engineering materials. *Biomaterials* 2005, 26(33), 6565–6578.

Doi, Y. *Microbial polyesters*, VCH: New York, 1990.

Ee, L. S.; Rajab, N. F.; Osman, A. B.; Sudesh, K.; Inayat-Hussain, S. H. Mutagenic and clastogenic characterization of poststerilized poly(3-hydroxybutyrate-*co*-4-hydroxybutyrate) copolymer biosynthesized by *Delftia acidovorans*. *J Biomed Mater Res A* 2009, 91(3), 786–794.

Foster, L. J. R.; Davies, S. M.; Tighe, B. J. Centrifugally spun polyhydroxybutyrate fibers: effect of process solvent on structure, morphology and cell response. *J Biomat Sci-Polym E* 2001, 12(3), 317–336.

Freier, T. Biopolyesters in tissue engineering applications. *Adv Polym Sci* 2006, 203, 1–61.

Gorenflo, V.; Steinbüchel, A.; Marose, S.; Rieseberg, M.; Scheper, T. Quantification of bacterial polyhydroxyalkanoic acids by Nile red staining. *Appl Microbiol Biotechnol* 1999, 51(6), 765–772.

Han, L.; Han, C.; Zhang, H.; Chen, S.; Dong, L. Morphology and properties of biodegradable and biosourced polylactide blends with poly(3-hydroxybutyrate-*co*-4-hydroxybutyrate. *Polym Composite* 2012, 33(6), 850–859.

Hoerstrup, S. P.; Sodian, R.; Daebritz, S.; Wang, J.; Bacha, E. A.; Martin, D. P.; Moran, A. M.; Guleserian, K. J.; Sperling, J. S.; Kaushal, S.; Vacanti, J. P.; Schoen, F. J.; Mayer, J. E. Functional living trileaflet heart valves grown *in vitro*. *Circulation* 2000, 102, 44–49.

Jendrossek, D.; Schirmer, A.; Schlegel, H. G. Biodegradation of polyhydroxyalkanoic acids. *Appl Microbiol Biotechnol* 1996, 46(5–6), 451–463.

Kahar, P.; Tsuge, T.; Taguchi, K.; Doi, Y. High yield production of polyhydroxyalkanoates from soybean oil by *Ralstonia eutropha* and its recombinant strain. *Polym Degrad Stabil* 2004, 83(1), 79–86.

Kang, C.; K. Kusaka, S.; Doi, Y. Structure and properties of poly(3-hydroxybutyrate-*co*-4-hydroxybutyrate) produced by *Alcaligenes latus*. *Biotechnol Lett* 1995, 17(6), 583–588.

Kessler, B.; Ren, Q.; De Roo, G.; Prieto, M. A.; Witholt, B. Engineering of biological systems for the synthesis of tailor-made polyhydroxyalkanoates, a class of versatile polymers. *Chimia* 2001, 55(3), 119–122.

Khanna, S.; Srivastava, A. K. Recent advances in microbial polyhydroxyalkanoates. *Process Biochem* 2005, 40(2), 607–619.

Kim, Y.; Lenz, R. Polyesters from microorganisms. In: *Biopolyesters*, Babel, W., Steinbuchel, A., Eds.; Springer: Berlin, 2001; 51–79.

Kobayashi, T.; Uchino, K.; Abe, T.; Yamazaki, Y.; Saito, T. Novel intracellular 3-hydroxybutyrate-oligomer hydrolase in *Wautersia eutropha* H16. *J Bacteriol* 2005, 187(15), 5129–5135.

Koyama, N.; Doi, Y. Miscibility of binary blends of poly[(R)-3-hydroxybutyric acid] and poly[(S)-lactic acid]. *Polymer* 1997, 38, 158–1593.

Lee, S. Y. Bacterial polyhydroxyalkanoates. *Biotechnol Bioeng* 1996, 49(1), 1–14.

Lee, W. H.; Azizan, M. N. M.; Sudesh, K. Effects of culture conditions on the composition of poly(3-hydroxybutyrate-*co*-4-hydroxybutyrate) synthesized by *Comamonas acidovorans*. *Polym Degrad Stabil* 2004, 84(1), 129–134.

Lee, W. H.; Loo, C. Y.; Nomura, C. T.; Sudesh, K. Biosynthesis of polyhydroxyalkanoate copolymers from mixtures of plant oils and 3-hydroxyvalerate precursors. *Bioresource Technol* 2008, 99(15), 6844–6851.

Li, G.; Liu, Y.; Li, D.; Zhang, L.; Xu, K. A comparative study on structure property elucidation of P3/4HB and PEG-based block polyurethanes. *J Biomed Mater Res A* 2012, 100(9), 2319–2329.

Li, R.; Zhang, H.; Qi, Q. The production of polyhydroxyalkanoates in recombinant *Escherichia coli*. *Bioresource Technol* 2007, 98(12), 2313–2320.

Li, Z. J.; Shi, Z. Y.; Jian, J.; Guo, Y. Y.; Wu, Q.; Chen, G. Q. Production of poly(3-hydroxybutyrate-*co*-4-hydroxybutyrate) from unrelated carbon sources by metabolically engineered *Escherichia coli*. *Metab Eng* 2010, 12(4), 352–359.

Loo, C. Y.; Sudesh, K. Biosynthesis and native granule characteristics of poly(3-hydroxybutyrate-*co*-3-hydroxyvalerate) in *Delftia acidovorans*. *Int J Biol Macromol* 2007, 40(5), 466–471.

Madison, L. L.; Huisman, G. W. Metabolic engineering of poly(3-hydroxyalkanoates): from DNA to plastic. *Microbiol Biol Rev* 1999, 63(1), 21–53.

Martin, D. P.; Williams, S. F. Medical applications of poly 4-hydroxybutyrate: a strong flexible absorbable biomaterial. *Biochem Eng J* 2003, 16, 97–105.

Matsusaki, H.; Abe, H.; Doi, Y. Biosynthesis and properties of poly(3-hydroxybutyrate-*co*-3-hydroxyalkanoates) by recombinant strains of *Pseudomonas* sp. 61–3. *Biomacromolecules* 2000, 1(1), 17–22.

Mergaert, J.; Webb, A.; Anderson, C.; Wouters, A.; Swings, J. Microbial degradation of poly(3-hydroxybutyrate) and poly(3-hydroxybutyrate-*co*-3-hydroxyvalerate) in soils. *Appl Environ Microbiol* 1993, 59(10), 3233–3238.

Mitomo, H.; Hsieh, W. C.; Nishiwaki, K.; Kasuya, K.; Doi, Y. Poly(3-hydroxybutyrate-*co*-4-hydroxybutyrate) produced by *Comamonas acidovorans*. *Polymer* 2001, 42(8), 3455–3461.

Mukai, K.; Yamada, K.; Doi, Y. Efficient hydrolysis of polyhydroxyalkanoates by *Pseudomonas stutzeri* YM1414 isolated from lake water. *Polym Degrad Stabil* 1994, 43(3), 319–327.

Opitz, F.; Schenke-Layland, K.; Cohnert, T. U.; Starcher, B.; Halbhuber, K. J.; Martin, D. P.; Stock, U. A. Tissue engineering of aortic tissue: dire consequence of suboptimal elastic fiber synthesis *in vivo*. *Cardiovasc Res* 2004, 63, 719–730.

Philip, S.; Keshavarz, T.; Roy, I. Polyhydroxyalkanoates: biodegradable polymers with a range of applications. *J Chem Technol Biotechnol* 2007, 82(3), 233–247.

Potter, M.; Steinbüchel, A. Poly(3-hydroxybutyrate) granule-associated proteins: impacts on poly(3-hydroxybutyrate) synthesis and degradation. *Biomacromolecules* 2005, 6(2), 552–560.

Qiu, H.; Li, D.; Chen, X.; Fan, K.; Ou, W.; Chen, K. C.; Xu, K. Synthesis, characterizations, and biocompatibility of block poly(ester-urethane) based on biodegradable poly(3-hydroxy-butyrate-*co*-4-hydroxybutyrate) (P3/4HB) and poly(ε-caprolactone). *J Biomed Mater Res A* 2013, 101(1), 75–86.

Rahayu, A.; Zaleha, Z.; Yahya, A. R. M.; Majid, M. I. A.; Amirul, A. A. Production of copolymer poly(3-hydroxybutyrate-*co*-4-hydroxybutyrate) through a one-step cultivation process. *World J Microb Biot* 2008, 24(11), 2403–2409.

Rao, U.; Kumar, R.; Balaji, S.; Sehgal, P. K. A novel biocompatible poly(3-hydroxy-*co*-4-hy-droxybutyrate) blend as a potential biomaterial for tissue engineering. *J Bioact Compat Pol* 2011, 26, 452–463.

Rao, U.; Sridhar, R.; Sehgal, P. K. Biosynthesis and biocompatibility of poly(3-hydroxybutyr-ate-*co*-4-hydroxybutyrate) produced by *Cupriavidus necator* from spent palm oil. *Biochem Eng J* 2010, 49(1), 13–20.

Rehm, B. H. A.; Steinbüchel, A. Biochemical and genetic analysis of PHA synthases and other proteins required for PHA synthesis. *Int J Biol Macromol* 1999, 25(1–3), 3–19.

Saito, Y.; Doi, Y. Microbial synthesis and properties of poly(3-hydroxybutyrate-*co*-4-hydroxy-butyrate) in *Comamonas acidovorans*. *Int J Biol Macromol* 1994, 16(2), 99–104.

Saito, Y.; Nakamura, S.; Hiramitsu, M.; Doi, Y. Microbial synthesis and properties of poly(3-hydroxybutyrate-*co*-4-hydroxybutyrate). *Polym Int* 1996, 39(3), 169–174.

Shinoka, T.; Breuer, C. K.; Tanel, R. E.; Zund, G.; Miura, T.; Ma, P. X.; Langer, R.; Vacanti, J. P.; Mayer, J. E. Tissue engineering heart valves: valve leaflet replacement study in a lamb model. *The Annals of Thoracic Surgery* 1995, 60S, S513-S516.

Singh, S.; Mohanty, A. K.; Sugie, T.; Takai, Y.; Hamada, H. Renewable resource based biocom-posites from natural fiber and polyhydroxybutyrate-*co*valerate (PHBV). *Bioplastic. Compos Part A – Appl* 2008, 39(5), S875-S886.

Sodian, R.; Sperling, J. S.; Daebritz, S.; Egozy, A.; Stock, U.; Mayer, J. E. Jr. Fabrication of trileaflet heart valve scaffold from a polyhydroxyalkanoate biopolyester for use in tissue en-gineering. *Tissue Eng* 2000, 6, 183–188.

Spiekermann, P.; Rehm, B. H. A.; Kalscheuer, R.; Baumeister, D.; Steinbüchel, A. A sensitive, viable-colony staining method using Nile red for direct screening of bacteria that accumulate polyhydroxyalkanoic acids and other lipid storage compounds. *Arch Microbiol* 1999, 171(2), 73–80.

Sridewi, N.; Bhubalan, K.; Sudesh, K. Degradation of commercially important polyhydroxyal-kanoates in tropical mangrove ecosystem. *Polym Degrad Stabil* 2006, 91(12), 2931–2940.

Steinbüchel, A.; Lütke-Eversloh, T. Metabolic engineering and pathway construction for bio-technological production of relevant polyhydroxyalkanoates in microorganisms. *Biochem Eng J* 2003, 16(2), 81–96.

Stock, U. A.; Sakamoto, T.; Hatsuoka, S.; Martin, D. P.; Nagashima, M.; Moran, A. M.; Moses, M. A.; Khalil, P. N.; Schoen, F. J.; Vacanti, J. P.; Mayer, J. E. Patch augmentation of the pulmonary artery with bioabsorbable polymers and autologous cell seeding. *J Thorac Cardiov Sur* 2000, 120, 1158–1167.

Sudesh, K.; Iwata, T. Sustainability of biobased and biodegradable plastics. *CLEAN* 2008, 36, 433–442.

Sudesh, K. Microbial polyhydroxyalkanoates (PHAs): an emerging biomaterial for tissue engineering and therapeutic applications. *Med J Malaysia* 2004, 59B, S55-S56.

Sudesh, K.; Abe, H.; Doi, Y. Synthesis, structure and properties of polyhydroxyalkanoates: biological polyesters. *Prog Polym Sci* 2000, 25, 1503–1555.

Sudesh, K.; Doi, Y. Molecular design and biosynthesis of biodegradable polyesters. *Polym Adv Technol* 2000, 11(8–12), 865–872.

Sudesh, K.; Fukui, T.; Taguchi, K.; Iwata, T.; Doi, Y. Improved production of poly(4-hydroxybutyrate) by *Comamonas acidovorans* and its freeze-fracture morphology. *Int J Biol Macromol* 1999, 25(1–3), 79–85.

Sun, Z. Y.; Ramsay, J. A.; Guay, M.; Ramsay, B. A. Fermentation process development for the production of medium-chain-length poly 3-hyroxyalkanoates. *Appl Microbiol Biotechnol* 2007, 75(3), 475–485.

Suriyamongkol, P.; Weselake, R.; Narine, S.; Moloney, M.; Shah, S. Biotechnological approaches for the production of polyhydroxyalkanoates in microorganisms and plants – a review. *Biotechnol Adv* 2007, 25(2), 148–175.

Taguchi, S.; Doi, Y. Evolution of polyhydroxyalkanoate (PHA) production system by "enzyme evolution": successful case studies of directed evolution. *Macromol Biosci* 2004, 4(3), 145–156.

Tang, Y.; Liang, D.; Lou, B. Crystallization and rheological behavior of poly(butylenessuccinate)/poly(3-hydroxybutyrate-*co*-4-hydroxybutyrate) blends. *Polym Mat Sci Eng* 2012, 28(6), 28–31.

Tsuge, T. Metabolic improvements and use of inexpensive carbon sources in microbial production of polyhydroxyalkanoates. *J Biosci Bioeng* 2002, 94(6), 579–584.

Valentin, H. E.; Dennis, D. Production of poly(3-hydroxybutyrate-*co*-4-hydroxybutyrate) in recombinant *Escherichia coli* grown on glucose. *J Biotechnol* 1997, 58(1), 33–38.

Vigneswari, S.; Majid, M. I. A.; Amirul, A. A. Tailoring the surface architecture of poly(3-hydroxybutyrate-*co*-4-hydroxybutyrate) scaffolds. *J Appl Polym Sci* 2012, 124(4), 2777–2788.

Vigneswari, S.; Nik, L. A.; Majid, M. I. A.; Amirul, A. A. Improved production of poly(3-hydroxybutyrate-*co*-4-hydroxbutyrate) copolymer using a combination of 1,4-butanediol and γ-butyrolactone. *World J Micro Biot* 2010, 26(4), 743–746.

Vigneswari, S.; Vijaya, S.; Majid, M. I. A.; Sudesh, K.; Sipaut, C. S.; Azizan, M. N. M.; Amirul, A. A. Enhanced production of poly(3-hydroxybutyrate-*co*-4-hydroxybutyrate) copolymer with manipulated variables and its properties. *J Ind Microbiol Biot* 2009, 36(4), 547–556.

Wang, K.; Wang, Y.; Zhang, R.; Li, Q.; Shen, C. Preparation and characterization of microbial biodegradable poly(3-hydroxybutyrate-*co*-4-hydroxybutyrate)/organoclay nanocomposites. *Polym Compos* 2012, 33(5), 838–842.

Wang, Y. W.; Mo, W.; Yao, H.; Wu, Q.; Chen, J.; Chen, G. Q. Biodegradation studies of poly(3-hydroxybutyrate-*co*-3-hydroxyhexanoate). *Polym Degrad Stabil* 2004, 85(2), 815–821.

Wiggam, M. I.; O'Kane, M. J.; Harper, R.; Atkinson, A. B.; Hadden, D. R.; Trimble, E. R.; Bell, P. M. Treatment of diabetic ketoacidosis using normalization of blood 3-hydroxybutyrate concentration as the endpoint of emergency management: a randomized controlled study. *Diabetes Care* 1997, 20(9), 1347–1352.

Williams, S. F.; Martin, D. P. Applications of PHAs in medicine and pharmacy. In: *Biopolymers: polyesters III, applications and commercial products*, Doi, Y., Steinbüchel, A., Eds.; WILEY-VCH: Germany, 2002, 91–127.

Williams, S. F.; Martin, D. P.; Horowitz, D. M.; Peoples, O. P. PHA applications: addressing the price performance issue. I. Tissue engineering. *Int J Biol Macromol* 1999, 25(1–3), 111–121.

Wu, Y.; Zhang, L.; Zeng, X.; Mai, L.; Wu, C.; Tang, S.; Yu, X. Preparation and degradation property of composite membrane of porous poly(3-hydroxybutyrate-*co*-4-hydroxybutyrate)/calcium metaphosphate. *J Clin Rehab Tissue Eng Res* 2012, 15(51), 9531–9534.

Xu, X. Y.; Li, X. T.; Peng, S. W.; Xiao, J. F.; Liu, C.; Fang, G.; Chen, K. C.; Chen, G. Q. The behavior of neural stem cells on polyhydroxyalkanoate nanofiber scaffolds. *Biomaterials* 2010, 31(14), 3967–3975.

Yang, G.; Cai, Z. J. Preparation and characterization of electrospun poly(3 -hydroxybutyrate-*co*-4-hydroxybutyrate)nanofiber menbranes. *Adv Mat Res* 2011, 1527, 332–334.

Yang, X.; Zhao, K.; Chen, G. Q. Effect of surface treatment on the biocompatibility of microbial polyhydroxyalkanoates. *Biomaterials* 2002, 23(5), 1391–1397.

Ying, T. H.; Ishii, D.; Mahara, A.; Murakami, S.; Yamaoka, T.; Sudesh, K.; Samian, R.; Fujita, M.; Maeda, M.; Iwata, T. Scaffolds from electrospun polyhydroxyalkanoate copolymers: fabrication, characterization, bioabsorption and tissue response. *Biomaterials* 2008, 29(10), 1307–1317.

Yoshie, N.; Azuma, Y.; Sakurai, M.; Inoue, Y. Crystallization and compatibility of poly(vinyl alcohol)/poly(3-hydroxybutyrate) blends: influence of blend composition and tacticity of poly(vinyl alcohol). *J Appl Polym Sci* 1995, 56(1), 17–24.

Yu, Z.; Yang, Y.; Zhang, L.; Ding, Y.; Chen, X.; Xu, K. Study on short glass fiber-reinforced poly(3-hydroxybutyrate-*co*-4-hydroxybutyrate) composites. *J Appl Polym Sci* 2012, 126(3), 822–829.

Zhang, K.; Mohanty, A. K.; Misra, M. Fully biodegradable and biorenewable ternary blends from polylactide, poly(3-hydroxybutyrate-*co*hydroxyvalerate) and poly(butylene succinate) with balanced properties. *ACS Appl Mat Interf* 2012, 4(6), 3091–310.

Zinn, M.; Witholt, B.; Egli, T. Occurrence, synthesis and medical application of bacterial polyhydroxyalkanoate. *Adv Drug Deliver Rev* 2001, 53(1), 5–21.

CHAPTER 12

ALGAL BIOTECHNOLOGY FOR BIOENERGY, ENVIRONMENTAL REMEDIATION AND HIGH-VALUE BIOCHEMICALS

MOHD AZMUDDIN ABDULLAH, SYED MUHAMMAD USMAN SHAH, ASHFAQ AHMAD, and HAMDY EL-SAYED

CONTENTS

12.1   INTRODUCTION

Algae are one of the most primitive life forms and most abundant on earth. They consist of one or more of eukaryotic cells containing chlorophyll. It may be a single cell, colonies, filament of cells, or as in the kelp, simple tissues. The cell doubling time is typically 1–2 days, and under optimal conditions can be as short as 6 hours for reported in *Chlamydomanas* species (Chen *et al.*, 2009), as compared to bacteria (20 min), animal cells (20 hours) and plant cells (2–4 days). Algae inhabit particularly the oceans, rivers, lakes, streams, ponds and swamps, but some species can be found in the soil and survive dry conditions for a long time. Because of such diverse ecological habitats ranging from seawater, freshwater or brackish water, they are equipped to flourish in extreme temperature and pHs such as at 80°C, in and around hot springs, or in the snow and ice of Arctic and Antarctic regions (Rosenberg *et al.*, 2008). In the tropics and subtropics, algae may be found on leaves, woods and stones, within or on plants and animals. Although for many years categorized as plants because of their photosynthetic ability, algae are now placed within the diverse kingdom Protista of eukaryotic, predominantly single-celled microscopic organisms (Hollar, 2012).

It is estimated that the upper limit of algal species in nature is about 10 million, and only a small portion of which is identified taxonomically (Figs. 12.1 and 12.2). Algae are classified into (1) microalgae, which include diatoms (*Bacillariophyceae*), green algae (*Chlorophyceae*), red algae (*Rhodophyceae*), yellow-green algae (*Xanthophyceae*), golden algae (*Chrysophyceae*), brown algae (*Phaeophyceae*), and Euglenoids; and (2) macroalgae (seaweeds or multicellular algae), commonly categorized into green (*Chlorophyta*), brown (*Phaeophyta*) and red algae (*Rhodophyta*). Most of the red and brown algae are marine species, though there are some rare freshwater species in each group. Blue-green algae are single-celled organisms that use photosynthesis to create food, do not have eukaryotic cells, and are not classified as algae, but as chlorophyll-containing cyanobacteria (Hollar, 2012).

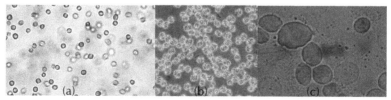

FIGURE 12.1   Locally isolated microalgal strains: (a) *Isochrysis galbana*, (b) *Tetraselmis* species, (c) *Nannochloropsis oculata*.

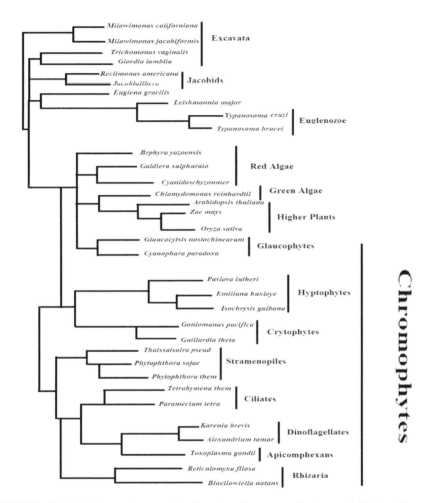

FIGURE 12.2 Phylogenetic tree showing the placement of algae and higher plants among some of their common protist neighbors (Larkum *et al.*, 2012).

Algae offer diverse spectrum of valuable products and environmental solutions such as food, nutritional compounds, omega-3 fatty acids, animal feed, energy sources (including jet fuel, aviation gas, biodiesel, gasoline, and bioethanol), organic fertilizers, biodegradable plastics, recombinant proteins, pigments, medicines, pharmaceuticals and vaccines. Red algae are the most important sources of many biologically active metabolites (Pienkos and Darzins, 2009). The productivity of algae can be enhanced through simple, efficient and economical reactor design, culturing, harvesting and extraction technique. The development in genetic and metabolic engineering is likely to have the greatest impact on improving the economics of microalgal productivity (Hankamer *et al.*, 2007).

12.2 ALGAL CULTIVATION

Most of commercial, large-scale outdoor microalgae cultivation are artificial open ponds as they are cheap to build and easy to operate and scale up (Brennan and Owende, 2010). The raceway ponds, usually lined with plastic or cement, are about 20 to 35 cm deep to ensure adequate exposure to sunlight. Paddle wheels provide motive force and keep algae suspended in water. The ponds are supplied with water and nutrients, and mature algae are continuously removed at one end. The raceway ponds holds relatively low capital and maintenance costs while circular ponds are less attractive because of expensive concrete construction, high energy consumption of stirring, the mechanical complexity of supplying CO_2 and inefficient land use (Chen *et al.*, 2009). There are, however, several disadvantages such as low productivity and biomass yield, high harvesting cost, water loss through evaporation, limited number of species that can be grown in ponds as they are vulnerable to contamination, lower efficiency of carbon dioxide use and temperature fluctuations due to diurnal variations as they are difficult to control in open ponds (Chisti, 2007; Chen *et al.*, 2009). Our study in comparing between small-scale and 30–300L open tank system for cultivation of *Pavlova lutheri* (Table 12.1) suggest that major challenges to address are efficient agitation for homogeneous distribution of cell suspension and nutrients, and getting representative sampling (Figs. 12.3 and 12.4).

TABLE 12.1 Comparison of kinetics between 250 mL and large-scale batch cultures of *Pavlova lutheri* (Shah *et al.*, 2013).

Experimental conditions	Final cell concentration, N_{max} (10^6cells. ml^{-1})	Final dry weight, X_{max} (gL^{-1})	Maximum biomass formation rate, X'_{max} ($gL^{-1}d^{-1}$)	Maximum specific growth rate, μ_{max} (d^{-1})	Doubling time, t_d (day)
250 mL	13.5	0.45	0.028	0.12	5.77
1 L	12.2	0.43	0.026	0.12	5.77
5 L	10.9	0.41	0.025	0.13	5.33
30 L	9.77	0.39	0.024	0.14	4.95
300 L	9.65	0.35	0.022	0.12	5.77

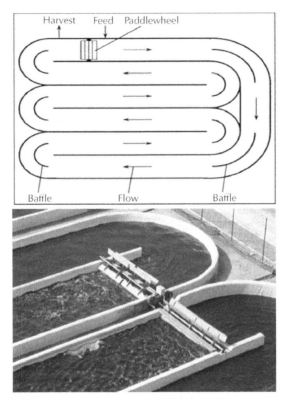

FIGURE 12.3 Arial view of a raceway pond (Chisti, 2007).

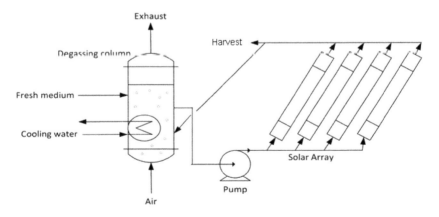

FIGURE 12.4 Tubular Horizontal Photobioreactor (Chisti, 2007).

The limitations of open pond systems lead to the development of enclosed photobioreactors (PBRs). There are two major types – tubular and plate types, with only the tubular reactor is used commercially. The enclosed structure, narrow light path, large illuminating area and relatively controllable environment, together with less contamination issues facilitate higher cell density in PBRs than in open pond system (Ugwu et al., 2008). Table 12.2 suggests that PBR is superior, capable of producing nearly 30 times more concentrated biomass, requiring significantly less surface area. The major drawbacks of PBRs for mass production of biofuels are high capital and operation costs, excessive energy demand for pumping, mixing and harvesting of culture and medium, the process gradients of pH, dissolved oxygen and CO_2 along the tubes, wall growth, fouling, hydrodynamic stress, and high expenses to scale up (Borowitzka, 2008; Ugwu et al., 2008). Techno-economic studies show that the financial feasibility of PBRs is substantially lower than for open ponds. In the base case, average total cost of lipid production is $31.61/gal for PBRs and $12.73/gal for open ponds (Richardson et al., 2012). Innovative hybrid PBR concept with reduced energy demand, higher biomass concentration, and lower production cost must be developed to improve the economics (Morweiser et al., 2010).

TABLE 12.2   Comparison of photobioreactor and raceway production methods (Chisti, 2007).

Variable	Photobioreactor	Raceway ponds
Annual biomass production (kg)	100,000	100,000
Volumetric productivity (kg $m^{-3}d^{-1}$)	1.535	0.117
Areal productivity (kg $m^{-2}d^{-1}$)	0.072	0.035
Biomass concentration in broth (kg m^{-3})	4	0.14
Dilution rate (d^{-1})	0.384	0.25
Area needed (m^2)	5681	7828
Oil yield (m^3ha^{-1})	58.7	42.6
Annual CO_2 consumption (kg)	183,333	183,333
System geometry	132 parallel tubes/unit; 80 m tubes; 0.06 m tube diameter	978 m^2/pond; 12 m wide, 82 m long, 0.30 m deep
Number of units	6	8

12.3 HARVESTING AND EXTRACTION

The economics of mass production by microalgae could be enhanced by improvements in cultivation technique and low energy, simple yet effective downstream processing. Harvesting and dewatering of small algal species in dilute suspensions at concentrations between less than 1 g/L (ponds) and 3–15 g/L (photobioreactors) are difficult and energy intensive. Dewatering to about 20–30% water content is necessary to reduce volume and weight, to minimize transportation and downstream costs and to extend the shelf-life of the microalgae concentrate (Grima *et al.*, 2003). Different physical, chemical, and biological methods can be applied depending on the type of algae, the requirements of the downstream processes, and the desired product quality. An initial step to aggregate microalgal cells and enhance the ease of sedimentation or centrifugal recovery, may involve autoflocculation and flocculation with alum, ferric chloride, chitosan, or hydrophobic absorbents and collection by dissolved air flotation, which thicken the material to 10% dry weight content (100 g/L) (Grima *et al.*, 2003).

Gravity sedimentation is economically attractive but requires substantial area and that the downstream processes and product targets are tolerant to coagulant contamination. High-speed continuous centrifugation may seem a preferred method as it is used commercially to harvest high-value metabolites for hatcheries and nurseries in aquaculture. The drawback is that strong gravitational and shear forces can damage cell structure (Harun *et al.*, 2010b), and require large capital investments, operating costs and high throughput processing of large quantities of water and algae (Johnson and Wen, 2010). Filtration by membrane, micropressure or vacuum filters are competitive for mechanical simplicity and availability in large unit sizes, but hampered by extensive operation costs and hidden preconcentration requirements (Harun *et al.*, 2010a). For large-scale microalgae such as *Chlorella*, *Scenedesmus* and *Spirulina*, spray-drying, drum-drying, freeze-drying, and sun-drying are used. Drying is required to achieve high biomass concentrations, and the costs climb steeply with incremental temperature and duration. Air-drying suits low-humidity climates, but require extra space and considerable time, while solar or wind may be subjected to climatic variation.

The concept of like substances dissolve in other substances with similar chemical properties is the basis behind the earliest and well-known cosolvent extraction procedure (Bligh and Dyer, 1959). Organic solvent extraction is a widely used method for lipid extraction from traditional oilseed plants, and different extraction systems have also been tested with algae cultures (Fajardo *et al.*, 2007). In order to maximize the lipid extraction efficiency, the organic solvent used has to match the lipid polarity profile in the cells. For industrial applications, the extraction solvents should be cheap, easy to remove, have low toxicity, insoluble in water, efficient in dissolving targeted components, and ideally recyclable.

Lipid extraction method using a mixture of chloroform and methanol [2:1 (v/v)] has been widely used for a variety of materials including animal or plant tissue, and microorganisms. With microalgae, chloroform extractables include hydrocarbons, carotenoids, chlorophylls, sterols; triacylglycerols, wax esters, long-chain alcohols, aldehydes and free fatty acids, and methanol extractables include phospholipids and traces of glycolipids, many of which are nonlipid compounds (Fajardo *et al.*, 2007). After extraction is complete, water is added to the cosolvent mixture until a two-phase system develops in which water and chloroform separate into two immiscible layers. The lipids mainly separate into the chloroform layer and can be recovered. Other combinations of cosolvents include hexane/isopropanol, dimethyl sulfoxide/petroleum ether, hexane/ethanol for microalgae and hexane/isopropanol (Park *et al.*, 2007). The hexane system has been promoted because hexane and alcohol will readily separate into two separate phases when water is added, thereby improving downstream separations. With greater need to address the green and environmental concerns, more intensive research must be developed in the use of more environmentally friendly supercritical liquid, ionic liquid or combination of physicomechanical disruption with water and reactive solvents such as hydrogen peroxide and organic acids (Nazir *et al.*, 2012).

12.4 BIOENERGY FROM MICROALGAE

Algae appear to represent the only means of renewable current generation of biofuels (Chisti, 2007; Li *et al.*, 2008; Schenk *et al.*, 2008). Microalgal biofuels have much lower impact on the environment and food supply than conventional biofuel-producing crops, have low viscosity and density, high calorific value, with properties more appropriate for biofuel than lignocellulosic materials due to their naturally high-lipid content, semisteady state production and suitability in variety of climates (Clarens *et al.*, 2010). Desirable characteristics of algal strains for biofuel production (Brennan and Owende, 2010) include: (1) robust and able to survive shear stresses common in PBRs; (2) able to dominate wild strains in open pond production systems; (3) high CO_2 sinking capacity; (4) limited nutrient requirements; (5) tolerant to a wide range of temperature resulting from the diurnal cycle and seasonal variations; (6) potential to provide valuable coproducts; (7) fast productivity cycle; (8) high photosynthetic efficiency; and (9) display self-flocculation characteristics.

Figure 12.5 shows microalgal conversion of resources to intermediates and fuels. If algal production could be scaled up to industrial capacity to meet the current fuel demand, less than 6 million hectares of land (less than 0.4% of arable land) would be needed worldwide, an achievable goal from global agriculture perspective. Most of efficient algal species are marine, thus discounting the need for freshwater in culturing phase (Gressel, 2008). The major challange remain in determining ways for downstream processing suitable for producing biofuel and other bioproducts (Christenson and Sims, 2011).

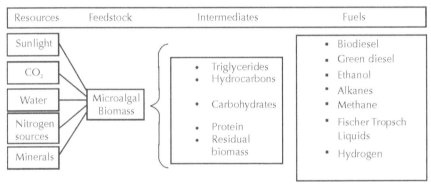

FIGURE 12.5   Fuel production options from microalgal cell components (Darzins *et al.*, 2010).

12.4.1   *BIODIESEL*

Biodiesel is typically produced from oleaginous crops such as soybean, sunflower, rapeseed, and from palm by mono-alcoholic transesterification process, in which triglycerides react with mono-alcohol (most commonly ethanol or methanol) with alkali, acid or enzyme catalysis (Hankamer *et al.*, 2007; Li *et al.*, 2008). Biodiesel production from photosynthetic algae, which grow on CO_2 has great potential as biofuel as it provides maximum net energy. The oil conversion into biodiesel is much less energy-intensive than methods for conversion to other fuels, making biodiesel the desired end-product from algae. This requires an integrated approach (Fig. 12.6) in selecting high-oil content strains, optimum cultivation conditions and cost effective methods of harvesting, oil extraction and conversion of oil to biodiesel. Several microalgal strains have been screened such as *Chlorella vulgaris*, *Spirulina maxima*,

Chlamydomonas reinhardtii, Nannochloropsis sp., *Neochloris oleoabundans,*
Scenedesmus obliquus, Nitzchia sp., *Schizochytrium* sp., *Chlorella protothe-*
coides and *Dunaliella tertiolecta* for the highest lipid production in terms of
quantity (combination of biomass productivity and lipid content) and quality
(fatty acid composition) for biodiesel production (Miao and Wu, 2006; Xu *et*
al., 2006; Chisti, 2007; Rodolfi *et al.*, 2009; Lopes da Silva *et al.*, 2009; Gouveia and Oliveira, 2009; Gouveia *et al.*, 2009; Morowvat *et al.*, 2010).

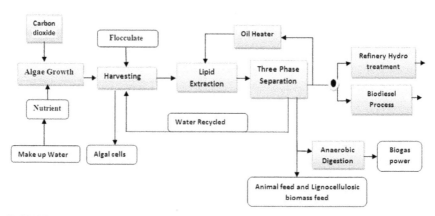

FIGURE 12.6   Algae lipid production process (Darzines *et al.*, 2010).

12.4.1.1   *NUTRITIONAL, PHYSIOLOGICAL AND ENVIRONMENTAL FACTORS*

Photoperiod, salinity and pH are important parameters for optimization of
microalgal productivity. As shown in our study with *Pavlova lutheri* (Table
12.3), photoperiod of 24 hours, at 35 ppt NaCl and pH 8–9 have improved the
cell growth and lipid content with maximum cell density of $13.3–14.1 \times 10^6$
cells mL^{-1}, dry weight of 0.45 gL^{-1} and lipid content of 35–36%. However,
the high doubling time of 5.33–5.77 days and μ_{max} of 012–0.13 d^{-1} may suggest the need to improve quality of strain and cultivation conditions. The lipid
contents at 26–36% DW are similar to that reported for *Nannochloropsis,*
Nannochloris and *Phaeodactylum tricornutum* (Griffiths and Harrison, 2009).
Biomass productivity of *E. oleoabundans* and *A. falcatus* has been reported
at 0.46 $gL^{-1}day^{-1}$ and lipid contents of 36% and 24% DW, respectively; while
several strains such as marine *Tetraselmis suecica* and the freshwater *C. so-*
rokiniana show high growth rates of 0.59 and 0.55 g L^{-1} day^{-1}, but with lower
lipid contents of 17% and 18% DW, respectively.

TABLE 12.3 Kinetics of cell growth and lipid production of *Pavlova lutheri* (Shah *et al.*, 2013).

Experimental conditions	Maximum biomass formation rate, X'_{max} ($gL^{-1}d^{-1}$)	Maximum specific growth rate, μ_{max} (d^{-1})	Doubling time, t_d (day)	Lipid content (%)
Photoperiod (h)				
24	0.028	0.12	5.77	35
12	0.018	0.11	6.30	30
Dark	0.007	0.10	6.93	15
NaCl (ppt)				
15	0.007	0.11	6.30	17
20	0.010	0.12	5.77	21
25	0.014	0.13	5.33	26
30	0.022	0.13	5.33	29
35	0.026	0.12	5.77	36
40	0.025	0.12	5.77	34
pH				
5	0.008	0.11	6.30	24
6	0.016	0.11	6.30	25
7	0.023	0.12	5.77	32
8	0.028	0.13	5.33	35
9	0.027	0.12	5.77	33
10	0.016	0.11	6.30	23

For *Pavlova* sp., optimal pH 8–9 is similar to *Chlorella vulgaris* at pH 7–8 (Cui *et al.*, 2010) and pH 8.2–8.7 for cyanobacteria (Kim *et al.*, 2007). Optimum salinity for *P. lutheri* at 30–40 ppt agrees well with that reported for *N. acceptata, N. pelliculosa* and *N. saprophila* (Griffiths and Harrison, 2009). The effects of low salinity level on growth retardation has also been observed in *N. gregari* (Morin *et al.*, 2008), and on lower lipid contents (18–19% DW) in nutrient replete conditions of *Chlamydomonas applanata* and *C. reinhardtii*

(Griffiths and Harrison, 2009). Different geographical strains have different preferences towards salinity. Optimal salinity may be a function of immediate conditions from which the strain is initially isolated. Species isolated at higher salinities will grow better at higher salinities and not as good at lower salinities (Cucchiari *et al.*, 2008). Higher lipid values (60.6 % DW) have been reported for the genus *Dunaliella* in salt concentration as high as 0.5 M NaCl (equivalent to seawater, or 35 psu) (Takagi *et al.*, 2006) and increases as NaCl increases. The highest lipid content of 50% for *S. obliquus* has been reported after 10 day stress period when NaCl at 1.0 M is used (Dujjanutat *et al.*, 2011).

Studies on *Ochromonas danica* and *Nannochloropsis salina* similarly indicate that cell growth and lipid content increase with increasing photoperiod (Verma *et al.*, 2010). The effects of light and photoperiod may at the same time influence cultivation temperature. Increased microalgal lipid productions to 26–36% at temperature 25–30°C for several microalgal species have been reported (Shah *et al.*, 2012) while reduced lipid production to 15–20% at extremes of low and high temperature has been observed in *Isochrysis galbana* and *Nanochloropsis* sp. (Fotoon *et al.*, 2011). Light is the driving force of photosynthesis, as well as for cell photo-acclimatization. Physiological properties of phytoplankton and photosynthetic organisms can be changed upon exposure to photoperiod and light intensity. Temperatures influence enzyme activity in glycolysis and the Krebs cycle and consequently the metabolism of carbon sources.

12.4.1.2   CONVERSION INTO BIODIESEL

Triacylglycerols are typically nonvolatile that transesterification with short alkyl moieties to manufacture biodiesel, methyl or ethyl residues is required. There are several chemical reaction steps: reversible hydrolysis where triglycerides are converted into diglycerides; diglycerides are converted into monoglycerides; and monoglycerides into free fatty acids and glycerol (as byproduct); followed by reesterification with short chain alcohol (ethanol or methanol), in the presence of a catalyst. Studies show that lipases are excellent for various vegetable oil conversions to methyl ester. With the molar ratio of oil to alcohol 3:1–6:1 and the conversion temperature of 35–50°C at atmospheric pressure, crude oil yield more than 90% can be achieved (Fjerbaek *et al.*, 2009; Mata *et al.*, 2010). If lipase is used, hydrolysis and esterification may take place concurrently, but the thermal liability of enzyme makes this possibility of a lesser interest for industrial scale. A promising alternative that may reduce processing costs is *in situ* transesterification. This process leads to conversion of fatty acids to alkyl esters right inside the biomass, thereby eliminating the solvent extraction step and facilitating the need for harvesting biomass and drying. Such method, which includes alcoholysis, leads up to 20% higher biodiesel yield than the conventional process and reduces wastes (Ehimen et al., 2010).

12.4.2 BIOETHANOL

There has been significant surge in research to investigate microalgal utilization as an advanced energy feedstock for bioethanol production (Rosenberg et al., 2008). The prospective of microalgae for bioethanol is based on the fact that 75% of algal complex carbohydrates could be hydrolyzed into a fermentable hexose monomer and the theoretical value is 80% from fermented yield of ethanol (Huntley and Redalje, 2007). Microalgae can assimilate cellulose, which can also be fermented to bioethanol. Absence of nonphotosynthetic supporting structures (roots, stems, leaves), means microalgae do not have to spend energy towards distribution and transportation of storage molecules between tissues like starch (Chen et al., 2009). In addition, ease of supplying nutrients at optimal levels with only fewer requirements and a well-mixed aqueous environment as compared to soil, have all favored microalgal cultivation.

The principle of ethanol production from microalgae consists of cultivation of microorganisms, cell harvesting, biomass preparation, fermentation and extraction of ethanol. Generally two methods are used, namely fermentation (biochemical process) and gasification (thermo-chemical process) (Singh and Gu, 2010). After oil extraction from the microalgal biomass, fermentation process ensues, using enzymes glucoamylase, α-amylase and yeast, bacteria or fungi for fermenting the sugars to ethanol, carbon dioxide and water (Dismukes et al., 2008). The fermentation process can be represented by:

$$C_6H_{12}O_6 + 2ADP + Pi \rightarrow 2C_2H_5OH + 2CO_2 + 2ATP$$

The biomass preparation can be carried out with mechanical process or enzymatic cell wall break down to make carbohydrates more available, as well as breaking down large molecules of carbohydrates. When cells are disrupted, the yeast Saccharomyces cerevisiae is added to the biomass and fermentation begins. In this way, sugar is converted into ethanol, and ethanol is purified by distillation (Amin, 2009). For microalgae cultivation, CO_2 can be recycled as a nutrient or residual biomass in the process of anaerobic digestion for methane production, such that in essence all organic matters are accounted for (Harun et al., 2010a, 2010b; Singh and Gu, 2010).

12.4.3 BIOMETHANE

Biogas is mainly composed of a mixture of methane (55–75%) and carbon dioxide (25–45%) produced by anaerobic microorganisms during anaerobic digestion. It involves the breakdown of organic matter to produce biogas with an energy content of about 20–40% of lower heating value of the feedstock. Anaerobic digestion process occurs in three sequential stages: hydrolysis, fermentation and methanogenesis. During hydrolysis, complex organic bio-polymers (e.g., carbohydrates, lipids and proteins) are hydrolyzed and broken down into soluble sugars. The fermentation carried out by bacteria converts sugars into alcohols, acetic acid, volatile fatty acids and gas containing H_2 and CO_2, which is primarily metabolized by methanogenesis into CH_4 (60–70%) and CO_2 (30–40%). The second main component, carbon dioxide, has to be removed before methane is used as fuel gas and converted to produce electricity (Hankamer et al., 2007; Holm-Nielsen, 2009).

Anaerobic digestion is suitable for high moisture content (80–90%) organic wastes, which can be useful for w*et al*gal biomass. Comparing microalgal biomass with the plant biomass that is conventionally used for the production of biogas, the former has the advantage that they grow in liquid medium and the space for cultivation is two-dimensional. Microalgae contain lower cellulose and no lignin, leading to good process stability and high conversion efficiencies (Vergara-Fernandez et al., 2008). A maximum methane production rate of 1.61 $LL^{-1}d^{-1}$ has been reported under mesophilic condition when algal sludge is mixed with 60% waste paper at 5 g VS L^{-1} d^{-1} loading rate (Yen and Brune, 2007). Digestion of algal cell walls by the addition of waste paper has a positive effect on the anaerobic digestion, an indication of the increased cellulase activity stimulated by specific nature of the waste paper. Furthermore, codigestion leads to the dilution of toxic compounds by maintaining them under toxic threshold level.

The conversion of microalgal biomass into biogas, can also recover energy through the extraction of lipids that can be used for biodiesel production (Brennan and Owende, 2010). The high energetic content of lipids makes microalgae attractive substrates for anaerobic digestion due to their higher gas production potential as compared to carbohydrates and proteins. The higher the lipids content of the cell, the higher the potential of methane yield (Cirne et al., 2007). The remaining biomass can be further processed into fertilizers, which promote sustainable agricultural practices and greater efficiencies, at reduced algal production costs.

12.4.4   BIOHYDROGEN

Hydrogen is an important fuel and is widely applied in fuel cells, liquefaction of coal and upgrading of heavy oils. Hydrogen can be produced by steam reformation of bio-oils, dark and photo fermentation of organic materials, and photolysis of water catalyzed by special microalgal species (Kapdan and Kargi, 2006; Ran *et al.*, 2006; Wang *et al.*, 2009). Cyanobacteria, which can be used for the production of hydrogen have been the subject of several reviews (Tamagnini *et al.*, 2007). Cyanobacteria are able to diverge the electrons emerging from the two primary reactions of oxygenic photosynthesis directly into the production of H_2, making them attractive for H_2 production from solar energy and water. In cyanobacteria, two natural pathways for H_2 production can be used: 1) H_2-production as a by-product during nitrogen fixation by nitrogenases; and 2) H_2-production directly by bidirectional hydrogenase (Angermayr *et al.*, 2009). Nitrogenases require ATP whereas bidirectional hydrogenases do not require ATP, hence making them more efficient for H_2-production with a much higher turnover.

Direct and indirect photoproduction of H_2 using microalgae, resort to reduced ferredoxin (Fd) as electron donor, coupled to the action of hydrogenases (Seibert *et al.*, 2008). In the direct pathway, photo-oxidation of water occurs, and both Photosystem I (PSI) and Photosystem II (PSII) play a role in supplying reductants (or electrons) to Fd via the photosynthetic electron transfer chain. Indirect pathway involves carbon oxidative metabolism (e.g., starch degradation), and NADP plastoquinone oxidoreductase and PSI activities are required to supply the reductants. Either electron source (i.e., water or starch) can be used, but the contribution of each depends on the type of strain, culture conditions, extent of damage of PSII and specific metabolic constraints (Posewitz *et al.*, 2009). PSII-dependent H_2 photoevolution involving water as a source of electrons produces 2:1 stoichiometric amount of H_2 to O_2.

Light absorption by PSI, and the ensuing electron transport raises the redox potential of such electrons to the redox equivalent of Fd and hydrogenase, thus permitting generation of molecular H_2 (Gibbs *et al.*, 1986). A hydrogenase (containing Fe as prosthetic group) is expressed during such incubation, and catalyzes light-mediated production of H_2 with high specific activity (Melis *et al.*, 2000). Light absorption by photosynthetic mechanism is essential for generation of hydrogens, as it brings about oxidation of water that releases electrons and protons, and facilitates endergonic transport of electrons to Fd. This Fd thus serves as physiological electron donor to the Fe-hydrogenase, that links enzyme to the electron transport chain in the chloroplasts of microalgae (Florin *et al.*, 2001). The activity of hydrogenase is only transient, a mere several seconds to a few minutes, as the light-dependent oxidation of water also entails release of O_2 that is an inhibitor of Fe-hydrogenase. Dark anaerobic incubation and the consequent induction of hydrogenase (Gfeller and Gibbs, 1984) produce electrons for the photosynthetic apparatus from catabolism of the endogenous substrate and the corresponding oxidative carbon metabolism. Electrons are fed to the photosynthetic electron transport chain between PSI and PSII, probably at the level of plastoquinone.

Expensive chemical and mechanical methods have been developed to remove O_2 produced by the photosynthetic activity of microalgae, through addition of O_2 scavengers or reductants and purging with inert gas (Borodin *et al.*, 2000). In the absence of sulfur but in the presence of light, *C. reinhardtii* decreases its PSII activity to the rate of O_2 uptake by respiration implying that microalgal cells consume internally all remaining O_2, and sufficiently fast to generate their own anaerobic microenvironment. Cells will induce (reversible) hydrogenase and produce H_2, for up to 4 days (Melis *et al.*, 2000). If sulfate is subsequently added to the spent cultures at high concentration, further cycles of cell growth and H_2 production will be observed. The presence of the PSII inhibitor 3-(3,4-dichlorophenyl)-1,1-dimethylurea (DCMU) generates 2:1 ratio of H_2 to CO_2, with sufficiently long dark anaerobic incubation. High rates of H_2 production will occur upon illumination of the microalgae in the presence of DCMU (Happe *et al.*, 1993).

12.5   ALGAE FOR WASTE TREATMENT

12.5.1   *REDUCTION OF NITROGEN, PHOSPHORUS AND HEAVY METALS*

Microalgal culture offers tertiary biotreatment of wastewater coupled with the production of potentially valuable biocompounds. Aquaculture systems involving microalgae production and wastewater treatment (e.g., of amino acids, enzyme, or food industries) are promising for microalgae growth in combination with biological cleaning. Green microalgal species such as *Ulva* spp. and *Monostroma* spp. (McHugh, 2003) could use inorganic nitrogen and phosphorus for growth. Microalgae can also moderate the effects of sewage effluent and industrial sources of nitrogenous waste such as those originating from fish aquaculture or water treatment but affecting biodiversity, therefore help to reduce the eutrophication in the aquatic environment. For this, microalgae must tolerate a wide variation in medium conditions (e.g., salinity, extreme pH and heat).

Some industrial processes such as plastic manufacturing and electroplating continuously release substantial amounts of heavy metals into the environment and some algal species are able to remove heavy metals, as well as some toxic organic compounds and therefore do not lead to secondary pollution (Abdel-Raouf et al., 2012). Algae readily take up heavy metals like cadmium from the environment and then induce a heavy metal stress response, which includes production of heavy metal binding factors and proteins. For binding of heavy metals such as copper, nickel, lead, zinc and cadmium, macroalgal species such as *Laminaria, Sargassum, Macrocystis, Ecklonia, Ulva, Lessonia,* and *Durvillaea* have been reported (McHugh, 2003). The removal rate of calcium, magnesium, ferum, mangan and aluminum varies among the microalgae species, ranging from 50 to 99% (Woertz et al., 2011). However, higher heavy metal levels hinder other main processes (e.g., photosynthesis, growth) and finally kill the cells (Siripornadulsil et al., 2002).

12.5.2 CARBON DIOXIDE AND FLUE GASES REMOVAL

Among various methods for CO_2 reduction (absorption, adsorption, chemical fixation), CO_2 mitigation by plants and photosynthetic microorganisms has attracted much attention as an alternative strategy because it leads to the production of biomass energy in the process of CO_2 fixation during photosynthesis. For photosynthetic organisms, water, nutrients and carbon dioxide are vital to growth, but the atmospheric CO_2 concentration limits the growth of these organisms. Thus a cheap source of CO_2 to fuel their photosynthetic process is needed (Wang et al., 2008). However, increased CO_2 capture by terrestrial plants is expected to contribute only 3–6% of fossil fuel emissions, largely due to slow growth rates. Microalgal growth rate is higher during exponential growth than terrestrial plants, with higher photosynthetic efficiencies and more proficient in capturing carbon (Li, 2008; Packer, 2009). Macroalgae have the ability to fix CO_2 and capture solar energy with efficiency of 10 to 50 times greater than that of terrestrial plants. Hence, biological CO_2 fixation by algal photosynthesis has been proposed as an economically feasible method (Skjanes, 2007).

Flue gases from industrial plants have the potential to be used as a feed to phototrophic microalgae and do not harmfully affect algal growth (Jeong et al.,

2003). The use of algae as carbon sequestrator is only considered feasible if they may also have additional value as chemicals and biofuel feedstocks rather than merely as a carbon sequester. Solar energy-driven CO_2 fixation technologies using macroalgae have been used to convert CO_2 in the stack gas from a thermal power station into energy-rich biomass. Since flue gases from industries such as steel-making plants and thermal power stations contain about 500 times higher concentration of CO_2 [10–20 % (v/v)] than that in the air, there may be inhibition of algal growth by high CO_2 concentration, requirement of large amount of nutrients such as nitrogen and phosphorus and low CO_2 conversion due to short gas retention time. Screening and selection of suitable algal strains having tolerance to high CO_2 concentration while producing lipid for subsequent biodiesel production, have been extensively carried out (Chisti, 2007). The desired microalgae strains should have the following characteristics: (1) high growth rate and biomass productivity, (2) high tolerance to trace amount of acidic components from flue gases such as NO_x and SOx, and (3) able to sustain growth even under extreme culture conditions (e.g., high temperature of water due to direct introduction of flue gases) (Brennan and Owende, 2010). Suitable species for CO_2-fixation include *Chlorella* sp., *Emiliania huxley*, *Spirulina platensis*, *Nannochloropsis* sp., and *Phaeodactylum*.

The challenge of limited availability of land for large scale CO_2-capturing from industrial or power plants by microalgae have to be overcome by sophisticated area-efficient techniques to recycle CO_2 by microalgae (Sydney *et al.*, 2010). There are specific parameters to be optimized such as how efficiently the microalgae could use CO_2 in order to avoid the release of excess CO_2 into the atmosphere, the design of effective closed and open system and irrigation, planting, fertilization and harvesting. Open system may contain various species of microalgae that will have different utilization efficiencies that may directly release excess CO_2 into the atmosphere, while closed systems allow for easy control of the CO_2 (Suali and Sarbatly, 2010). If the purpose of algae cultivation is to sequester industrial CO_2 outputs, it also has to take into account that during night time and cloudy days, algal reproduction rates slow down and thus take up less CO_2. This would require the installation of gas storage facilities to cope up with the influx of CO_2 at night.

A pilot scale system is successfully developed to culture microalgae using industrial flue gases where *Scenedesmus obliquus* tolerates high concentration of CO_2 up to 12% (v/v) with optimal removal efficiency of 67% (Li *et al.*, 2011). CO_2 tolerance of *Chlorella vulgaris* is enhanced by gradual increase of CO_2 concentration, while *S. obliquus*, *Chlorella kessleri* and *Spirulina* sp. have exhibited good tolerance to high CO_2 contents (up to 18% CO_2) indicating their great potentials for CO_2 fixation from CO_2-rich streams (Lehmann, 2007). CO_2 consumption rate of 549.90 mg/L day for a maximum *S. obliquus* biomass productivity and lipid productivity of 292.50 mg/L day and 78.73 mg/L day (38.9% lipid content per dry weight of biomass), respectively, have been reported using two-stage cultivation system with 10% CO_2 (Ho *et al.*, 2010).

12.6   HIGH-VALUE BIOCHEMICALS

There are three areas of research in aquatic natural products that have emerged in the last 3 decades – toxins, bioproducts and chemical ecology. With more than 15,000 novel compounds chemically determined, algae is a promising group to provide novel biochemically active substances (Mayer and Hamann 2004, 2005; Singh *et al.*, 2005; Blunt *et al.*, 2005). Algae have developed significant level of structural–chemical diversity, from different metabolic pathways to survive in a competitive freshwater and marine environment (Barros *et al.*, 2001; Puglisi *et al.*, 2004). These include pigments, antioxidants, β-carotenes, polysaccharides, triglycerides, fatty acids, vitamins, and biomass. Species such as Cyanobacteria *Phormidium cebennse*, *Oscillatoria raciborskii*, *Scytonema burmanicum*, *Calothrix elenkinii*, and *Anabaena variabilis*, show anti HIV-1 activity, and give positive tests for the presence of sulfolipids. Hydrocolloids, alginate, agar, and carrageenan, produced from seaweeds are largely used as viscosity-modifying agents in foods and pharmaceuticals. As they are largely used as bulk commodities in as varied industrial sectors as pharmaceuticals, cosmetics, nutraceuticals, functional foods, and biofuels, readily available supply of extracts, fractions or pure compounds are of prime importance (Dos Santos *et al.*, 2005).

12.6.1   PIGMENTS

Algal classification into brown (*Phaeophyceae*), red (*Phaeophyceae*), and green algae (*Chlorophyceae*) are based on their natural pigments contents (Khan *et al.*, 2010) that they are considered potential sources of nonanimal natural pigments. Marine algae contain three basic classes of natural pigments: chlorophylls, carotenoids and phycobiliproteins. Natural pigments are used not only as colorants, but also as antioxidant, anticancer, antiinflammatory, antiobesity, neuroprotective and antiangiogenic activities (Pangestuti and Kim, 2011).

12.6.1.1   CAROTENOIDS

Carotenoids are colored lipid-soluble compounds found in higher plants and algae, as well as animals, fungi, and bacteria. They are responsible for the red, orange and yellow colors of plant leaves, fruits, and flowers, as well as the color of feathers, crustacean shells, fish flesh and skin (Negro and Garrido-Fernández, 2000). The chemical structure of more than 600 different carotenoids is derived from a 40-carbon polyene chain, which can be considered as the backbone of the molecule. This gives carotenoids their typical molecular structure, light-absorbing characteristics and chemical properties. The chain may be terminated by cyclic groups (rings) and can be complemented with oxygen-containing functional groups. The hydrocarbon carotenoids are named carotenes, whereas, xanthophylls are oxygenated derivatives. In the latter, oxygen can be present as OH groups (as in lutein), as oxi-groups (as in cantaxanthin), or in a combination of both (as in astaxanthin) (Higuera-Ciapara et al., 2006). β-carotene (Fig. 12.7) and astaxanthin have been used commercially more than zeaxanthin, lycopene and bixin (Del Campo et al., 2000).

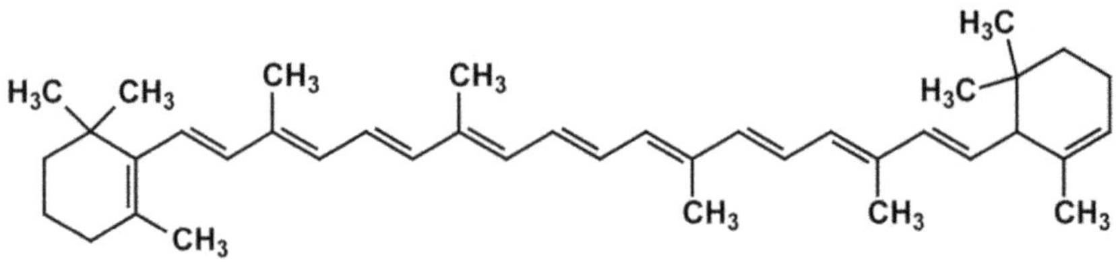

FIGURE 12.7   Structure of β-Carotene.

The most important uses of carotenoids are as natural food colorants (e.g., orange juice), and as additive for animal feed (poultry, fish) and cosmetics (Del Campo et al., 2000). Carotenoids are known to possess antioxidative, anticancer, antiviral, and antiinflammatory activities (Rao and Rao, 2007), and reduce risk for certain types of immune diseases such as asthma and atopic dermatitis. Dunaliella salina a halophilic green biflagellate microalga has long been recognized as a source of β-carotene (Ben-Amotz, 1999). The nutritional and therapeutic relevance of carotenoids is due to the ability to act as provitamin A, which can be converted into vitamin A (Gouveia and Empis, 2003; García-González et al., 2005). Studies suggest that humans fed on a diet high in β-carotene from Dunaliella maintains higher than average levels of serum carotenoids, and reduces incidence of cancer and degenerative diseases

(Ben-Amotz, 1999). *Haematococcus pluvialis* for astaxanthin is now cultivated at large scale (Olaizola and Huntley, 2003) and *Scenedesmus* cultivation
is already established at pilot scale to produce lutein-rich microalgal biomass
(Shi *et al.*, 2006).

12.6.1.2 CHLOROPHYLLS

Most algae when cultured under optimum conditions contain about 4% dry
weight of chlorophyll. Cyanobacteria typically contain chlorophyll-*a* while
species of green algae mostly have chlorophyll-*b* (Dring, 1991; Rosenberg *et
al.*, 2008). *Chlorella* is reported to have high amount of chlorophyll. Chlorophyll has found application as pigment ingredient in food additive industry
(Humphrey, 2004). The chelating agent activity of chlorophyll can be used in
ointment for liver recovery and ulcer treatment. Moreover, it repairs cells, increases hemoglobin in blood and as cell growth promoter. Chlorophyll-*a* and
related compounds derived from *Phaeophyta* have antioxidant activities in
methyl linolenate systems (Le Tutour *et al.*, 1998). Strong antioxidant activity
of *Enteromorpha prolifera* is attributed to chlorophyll derivatives, pheophor-
bide, rather than phenolic compounds (Cho *et al.*, 2011). Cancer preventive
effects of chlorophyll and its derivatives have been extensively studied, with
particular emphasis on *in vitro* antimutagenic effect against numerous dietary
and environmental mutagens (Ferruzzi and Blakesless, 2007).

12.6.1.3 PHYCOBILIPROTEINS

Phycobiliproteins are water-soluble fluorescent proteins used as accessory
or antenna pigments for photosynthetic light collection, which absorb visible
light energy (450–650 nm) (Sousa *et al.*, 2006). They are main photorecep-
tor for photosynthesis in red algae, cyanobacteria and cryptomonads (Glazer,
1994). In several algae, to optimize the capture of light and transfer of energy, phycobiliproteins are arranged in subcellular structures called phyco-
bilisomes, which allow the pigments to be geometrically arranged. The colors
of the phycobiliproteins arise from the presence of covalently attached prosthetic groups, bilins, which are linear tetrapyrroles derived biosynthetically
from heme via biliverdin. The three major categories of phycobiliproteins are
phycocyanins, allophycocyanins and phycoerythrins, the latter being the most
abundant phycobiliproteins found in many red algae species (Glazer, 1994).
Phycobiliproteins show antioxidative properties and used as a natural dye and
colorant, and in the cyanobacterial based foods and health foods (C-phycocy-

anin) and cosmetic industry (C-phycocyanin and R-phycoerythrin) (Eriksen, 2008). The therapeutic potential of R-phycocyanin (R-PC), a novel phycobili-protein, against allergic airway inflammation has been shown with decreased endocytosis and augmentation of IL-12 production in mouse (Chang *et al.*, 2011). *Spirulina* and its phycobiliprotein are useful for regulating allergic in-flammatory responses (Sekar and Chandramohan, 2008).

12.6.2 POLYUNSATURATED FATTY ACIDS

Polyunsaturated fatty acids (PUFAs) are essential for human development and physiology (Hu *et al.*, 2008) to reduce the risk of cardiovascular disease (Ruxton *et al.*, 2007); and for antioxidant, antiatherosclerotic, antiinflamma-tory, and immunoregulatory activities (Wojenski *et al.*, 1991; Calder, 1998; Kim *et al.*, 2010). Marine algae are important sources of essential fatty acids (Kay, 1991; Sanchez-Machado *et al.*, 2002) with applications such as addi-tives for infant milk formula and chicken feed to produce omega-3 enriched eggs. Docosahexaenoic acid (DHA) (Fig. 12.8) is the commercially available algal PUFA (Spolaore *et al.*, 2006). Marine *Schizochytrium mangrovei* micro-algae are reported to have the main component of DHA in a range of 33–39% of total fatty acid (Jiang *et al.*, 2004). Other species include *Crypthecodinium cohnii* (19.9%), *Amphidium caryerea* (17.0%) and *Thrautocytrium aureum* (16.1%) (Vazhappilly and Chen, 1998). *Isochrysis galbana* also produce sig-nificant amount of DHA with specific productivity around 0.16 g $L^{-1}d^{-1}$ (Patil *et al.*, 2007). Several species have demonstrated industrial production poten-tial of eicosapentaenoic acid (EPA) from *Porphyridium purpureum*, *Phaeo-dactylum tricornutum*, *Isochrysis galbana*, *Nannochloropsis* sp. and *Nitzschia laevis* (Robles *et al.*, 1998; Chini Zittelli *et al.*, 1999; Molina Grima *et al.*, 2003; Wen and Chen, 2003).

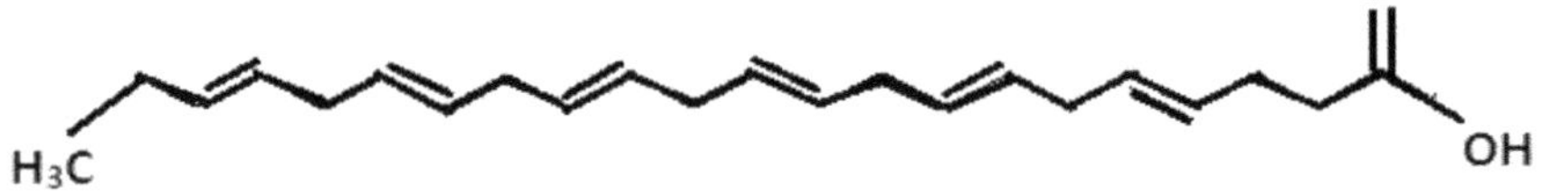

FIGURE 12.8   Docosahexaenoic acid (DHA).

12.6.3 POLYSACCHARIDES

12.6.3.1 HYDROCOLLOIDS

Hydrocolloids are high molecular weight, main structural components of seaweed cell walls and may be involved in recognition mechanisms between seaweeds and pathogens (Potin *et al.*, 1999). Algal polysaccharides have exhibited antioxidant, antiviral, antitumor and anticoagulant activities (Mayer and Lehmann, 2001; Mayer and Hamann, 2004; Smit, 2004). Porphyran of red algae *Porphyra tenera* and *P. yezoensis* have been reported to inhibit the contact hypersensitivity reaction induced by 2,4,6-trinitrochlorobenzene via decreasing the serum level of IgE in Balb/c mice (Ishihara *et al.*, 2005). Fractions of polysaccharides from red extracts inhibit herpes simplex virus (HSV) and other viruses (Burkholder and Sharma, 1969; Deig, 1974; Ehresmann *et al.*, 1977; Richards *et al.*, 1978). Sulfated polysaccharide, with fucose as the main sugar component, isolated from hot water extract of brown algae, *Sargassum horneri* (Turner) C. Agardh demonstrates potent antiviral activity against HSV-1(Hoshino *et al.*, 1998). Polysaccharides isolated from the brown seaweed *Fucus vesiculosus* show anti-HIV activity (Beress *et al.*, 1993; Moen and Clark, 1993); and galactan sulfate isolated from red seaweed *A. tenera*, has an IC_{50} of 0.5–0.6 g mL^{-1} against HIV-1 and HIV-2 (Witvrouw *et al.*, 1994). Extracellular sulfated polysaccharides of marine microalgae *Cochlodinium polykrikoides* inhibit cytopathic effect of influenza virus types A and B (Hasui *et al.*, 1995).

12.6.3.2 AGAR, CARRAGEENAN AND ALGINATE

The most extracted polysaccharides from seaweeds are agar, carrageenan and alginate, which are of interest to food and cosmetic industries due to their gelling, viscosifying and emulsifying properties. Agar and carrageenan are sulfated polysaccharides mainly extracted from Rhodophyceae; while alginate, a binary polyuronide made up of mannuronic acid and guluronic acid, is extracted from Phaeophyceae (Cardozo *et al.*, 2007). Agar is the generic name for seaweed galactans containing α-(1,4)-3,6-anhydro-L-galactose and β-(1–3)-D-galactose residues with a small amount of sulfate esterification, typically up to 6% (w/w) (Hammingson *et al.*, 1996). Agar is produced mainly from red algae (primarily the *Gracilaria* genus) (Marinho-Soriano, 2001). Low quality agar is used in food products and the industrial applications include paper sizing/coating, adhesives, textile printing/dyeing, castings, and impressions. Medium quality agar is used as the gel substrate in biological culture media or as bulking agents, laxatives, suppositories, capsules, tablets and anticoagulants in pharmaceutical products. The most highly purified and upper market types (the neutral fractions called agarose) are used for separation in molecular biology (electrophoresis, immunodiffusion and gel chromatography) and as anticancer agent since it can induce apoptosis in cancer *in vitro* (Chen *et al*., 2004).

Carrageenan is a polysaccharide, which is built up from D-galactopyranose units only (Goncalves *et al*., 2002). It is derived from red algae especially from *Kappaphycus alvarezii* found throughout the coasts of North America and Europe, and grown in large scale for supply to the food industries. Natural carrageenans are mixtures of different sulfated polysaccharides with different composition. Carrageenans are ideal food additives with gelling and emulsifying properties ranging from soft slime to brittle gel that one could nearly walk upon (Van de Velde *et al*., 2004). They are more widely used than agar because of their thickening and suspension properties as emulsifiers/stabilizers in numerous foods, especially milk-based products. Carragenans have shown antitumor, antiviral, anticoagulant and immunomodulation activities (Schaeffer and Krylov, 2000; Zhou *et al*., 2005).

Alginate is the common name given to a family of linear polysaccharides containing 1,4-linked β-D-mannuronic and α-L-guluronic acid residues arranged in a nonregular, blockwise order along the chain (Andrade *et al*., 2004). The alginates are salts of alginic acid, and the sodium salts in particular are also known as algin. The main sources of alginates are brown macroalgae (Phaeophyceae), mainly from *Laminaria, Macrocystis*, and *Ascophyllum* species, and generally recovered from the macroalgal biomass by extracting the insoluble alginic acid salts with hot alkali (sodium carbonate). Sodium alginate is then separated from insoluble seaweed residue by filtration and then purified (McHugh, 1987). There are several applications in the textile industry, and alginate from brown seaweed is widely used in the food and pharmaceutical industries to chelate metal ions and to form highly viscous solution. The production of different alginate oligosaccharides with lyases is required for the development of a more functional alginate for industrial use (Yamasaki *et al*., 2005). The quality and content of agar depend on its specific physicochemical characteristics, and closely related to environmental parameters (Daugherty and Bird, 1988), growth and reproductive cycle. Agar extracted from *Gracilaria gracilis* and *Gracilaria bursa-pastoris* show maximum yield during spring (30%) and summer (36%), while minimum yield is observed during autumn (19%) and winter (23%), respectively (Marinho-Soriano, 2001).

The gelling temperature also shows significant seasonal variation for both species although agar extracted from *G. gracilis* exhibits better qualities than *G. bursa-pastoris* and a better candidate for industrial use (Marinho-Soriano and Bourret, 2003).

12.7   PRODUCTIVITY ENHANCEMENT

12.7.1   STRAIN SELECTION AND IMPROVEMENT

Algal biotechnology breakthroughs enable new approaches to generate algae with desirable properties for the production of biofuels and bioproducts. There are now intensive global research efforts aimed at increasing and modifying the accumulation of lipids, alcohols, hydrocarbons, polysaccharides, and other bio-based compounds in algae through genetic engineering. The strategies to introduce, delete, disrupt and modify genes or gene expression in particular algal species have been developed to enhance physiological properties and to optimize algal production systems. Mutagenesis, selectable marker genes and transformation techniques are among genetic tools developed. Mutagenesis and characterization of mutants is pertinent to understand gene function and to generate desired strains, which can be affected by spontaneous mutants arising from errors in DNA replication, or directed evolution with rationale understanding of the changes made. Naturally occurring mutations necessitate large amount of screening, while standard chemical or UV-based mutagenesis may readily generate mutants but is hampered by the introduction of multiple mutations in a genome and the difficulty in mapping the locus responsible for the phenotype.

Targeted or tagged mutagenesis offers the advantage of simplified identification of the mutated gene. Targeted approaches rely on homologous recombination (if the native gene is to be entirely replaced), or introduction of a modified copy of the gene that is inserted elsewhere into the genome. Tagging can be accomplished by introducing a selectable marker randomly into the genome or through the use of transposons (Adams *et al.*, 2005). Typically, this is accomplished by introducing an antibiotic resistance gene as a selectable marker along with the DNA of interest on an extrachromosomal plasmid. Several antibiotic markers have been developed for microalgae including resistance to neomycin, kanamycin, zeocin and nourseothricin (Poulsen *et al.*, 2006). A common method for introducing DNA into algal cells is the biolistic (gene gun) approach, which is useful for both nuclear and chloroplast transformation. Other methods include electroporation, vortexing with glass beads or silicon carbide whiskers (Hallmann, 2007).

Gene expression control elements or transcriptional regulators can modulate the levels of mRNA, which can subsequently affect algae traits. Frequently, transgenes are overexpressed by using strong control elements, but considering the need for balance in cellular metabolism, intermediate, slightly elevated, or reduced levels of expression may be desirable. Control element strength can be evaluated by monitoring mRNA levels by quantitative PCR or high throughput microarrays. Inducible and repressible promoters that can be activated by simple manipulations are desirable, allowing for precise control over the timing of gene expression (Poulsen and Kroger, 2005). RNA Interference (RNAi) is a tool to down-regulate gene expression, especially in the study of polyploid organisms or when dealing with redundant genes where traditional genetic manipulations are difficult. RNAi operates through double-stranded RNAs that are cut down to small sizes and used to target suppression of specific genes by base pairing. RNAi can inhibit transcription and control translation by either cleaving specific mRNAs or sequestering them away from the ribosome (Ohnuma *et al.*, 2009). Tagging proteins with fluorescent markers is useful in determining their intracellular location and can provide semiquantitative evaluation of their abundance in a simple measurement. This could be useful in monitoring intracellular metabolic processes associated with biofuel precursor production. Green fluorescent protein and its derivatives are the most widely used and versatile protein tags (Regoes and Hehl, 2005).

12.7.2   METABOLIC ENGINEERING

Metabolic engineering brings rationality to strain improvement – from empirical approach through mutagenesis and selection, to a more directed improved productivity through the modification of specific biochemical(s) or the introduction of new one(s), via molecular biology, physiology, bioinformatics, computer modeling and control engineering. It comprises a synthesis step that introduces new pathways and genetic controls; an analysis step to elucidate the properties of metabolic reaction networks and the evaluation of the recombinant physiological state via metabolic flux determination (Abdullah *et al.*, 2005; Abdullah *et al.*, 2008). It is a systems-based analysis – identifying the reaction and/or transport bottlenecks, thermodynamic feasibility, pathway flux distribution and flux control. Metabolic pathway engineering assists to estimate the enzyme activities in the metabolic network that can be upregulated or downregulated for enhanced lipid productivity. In recent years, studies on the lipid biochemistry of algae have shifted from experiments with a few model organisms to encompass a much larger number of often-unusual algae. This has led to the discovery of new compounds, including major membrane components, as well as the elucidation of lipid signaling pathways. Attempts have been made to discover genes that code for expression of various proteins involved in the production of very long-chain PUFA such as arachidonic, eicosapentaenoic and docosahexaenoic acids.

Green alga *Chlamydomonas reinhardtii* has been the focus of most molecular and genetic phycological research. Nuclear, mitochondrial, and chloroplast genomes from several microalgae have been sequenced, and several more are being sequenced (Radakovits *et al.*, 2010). Most of the tools for the expression of transgenes and gene knockdown have been developed for and are specific to this species. However, the tools are now also being rapidly developed for diatoms and other algae that are of greater interest for industrial applications. Some of the earliest works on fatty acid biosynthesis are carried out with the green alga, *C. vulgaris* and such organisms continue to offer great possibilities for elucidating the metabolism and functions of acyl lipids. Biosynthesis of glycerolipids has been investigated in *Chlorella kessleri* with special emphasis on the fatty acid distribution at the *sn*-1 and *sn*-2 positions in membrane lipids (Guschina and Harwood, 2006).

12.7.2.1 CARBOHYDRATE METABOLISM

The most common source of metabolic energy among living organisms is sugar. Sugar catabolic pathways are active mainly during the dark phase of the light–dark cycle. These pathways are responsible for producing NAD(P)H and other biosynthetic metabolites involved in the normal cellular functions. The major route of glucose degradation is the oxidative pentose phosphate pathway (OPP) cycle and is considered as the main CO_2 fixation mechanism in cyanobacteria. During the first step of Calvin cycle (dark phase of photosynthesis), CO_2 is assimilated through the carboxylation of ribulose-1,5-biphosphate (RuBP), to form 3-phosphoglycerate (3PG), which then forms glucose-6-phosphate (G6P) through gluconeogenesis (Quintana *et al.*, 2011). The key enzymes in the oxidation of G6P through the OPP cycle are G6P dehydrogenase and 6-phosphogluconate dehydrogenase. From metabolic regulation point of view, G6P dehydrogenase controlled at the level of gene expression may need further attention as low RuBP levels significantly reduce this enzyme activity (Kaplan *et al.*, 2008). The enzyme ADP-glucose diphosphorylase

(AGPase encoded by the *agp* gene, also known as *glgC*) controls glycogen synthesis in bacteria in an ATP-dependent reaction and seems to be regulated by 3PG (activator) and Pi (inhibitor) (Ballicora *et al.*, 2003).

$$GlclP + ATP \leftrightarrow ADP\text{-}Glc + PPi$$

12.7.2.2   *LIPID METABOLISM*

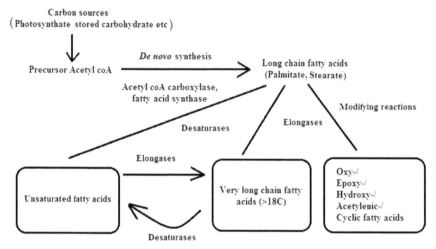

FIGURE 12.9   Fatty acid biosynthesis in plants (Harwood, 1996).

Figure 12.9 shows fatty acid biosynthesis in plants. Microalgae can fix CO_2 into sugars using energy from the sun, and fixed sugars are further processed to produce acetyl-CoA (coenzyme A), with more than one pathway may contribute to maintain the acetyl-CoA pool. Acetyl-CoA provided by photosynthesis serves as the precursor for fatty acid synthesis in the chloroplast. Fatty acids are the building blocks of many cellular lipid types including triacylglycerols. The first committed step of fatty acid synthesis is catalyzed by a multifunctional enzyme complex, that is, acetyl CoA carboxylase (ACCase), which produces malonyl-CoA from acetyl CoA and bicarbonate. Before being further used by the fatty acid synthase machinery, the malonyl group is transferred from CoA to ACP (acetyl carrier protein) catalyzed by a malonyl-CoA:acyl carrier protein malonyltransferase. The common 16- or 18-carbon fatty acids are formed by a series of two-carbon chain elongating reactions catalyzed by a multisubunit enzyme named fatty acid synthase (FAS). Fatty acid synthesis requires stoichiometric amount of ATP, acetyl CoA and NADPH for each two carbon added to the growing acyl chain. Photosynthetic reactions are thus essential not only in providing carbon source but also in generating reducing power (NADH and NADPH) and energy (ATP) for fatty acid synthesis. For most algae species, the final acyl chains emerging from the chloroplast are 16- or 18- carbons in length. The chain-elongating reaction is terminated by the action of acyl-ACP thioesterase (fats). The specificity of this enzyme usually determines the final chain length of the product. The released free fatty acids can cross the plastid envelop membrane where they are esterified to CoA via another enzymatic reaction catalyzed by long chain acyl-CoA synthases (LACS) (Goodson *et al.*, 2011).

Lipids, in the form of triacylglycerides, typically provide storage function in the cell that enables microalgae to tolerate adverse environmental conditions. Essentially, algal biomass and triacylglycerides compete for photosynthetic assimilation and a reprogramming of physiological pathways is required to stimulate lipid biosynthesis (Sharma *et al.*, 2012). Fatty acid and protein biosynthetic pathways possess phosphoenolpyruvate (PEP) as a common substrate. Thus, when PEP is converted to oxaloacetate (OAA) by PEP carboxylase (PEPC), it enters into protein synthesis and is directed to fatty acid synthesis when transformed to malonyl-CoA. PEP is converted to pyruvate-by-pyruvate kinase, and then by pyruvate dehydrogenase in a second reaction to form acetyl-coenzyme A (acetyl-CoA). In addition, pyruvic acid can be converted to alanine and thus participates in protein metabolism (Quintana *et al.*, 2011). On the other hand, acetyl-CoA can be converted to malonyl-CoA in a rate-limiting reaction catalyzed by acetyl-CoA carboxylase, which is the first step towards fatty acid synthesis. High concentrations of acetyl-CoA or free fatty acids stimulate PEPC activity in *E. coli*. In certain cyanobacterial strains, the increased levels of acetyl-CoA do not influence PEPC activity. PEPC in *Synechococcus vulcanus* is strongly activated by fructose-1,6-diphosphate while aspartate acts as a strong suppressor and has been reported to reduce the PEPC activity in *Coccochloris peniocystis*. PEPC seems to divert the carbon flux away from fatty acid biosynthesis. Thus, the antisense expression of PEPC-coding gene (ppc), in *Synechococcus* sp., has led to a lipid content increase (Song *et al.*, 2008).

12.7.2.3   CAROTENOID PATHWAY ENGINEERING

Carotenogenesis pathways and their enzymes are mainly investigated in cyanobacteria and terrestrial plants among oxygenic phototrophs. Algae have common pathways with terrestrial plants and also the additional algae-specific pathways proposed solely based on the chemical structures of carotenoids. Some common carotenogenesis genes in algae are suggested from homology of the known genes (Bertrand, 2010), but most genes and enzymes for algae-specific pathways are still unknown. All carotenoids in oxygenic phototrophs are dicyclic carotenoids: β-carotene, α-carotene and their derivatives, and are all derived from lycopene except for myxol glycosides and oscillol diglycosides in cyanobacteria, which are monocyclic and acyclic carotenoids, respectively. Lycopene is cyclized into either β-carotene through γ-carotene, or α-carotene through γ-carotene or δ-carotene. Three distinct families of lycopene cyclases have been identified in carotenogenetic organisms. One large family contains CrtY in some bacteria except cyanobacteria, and CrtL (CrtL-b, Lcy-b) in some cyanobacteria and plants. Lycopene ε-cyclases (CrtL-e, Lcy-e) from plants and lycopene β-monocyclases (CrtYm, CrtLm) from bacteria are also included. Their amino acid sequences exhibit significant five conserved regions and have an NAD(P)/FAD-binding motif.

Isopentenyl pyrophosphate (IPP), a C5-compound, is the source of isoprenoids, terpenes, quinones, sterols, phytol of chlorophylls, and carotenoids. There are two known independent pathways of IPP synthesis: the classical mevalonate (MVA) pathway and the alternative, nonmevalonate, 1-deoxy-D-xylulose-5-phosphate (DOXP) pathway. In the MVA pathway, acetyl-Coenzyme A is converted to IPP through MVA, and the enzymes and genes are well studied. The pathway is found in plant cytoplasm, animals and some bacteria. The DOXP pathway found in the 1990s, proposes that pyruvate and glycelaldehyde are converted to IPP (Miziorko, 2011). Three enzymes from *Rhodophyta*, *Cyanidioschyzon merolae*, and Chlorophyceae, *Dunaliella salina* and *Haematococcus pluvialis*, are functionally confirmed (Cunningham *et al.*, 2007). Under stressful environment such as high light intensity, UV irradiation and nutrition stress, some Chlorophyceae such as *Haematococcus*, *Chlorella* and *Scenedesmus*, accumulate ketocarotenoids, canthaxanthin and astaxanthin, which are synthesized by combining CrtR-*b* and β-carotene ketolase (CrtW, BKT) (Lemoine and Schoefs, 2010).

12.7.3 TRANSCRIPTOMICS, PROTEOMICS AND METABOLOMICS

Systems biology approaches such as transcriptomics and proteomics have become essential for understanding how microorganisms respond and adapt to changes in their physical environment and offer additional possibility of identifying differentially expressed genes and proteins that are either directly involved in lipid biosynthesis and degradation or that are coordinately regulated. The identification of key regulatory genes and their proteins such as transcription factors, kinases and phosphatases, and their over or under expression in transgenic cells can efficiently alter the whole physiological pathways. The application of high-throughput approaches could accelerate research on algal-derived biofuels by providing the framework for hypothesis-based strain improvement programs, built on an improved fundamental understanding of the specific pathways and regulation of networks involved in algal oil production (Guarnieri *et al.*, 2011). Transcriptome sequencing is a more efficient approach for obtaining microalgae functional genomics information, targeting only coding DNA. This reduced sequencing requirement coupled with the rapidly evolving next-generation sequencing methods, can result in high transcriptome coverage depth and facilitates the *de novo* assembly of transcriptomes from species where full genomes do not exist. The sequencing of genes of importance and *de novo* transcriptome assembly for *Dunaliella tertiolecta*, have been identified, opening up avenue for more rapid and creation of transcriptomes to enable researchers to focus on organisms of direct biofuels interest (Parchman *et al.*, 2010).

Proteomics using large two-dimensional polyacrylamide gel electrophoresis (2-DE) allows multiple expressed proteins to be separated and mapped, providing a convenient and powerful method for monitoring variations at translational levels (Kim *et al.*, 2008). Proteomic technology is useful for investigating the correlation between expressed proteins and behavioral and physiological phenotypes during biological cycles. However, poor reproducibility of the results caused by heavy polysaccharides in algae may have hampered the use of these techniques in the field of marine biology. Rapid progress has been made in 2-DE methodology for marine algae where proteomic analysis done for species recognition of several harmful bloom-related dinoflagellates (Chan *et al.*, 2006). More recent studies have shown that 2-DE results are useful for analysis of the phylogenetic relationships among closely related species of algae (Kim *et al.*, 2008). Protein profiles of a common mixotrophic dinoflagellate, *Prorocentrum micans*, growing autotrophically and mixotrophically (fed on the cryptophyte *Rhodomonas salina*) suggest that comparative proteomics may be a useful tool for analysis of the mixotrophism of dinoflagellates. With more standardized methods of protein extraction and consistent culture conditions, this new tool could have a profound influence on studies of the cellular responses of dinoflagellates to environmental changes (Shim *et al.*, 2011). Transcriptomics and proteomics are highly dependent upon available genomic sequence data, and the lack of these data has hindered the pursuit of such analyzes for many oleaginous microalgae. A strategy has been developed with which to bypass the necessity for genomic sequence information by using the transcriptome as a guide. The triacylglycerol biosynthetic pathway in the unsequenced oleaginous microalga, *Chlorella vulgaris* indicates an upregulation of both fatty acid and triacylglycerol biosynthetic machinery under oil-accumulating conditions, and demonstrates the utility of a *de novo* assembled transcriptome as a search model for proteomic analysis of an unsequenced microalga (Guarnieri *et al.*, 2011).

Metabolomics aims at identifying and quantifying all the metabolites present in a biological sample at an unbiased level (Dettmer *et al.*, 2007). Metabolomics has successfully been applied in environmental studies and in stress physiology for studying nutrient depletion in algae, freezing tolerance of plants or temperature stress (Guy *et al.*, 2008). Genome-scale model-driven algal metabolic design assists production optimization of the compounds of interest by analyzing the utilization of metabolites in the complex, interconnected metabolic networks. Network analysis can direct microbial development efforts towards successful strategies and enable quantitative fine-tuning of the network for optimal product yields while maintaining the robustness of the production. The application of such analytical approaches to algal systems is limited to date, although metabolic network analysis can improve understanding of algal metabolic systems and play an important role in accelerating the adoption of new biofuel technologies (Schmidt *et al.*, 2010).

12.7.4   BIOINFORMATICS

Sequenced genomes are an essential basis of information for the interpretation of transcriptomic and proteomic data. With the development of more powerful sequencing methods, at reduced cost and more coverage obtained in a shorter period of time, obtaining a genome sequence is a prerequisite for rationale strain improvement for biofuels research. Bioinformatics analysis of sequenced genomes, especially at the basic level of gene annotation, will be essential to make sequence data usable. Comparative genomics approaches between related organisms and organisms that carry out similar functions can also help assign gene function and identify metabolic pathways of interest. Access to microalgal genome sequences facilitates genetic manipulation, and the availability of rapid large-scale sequencing technology speeds up microalgal research.

Except for cyanobacteria, for, which over 20 completed genome sequences are available, the nuclear genomes of only a handful of microalgal species have been fully or partially sequenced to date. These species include unicellular green algae (*C. reinhardtii*, *Volvox carteri*), a red alga (*Cyanidioschizon merolae*), several picoeukaryotes (*Osteococcus lucimarinus*, *Osteococcus tauris*, *Micromonas pussilla*, *Bathycoccus* sp.), a pelagophyte (*Aureococcus annophageferrens*), a coccolithophore (*Emiliania huxleyi*), and several diatoms (*Phaeodactylum tricornutum*, *Thalassiosira pseudonana*, *Fragilariopsis cylindrus*) (Radakovits *et al.*, 2010). Genome size in algae can vary substantially, even in closely related species (Connolly *et al.*, 2008). One reason for this variation is likely due to be the accumulation of repeated sequences in the larger genomes (Hawkins *et al.*, 2006). Even though new sequencing technologies readily enable accumulation of data for large genomes, assembly of such data (especially with short read lengths) can be more challenging in repeat-laden genomes.

12.8 CONCLUSION AND FUTURE OUTLOOK

Algal systems are gaining increasing attention to meet the demand for improved systems of bioenergy production, environmental remediation and high value biochemicals. Various high-value chemical compounds can be the target compounds for optimization – lipids, triglycerides, fatty acids, polysaccharides, alcohols, hydrocarbons, pigments, antioxidants, β-carotenes, vitamins, and biomass for industrial sectors such as pharmaceuticals, nutraceuticals, functional foods and biofuels. Change in temperature, pH and salinity can induce lipids, though may be difficult to regulate on a large-scale cultivation system. In bioremediation, microalgae offer tertiary biotreatment of industrial wastewater or flue gases. Use of biorefinery concept and advances in photobioreactor engineering especially the hybrid tubular photobioreactors could make the biodiesel production from microalgal biomass more economical. Producing low-cost microalgal bioenergy and high-value biocompounds require primarily improvements to algal biology through genetic and metabolic engineering for overproduction. This hinges upon utilization of transcriptomics, genomics and metabolomics to screen for and develop new strains. The promising opportunity to use transgenic algae derived from a fast growing species, such as *Chlamydomonas*, the use of transgenics, has already resulted in several business start-ups in this field during the last few years. This is only the beginning of this field of research and much remains to be achieved to optimize the full potential of algae.

KEYWORDS

- **Algal cultivation**
- **Biodiesel**
- **Bioenergy**
- **Bioethanol**
- **Biohydrogen**
- **Bioinformatics**
- **Biomethane**
- **Environmental remediation**
- **High-value biochemicals**
- **Metabolic engineering**
- **Metabolomics**
- **Microalgae**
- **Proteomics**
- **Transcriptomics**
- **Waste treatment**

REFERENCES

Abdel-Raouf, N.; Al-Homaidan, A. A.; Ibraheem, B. M. Microalgae and wastewater treatment. *Saudi J Biol Sci* 2012, 19, 257–275.

Abdullah, M. A.; Lajis, N. H.; Ali, A. M.; Marziah, M.; Sinskey, A. J.; Rha, C. K. Issues in plant cell culture engineering for enhancement of productivity. *Dev Chem Eng Min Proc* 2005, 13(5/6), 573–587.

Abdullah, M. A.; Anisa, U. R.; Rahmah, A.; Sinskey, A. J.; Rha, C. K. Cell engineering and molecular pharming for biopharmaceuticals. *The Open Med Chem J* 2008, 2, 49–61.

Adams, J. E.; Colombo, S. L.; Mason, C. B.; Ynalvez, R. A.; Tural, B.; Moroney, J. V. A mutant of *Chlamydomonas reinhardtii* that cannot acclimate to low CO_2 conditions has an insertion in the Hdh1 gene. *Functional Plant Biol* 2005, 32, 55–66.

Amin, S. Review on biofuel oil and gas production processes from microalgae. *Energy Convers Manage* 2009, 50, 1834–1840.

Andrade, L. R.; Salgado, L. T.; Farina, M.; Pereira, M. S.; Mourao, P. A. S.; Amado-Filho, G. M. Ultrastructure of acidic polysaccharides from the cell walls of brown algae. *J Struct Biol* 2004, 145, 216–225.

Angermayr, S. A.; Hellingwer, K. J.; Lindblad, P.; Teixeira de Mattos, M. J. Energy biotechnology with Cyanobacteria. *Curr Opin Biotechnol* 2009, 20, 257–263.

Ballicora, M. A.; Iglesias, A. A.; Preiss, J. ADP-glucose pyrophosphorylase, a regulatory enzyme for bacterial glycogen synthesis. *Microbiol Mol Biol Rev* 2003, 67, 213–225.

Barros, M. P.; Pinto, E.; Colepicolo, P.; Pedersén, M. Astaxanthin and peridinin inhibit oxidative damage in Fe^{2+} loaded liposomes: scavenging oxyradicals or changing membrane permeability. *Biochem Biophys Res Commun* 2001, 288, 225–232.

Ben-Amotz, A. *Dunaliella* β-carotene: from science to commerce. In *Enigmatic microorganisms and life in extreme environments*, Seckbach, J., Ed.; Kluwer Academic Publishers: Netherlands, 1999, 401–410.

Beress, A.; Wassermann, O.; Tahhan, S.; Bruhn, T.; Beress, L.; Kraiselburd, E. N.; Gonzalez, L. V.; de Motta, G. E.; Chavez, P. I. A new procedure for the isolation of antiHIV compounds (polysaccharides and polyphenols) from the marine alga *Fucus vesiculosus*. *J Nat Prod* 1993, 56, 478–488.

Bertrand, M. Carotenoid biosynthesis in diatoms. *Photosynth Res* 2010, 106, 89–102.

Bligh, E. G.; Dyer, W. J. A rapid method of total lipid extraction and purification. *Can J Biochem Physiol* 1959, 37, 911–917.

Blunt, J. W.; Copp, B. R.; Munro, M. H. G.; Northcote, P. T.; Prinsep, M. R. Marine natural products. *Nat Prod Rep* 2005, 22, 15–61.

Borodin, V. B.; Tsygankov, A. A.; Rao, K. K.; Hall, D. O. Hydrogen production by *Anabaena variabilis* PK84 under simulated outdoor conditions. *Biotechnol Bioeng* 2000, 69(5), 478–485.

Borowitzka, M. A. Marine and halophilic algae for the production of biofuels. *J. Biotechnol* 2008, 136(1), S7.

Brennan, L.; Owende, P. Biofuels from microalgae – a review of technologies for production, processing, and extractions of biofuels and coproducts. *Renew Sustain Energy Rev* 2010, 14, 557–577.

Burkholder, P. R.; Sharma, G. M. Antimicrobial agents from the sea. *Loydia* 1969, 32, 466–483.

Calder, P. C. Immunoregulatory and antiinflammatory effects of n-3 polyunsaturated fatty acids. *Braz J Med Biol Res* 1998, 31, 467–490.

Cardozo, K. H. M.; Guaratini, T.; Barros, M. P.; Falcão, V. R.; Tonon, A. P.; Lopes, N. P.; Campos, S.; Torres, M. A.; Souza, A. O.; Colepicolo, P.; Pinto, E. Metabolites from algae with economical impact. *Comparative Biochem Physiol* 2007, 146, 60 78.

Chan, L. L.; Sit, W. H.; Lam, P. K. S.; Hsieh, D. P. H.; Hodgkiss, I. J.; Wan, J. M. F.; Ho, A. Y. T.; Choi, N. M. C.; Wang, D. Z.; Dudgeon, D. Identification and characterization of a "biomarker of toxicity" from the proteome of the paralytic shellfish toxin-producing dinoflagellate *Alexandrium tamarense* (Dinophyceae). *Proteomics* 2006, 6, 654–666.

Chang, C. J.; Yang, Y. H.; Liang, Y. C.; Chiu, C. J.; Chu, K. H.; Chou, H. N. A novel phycobiliprotein alleviates allergic airway inflammation by modulating immune responses. *Am J Respir Crit Care Med* 2011, 183, 15–25.

Changwei, H.; Mei, L.; Jianlong, L.; Qin, Z.; Zhili, L. Variation of lipid and fatty acid compositions of the marine microalga *Pavlova viridis* (Prymnesiophyceae) under laboratory and outdoor culture conditions. *World J Microbiol Biotechnol* 2008, 24, 1209–1214.

Chen, P.; Min, M.; Chen, Y.; Wang, L.; Li, Y.; Chen, Q.; Wang, C.; Wan, Y.; Wang, X.; Cheng, Y.; Deng, S.; Hennessy, K.; Lin, X.; Liu, Y.; Wang, Y.; Martinez, B.; Ruan, R. Review of the biological and engineering aspects of algae to fuels approach. *Int J Agric Biol Eng* 2009, 2, 1–30.

Chen, Y. H.; Tu, C. J.; Wu, H. T. Growth-inhibitory effects of the red alga *Gelidium amansii* on cultured cells. *Biol Pharm Bull* 2004, 27, 180–184.

Chini Zittelli, G.; Lavista, F.; Bastianini, A.; Rodolfi, L.; Vincenzini, M.; Tredici, M. R. Production of eicosapentaenoic acid by *Nannochloropsis* sp. cultures in outdoor tubular photobioreactors. *J Biotechnol* 1999, 70, 299–312.

Chisti, Y. Biodiesel from microalgae. *Biotechnol Adv* 2007, 25, 294–306.

Cho, M. L.; Lee, H. S.; Kang, I. J.; Won, M. H.; You, S. G. Antioxidant properties of extract and fractions from *Enteromorpha prolifera*, a type of green seaweed. *Food Chem* 2011, 127, 999–1006.

Christenson, L.; Sims, R. Production and harvesting of microalgae for wastewater treatment, biofuels, and bioproducts. *Biotechnol Adv* 2011, 29(6), 686–702.

Cirne, D. G.; Paloumet, X.; Bjornsson, L.; Alves, M. M.; Mattiasson, B. Anaerobic digestion of lipid-rich waste effects of lipid concentration. *Renew Energy* 2007, 32, 965–975.

Clarens, A. F.; Resurreccion, E.; White, M.; Colosi, A. Environmental life cycle comparison of algae to other bioenergy feedstocks. *Environ Sci Technol* 2010, 44, 1813–1819.

Connolly, J. A.; Oliver, M. J.; Beaulieu, J. M.; Knight, C. A.; Tomanek, L.; Moline, M. A. Correlated evolution of genome size and cell volume in diatoms (bacillariophyceae). *J Phycol* 2008, 44(1), 124–131.

Cucchiari, E.; Guerrini, F.; Penna, A.; Totti, C.; Pistocchi, R. Effect of salinity, temperature, organic and inorganic nutrients on growth of cultured *Fibrocapsa japonica* (Raphidophyceae) from the northern Adriatic Sea. *Harmful Algae* 2008, 7, 405–414.

Cunningham, S. A.; Kanzow, T.; Rayner, D.; Molly, O.; Johns, B. W. E.; Marotzke, J.; Hannah, R.; Longworth, T. Temporal variability of the Atlantic meridional overturning circulation at 26.5 N. *Science* 2007, 317(5840), 935–938.

Daugherty, K. B.; Bird, T. K. Salinity and temperature effects on agar production from *Gracilaria verrucosa* strain G-16. *Aquaculture* 1988, 75, 105–113.

Deig, E. F. Inhibition of herpes virus replication by marine algae extracts. *Antimicrob Agents Chemother* 1974, 6, 524–525.

Del Campo, J. A.; Moreno, J.; Rodríguez, H.; Vargas, M. A.; Rivas, J.; Guerrero, M. G. Carotenoid content of chlorophycean microalgae: factors determining lutein accumulation in *Muriellopsis* sp. (Chlorophyta). *J Biotechnol* 2000, 76, 51–59.

Dettmer, K.; Aronov, P. A.; Hammock, B. D. Mass spectrometry-based metabolomics. *Mass Spectrometry Reviews* 2007, 26, 51–78.

Dismukes, G. C.; Carrieri, D.; Bennette, N.; Ananyev, G. M.; Posewitz, M. C. Aquatic phototrophs: efficient alternatives to land-based crops for biofuels. *Curr Opin Biotech* 2008, 19(3), 235–240.

Dos Santos, M. D.; Guaratini, T.; Lopes, J. L. C.; Colepicolo, P.; Lopes, N. P. Plant cell and microalgae culture. In *Modern biotechnology in medicinal chemistry and industry*, Taft, C. A.; Research Signpost: Kerala, India, 2005.

Dring, J. M. *The biology of marine plants*, 2nd Ed.; Cambridge University Press: Cambridge, 1991.

Dujjanutat, A.; Kaewkannetra, P. Effects of wastewater strength and salt stress on microalgal biomass production and lipid accumulation. *World Acad Sci Eng Technol* 2011, 60, 1163–1168.

Ehimen, E. A.; Sun, Z. F.; Carrington, C. G. Variables affecting the *in situ* transesterification of microalgae lipids. *Fuel* 2010, 89, 677–684.

Ehresmann, D. W.; Dieg, E. F.; Hatch, M. T.; Di Salvo, L. H.; Vedros, N. A. Antiviral substances from *California* marine algae. *J Phycol* 1977, 13, 37–40.

Eriksen, N. Production of phycocyanin a pigment with applications in biology, biotechnology, foods and medicine. *Appl Microbiol Biotechnol* 2008, 80, 1–14.

Fajardo, A. R.; Cerdain, L. E.; Medina, A. R.; Fernandex, F. G. A.; Grima, E. M. Lipid extraction from the microalgae *Phaeodactylum tricornutum*. *Eur J Lipid Sci Technol* 2007, 109, 120–126.

Ferruzzi, M. G.; Blakeslee, J. Digestion, absorption and cancer preventative activity of dietary chlorophyll derivatives. *Nutr Res* 2007, 27, 1–12.

Fjerbaek, L.; Christensen, K. V.; Norddahl, B. A. Review of the current state of biodiesel production using enzymatic transesterification. *Biotechnol Bioeng* 2009, 102, 1298–315.

Florin, L.; Tsokoglou, A.; Happe, T. A novel type of Fe-hydrogenase in the green alga *Scenedesmus obliquus* is linked to the photosynthetical electron transport chain. *J Biol Chem* 2001, 276, 6125–6132.

Fotoon, A. Q.; Sayegh, A.; David, J. S.; Montagnes, S. Temperature shifts induce intraspecific variation in microalgal production and biochemical composition. *Bioresour Technol* 2011, 102, 3007–3013.

García-González, M.; Moreno, J.; Manzano, J. C.; Florencio, F. J.; Guerrero, M. G. Production of *Dunaliella salina* biomass rich in 9-*cis*-β-carotene and lutein in a closed tubular photobioreactor. *J Biotechnol* 2005, 115, 81–90.

Gfeller, R. P.; Gibbs, M. Fermentative metabolism of *Chlamydomonas reinhardtii*: I. Analysis of fermentative products from starch in dark-light. *Plant Physiol* 1984, 75, 212–218.

Gibbs, M.; Gfeller, R. P.; Chen, C. Fermentative metabolism of *Chlamydomonas reinhardtii*: III. Photoassimilation of acetate. *Plant Physiol* 1986, 82, 160–166.

Glazer, A. N. Phycobiliproteins A family of valuable widely used fluorophores. *J Appl Phycol* 1994, 6, 105–112.

Goncalves, A. G.; Ducatti, D. R.; Duarte, M. E.; Noseada, M. D. Sulfated and pyruvylated disaccharide alditols obtained from a red seaweed galactan: ESIMS and NMR approaches. *Carbohydr Res* 2002, 337, 2443–2453.

Goodson, C.; Roth, R.; Wang, Z. T.; Goodenough, U. Structural correlates of cytoplasmic and chloroplast lipid body synthesis in *Chlamydomonas reinhardtii* and stimulation of lipid body production with acetate boost. *Eukaryotic Cell* 2011, 10, 1592–1606.

Gouveia, L.; Empis, J. Relative stabilities of microalgal carotenoids in microalgal extracts, biomass and fish feed: effect of storage conditions. *Innov Food Sci Emerg Technol* 2003, 4, 227–233.

Gouveia, L.; Marques, A. E.; Lopes da Silva, T.; Reis, A. *Neochloris oleabundans* UTEX 1185: a suitable renewable lipid source for biofuel production. *J Ind Microbiol Biotechnol* 2009, 36, 821–826.

Gouveia, L.; Oliveira, A. C. Microalgae as a raw material for biofuels production. *J Ind Microbiol Bioethanol* 2009, 36, 269–274.

Gressel, J. Transgenics are imperative for biofuel crops. *Plant Science* 2008, 174, 246–263.

Griffiths, M.; Harrison, S. Lipid productivity as a key characteristic for choosing algal species for biodiesel production. *J Appl Phycol* 2009, 21, 493–507.

Grima, M. E.; Belarbi, E. H.; Fernandez, F. G. A. Recovery of microalgal biomass and metabolites: process options and economics. *Biotechnology Advances* 2003, 20, 491–515.

Guarnieri, M. T.; Nag, A.; Smolinski, S. L.; Darzins, A.; Seibert, M. Examination of triacylglycerol biosynthetic pathways via *de novo* transcriptomic and proteomic analyzes in an unsequenced microalgae. *PLoS One* 2011, 6(10), 25851.

Guschina, I. A.; Harwood, J. L. Lipids and lipid metabolism in eukaryotic algae. *Progress in Lipid Research* 2006, 45(2), 160–186.

Guy, C.; Kaplan, F.; Kopka, J.; Selbig, J.; Hincha, D. K. Metabolomics of temperature stress. *Physiologia Plantarum* 2008, 132, 220–235.

Hallmann, A. Algal transgenics and biotechnology. *Transgenic Plant J* 2007, 1(1), 81–98.

Hammingson, J. A.; Furneaux, R. H.; Murray-Brown, H. V. Biosynthesis of agar polysaccharides in *Gracilaria chilensis* Bird, McLachlan Oliveira. *Carbohydr Res* 1996, 287, 101–115.

Hankamer, B.; Lehr, F.; Rupprecht, J.; Mssgnug, J. H.; Posten, C.; Kruse, O. Photosynthetic biomass and H_2 production by green algae: from bioengineering to bioreactor scale-up. *Physiol Plant* 2007, 131, 10–21.

Happe, T.; Naber, J. D. Isolation, characterization and N-terminal amino acid sequence of hydrogenase from the green alga *Chlamydomonas reinhardtii*. *Eur J Biochem* 1993, 214, 475–481.

Harun, R.; Singh, M.; Forde, G. M.; Danquah, M. K. Bioprocess engineering of microalgae to produce a variety of consumer products. *Renew Sust Energy Rev* 2010a, 14, 1037–1047.

Harun, R.; Jason, W. S. Y.; Cherrington, T.; Danquah, M. K. Exploring alkaline pretreatment of microalgal biomass for bioethanol production. *Appl Energy* 2010b, 88, 3464–3467.

Harwood, J. L. Recent advances in the biosynthesis of plant fatty acids. *Biochim Biophys Acta* 1996, 1301, 7–56.

Hasui, M.; Matsuda, M.; Okutani, K.; Shigeta, S. *In vitro* antiviral activities of sulfated polysaccharides from a marine microalga (*Cochlodinium polykrikoides*) against human immunodeficiency virus and other enveloped viruses. *Int J Biol Macromol* 1995, 17, 293–297.

Hawkins, J. S.; Kim, H.; Nason, J. D.; Wing, R. A.; Wendel, J. F. Differential lineage-specific amplification of transposable elements is responsible for genome size variation in *Gossypium*. *Genome Research* 2006, 16(10), 1252–1261.

Higuera-Ciapara, I.; Félix-Valenzuela, L.; Goycoolea FM. Astaxanthin: a review of its chemistry and applications. *Crit Rev Food Sci Nutr* 2006, 46, 185–196.

Ho, S. H.; Chen, W. M.; Chang, J. S. *Scenedesmus obliquus* CNW-N as a potential candidate for CO_2 mitigation and biodiesel production. *Bioresour Technol* 2010, 101, 8725–8730.

Hollar, S. *A closer look at bacteria, algae and protozoa.* Britannica Educational Publishers: New York, 2012.

Holm-Nielsen, J. B.; Seadi, T. A.; Oleskowicz-Popiel, P. The future of anaerobic digestion and biogas utilization. *Bioresource Technology* 2009, 100, 5478–5484.

Hoshino, T.; Hayashi, T.; Hayashi, K.; Hamada, J.; Lee, J. B.; Sankawa, U. An antivirally active sulfated polysaccharide from *Sargassum horneri* (Turner) C. Agardh. *Biol Pharm Bull* 1998, 21, 730–734.

Hu, C.; Li, M.; Li, J.; Zhu, Q.; Liu, Z. Variation of lipid and fatty acid compositions of the marine microalga *Pavlova viridis* (Prymnesiophyceae) under laboratory and outdoor culture conditions. *World J Microbiol Biotechnol* 2008, 24, 1209–1214.

Humphrey, A. M. Chlorophyll as a color and functional ingredient. *J Food Sci* 2004, 69, 422–425.

Huntley, M. E.; Redalje, D. G. CO_2 Mitigation and renewable oil from photosynthetic microbes: a new appraisal. *Mitig Adapt Strateg Glob Change* 2007, 12, 573–608.

Ishihara, K.; Oyamada, C.; Matsushima, R.; Murata, M.; Muraoka, T. Inhibitory effect of porphyran, prepared from dried Nori, on contact hypersensitivity in mice. *Biosci Biotechnol Biochem* 2005, 69, 1824–1830.

Jeong, M. L.; Gillis, J. M.; Hwang, J. Y. Carbon dioxide mitigation by microalgal photosynthesis. *Bull Korean Chem Soc* 2003, 24(12), 1763–1766.

Jiang, Y.; Fan, K. W.; Wong, R. T. Y.; Chen, F. Fatty acid composition and squalene content of the marine microalga *Schizochytrium mangrove. J Agric Food Chem* 2004, 52, 1196–2000.

Johnson, M.; Wen, Z. Development of an attached microalgal growth system for biofuel production. *App Microbiol Biotechnol* 2010, 85(3), 525–534.

Kapdan, I. K.; Kargi, F. Bio-hydrogen production from waste materials. *Enzyme Microb Technol* 2006, 38, 569–582.

Kaplan, A.; Hagemann, M.; Bauwe, H.; Kahlon, S.; Ogawa, T. Carbon acquisition by cyanobacteria: mechanisms, comparative genomics and evolution. In *The Cyanobacteria: molecular biology, genomics and evolution*, Herrero, A., Flores, E., Eds.; Caister Academic Press: Poole, 2008; 305–334.

Kay, R. A. Microalgae as food and supplement. *Crit Rev Food Sci Nutr* 1991, 30, 555–573.

Khan, S.; Kong, C.; Kim, J.; Kim, S. Protective effect of *Amphiroa dilatata* on ROS induced oxidative damage and MMP expressions in HT1080 cells. *Biotechnol Bioprocess Eng* 2010, 15, 191–198.

Kim, C. J.; Jung, Y. H.; Oh, H. M. Factors indicating culture status during cultivation of *Spirulina* (Arthospira) *platensis. J Microbiol* 2007, 45(2), 122–127.

Kim, G. H.; Shim, J. B.; Klochkova, T. A.; West, J. A.; Zuccarello, G. C. The utility of proteomics in algal taxonomy: *Bostrychia radicans/B. moritziana* (Rhodomelaceae, Rhodophyta) as a model study. *J Phycol* 2008, 44, 1519–1528.

Kim, J. A.; Kong, C. S.; Kim, S. K. Effect of *Sargassum thunbergii* on ROS mediated oxidative damage and identification of polyunsaturated fatty acid components. *Food Chem Toxicol* 2010, 48, 1243–1249.

Larkum, A. W. D.; Ross, I. L.; Kruse, O.; Hankamer B. Selection, breeding and engineering of microalgae for bioenergy and biofuel production. *Trends in Biotechnology* 2012, 30, 198–205.

Le Tutour, B.; Benslimane, F.; Gouleau, M.; Gouygou, J.; Saadan, B.; Quemeneur, F. Antioxidant and prooxidant activities of the brown algae, *Laminaria digitata, Himanthalia elongata, Fucus vesiculosus, Fucus serratus* and *Ascophyllum nodosum. J Appl Phycol* 1998, 10, 121–129.

Lemoine, Y.; Schoefs, B. Secondary ketocarotenoid astaxanthin biosynthesis in algae: a multifunctional response to stress. *Photosynth Res* 2010, 106, 155–177.

Li, F. F.; Yang, Z. H.; Zeng, R.; Yang, G.; Chang, X.; Yan, J. B. Microalgae capture of CO_2 from actual flue gas discharged from a combustion chamber. *Ind Eng Chem Res* 2011, 50, 6496–6502.

Li, Y.; Horsman, M.; Wu, N.; Lan, C. Q.; Dubois-Calero, N. Biofuels from microalgae. *Biotechnol Prog* 2008, 24, 815–820.

Lopes da Silva, T.; Reis, A.; Medeiros, R.; Oliveira, A. C.; Gouveia, L. Oil production towards biofuel from autotrophic microalgae semicontinuous cultivations monitorized by flow cytometry. *Appl Biochem Biotechnol* 2009, 159, 568–578.

Marinho-Soriano, E. Agar polysaccharides from *Gracilaria* species Rhodophyta (Gracilariaceae). *J Biotechnol* 2001, 89, 81–84.

Marinho-Soriano, E.; Bourret, E. Effects of season on the yield and quality of agar from *Gracilaria* species (Gracilariaceae Rhodophyta). *Bioresour Technol* 2003, 90, 329–333.

Mata, T. M.; Martins, A. A.; Caetano, N. S. Microalgae for biodiesel production and other applications: a review. *Renew Sustain Energy Rev* 2010, 14, 217–232.

Mayer, A. M. S.; Hamann, M. T. Marine pharmacology in 2000: marine compounds with antibacterial, anticoagulant, antifungal, antiinflammatory, antimalarial, antiplatelet, antituberculosis, and antiviral activities; affecting the cardiovascular, immune, and nervous system and other miscellaneous mechanisms of action. *Mar Biotechnol* 2004, 6, 37–52.

Mayer, A. M. S.; Hamann, M.T. Marine pharmacology in 2001–2002: marine compounds with anthelmintic, antibacterial, anticoagulant, antidiabetic, antifungal, anti inflammatory, antimalarial, antiplatelet, antiprotozoal, antituberculosis, and antiviral activities; affecting the cardiovascular, immune and nervous systems and other miscellaneous mechanisms of action. *Comp. Biochem Physiol, C Toxicol Pharmacol* 2005, 140, 265–286.

Mayer, A. M. S.; Lehmann, V. K. B. Marine pharmacology in 1999: antitumor and cytotoxic compounds. *Anticancer Res* 2001, 21, 2489–2500.

McHugh, D. J. *Production and utilization of products from commercial seaweeds*. FAO: Rome, 1987.

McHugh, D. J. A guide to the seaweed industry. *FAO Fisheries Technical Papers* 2003, 441, 1–105.

Melis, A.; Zhang, L.; Forestier, M.; Ghirardi, M. L.; Seibert, M. Sustained photobiological hydrogen gas production upon reversible inactivation of oxygen evolution in the green alga *Chlamydomonas reinhardtii*. *Plant Physiol* 2000, 122, 127–136.

Miao, X.; Wu, Q. Biodiesel production from heterotrophic microalgal oil. *Bioresour Technol* 2006, 97, 841–846.

Miziorko, H. M. Enzymes of the mevalonate pathway of isoprenoid biosynthesis. *Arch Biochem Biophys* 2011, 505, 131–143.

Moen, L. K.; Clark, G. F. A novel reverse transcriptase inhibitor from *Fucus vesiculosus*. *Int Conf AIDS* 1993, 9, 145–161.

Molina, Grima, E.; Belarbi, E. H.; Acién Fernández, F. G.; Robles, M. A.; Chisti, Y. Recovery of microalgal biomass and metabolites: process options and economics. *Biotechnol Adv* 2003, 20, 491–515.

Morin, S.; Coste, M.; Delmas, F. A comparison of specific growth rates of periphytic diatoms of varying cell size under laboratory and field conditions. *Hydrobiologia* 2008, 614, 285–297.

Morowvat, M. H.; Rasoul-Amini, S.; Ghasemi, Y. *Chlamydomonas* as a "new" organism for biodiesel production. *Bioresour Technol* 2010, 101, 2059–2062.

Morweiser, M.; Kruse, O.; Hankamer, B. Developments and perspectives of photobioreactors for biofuel production. *Appl Microbiol Biotechnol* 2010, 87(4), 1291–1301.

Nazir, M.S.; Wahjoedi, B.A.; Yussof, A.W.; Abdullah, M.A. Ecofriendly extraction and characterization of cellulose from Oil palm empty fruit bunches. *BioResources* 2013, 8(2), 2161–2172.

Negro, J. J.; Garrido-Fernández, J. Astaxanthin is the major carotenoid in tissues of white storks (*Ciconia ciconia*) feeding on introduced crayfish (*Procambarus clarkii*). *Comp Biochem Physiol Part B Biochem Mol Biol* 2000, 126, 347–352.

Ohnuma, M.; Misumi, O.; Fujiwara, T.; Watanabe, S.; Tanaka, K.; Kuroiwa, T. Transient gene suppression in a red alga, *Cyanidioschyzon merolae* 10D. *Protoplasma* 2009, 236(1), 107–112.

Olaizola, M.; Huntley, M. E. Recent advances in commercial production of astaxanthin from microalgae. In *Biomaterials and bioprocessing*, Fingerman, M., Nagabhushanam, R., Eds.; Science Publishers: Boca Raton, 2003; 143–164.

Packer, M. Algal culture of carbon dioxide; biomass generation as a tool for greenhouse gas mitigation with reference to New Zealand energy strategy and policy. *Energy Policy* 2009, 37, 3428–3437.

Pangestuti, R.; Kim, S. K. Biological activities and health benefit effects of natural pigments derived from marine algae. *J Funct Foods* 2011, 3, 255–266.

Parchman, T. L.; Geist, K. S.; Grahnen, J. A.; Benkman, G. W.; Buerkle, C. A. Transcriptome sequencing in an ecologically important tree species: assembly, annotation, and marker discovery. *BMC Genomics* 2010, 11, 180.

Park, P. K.; Kim, E. Y.; Chu, K. H. Chemical disruption of yeast cells for the isolation of carotenoid pigments. *Sep Purif Technol* 2007, 53(2), 148–152.

Patil, V.; Kallqvist, T.; Olsen, E.; Vogt, G.; Gislerod, H. R. Fatty acid composition of 12 microalgae for possible use in aquaculture feed. *Aqua Int* 2007, 15, 1–9.

Pienkos, P. T.; Darzins, A. The promise and challenges of microalgal derived biofuels. *Biofuels Bioprod. Bioref.* 2009, 3, 431–440.

Posewitz, M. C.; Dubini, A.; Meuser, J. E.; Seibert, M.; Ghirardi, M. L. Hydrogenases, hydrogen production and anoxia. In *The Chlamydomonas sourcebook*, Vol. 2, Stern, D. B., Ed.; Academic Press: Massachusetts, 2009; 217–255.

Potin, P.; Bouarab, K.; Kupper, F.; Kloareg, B. Oligosaccharide recognition signals and defense reactions in marine plant-microbe interactions. *Curr Opin Microbiol* 1999, 2, 276–283.

Poulsen, N.; Chesley, P. M.; Kröger, N. Molecular genetic manipulation of the diatom *Thalassiosira pseudonana* (bacillariophyceae). *J Phycol* 2006, 42(5), 1059–1065.

Poulsen, N. N.; Kroger, B. A new molecular tool for transgenic diatoms. *The FEBS Journal* 2005, 272, 3413–3423.

Puglisi, M. P.; Tan, L. T.; Jensen, P. R.; Fenical, W. Capisterones A and B from the tropical green alga *Penicillus capitatus*: unexpected antifungal defenses targeting the marine pathogen *Lindra thallasiae*. *Tetrahedron* 2004, 60, 7035–7039.

Quintana, N.; Van der Kooy, F.; Van de Rhee, M. D.; Voshol, G. P.; Verpoorte, R. Renewable energy from Cyanobacteria: energy production optimization by metabolic pathway engineering. *Appl Microbiol Biotechnol* 2011, 91, 471–490.

Radakovits, R.; Jinkerson, R. E.; Darzins, A. I.; Posewitz, M. C. Genetic engineering of algae for enhanced biofuel production. *Eukaryotic Cell* 2010, 9(4), 486–501.

Ran, C. Q.; Chen, Z. A.; Zhang, W.; Yu, X. J.; Jin, M. F. Characterization of photobiological hydrogen production by several marine green algae. *Wuhan Ligong Daxue Xuebao* 2006, 28(2), 258–263.

Rao, A. V.; Rao, L. G. Carotenoids and human health. *Pharmacol* 2007, 55, 207–216.

Regoes, A.; Hehl, A. B. SNAP-tag™ mediated live cell labeling as an alternative to GFP in anaerobic organisms. *Biotechniques* 2005, 39(6), 809.

Richards, J. T.; Glasgow, L. A.; Overall, J. C. Jr.; Deig, E. F.; Hatch, M. T. Antiviral activity of extracts from marine algae. *Antimicrob Agents Chemother* 1978, 14, 24–30.

Richardson, J. W.; Johnson, M. D.; Outlaw, J. L. Economic comparison of open pond raceways to photo bio-reactors for profitable production of algae for transportation fuels in the Southwest. *Algal Res* 2012, 1, 93–100.

Robles, M.; Molina, A.; Grima, E.; Giménez, A.; González, M. J. Downstream processing of algal polyunsaturated fatty acids. *Biotechnol Adv* 1998, 16, 517–580.

Rodolfi, L.; Zitelli, G. C.; Bassi, N.; Padovani, G.; Biondi, N.; Bonini, G.; Tredici, M. R. Microalgae for oil: strain selection, induction of lipid synthesis and outdoor mass cultivation in a low-cost photobioreactor. *Biotech Bioeng* 2009, 102, 100–112.

Rosenberg, J. N.; Oyler, G. A.; Wilkinson, L.; Betenbaugh, M. J. A green light for engineered algae redirecting metabolism to fuel a biotechnology revolution. *Curr Opin Biotechnol* 2008, 19, 430–436.

Ruxton, C. H. S.; Reed, S. C.; Simpson, M. J. A.; Millington, K. J. The health benefits of omega-3 polyunsaturated fatty acids: a review of the evidence. *J Human Nutr Diet* 2007, 20, 275–285.

Sanchez-Machado, D. I.; Lopez-Cervantes, J.; Lopez-Hernandez, J.; Paseiro-Losada, P. Fatty acids, total lipid, protein and ash contents of processed edible seaweeds. *Food Chem* 2002, 85, 439–444.

Schaeffer, D. J.; Krylov, V. S. Anti-HIV activity of extracts and compounds from algae and cyanobacteria. *Ecotoxicol Environ Saf* 2000, 45, 208–227.

Schenk, P. M.; Skye, R.; Hall, T.; Stephens, E.; Marx, U. C.; Mussgnug, J. H.; Posten, C.; Kruse, O.; Hankamer, B. Second generation biofuels: high-efficiency microalgae for biodiesel production. *Bioenergy Res* 2008, 1, 20–43.

Schmidt, B. J.; Lin-Schmidt, X.; Chamberlin, A.; Salehi-Ashtiani, K.; Papin, J. A. Metabolic systems analysis to advance algal biotechnology. *Biotechnol J* 2010, 5, 660–670.

Seibert, M.; King, P. W.; Posewitz, M. C.; Melis, A.; Ghirardi, M. L. Photosynthetic water splitting for hydrogen production. In *Bioenergy*, Wall, J., Hardwood, C. S., Demain, A. L., Eds.; ASM Press: Washington, 2008; 273–291.

Sekar, S.; Chandramohan, M. Phycobiliproteins as a commodity: trends in applied research, patents and commercialization. *J Appl Phycol* 2008, 20, 113–136.

Sharma, K. K.; Schuhmann, H.; Schenk, P. M. High lipid induction in microalgae for biodiesel production. *Energies* 2012, 5, 1532–1553.

Shi, X.; Zhengyun, W.; Chen, F. Kinetic model of lutein production by heterotrophic *Chlorella* at various pH and temperature. *Mol Nutr Food Res* 2006, 50, 763–768.

Shim, J.; Klochkova, T. A.; Han, J. W.; Kim, G. H.; Yoo, Y. D.; Jeong, H. J. Comparative proteomics of the mixotrophic dinoflagellate *Prorocentrum micans* growing in different trophic modes. *Algae* 2011, 26(1), 87–96.

Singh, J.; Gu, S. Commercialization potential of microalgae for biofuels production. *Renew Sust Energy Rev* 2010, 14(9), 2596–2610.

Singh, S.; Kate, B. N.; Banerjee, U. C. Bioactive compounds from cyanobacteria and microalgae: an overview. *Crit Rev Biotechnol* 2005, 25, 73–95.

Siripornadulsil, S.; Traina, S.; Verma, D. P.; Sayre, R. T. Molecular mechanisms of proline-mediated tolerance to toxic heavy metals in transgenic microalgae. *Plant Cell* 2002, 14, 2837–2847.

Skjanes, K.; Lindblad, P.; Muller, J. BioCO$_2$ – a multidisciplinary, biological approach using solar energy to capture CO$_2$ while producing H$_2$ and high value products. *Biomol Eng* 2007, 24, 405–413.

Smit, A. J. Medicinal and pharmaceutical uses of seaweed natural products: a review. *J Appl Phycol* 2004, 16, 245–262.

Song, D.; Hou, L.; Shi, D. Exploitation and utilization of rich lipids-microalgae, as new lipids feedstock for biodiesel production – a review. *Sheng Wu Gong Cheng Xue Bao* 2008, 24, 341–348.

Sousa, I.; Batista, A. P.; Raymundo, A.; Empis, J. Rheological characterization of colored oil-in-water food emulsions with lutein and phycocyanin added to the oil and aqueous phases. *Food Hydrocolloids* 2006, 20, 44–52.

Spolaore, P.; Joannis-Cassan, C.; Duran, E.; Isambert, A. Commercial applications of microalgae. *J Biosci Bioeng* 2006, 101, 87–96.

Suali, E.; Sarbatly, R. Potential of CO$_2$ utilization by microalgae in Malaysia. *Int J Global Environ Issues* 2010, 12(2), 150–160.

Sydney, E. B.; Sturm, W.; Cesar de Carvalho, J.; Thomas-Soccol, V.; Larroche, C.; Pandey, A.; Soccol, C. R. Potential carbon dioxide fixation by industrially important microalgae. *Bioresour Technol* 2010, 101, 5892–5896.

Takagi, M.; Karseno, T.; Yoshida, T. Effect of salt concentration on intracellular accumulation of lipids and triacylglyceride in marine microalgae *Dunaliella* cells. *J Biosci Bioeng* 2006, 101, 223–226.

Tamagnini, P.; Leitao, E.; Oliveira, P.; Ferriera, D.; Pinto, F.; Harris, D. J.; Heidorn, T.; Lindblad, P. Cyanobacterial hydrogenases: diversity, regulation and applications. *FEMS Microbiol Rev* 2007, 31, 692–720.

Ugwu, C. U.; Aoyagi, H.; Uchiyama, H. Photobioreactors for mass cultivation of algae. *Bioresour Technol* 2008, 99, 4021–4028.

Van de Velde, F.; Pereira, L.; Rolleman, H. S. The revised NMR chemical shift data of carrageenans. *Carbohydr Res* 2004, 339, 2309–2313.

Vazhappilly, R.; Chen, F. Eicosapentaenoic acid and docosahexaenoic acid production potential of microalgae and their heterotrophic growth. *J Am Oil Chem Soc* 1998, 75, 393–397.

Vergara-Fernandez, A.; Vargas, G.; Alarcon, N.; Velasco, A. Evaluation of marine algae as a source of biogas in a two-stage anaerobic reactor system. *Biomass Bioenerg* 2008, 32, 338–344.

Verma, N. M.; Mehrotra, S.; Shukla, A.; Mishra, B. N. Prospective of biodiesel production using microalgae as the cell factories: a comprehensive discussion. *Afr J Biotechnol* 2010, 9(10), 1402–1411.

Wang, B.; Li, Y.; Wu, N.; Lan, C. Q. CO_2 biomitigation using microalgae. *Appl Microbiol Biotechnol* 2008, 79, 707–718.

Wang, C.; Li, H.; Wang, Q.; Ping, W. Effect of pH on growth and lipid content of *Chlorella vulgaris* cultured in biogas slurry. *Chinese J Biotechnol* 2010, 26(8), 1074–1079.

Wen, Z. Y.; Chen, F. Heterotrophic production of eicosapentaenoic acid by microalgae. *Biotechnol Adv* 2003, 21, 273–294.

Witvrouw, M.; Este, J. A.; Mateu, M. E.; Reymen, D.; Andrei, G.; Snoeck, R.; Ikeda, S.; Pauwels, R.; Vittori Bianchini, N.; Desmyter, J.; De Clercq, E. Antiviral activity of a sulfated polysaccharide extracted from the red seaweed *Aghardhiella tenera* against human immunodeficiency virus and other enveloped viruses. *Antiviral Chem Chemother* 1994, 5, 297–303.

Woertz, I.; Feffer, A.; Lundquist, T.; Nelson, Y. Algae grown on dairy and municipal wastewater for simultaneous nutrient removal and lipid production for biofuel feedstock. *J Environ Eng* 2011, 135(11), 1115–1123.

Wojenski, C. M.; Silver, M. J.; Waker, J. Eicosapentaenoic acid ethyl ester as an antithrombotic agent: comparison to an extract of fish oil. *Biochim Biophys Acta* 1991, 1081, 33–38.

Xu, H.; Miao, X.; Wu, Q. High quality biodiesel production from a microalga *Chlorella* protothecoides by heterotrophic growth in fermenters. *J Biotechnol* 2006, 126, 499–507.

Yamasaki, M.; Ogura, K.; Hashimoto, W.; Mikami, B.; Murata, K. A structural basis for depolymerization of alginate by polysaccharide lyase family 7. *J Mol Biol* 2005, 352, 11–21.

Yen, H. W.; Brune, D. E. Anaerobic codigestion of algal sludge and waste paper to produce methane. *Bioresour Technol* 2007, 98, 130–134.

Zhou, G. F.; Xin, H.; Sheng, W.; Sun, Y.; Li, Z.; Xu, Z. *In vivo* growth inhibition of S180 tumor by mixture of 5-Fu and low molecular lambda carrageenan from *Chondrus ocellatus*. *Pharmacol Res* 2005, 51, 153–157.

Zhou, M. L.; Shao, J. R.; Tang, Y. X. Production and metabolic engineering of terpenoid indole alkaloids in cell cultures of the medicinal plant *Catharanthus roseus* (L.) G. Don (Madagascar periwinkle). *Biotechnol Appl Biochem* 2009, 52, 313–323.

CHAPTER 13

MOLECULAR DOCKING: A PRACTICAL APPROACH FOR PROTEIN INTERACTION ANALYSIS

NEIL ANDREW D. BASCOS

CONTENTS

13.1   INTRODUCTION

Recent developments in sequencing technologies and biophysical analysis have led to a torrential rate of data acquisition. Computational methods provide systematic ways to sift through the flood of data for nuggets of golden information (Baxevanis and Ouellette, 2001). This review focuses on three computational methodologies for the study of protein structure and the protein interactions involved in complex formation and ligand binding.

13.1.1   HOMOLOGY MODELLING

Christian Anfinsen's classical work on the denaturation and renaturation of ribonuclease revealed the importance of the amino acid sequence in defining the three-dimensional structure of polypeptides (Anfinsen *et al.*, 1954). This discovery provided the experimental basis for *de novo* protein structure prediction, that is, the prediction of how a protein folds into its final functional form based on its amino acid sequence. Several algorithms have been developed for *de novo* protein structure prediction. The accuracy of their predicted structures compared to experimental observations is assessed in the biannual competition for the critical assessment of structural prediction (CASP). While recent developments in algorithm design and parallel processing have raised the accuracy of these predictions, the solution to the "protein-folding problem" is far from complete. In the absence of this complete solution, other methods are tapped for the prediction of protein structure. Perhaps the most common method used is that of homology modeling.

Homology modeling is the process of predicting the structure of a protein based on the relatedness of its amino acid sequence compared to previously defined protein structures. This method relies on Anfinsen's theorem regarding the importance of amino acid sequence for three-dimensional structures and the observed conservation of specific structural features in proteins of similar function. Homology modeling requires at least three pieces of information. Firstly, the amino acid sequence of the query protein. Secondly, similar amino acid sequences in the data repositories. And lastly, a defined structure for these similar sequences, on which the structural model can be based. Homology modeling involves two processes. An initial sequence alignment is required to define homologous (related) proteins within a database. Subsequently, the aligned query sequence may be fit using defined structures of homologous proteins as scaffolds. It is possible that the query sequences contain several sequence stretches that align with different structural domain types. In these cases, homology models can be generated for each of the domains separately prior to a prediction of their association to form the complete protein structure. Proteins are believed to attain their native/functional features through the minimization of the polypeptides' free energy (Rose *et al.*, 2006). Validation of generated homology models is commonly done using molecular dynamics simulations for energy minimization.

13.1.2 PROTEIN–PROTEIN INTERACTIONS

Molecular Docking Simulations are used to predict the mechanisms of protein action in coordination with other proteins and cofactors. Molecular Docking simulations generally test the interaction between one target and one probe molecule. These molecules are usually composed of either a target protein and a probe protein, or a target protein and a probe small molecule ligand (Meng *et al.*, 2011; Mukesh and Rakesh, 2011). Molecular docking algorithms are used to investigate protein interactions in order to determine the mechanisms of their action. Through these techniques, information may be gathered on the orientation of the interacting partners, as well as their binding affinities. Binding partner orientation may be important in the generation of functional domains such as binding surfaces in heterodimeric receptors (Xiong *et al.*, 2001). Binding affinities of partner proteins define the probability of their association in the presence of competitors and other cofactors (Eisenmesser *et al.*, 2005). Standard methodologies for molecular docking assign one of the binding partners as a target, and one a probe. By convention, larger proteins are assigned to be targets and smaller proteins are assigned as probes. This convention provides a more efficient allocation of computational resources by setting the larger target molecules to be immobile during the docking process, while the smaller probe molecule is moved to test the different binding orientations (Krippahl *et al.*, 2003).

Target and probe interactions may be tested using three types of algorithms: hard docking, induced-fit and flexible docking protocols (Meng *et al.*, 2011). Hard docking algorithms assume a classical "lock and key" association for the binding partners (Mukesh and Rakesh, 2011). Best-fit docks of rigid targets and probes are determined based on complementarities of shape and surface properties. Induced-fit docking protocols simulate flexibility in the probe molecule. These algorithms were developed to account for observed variations in protein structure when bound as part of complexes/reaction centers. Bound proteins were observed to adopt changed conformations, induced to fit the binding site (Mukesh and Rakesh, 2011). These changes in protein structure have been associated with their abilities to modulate enzymatic reactions (Landry, 2003). The structure of the probe may be varied to determine conformations of optimal association. Variations of probe structure involve rotations in defined torsion angles (Meng *et al.*, 2011; Mukesh and Rakesh, 2011).

Flexible docking protocols simulate binding with variable torsion angles for both the target and probe molecules. Flexibility in both target and probe is meant to provide a more accurate representation of *in vivo* docking, where both target and probe molecules may adjust their conformations to facilitate their interaction. The simulation of target molecule flexibility makes these protocols more computationally intensive than the other types (Meng *et al.*, 2011; Mukesh and Rakesh, 2011). All three docking protocols provide inferred dock structures based on target and probe interactions. The probabilities of each dock are calculated based on free energy levels and binding affinity. Structures docked with higher binding affinity and/or lower free energy levels have an increased probability of occurrence. However, previous knowledge of the system, including information on factors that are observed to affect the interaction (e.g., electrostatics, hydrophobic interactions) may be used to better screen the results (Krippahl *et al.*, 2003).

13.1.3 PROTEIN–LIGAND INTERACTIONS

Protein–ligand docking simulations provide perhaps the most lucrative prospects for the use of molecular docking. This protocol may be used to investigate the binding of small molecule ligands unto their receptors and is commonly used for the discovery of novel drug treatments (Goodsell and Olson, 1990; Meng *et al.*, 2011; Mukesh and Rakesh, 2011). Similar to protein-protein docking, protein–ligand docking simulations require the assignment of a target, and a probe molecule. The small molecule ligand is most often assigned as the probe. Protein–ligand binding simulations may use hard docking, induced-fit and flexible docking protocols. The small size of the ligands allows their efficient use of the more computationally extensive induced-fit and flexible docking protocols.

Protein–Ligand docking simulations provide probable ligand docking sites and binding affinities (Goodsell and Olson, 1990). These protocols may be used to screen ligand repositories to identify samples that will bind the target enzyme/receptor. Ligands found to have similar binding sites as known drugs have the potential for use as competitors. The probability of successful competition may be assessed with their binding affinities. Greater binding affinity of one ligand suggests its predominance in a competitive reaction (Eisenmesser *et al.*, 2005). Ligands found to bind at alternative sites may have the potential to affect the receptor through other means. While these may not be direct competitors for known drug binding site, they may still serve as allosteric regulators of the enzyme/receptor's function.

13.2 MOLECULAR DOCKING RESOURCES

Several software packages have been developed for each of the docking protocols. Some commonly used examples are provided in Table 13.1. As with any experimental system, these simulation programs are best used with a clear understanding of their capabilities and limitations. Most of these programs come with tutorials and sample data that can be used to familiarize new users with their system. Familiarization with the different software will allow researchers to adapt them for use in their own fields of study. The following section provides basic examples of how the different docking protocols may be used in the analysis of protein structure and interactions.

TABLE 13.1   Molecular docking resources.

Experimental type	Software	Reference
Homology Modelling	Deepview/SwissPDB Viewer	Guex *et al.*, 1995
	VMD	Humphrey *et al.*, 1996
Protein-Protein Interactions	Chemera:BiGGER	Krippahl *et al.*, 2003
	FT Dock	Gabb *et al.*, 1997
Protein-Ligand Interactions	AutoDock Vina	Goodsell and Olson, 1990
	FlexX	Rarey *et al.*, 1996
	GOLD	Verdonk *et al.*, 2003
	DOCK	Kuntz *et al.*, 1982
	FLOG	Miller *et al.*, 1994

13.2.1 *PROTEIN STRUCTURE PREDICTION*

Structures may be predicted for query proteins based on the structures of homologous/related proteins. Structure prediction may be based on regions of high sequence similarity between the compared proteins. The polypeptide sequence of the query protein may be fit unto the reference protein structure in regions of high similarity. Energy minimization and molecular dynamics simulations may be used to predict the structure of the dissimilar regions.

1. Acquire the amino acid sequence of the query protein.
2. Search for probable homologs of the query protein (In Genbank (www.ncbi.nlm.nih.gov/genbank), this may be done using the Basic Local Alignment Search Tool (BLAST)).
3. Acquire the amino acid sequences of the homologous proteins.
4. Search and acquire available molecular structure files of the returned homologous proteins (search the protein data bank (www.pdb.org) for structure files of the BLAST results).
5. Conduct pairwise sequence alignments of the query protein and homologous proteins with defined structures (reference proteins).
6. Determine regions of sequence similarity query and reference proteins.
7. Acquire a molecular structure viewer (e.g., DeepView/SwissPDB Viewer (Guex et al., 1995) or VMD (Humphrey et al., 1996)).
8. Fit query protein sequences unto the reference protein structures in the defined regions of sequence similarity. In DeepView, several options are available for this function under the Fit tab. The most simple is the use of Magic Fit. Alternatively, regions of similarity can be selected individually, and fit unto the corresponding sections of the protein structure.
9. Minimize the energy of the fit protein structure. In DeepView, this can be done using the Energy Minimization tool, which may be accessed through the Tools tab. This structure may be defined as a homology model of the query protein.
10. Molecular dynamics simulations for equilibration/energy minimization may be done on the homology model. The resulting structures after the equilibration can be compared to determine areas of variation between the model and the reference. Additionally, a comparison of the residue root mean square deviations for the two protein forms provides a measure of their respective flexibilities (optional).

13.2.2 SITE DIRECTED MUTAGENESIS

A very basic, but, useful application for homology modeling is the prediction of the effect of single mutations on a protein's structure. This technique allows the prediction of changes in a protein's active site due to the point mutation. This technique is best used with proteins that have available molecular structure files. In the absence of previously defined structures, homology models of the unmutated and mutated protein forms may be generated and compared.

1. Acquire a molecular structure viewer (e.g., DeepView/SwissPDB Viewer (Guex et al., 1995) or VMD (Humphrey et al., 1996)).
2. Acquire molecular structure files of the unmutated protein from a database (e.g., Protein Data Bank at www.pdb.org).
3. Open the molecular structure file (e.g., pdb file) using the molecular viewer.
4. Mutate the desired residue. In DeepView, you can choose the desired residue with the mouse and change it to the desired amino acid using the Mutate icon.
5. Minimize the energy of the mutated protein. In DeepView, this can be done using the Energy Minimization tool, which may be accessed through the Tools tab.
6. Compare the unmutated and mutated protein forms to determine areas of variation.
7. Molecular dynamics simulations for equilibration/energy minimization may be done on both the unmutated and mutated protein forms. The resulting structures after the equilibration can be compared to determine areas of variation. Additionally, a comparison of the residue root mean square deviations for the two protein forms provides a measure of their respective flexibilities (optional).

13.2.3   PROTEIN COMPLEX FORMATION

The interaction between proteins can be simulated using molecular docking protocols. These predictions provide information on probable orientations and affinities of binding. Certain docking algorithms also provide information on the effect of different factors (e.g., electrostatics, hydrophobics, geometry, etc.). Experimental data on the binding interaction can be used to assign the appropriate weights of the factors in the predicted results (e.g., previous studies showing the importance of electrostatics in the interaction may be the basis for using this factor in sorting the generated docks). The following methodology is based on the use of the BiGGER docking software of the Chemera Tool for Biomedical Research (Krippahl et al., 2003) The BiGGER program generates a user-defined number of docks that may be sorted based on geometric complementarity, electrostatics, side-chain interactions, hydrophobic interactions and a total global score. A particular factor from these choices may be chosen as the basis for dock predominance.

1. Acquire molecular structure files of the interacting proteins from a database (e.g., Protein Data Bank at www.pdb.org).
2. Assign the larger protein as the target.
3. Assign the smaller protein as the probe.
4. Run the docking simulation.
5. Retrieve the datafiles.
6. Sort the returned docks based on appropriate factors that have been documented to affect the binding.
7. Compare the interactions observed in the generated docks with previous experimental data.

13.2.4   DRUG COMPETITOR SEARCH

Screens for possible ligands of known drug targets may be done using molecular dynamics simulations. The generated data provides information on the possible orientations and binding affinities of the test ligands when bound to the enzyme/receptor. Relative strengths of binding affinity may provide insights on a ligand's potential as a competitor for a binding site. Observed binding sites for the ligand may also provide information on its role in either direct or allosteric regulation of the receptor function. The following methodology is based on the use of the Autodock Vina (Goodsell and Olson, 1990) for modeling protein–ligand interactions.

1. Acquire a molecular structure file for the enzyme/receptor protein from a database (e.g., Protein Data Bank at www.pdb.org).
2. Acquire a molecular structure file for the ligand from a database.
3. Generate PDBQT files for both the receptor and the ligand; this adds the appropriate charged groups (e.g., missing H) to the molecules in preparation for docking. This process also assigns the rotatable torsion angles of the ligand that will be varied during the docking process to optimize the interactions.
4. Define the search space for ligand binding.
5. Write a config file that contains the information required to run Autodock Vina using the command line (optional).
6. Run Autodock Vina using the command line. Include all the necessary information as part of the command.
7. Analyze generated docks; Autodock provides an output of 9 generated docks. The binding affinities of each of these docks are provided along

with their difference in position relative to the top dock. A comparison of the generated docks for two possible ligands will reveal their competition for a particular site, as well as the probability of one's predominance over the other.

13.3 CONCLUSION

Molecular docking protocols allow the simulation of protein structures and protein interactions. Through these techniques, researchers may prescreen experimental conditions for those that have greater potential returns. This advantage is particularly relevant for researchers that have limited funds for experimental trials. Continued development of this field promises greater opportunities for researchers of all (financial) backgrounds to contribute to the pool of knowledge. In addition, the refinement of data acquisition afforded by these techniques holds the potential for an increase in the efficiency with, which research is conducted.

KEYWORDS

- **Drug competitor**
- **Homology**
- **Ligand**
- **Molecular docking**
- **Mutagenesis**
- **Protein interaction**

REFERENCES

Anfinsen, C. B.; Redfield, R. R.; Choate, W. L.; Page, J.; Carroll, W. Studies on the gross structure, cross-linkages, and terminal sequences in ribonuclease. *J Biol Chem* 1954, 207(1), 201–210.

Baxevanis, A. D.; Ouellette, B. F. F. *Bioinformatics: a practical guide to the analysis of genes and proteins*, 2nd Ed.; John Wiley and Sons: New York, 2001.

Eisenmesser, E. Z.; Millet, O.; Labeikovsky, W.; Korshnev, D. M.; Wolf-Watz, M.; Bosco, D. A.; Skalicky, J. J.; Kay, L. E.; Kern, D. Intrinsic dynamics of an enzyme underlies catalysis. *Nature* 2005, 438(7064), 117–121.

Gabb, H. A.; Jackson, R. M.; Sternberg, M. J. Modelling protein docking using shape complementarity, electrostatics and biochemical information. *J Mol Biol* 1997, 272(1), 106–120.

Goodsell, D. S.; Olson, A. J. Automated docking of substrates to proteins by simulated annealing. *Proteins* 1990, 8(3), 195–202.

Guex, N.; Peitsch, M. C. SWISS-PDBVIEWER: a fast and easy to use pdb viewer for Macintosh and PC. *Protein Data Bank Quaterly Newsletter* 1996, 77, 7.

Humphrey, W.; Dalke, A.; Schulten, K. VMD – visual molecular dynamics. *J Mol Graph* 1996, 14(1), 33–38.

Krippahl, L.; Moura, J. J.; Palma, P. N. Modelling protein complexes with BiGGER. *Proteins* 2003, 52(1), 19–23.

Kuntz, I. D.; Blaney, J. M.; Oatley, S. J.; Langridge, R.; Ferrin, T. E. A geometric approach to macromolecule-ligand interactions. *J Mol Biol* 1982, 161(2), 269–288.

Landry, S. J. Structure and energetics of an allele-specific genetic interaction between dnaJ and dnaK: correlation of nuclear magnetic resonance chemical shift perturbations in the J-Domain of Hsp40/DnaJ with binding affinity for the ATPase domain of Hsp70/DnaK. *Biochemistry* 2003, 42(17), 4926–4936.

Meng, X. Y.; Zhang, H. X.; Mezei, M.; Meng, C. Molecular Docking: a powerful approach for structure based drug discovery. *Curr Comput Aided Drug Des* 2011, 7(2), 146–157.

Miller, M. D.; Kearsley, S. K.; Underwood, D. J.; Sheridan, R. P. FLOG: a system to select 'quasi-flexible' ligands complementary to a receptor of known three-dimensional structure. *J Comput Aided Mol Des* 1994, 8(2), 153–174.

Mukesh, B.; Rakesh, K. Molecular docking: a review. *Int J Res Ayurv Pharm* 2011, 2(6), 1746–1751.

Rarey, M.; Kramer, B.; Lengauer, T.; Klebe, G. A fast flexible docking method using an incremental construction algorithm. *J Mol Biol* 1996, 261(3), 470–489.

Rose, G. D.; Fleming, P. J.; Banavar, J. R.; Maritan, A. A backbone-based theory of protein folding. *PNAS* 2006, 103(45), 16623–16633.

Verdonk, M. L.; Cole, J. C.; Hartshorn, M. J.; Murray, C. W.; Taylor, R. D. Improved protein–ligand docking using GOLD. *Proteins* 2003, 52(4), 609–623.

Xiong, J. P.; Stehle, T.; Diefenbach, B.; Zhang, R.; Dunker, R.; Scott, D.; Joachimiak, A.; Goodman, S.; Arnaout, M. A. Crystal structure of the extracellular segment of integrin $\alpha_v\beta_3$. *Science* 2001, 294, 339–345.

APPLICATION OF BIOTECHNOLOGY AND BIOINFORMATICS IN DRUG DESIGNING AND DISCOVERY

A. K. M. MAHBUB HASAN, SAJIB CHAKROBARTY, RAJIB CHAKROVORTY, and A. H. M. NURUN NABI

CONTENTS

14.1 INTRODUCTION

The process of drug discovery involves the identification of candidates, synthesis, characterization, screening and finally, the assays of their therapeutic efficacy. The history of drug discovery is as old as that of human being. The history of drug discovery goes back in parallel as mankind itself. In ancient time, majority of human societies have used drugs for their treatment. Through simple trial and error approach, man discovered that berries, roots, leaves and barks could be used for medicinal purposes to alleviate symptoms of illness. Scientists have found evidence of using herbal mixtures by the Stone Age humans from Caribbean island of Carriacou. These types of drugs were discovered in South America between 100 BC and 400 BC. In that age, the drug discovery was also accompanied with drug delivery system in the form of ceramic bowls, tubes for inhaling drug fumes or powders. Even Indian subcontinent has a very rich history of using drugs. Archeologists and anthropologists helped to dig out the history of drugs, whereas biochemists, biotechnologists, molecular biologists, pharmacists, geneticists and computational biologists have contributed in designing and developing new drugs. While people during 19th and early centuries used crude plants or their extracts as folk medicine, modern society is using semisynthetic and synthetic drugs along with purified herbal medicines. Isolation of biologically active organic molecules started in the latter part of the 19th century in relatively pure form for medicinal use. Salicylic acid (precursor of aspirin from willow bark), morphine, codeine (from opium poppy), antimalarial agent quinine (cinchona, china bark) are among the first naturally occurring drugs that made a breakthrough in therapeutics. Therapeutic drugs have made a major contribution in healing the life threatening diseases and thus, increase average life expectancy almost all over the world.

Major advances in synthetic chemistry and biochemistry provided further momentum in the area of therapeutic agents. Aspirin, the first synthetic pharmaceutical drug, was synthesized in the latter half of the 19th century. But its recognition as the universal painkiller came in early 1900s and this discovery initiated the era of therapeutic agents. The milestone discoveries of drugs in 1930s and 1940s include synthetic sulfa drugs, the natural antibiotic penicillin (by Alexander Fleming from *Penicillium notatum*), the semisynthetic antibiotic, tetracycline (from natural *Streptomyces aureofaciens* by Benjamin Duggar) and streptomycin (aminoglycoside from *Streptomyces griseus* by Salman Waksman). The importance of vitamins and diseases caused by their deficiencies was also being uncovered during this period. Further advancements in technologies such as X-ray crystallography, NMR spectroscopy and mass spectrometry; developments and improvements of electrophoresis, ultracentrifugation, high performance liquid chromatography (HPLC) have contributed to the discovery of additional chemicals with therapeutic activities and also, to the development of vaccines. However, discovery of drugs got a new rhythm and pace with the revelation of the advanced techniques of molecular biology. Watson and Crick's discovery of double helix structure of DNA in 1953 (based on X-ray diffraction work of Rosalind Franklin and Maurice Wilkins) marked the initiation of designing drugs on the basis of molecular modeling. Besides helping to uncover the mechanism of DNA replication, transcription and translation, this great discovery also assisted in gradual development of the better understanding of viral genome replication. This proved to be a critical work for the antiviral drug discovery in subsequent decades due to the beginning of the identification of molecular targets in the viral replication cycle. The commercial potential of molecular biology and biotechnology in drug discovery was first recognized in the mid of 1970s (Duenas-Gonzalez, 2010) along with DNA sequencing, recombinant DNA technology and DNA cloning techniques. Biotechnology can be defined with respect to two main drug-discovery applications: supporting technologies used in the development of traditional, small-molecule drugs and the development of protein therapeutics, such as insulin, erythropoietin and monoclonal antibodies. The polymerase chain reaction (PCR) of the 1980s resulted in major advances in biotechnology that have had significant impact in drug discovery. Also, this industry grew rapidly after the approval of recombinant insulin in 1982.

In the 1990s, combinatorial chemistry, molecular modeling and bioinformatics have further revolutionized the discovery of newer generation drugs based on genomics and proteomics background. It has been of great importance to develop fast and accurate target identification and prediction method for the discovery of targeted drugs, construction of drug-target interaction network as well as the analysis of small molecule regulating network. Computational tools offer the advantage of delivering new drug candidates more quickly and at a lower cost. Major roles of computation in drug discovery are: virtual screening and *de novo* design, *in silico* ADME/T prediction and advanced methods for determining protein–ligand binding and structure based drug design. In recent years, a discipline called 'Systems Biology' (Kitano, 2002) is giving new hopes to the researchers by providing a broader perspective as well as solid platform to identify novel drug targets, which might make drug discovery and drug testing better, faster and cheaper using large scale data of genomics, proteomics and metabolomics (Fig. 14.1). It could lead to the identification of novel drug targets or assist with the assessment or pre-dicting dosage regimen of drug compounds. For example, use of advanced computer simulated models, novel drug targets might be identified by analyz-ing the key step controlling the disease-related pathway. By modeling organ specific metabolic stress responses, drug toxicity could, one day, be evaluated quickly and economically.

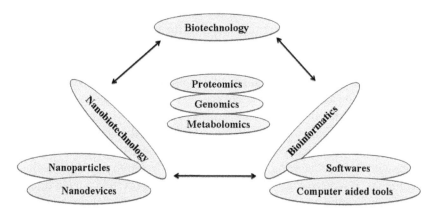

FIGURE 14.1 Relationship of the most powerful approaches used in drug designing and development in this modern world. It is important to bring together the nanoengineering, mathematical modeling and biomedical sciences to develop nanotechnology-based therapeutics and diagnostic platforms that address unmet medical needs.

Development of multiorgan and multitissue pharmacokinetic simulated model, it could be possible to assess the pattern of drug distribution and its metabolism. Drug dosage regimen system could be established using different concentrations or dose frequencies along with computer-aided modeled tumor responses (Bugrim *et al.*, 2004; Cho *et al.*, 2006; Kumar *et al.*, 2006). Argu-ably, these kinds of computer-aided programs are required to be validated through wet-lab experiments. However, they could at least direct towards a valid proposition or shorten the experimental time and efforts. Nanotechnol-ogy has also emerged as another promising area, which could be applied in drug discovery, drug delivery and pharmaceutical manufacturing. Nanotech-nology is the utilization of materials, devices and systems by controlling mat-ter at the nanometer scale (Kewal, 2005). Analyzes of signaling pathways through nanobiotechnology techniques might provide new insights into the disease, which would help to understand mechanism of action of drug and assist in recognizing new biomarkers. This might lead to new approaches for drug discovery. Nanobiotechnology not only involves nanoparticles in drug discovery (a nanosubstance called fullerenes is a drug candidate (Wilson, 2002)) but also includes several fundamental nanotechnologies (Venne *et al.*, 2005) and nanodevices all of which have equal contribution in discovering new drug (Chen *et al.*, 2003). In the case of target identification and validation, both nanodevices (e.g., use of nanotube electronic biosensors in proteomics) and nanotechnologies are used for imaging single biomolecule under physiological conditions for examining molecular reactions and interactions (Legleiter *et al.*, 2004). Moreover, biosensor (a nanodevice) and surface plasmon resonance technique (a nanotechnology) are widely used in lead identification (Grimm *et al.*, 2004).

Although modern technologies and methods have brought an enormous transformation in the discovery and developmental processes that involve several major steps for successful new drugs designing (Fig. 14.2 and Table 14.1), this path still takes a long way for reaching the final goal. It is evident that a major multidisciplinary approach is needed for the discovery and development of sensitive and effective drugs. Role of biochemistry, particularly pathobiochemistry and pharmacobiochemistry has been recognized in the development of a new and more potent medicine. This chapter will discuss about the interplay of biotechnology, bioinformatics including computational biology, protein folding and modeling along with mathematical models in designing, discovery, and development of new drugs in the following sections.

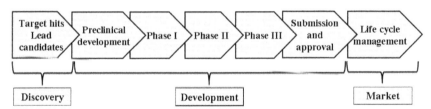

FIGURE 14.2   Discovery of drug takes the path from understanding a disease to bringing a safe and effective treatment to patients that involves not only plenty of time (10 to 15 years) with several steps and millions of billions of dollars (800 million to 1.0 billion USD) but also bringing people with diverse background together for a common goal. Multidisciplinary experts with knowledge in biochemistry, molecular biology, genetics, biotechnology, nanotechnology and bioinformatics should be associated in this process. Their collective effort should provide the platform to create sophisticated new drugs that will act with great precision.

TABLE 14.1 Description of major steps of how drugs can be designed, synthesized and marketed after approval of appropriate agency.

Major steps	Description
Pre-discovery to understand the disease or program selection	Researchers from companies, academic institutes and government contribute to the knowledge based understanding of the disease to be treated as well as the possible and the fundamental reason of the diseases. This takes years even with the modern facilities of technologies and tools; very often it ends with frustrating negative results. It is very important to understand how genes are altered and their respective protein product act in living cells and tissues and, how the mutant products affect the patient.
Target identification	Once the underlying cause is identified, the specific gene or protein should be chosen to be target for drug.
Validation of the drug target	This step involves testing of the drug target and confirmation of its role in disease by using cell culture and animal models. This actually indicates whether the course looks promising or it will lead to the dead ends.
Assay development and find a "lead compound"	Assay development fulfills several important criteria including relevance, robustness, practicality, feasibility, automation, cost, and quality. Random screening of plants, animals, microbial and synthetic sources is required for the identification of a promising new chemical entity using the developed assay systems that could become the drug called the lead compound. This could ultimately act on a target compound to alter the disease course. There are few ways for finding the lead compounds, of them bacteria and moldy plants offer many substances that could fight against diseases. By virtue of the knowledge of biotechnology, scientists can proceed towards the genetically engineered living systems to produce disease fighting biologically active molecules. Using sophisticated computer modeling, advanced robotics and simulation *de novo* programs, scientists can predict and test what kind of compound could be the most useful and/or potent as the most promising lead compounds. The high throughput screening (HTS) process is the most powerful one to search the lead compounds that run in a parallel fashion in a multi-well assay plates.

TABLE 14.1   *(Continued)*

Major steps	Description
Lead optimization and identification of a drug candidate	Promising lead compound that is specified from the above stages could then be modified structurally to obtain new properties (e.g. augmented binding affinity) for increasing its efficacy and potency. New techniques have revolutionized the way of lead optimization. Hundreds of analogues are produced; biologists test these compounds in biological system and chemists use this information for further modifications to identify the drug candidate; researchers use to start formulation of the lead compound as a candidate drug by adding the inactive ingredients that will hold it and deliver it to the specific site, way of taking the drug (whether the dug should be taken orally, by injection or inhaling).
Preclinical trials	Preclinical tests are performed *in vitro* in the laboratory and *in vivo* in animal models to determine if the drug is safe for human testing. During this stage, making of large quantities of drugs for clinical trials should also be considered. For clinical trials, number of subjects should be considered carefully because more subjects in the study would generate more statistically significant and acceptable data.
Clinical trials	The developed drug should get approval from the respective drug administration before clinical trials. Clinical trials in human require three steps; of them first one is to perform initial human testing of candidate drug in a small number, comprising of 20 to 100 healthy volunteers. The second step is to include 100 to 500 patients of specific disease for which the candidate drug has been developed and finally, the last step is to trial a large number (1000-5000) of patients to determine safety and efficacy of the drug.
Release of the drug	The drug should be released into the market (NDA/FDA approval).
Follow-up monitoring	There should be a monitoring system upon the utilization of the drug to collect information about the safety and efficacy of the drug on the particular group of patients.

14.2   BIOTECHNOLOGY IN DRUG DESIGN AND DISCOVERY

Recombinant DNA is a form of designed DNA that is isolated from natural host or created by combining two or more sequences that would not normally occur together. In terms of genetic engineering, it is created through the introduction of specific DNA sequence of interest into the already existing plasmids of bacteria or into an artificial DNA vector, to code for desired product for a specific purpose such as antibiotic resistance or disease protection. This is the basis of recombinant DNA technology, also called genetic engineering. A recombinant protein therefore is one that is derived from recombinant DNA technology (Hansson *et al.*, 2000; Mitchell, 2002; Steinberg and Raso, 1998). Molecular biology is now practiced as an engineering discipline, called bioprocess engineering, and spawned an entire industrial field of biotechnology. Recombinant technologies allow the production of large quantities of useful medicinal products, which are difficult to prepare from natural sources or are naturally unavailable. These products include hepatitis B vaccine, interferons, tumor necrosis factor, insulin, growth hormone, erythropoietin, tissue plasminogen activator, and recently the monoclonal antibodies. They have extensive implications in the prevention, diagnosis and/or treatment of many human diseases such as hepatitis, cancer, diabetes and myocardial infarction. The chemical nature and purity of recombinant DNA compounds will have to be assured, animal toxicity testing will be required, and sufficient data on clinical safety and sensitivity will have to be gathered for legitimate approval (Bhogal and Balls, 2008; Lasagna, 1986).

Recent developments in the field of molecular biology enable identification, analysis and cloning of protein coding genes for large-scale amplification. Besides naturally occurring genes, it is not possible to construct novel genes with desired properties such as coding for modified products, possessing enhanced biological activities and/or diminished undesirable properties, as well as entirely new designed substances. This is usually practiced in bacteria, but other systems involving yeasts or continuously growing transformed mammalian cell lines have recently been developed. Using transgenic approaches and transient expression in whole plants or plant cell culture, a variety of recombinant subunit vaccine candidates, therapeutic proteins and even monoclonal antibodies have now been produced with high yield (Yusibov *et al.*, 2011). Some of these products have been tested in early phase of clinical trials, and have been shown to be safe and effective. Examples of these products include mucosal vaccines for diarrheal diseases, hepatitis B and rabies; injectable vaccines for non-Hodgkin's lymphoma, H1N1 and H5N1 strains of influenza A virus and Newcastle disease in poultry; and topical antibodies for the treatment of dental caries and HIV (Yusibov *et al.*, 2011). The plant-based expression systems have some advantages, for example, low upstream costs, lack of human or animal pathogens, and ability to produce target proteins with desired structures and safe biological functions. More interestingly, Zebrafish

(*Danio rerio*) are now being used as prominent and a complete animal model for *in vivo* drug discovery and development (Chakraborty *et al.*, 2009). One should carefully consider that naturally occurring genes, when engineered and expressed for functional use in foreign hosts, may deviate structurally, biologically or immunologically from their natural counterparts. Such products may be designed with enhanced biological features and/or diminished undesirable effects compared with their natural counterparts in experimental or human subjects. Most of the medically important molecules are difficult to isolate from natural source and are often contaminate with other toxin and this forces the biological scientists to design the biological drug (Fig. 14.3).

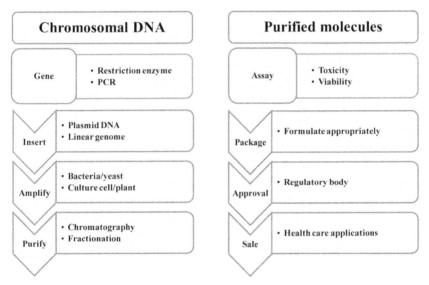

FIGURE 14.3   A DNA fragment (more specifically gene) of interest is isolated directly from natural source cutting by restriction enzyme or synthesized artificially using polymerase chain reaction techniques. This DNA fragment is inserted into DNA plasmid or live vector for example viral or yeast genome to multiply them in number of copies into suitable cellular host. This cell is harvested to produce the desired product in large scale. The product is purified using the appropriate chromatographic techniques and any other fractionation strategy depending on their solubility properties. Preclinical and clinical trials are important to assay their possible toxicity in human health and check the viability to store them. Packaging to maintain its quality is thus necessary followed by the approval from the drug regulatory body of the respective country. Finally, the drug is sold in the market for end user health care.

The advances in recombinant DNA technology have the potential to alter the drug-development process profoundly. Modern well-equipped pharmaceutical industry needs to adjust its research efforts by developing new research laboratories. Academic researchers in the universities, institutes and research centers are playing a vital role in developing the new biotechnology, supplying most of the basic scientific knowledge and the initial supply of the scientific work force. The recent progress of thinking for scientific training from the government to the pharmaceutical industry has inspired the improvement of pharmaceutical based research in biotechnology. DNA recombinant technology has already made a significant impact on medical practice and the pharmaceutical industry (Alper, 2003; Heinzelmann, 2011; Neumann and Neumann-Staubitz, 2010). The technology has allowed the emergence of new and better methods for manufacturing existing drug products, for example, recombinant human insulin and recombinant human growth hormone. The first successful treatment using animal derived insulin was tried about 90 years ago by Best (Best and Scott, 1923). In 1982 human insulin, the first recombinant product is permitted for use as medicine (Johnson, 1983). Later, it was described that recombinant IL-1a directly activates neutrophil function by stimulating hydrogen peroxide production and that IL-1a *in vivo* may augment the response of neutrophils to other stimulators such as foreign bodies (Ozaki *et al.*, 1987). Recombinant human interferon-gamma recovers partially the defective phagocyte microbicidal function in patients with chronic granulomatous disease (CGD) in childhood (Sechler *et al.*, 1988). Sometimes, antisera and monoclonal antibodies (mAbs) can be difficult to produce economically at large scale. Moreover, they usually require long development times and have some problems associated with safety issues especially in case of neurotoxin. New therapeutic strategies to develop and prepare antitoxins are urgently needed. Recently, it was reported that a single recombinant heterodimeric binding agent consisting of two high-affinity Botulinum neurotoxin (BoNT) binding heavy-chain-only Ab VH (VHH) agents and two epitopic tags, coadministered with an antitag mAb, protected mice from lethality with an efficacy equivalent to conventional BoNT antitoxin serum (Mukherjee *et al.*, 2012). This strategy was sufficient to protect mice from BoNT intoxication.

The potential applications of nanotechnology in life sciences, particularly nanobiotechnology, are included in drug discovery. Nanotechnology is an emerging branch for designing tools and devices of size 1 to 100 nm with specific function at the cellular, molecular and atomic levels. Nanotechnologies, including nanoparticles and various nanodevices such as nanobiosensors and nanobiochips are being used to improve drug discovery (Domingo-Espin *et al*., 2011; Jain, 2009). Nanomaterials are increasingly used in diagnostics, imaging and targeted drug delivery. The concept of employing nanotechnology in biomedical research and clinical practice is best known as nanomedicine that could potentially make a major impact to human health. Nanotechnology will assist the integration of molecular diagnostics/imaging with therapeutics and ultimately will facilitate the development of personalized medicine, that is, prescription of specific medications best suited for an individual (Bhogal and Balls, 2008; Teli *et al*., 2010). RNA interference (RNAi) has created excitement in clinical sciences because this principle can be used for gene-specific therapeutic activities that target the mRNAs of disease-related genes such as oncology, neurology, endocrinology and infectious diseases and thus play the pivotal role in drug discovery (Appasani, 2004; Gomase and Tagore, 2008). Time-consuming and expensive large-scale experimental approaches are progressively replaced by prediction-driven investigations with the help of computer-aided biotechnology (Chanumolu *et al*., 2012; Vivona *et al*., 2008). DrugBank (http://www.drugbank.ca) is now a highly annotated resource that continuously combines detailed drug data with comprehensive drug target and drug action information since its first release in 2006 (Wishart *et al*., 2008). It is used to facilitate *in silico* drug target discovery, drug design, drug docking or screening, drug metabolism prediction, drug interaction prediction and general pharmaceutical education. More importantly, DrugBank is now equipped with the information of biotech drugs. Significantly, more protein target data has also been added to the database, with the latest version of DrugBank containing three times as many nonredundant protein or drug target sequences (Wishart *et al*., 2008).

14.3   BIOINFORMATICS IN DRUG DESIGN

Despite the rapid development of the sophisticated methods, drug discovery still remains a complex and time-consuming process. In the conventional workflow, it typically starts with the target and lead compound identification, subsequent optimization of the lead compound followed by *in vitro* and *in vivo* studies including animal model, preclinical and clinical trials. In pregenomic era, potential drug candidates had to be chemically synthesized and then analyzed for subsequent properties like efficacy, pharmacokinetics, toxicity, and allergic properties and so on, which is equally time consuming and labor intensive. But with the advancement of genomics, proteomics and bioinformatics are driven by high-throughput techniques and powerful algorithms.

As a result, the whole paradigm of drug discovery has been revolutionized. The vast amount of information obtained from numerous high throughput screening (HTS) studies based on genomics and proteomics background, not only speed up the small molecule drug discovery but at the same time open a whole new dimension of targeted therapy and molecular medicine where a specific proteins or genes responsible for the disease is targeted via different approaches. Bioinformatics has a wide spectrum of vital applications throughout various steps of drug discovery beginning from drug target identification, followed by drug-target interaction, optimization of lead compound, quantitative structure-activity relationship (QSAR) analysis and toxicity prediction (Fig. 14.4).

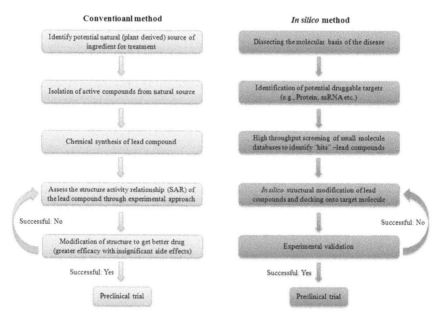

FIGURE 14.4   Major steps in conventional and *in silico* methods for drug design.

14.3.1   DRUG TARGET IDENTIFICATION THROUGH BIOINFORMATICS: IN SEARCH OF DRUGGABLE GENOME

The term "druggable genome" portrayed as the subset of human genome that potentially can serve as target for therapeutic intervention, has now drawn much of the attention of scientists and pharmaceutical companies. Bioinformatics has shown enormous potential in identifying drug target with high confidence in relatively short time. For instance in an attempt to outline the druggable genome, Hopkins and Groom classified the amino acid sequences of drug binding domains of existing drug targeted proteins into 130 families where only six protein families [G-protein coupled receptors (GPCRs), serine/threonine and tyrosine protein kinases, zinc metallopeptidases, serine proteases, nuclear hormone receptors and phosphodiesterases], comprises of the half of the total number of drug targets (Hopkins and Groom, 2002). They extended the list of potential drug targets up to 3051 by predicting the similar drug binding domains in human genome (Fig. 14.5) through the utilization of Interpro domain analysis tool (http://www.ebi.ac.uk/interpro/index.html) (Hopkins and Groom, 2002).

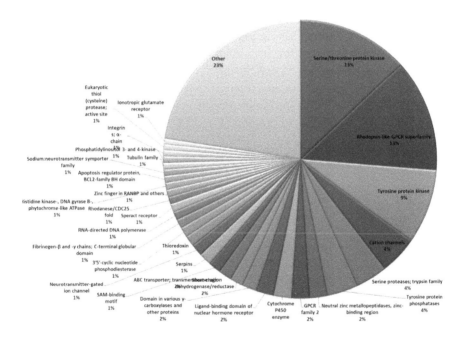

FIGURE 14.5 Distribution of predicted "druggable" targets in human genome and most of data were regenerated from Hopkins and Groom (2002) study. Out of 130 proteins families, the families that harbor 20 proteins or more were taken as separate group and rest of the families were considered as "others" as a whole.

Although the foundation of the druggable genome has been established, few remaining challenges demand for coordinated and systematic bioinformatics studies to exploit the full potential of the emerging concept of druggable genome. Human genome has not been fully understood yet as researchers is beginning to learn the vital role of regulatory mechanism through noncoding RNAs (e.g., microRNAs, shRNAs) and epigenetic factors in the pathogenesis of various diseases including cancer. Therefore, targeting only one gene may not be as helpful as it was expected. Another challenge would be to cope with the complexity of human proteome due to numerous posttranslational modifications, functional assembly formation and alternative splicing events.

Whatever the challenges, if it can assume that druggability is an inherent property of the protein, the bioinformatics analysis of a potential drug target could be beneficial. Although this approach is still in its infancy, it holds the promise to be a powerful weapon to combat complex and multidimensional diseases like cancer and diabetes. Since most small molecule drugs tend to bind to distinct regions of functional relevance to which endogenous molecules interact, the structural information of the potential target proteins can be investigated to elucidate the drug-binding region. There are numerous bioinformatics databases that serve as the archive of secondary, tertiary, globular structures and domains of human proteome. The most prominent among them is Protein Data Bank (PDB) a repository of atomic coordinates along with other structural information of proteins (Fig. 14.6). (http://www.rcsb.org/pdb/home/home.do).

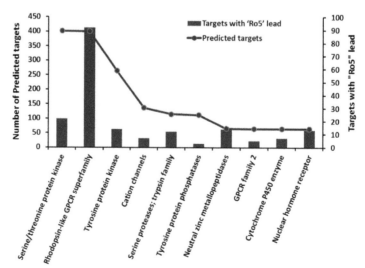

FIGURE 14.6 Comparison between the number of predicted targets and "Ro5" leads. Here Ro5 leads represents the compounds that satisfy the rule-of-five' parameters proposed by Lipinski *et al.* (2001).

All the protein-structure coordinates deposited in PDB are obtained through experimental methods such as X-ray crystallography, NMR spectroscopy, and cryo-electron microscopy and can be visualized by number of visualization tool such as PyMol. As of July 31, 2012 a total of 83407 structures (though atomic coordinates are not available for all of them) are deposited for human in PDB including various oncogenes, transport proteins, ion channels, membrane proteins, enzymes, which may serve as drug targets (Fig. 14.7). There are also databases specialized for drug targets that provides the structural information of the potential drug targets to the users. Two such databases are potential drug target database (PDTD) and therapeutic drug database (TDD) harboring 841 and 2025 potential drug target structures respectively.

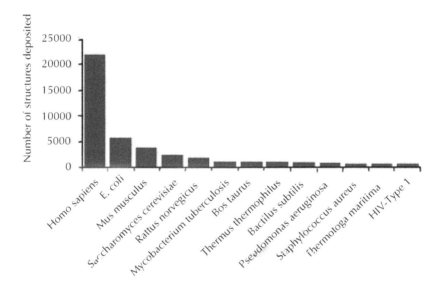

FIGURE 14.7 Comparative analysis of number of the structures deposited in PDB from various organisms.

14.3.2 STRUCTURAL BIOINFORMATICS LINKS LINEAR SEQUENCES TO DOMAINS AND 3D STRUCTURE

Despite the enormous growth of the protein structural databases, a vast majority of proteins structures in human genome is still remains as elusive and unresolved largely due to the varieties of biochemical properties of proteins such *in vitro* solubility, complex assembly formation, stability and others. For instance, membrane proteins, representing an attractive target for drugs, include ion channels and transporters of small molecules, and receptors encompassing various branches of cellular processes. However, for the majority of membrane proteins, the lack of the structural information hinders its potential to be effectively used as drug targets (Hambly *et al.*, 2006). Structural bioinformatics provides the unique opportunity to predict the 3D structure by powerful computational algorithms backed by robust statistical tests. Structure prediction methods can be divided into two broad categories: template based modeling (TBM) and '*ab initio*' or '*de novo*' modeling. Both these methods have their own advantages and disadvantages. Template based modeling is relatively accurate but needs a structurally resolved homologous protein as a starting point or template. TBM, alternatively called homology modeling, is based on the assumption that protein folding pattern is more conserved compared to liner sequence in the course of evolution and therefore query protein can be modeled with certain degree of confidence by mimicking the folding of a homologous template (Zhang, 2008). I-TASSER, a server system that develops 3D models by multiple-threading alignments, is one of the most popular among the various homology based protein 3-D structure and functions prediction tools (Roy *et al.*, 2010).

Other online and stand-alone tools for homology modeling can be found on the expassy proteomic tools repository. Unfortunately, suitable template with acceptable sequence similarity is not available for many proteins. This problem can be coped with the '*ab initio*' or '*de novo*' modeling approach based on the physical principles rather than previously solved template structures. However, the prediction accuracy can be compromised in case of '*de novo*' modeling approach (Nayeem *et al.*, 2006). In many cases, scientists may opt to look for drug binding domain instead of the whole 3D structure of the target proteins. There are multiple databases available online facilitating the identification of desired domains in user provided list of query protein sequences by various algorithms such as Hidden Markov model (HMM). The prominent examples of this domain databases include Pfam, Conserved Domain Database (CDD), Simple Modular Architecture Research tool (SMART) and InterProScan. A comparative picture of the human proteome data deposited in Pfam and InterPro demonstrates that 59,052 sequences and 93,527 domains are deposited in Pfam with 72% coverage while in InterPro the respective data are 25,635 and 60,698 with 73.8% coverage. The description and web link of some important online databases and tools important for drug discovery are now summarized in Table 14.2.

TABLE 14.2 Description and web link of some important online databases and tools important for drug discovery.

Database name	Description	Web link
Protein data bank (PDB)	The PDB archive is a repository of atomic coordinates and other information describing proteins and other important biological macro-molecules.	http://www.rcsb.org/pdb/home/home.do
Potential drug target database (PDTD)	PDTD is a comprehensive, web-accessible database of drug targets with known 3D-structures. PDTD contains 1207 entries covering 841 known and potential drug targets with structures from the Protein Data Bank (PDB).	http://www.dddc.ac.cn/pdtd/
Therapeutic drug database (TDD)	This database currently contains 2,025 targets, including 364 successful, 286 clinical trial, 44 discontinued and 1,331 research targets; 17,816 drugs, including 1,540 approved, 1,423 clinical trials, 14,853 experimental drugs and 3,681 multi-target agents.	http://bidd.nus.edu.sg/group/cjttd/TTD_HOME.asp
InterPro	InterPro is an integrated database of protein signatures and automatic annotation of proteins and genomes. InterPro classifies sequences at superfamily, family and subfamily levels, predicting the occurrence of functional domains, repeats and important sites.	http://www.ebi.ac.uk/interpro/index.html
PyMOL	PyMOL is a user-sponsored molecular visualization system on an open-source foundation.	http://www.pymol.org/
I-TASSER	I-TASSER server is an Internet service system for protein structure and function predictions. 3D models are built based on multiple-threading alignments by LOMETS and iterative TASSER assembly simulations.	http://zhanglab.ccmb.med.umich.edu/I-TASSER/

TABLE 14.2 *(Continued)*

Database name	Description	Web link
Pfam	The Pfam database is a large collection of protein families, each represented by multiple sequence alignments and hidden Markov models (HMMs)	http://pfam.sanger.ac.uk/
CDD	CDD is a protein annotation resource that consists of a collection of well-annotated multiple sequence alignment models for ancient domains and full-length proteins.	http://www.ncbi.nlm.nih.gov/Structure/cdd/cdd.shtml
SMART	SMART (a Simple Modular Architecture Research Tool) allows the identification and annotation of genetically mobile domains and the analysis of domain architectures. More than 500 domain families found in signaling, extracellular and chromatin-associated proteins are detectable.	http://smart.embl-heidelberg.de/help/smart_about.shtml
PubChem	PubChem is organized as three linked databases within the NCBI's Entrez information retrieval system. These are PubChem Substance, PubChem Compound, and PubChem BioAssay.	http://pubchem.ncbi.nlm.nih.gov/about.html
ChEMBL	ChEMBL is a database of bioactive drug-like small molecules, it contains 2-D structures, calculated properties (e.g. logP, Molecular Weight, Lipinski Parameters, etc.) and abstracted bioactivities (e.g. binding constants, pharmacology and ADMET data).	https://www.ebi.ac.uk/chembl/

TABLE 14.2 *(Continued)*

Database name	Description	Web link
PDBeChem	Dictionary of chemical components (ligands, small molecules and monomers) referred in PDB. It provides comprehensive search facilities for finding a particular component, or determining components in structure entries or *vice versa*.	http://www.ebi.ac.uk/ pdbe-srv/pdbechem/
ChemBank	ChemBank stores an increasingly varied set of cell measurements derived from, among other biological objects, cell lines treated with small molecules. Analysis tools are available and are being developed that allow the relationships between cell states, cell measurements and small molecules to be determined.	http://chembank. broadinstitute.org/
Cambridge Structural Database	A non-profit company and a registered charity the CCDC compiles and distributes the Cambridge Structural Database (CSD), the world's repository of experimentally determined organic and metal-organic crystal structures.	http://www.ccdc.cam. ac.uk/products/csd/

14.3.3 IN SILICO SCREENING OF POTENTIAL SMALL MOLECULES: CHEMINFORMATICS COMES OF AGE

With better understanding of intricate cellular pathways combined with powerful bioinformatics algorithms, the whole new era of drug discovery has begun and marked by the prediction of large number of druggable targets. The growing number of potential drug targets necessitates the development of libraries of large number of small drug-like molecules. To cope with the growing demand of new drugs, scientists employed combinatorial chemistry (CC) approach to produce large arrays of new drug-like molecules from existing building blocks by both solution- and solid-phase CC strategies (Hall *et al.*, 2001). Although combinatorial chemistry had the potential to yield large number of small molecules, only few of them appeared as probable drug candidates. It was believed that lack of diversity within the compounds could be a plausible explanation of this. However, the scientific community soon realized that management and comparative analysis of the data regarding different properties of the candidate molecules such as structure, bioactivity, biochemical properties, toxicity, and *in vitro* assays could result in designing better drugs. Subsequently, this realization leads to the concept of cheminformatics (Xu and Hagler, 2002). Cheminformatics provides a platform for the end users to analyze, compare and manage the small molecule libraries to search for potential bioactive drug like molecules. Among the freely available cheminformatics databases, PubChem, ChEMBL, PDBeChem, ChemBank and the Cambridge Structural Database are prominent. Pubchem is composed of three interconnected databases-PubChem Substance, PubChem Compound and PubChem BioAssay. PubChem substance database harbors detailed information regarding the chemical structures, description, external links including to PubMed, protein 3D structures, and biological screening results while the structural information is stored within PubChem Compound in clustered manner and cross-referenced by identity and similarity groups. The data obtained from BioActivity screens of chemical substances are stored in PubChem substance. It provides searchable descriptions of each BioAssay, including descriptions of the conditions and readouts specific to a screening protocol. A chemical diverse library may contain both drug and nondrug like compounds.

Therefore, algorithms have been developed to predict the drug like properties of the potential compounds based on the "Rule of 5" proposed by Lipinski *et al.* (Lipinski *et al.*, 2001). According to the rule of 5, absorption of permeation of a compound is likely to be poor if there are more than 5 hydrogen-bond donors, 10 hydrogen bond acceptors, the molecular weight (MWT) is greater than 500 and the calculated Log P (CLogP) (a measure of lipophilicity) is greater than 5 (Lipinski *et al.*, 2001). In recent years, a number of algorithms based on this Rule of 5 have been developed to predict the drug like properties in small compounds.

14.3.4 CHEMICAL GENOMICS – THE WEAPON OF CHOICE IN FUTURE

Pregenomics era has changed the whole landscape of drug discovery. With the burgeoning of huge "omics" data a revolutionary transition from conventional drug discovery approach to rational based targeted drug design are being occurred. Genome sequence information, together with functional genomics, can potentially provide the platform to identify large number of novel drug targets by revealing the "Druggable genome." On the other hand, cutting edge technologies in synthetic chemistry coupled with high throughput screening methods has now enabled to develop more target-specific lead compounds. The term "Chemical Genomics" has been coined that encompasses the genomic technologies enabling the chemical genetic approaches to study the ligand-target interaction in the context of living cell as a whole. Although the whole concept of rationale based drug design is still in its infancy compared to traditional drug discovery practice, it is showing much promise and can be our weapon of choice to combat complex diseases like cancer. Chemical genomics along with computational drug discovery methods can prioritize in wet lab experiments by setting up virtual screenings and predictions. Since the whole process of drug discovery has become a multidisciplinary entity, equal contribution from chemiinformatics, bioinformatics, genomics and proteomics is crucial to the success of this cutting edge field.

14.4 PROTEIN FOLDING AND COMPUTATIONAL DRUG DESIGNING

Proteins, also known as building blocks of life, perform different biological tasks through interaction with other molecules. It is required to maintain three-dimensional structures of proteins for their stability along with enough flexibility for performing biological functions. The problem of protein folding is to understand how a protein molecule of specified amino acid sequence ends up in a unique configuration that determines its biological function (Maddox, 1994). Knowledge of the 3D structure of proteins is necessary to understand their biological functions and this will lay the foundation for the comprehensive understanding of the biochemical and cellular functions of an organism and thus, also for applications like drug design. In the last 50 years, molecular biology has become more quantitative. Molecular basis of several diseases have been elucidated after the completion of human genome project. Gene sequences of several proteins have helped to understand their expression pattern. Genome wide association studies, single nucleotide polymorphisms have given access to the specific codon i.e., amino acid responsible for the misfolding of a disease-associated protein for which efficacy and potency of same drug may vary person to person. However, information about the gene sequences is not enough or is of little help in understanding the function of the corresponding protein, in manipulating its function and in designing drugs to act on it. This is why one should have three-dimensional structure of protein and once the conformation is known one can attempt designing a drug molecule.

The primary objective of most pharmaceutical chemistry is to generate new compounds that can modulate disease processes and most of these compounds are enzymes, that is, protein in nature and such compounds inhibit enzymatic reaction by capping the substrate binding sites. Only a small percentage of orally administered drugs have molecular masses above 500 daltons. The smaller compounds designed so far are deemed to be the most effective as drug if their properties facilitate not only interaction but also alternation of the function of given biological molecules. However, equal importance should be given to the fact that these compounds do not interact with most of the other molecules to generate potentially adverse side effects. With the advancement of combinatorial methods huge compilation of new components has become possible in a short period of time (Houghten, 2000; Schreiber, 2000).

Together with rapid screening methods, endless number of potentially active components is becoming available and, as our knowledge is expanding towards the understanding of the most complex biological events at the molecular level, we can increasingly use rational arguments in designing potential therapies and new potential molecules to test or screen (Bleicher *et al.*, 2003). Proteins bind to other molecules to stimulate or inhibit biological functions. The associates with which proteins bind are commonly known as ligands and these can be metal ions, small organic/inorganic molecules, or macromolecules like proteins or nucleic acids. The receptors on the membrane are generally flexible biological molecules and their dynamic behavior of a receptor has long been recognized as a complicating factor in computational drug design.

The use of a single, rigid protein structure usually from a high-quality X-ray crystal structure still is the standard in most applications (Zheng and Kyle, 1996). Advanced methods have been introduced to aid more accurate description of protein flexibility and its influence on ligand recognition. This has been aided by the exponential growth in the speed of computer processors, available RAM, and disk capacity and at the same time rapid decrease of the cost to acquire these. For a successful protein–ligand interaction, involvement of key residues is important in partner recognition that will also determine the affinity for the binding of ligand to its specific receptor. Identification of these key residues is essential for understanding protein's function, analyzing molecular interactions and guiding further experimental procedures (Rausell *et al.*, 2010). Although the experimental determination provides the most accurate assignment of the binding locations, the procedure can be time and labor intensive.

Process of determining the functional sites in proteins using computational approaches could be classified into sequence-and structure based methods (Capra and Singh, 2007; Pei and Grishin, 2001; Valdar, 2002; Wang et al., 2008). Of these two methods, sequence-based methods enjoy an advantage because functionally important residues are preferentially conserved during the evolution through natural selection. In many cases, however, the sequence or evolutionary conservation of residues does not necessarily translate into their involvement in ligand binding, as these residues may play a structural role in maintaining the global scaffold. Water molecules are critical factor for structure based drug design (Wong and Lightstone, 2011). There are, however, several approaches that could be used to identify the ligand binding site within a protein molecule: (i) ligand binding in proteins is like "an insertion of key into a lock" hypothesized by Emil Fisher (Fischer, 1894); (ii) shape and physiochemical complementarity are often used to detect concave pockets on protein's surface (Huang and Schroeder, 2006; Le Guilloux et al., 2009; Weisel et al., 2007); (iii) homology modeling, homologous proteins with similar global topology often bind similar ligands using a conserved set of residues (Russell et al., 1998); (iv) there are other methods that use calculated interaction energies (Laurie and Jackson, 2005) or protein structure dynamics (Landon et al., 2008; Lin et al., 2002) to examine the click of "lock and key." Accordingly, many contemporary methods use both geometric match and evolutionary information to identify binding site pockets and residues. Some of them use known protein–ligand complexes as templates (Tseng and Li, 2011; Wass et al., 2010; Xie and Bourne, 2008), whereas others use purely sequence-based homology information (Capra et al., 2009; Huang and Schroeder, 2006). Generally protein functions are predicted on the basis of sequence similarity. In these cases, the query sequence is screened mostly with the database by using sequence alignment tools such as BLAST (Altschul et al., 1990) and PSI-BLAST (Altschul et al., 1997) to search for the model sequence that has high degree of similarity. One or more model sequences can exhibit a sufficiently high level of similarity to the query sequence. If the homology is above 40% and functionally important motifs are conserved then we can hypothesize that the query sequence has a function that is quite similar to that of the model sequence. If we have knowledge regarding the functions of the model sequences, then it would be easy to draw some conclusions about the probable biological functions of the query sequence. As the level of similarity decreases, the conclusions on function that can be drawn from sequence similarity become less and less reliable. COGS (Tatusov *et al.*, 1997, 2000), ProDom (Corpet *et al.*, 2000), PFAM (Bateman *et al.*, 2000), SMART (Schultz *et al.*, 2000), PRINTS (Attwood *et al.*, 2000), Blocks (Henikoff *et al.*, 1999), ProtoMap (Yona *et al.*, 2000), InterPro (collective database of Pfam, PROSITE, PRINTS, ProDom, SWISSPROT+TREMBL available in http://www.ebi.ac.uk/interpro/) are several databases available on the web (Table 14.2) to classify functionally or structurally related proteins into different groups and these classifications may be helpful in predicting function of a particular protein. However, structure of a particular protein do play vital role in determining its functions and thus, prediction of the structure of a protein is the utmost important part in designing a drug.

For predicting protein structure, homology-based modeling has become a powerful tool. All proteins in a given protein structure database are tried and each template is ranked using scoring functions. The score reflects the likelihood that the query sequence assumes the template structure. The selection of a suitable template protein is often done via protein threading that helps to provide evidence of possible homology of query proteins with distantly related protein sequences, to detect possible homology in cases where sequence methods fail and to improve structural models for the query sequence via structurally more accurate alignments. Protein threading can be done mostly using methods based on hidden Markov models (Bateman *et al.*, 1999; Park *et al.*, 1998) and dynamic programming methods (Bowie *et al.*, 1991; Luthy *et al.*, 1992) based on profiles. Structural genomics helps to chart the protein structure space efficiently while functional annotations and/or assignment are made afterwards. Once a map of the protein structure space is available, this knowledge should provide additional insights about the function of protein inside the cell and other interacting partner molecules. Aspects of protein structure that are useful for drug design studies typically have to involve three-dimensional structure. Predicting the secondary structure of the protein is not sufficient. Even the similarity of the three-dimensional structures of two proteins cannot be taken as an indication for a similar function of these proteins. If the spatial shape of the site of the protein is known, to, which the drug is supposed to bind, then docking methods can be applied to select suitable lead compounds that have the potential of being refined to drugs. The speed of a docking method determines whether the method can be employed for screening compound databases in the search for drug leads. A docking method that takes a minute per instance can be used to screen up to thousands of compounds on a PC or hundreds of thousands of drugs on a suitable parallel computer. Docking methods that take the better part of an hour cannot be suitably employed for such large scale screening purposes. In order to screen really large drug databases with several hundred thousand compounds, docking methods that can handle single protein/drug pairs within seconds are needed. The high conformational flexibility of small molecules and the delicate structural changes in the protein-binding pocket upon docking (induced fit) are major complications in docking. Fast docking tools have been acclimatized to screen combinatorial drug libraries to provide a carefully selected set of molecular building blocks together with a small set of chemical reactions that link the modules. In this way, billions of diversified molecules could be available theoretically from a small set of reactants using combinatorial library. High-throughput screening (HTS) could be an alternative technique to docking for finding a lead compound. This is a laboratory method that allows screening thousands of components to the target protein in a day based on their binding affinity. This has an advantage over docking method is that it does not have to deal with the problem of insufficiently powerful computer models at the expense of high laboratory costs.

Now-a-days, the peptide drugs with their high specificity and low toxicity profile overcome the unattractive pharmacological properties of native peptides and protein fragments. Peptide vaccines to viral infections and antibacterial peptides led the way in clinical development. But recently many other diseases have been targeted, for example AIDS, cancer, and Alzheimer's disease (Otvos, 2008). A class of cationic amphiphilic peptides with short sequences, $G(IIKK)_n I-NH_2$ ($n = 1–4$), have been proposed that can effectively kill Gram-positive and Gram-negative bacteria (Hu *et al.*, 2011). Thus, such peptide offers new opportunities in the development of cost-effective and highly selective antimicrobial and antitumor peptide-based treatments. The challenges encountered by orally administered peptide and protein drugs, and the nature of lymphatic absorption after subcutaneous administration will need to be considered in designing of those drugs (Ferrer-Miralles *et al.*, 2011; Lin, 2009).

14.5 MATHEMATICS AND COMPUTATIONAL SCIENCE IN DRUG DESIGNING

Mathematics, another branch of pure science besides physics and biology, is somewhat different and strangely more overarching. Richard Feynman (noble prize winner) once remarked in one of his lectures that nature speaks through the language of mathematics and hence, in order to understand nature, one has to get familiar with the language of mathematics. Mathematics provides an abstract language and rules that do not necessarily explain any physical phenomenon. However, these abstract language and its rules, when applied to observations from nature, specifies the working of the nature. Though physics has been using the prowess of mathematics for very long, biology is catching up with it slowly (the slow nature of biology in using mathematics is not because it is lacking in any sense rather there was not enough observation to fit the mathematical rules into). Mathematics has been pivotal in studying of genes, proteins and other biochemical networks. It facilitates modeling of the behavior, defining working boundaries and then studying the system analytically to generate new hypothesis. Biological science progresses when these hypotheses are tested and validated (or rejected).

Studies of signal conduction, ion transport, mechanism of biochemical reaction via enzymes and as a whole the study of biology as a complex system-use tools of mathematics and have been useful to advance the fields beyond what was possible by biology alone. One distant cousin of mathematics is the computational science. With the advent of computers and other related technologies (such as powerful graphical processors, distributed computing, and large data storages), modern biologists can analyze large datasets and information in a short amount of time. We have been experiencing the combined power of mathematics and computational science in biology-for example, while the first genome took 10 years to decode, today we have started speculating about completing it within a few weeks. This remarkable achievement is made possible by every increasing power of computation available and by applying mathematical theories coupled with advancement of biological knowledge. Beyond the examples given, mathematics and computational science are no doubt playing a major role in biology in several other aspects. In this text, practical use of mathematics in the field of drug designing will be considered and emphasized. This section will not explain any specific mathematical methodology for designing drugs rather will explore the fields of drug development in detail. The background will give us an idea why the power of mathematics is critical for this field and how is mathematics assisting the current process of drug development.

Any business textbook will tell us to identify risks in every step and then, decide the most appropriate step after considering the risks. This is not an impossible task and, strangely, this is one of the major roles of mathematics and computational science in the drug development phases. Effectively, the cost (and thereby risk) for designing a drug can be reduced by even by 50% (Taft *et al.*, 2008). How does mathematics and computation achieve this feat? The underlying biology is revealed by the experiments carried out by the laboratories around the world. The collective understanding of underlying biological principle is modeled mathematically. By modeling, we mean the language for expressing the working of the biological system changes. However, this simple step now gives us the power to "analyze" the system through the use of mathematical logic (Fig. 14.8). For example, the process of signal transduction process involves (naively) complicated pathways involving biochemical as agents and gradient of these agents as drivers for various events. This process can be "modeled" as a network of nodes where each node is equivalent to agents in the actual process and the interaction between nodes are represented as edges (the edges can be again directional or directionless). The level of interaction can be modeled as "weight" and the value of the weight of the weight can be calculated from a set of gradient equations (differential calculus).

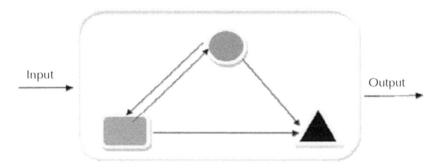

FIGURE 14.8   Biological processes modeled as a network of nodes and edges. The nodes (the shapes) are interacting with each other and the direction of interaction is evident from the direction of the arrows (edges). The boundary of the complete process is contained within the box – however, the model also takes external input and generates output – effectively communicating with other systems (in general the number of input and output can be more and unequal to each other).

Once the mathematical representation is available, the biological system can be analyzed purely based on the mathematical models and some additional parameters (such as initial concentration of molecules). In this way, it is possible a number of various scenario without conducting a laboratory experiment for each of those scenarios. For example, in Fig. 14.8, a possible variation can be made by removing the edge between the circle and the triangle-biological equivalent of inhibiting that path of the interaction. The system can be "run" on computers with varying initial concentration of different bio-agents involved and the results can be verified in a very short period of time.

In this way, the amounts of experiments (*in silico*) that can be done are huge. The reader can probably now take a guess on how this can be applied to drug designing – the *in silico* experiments can guide to a candidate molecule (lead candidate) with certain properties such as inhibiting the interaction between the circular and triangular agents as a potential drug (Fig.14.8). In this approach of system level drug designing, the use of mathematics and computational power is increasing rapidly while cost of such power is ever decreasing as compared to wet-lab experiments. Besides system level *in silico* approach, it is now possible to target the molecular level. Due to the vast amount of data on protein (or gene) sequence and structure, it is possible to predict the specific structure of the drug or target proteins (Kuroda *et al.*, 2012; Patschull *et al.*, 2012; Speck-Planche *et al.*, 2012). In these cases, it is even possible to simulate and predict the effects of the candidate drug at the molecular level and to compare (for example) the efficacy of several candidate drugs.

One other important aspect of drug discovery or designing is to understand the "side-effect" of the proposed drug. In the current paradigm, this can only be revealed fully during the clinical trial and definitely not before preclinical trials (the trials on animals sometime can reveal side effects but cannot be guaranteed). It may even be possible to discover critical side effects only after the drug is released in the market, in which case, it becomes immensely costly the pharmaceutical company. In recent times, some headway has been made to predict side effects (Kolaja, 2012; Lounkine *et al.*, 2012). It is to be noted that these latest research direction is in infancy. However, the results are already promising. In near future it may be possible to predict the side effect and, especially, detect adverse side effect a priori. This will be another major step in reducing drug development cost. The use of mathematics and computational science is immense in the field of fundamental biology. The importance of molecular (DNA, RNA, protein) structure is becoming ever more critical. Mathematical modeling, backed by the ever-increasing power of computers, is proving to be a power tool for biologists. The contribution of mathematics has now started to be felt in the practical applications and drug designing (or discovery) is one such major applications. Besides being a core theoretical tool, mathematical science is helping the drug industry design/test the drugs (or lead components) in a very cost effective manner. This is an active area and by no means complete at present. There are ample opportunities for future research in this domain. While the mathematical modeling is escalating the next biological discovery, new biology knowledge is also influencing new models. This mutual enforcement will continue and we will see some remarkable result. The end result will be a set of much better drug at a much lower price for every society.

14.6 CONCLUSION

Access to the genomic data has increased and become very easy. As a result, their large-scale expression data has opened new possibilities for the search for target proteins. Based on this development, many pharmaceutical companies have made large-scale investment in new technologies. For interpreting the generated data from the respective screening experiments, appropriate tools in bioinformatics are playing critical supportive role. Specifically, methods are required to identify interesting differentially expressed genes and to predict the function and structure of putative target proteins from differential expression data generated in an appropriate screening experiment. Discovery and development of a new drug needs to complete a long journey for, which association of multi disciplinary expertise, money and time are required. However, bioinformatics and mathematics along with the knowledge of genomics, proteomics, metabolomics has reformed the area of drug designing and development by reducing time, cost and manpower (Fig. 14.9). The days

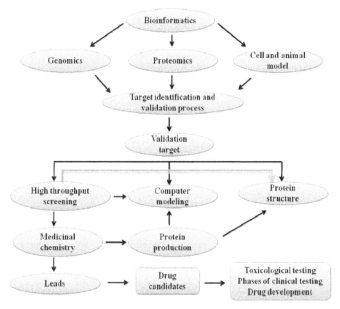

FIGURE 14.9 Paths of bioinformatics that involve in the discovery of lead compound followed by new drug candidates.

of being passive benefiters for developing worlds are now over and bioinformatics can help to open door in this region of the world. This area may also be helpful for developing countries to build up their own local research capacities by promoting collaboration via sharing data, expertise and resources. This could facilitate generation of trained manpower who would ultimately hunt the desire for achieving the goals.

KEYWORDS

- **Bioinformatics**
- **Biotechnology**
- **Chemical genomics**
- **Cheminformatics**
- **Computational science**
- **Drug design**
- **Drug discovery**
- **Protein folding**

REFERENCES

Alper, J. Drug development. Biotech thinking comes to academic medical centers. *Science* 2003, 299, 1303–1305.

Altschul, S. F.; Gish, W.; Miller, W.; Myers, E. W.; Lipman, D. J. Basic local alignment search tool. *J Mol Biol* 1990, 215, 403–410.

Altschul, S. F.; Madden, T. L.; Schaffer, A. A.; Zhang, J.; Zhang, Z.; Miller, W.; Lipman, D. J. Gapped BLAST and PSI-BLAST: a new generation of protein database search programs. *Nucleic Acids Res* 1997, 25, 3389–3402.

Appasani, K. RNA interference: from biology to drugs and therapeutics. *RNA Biol* 2004, 1, 118–121.

Attwood, T. K.; Croning, M. D.; Flower, D. R.; Lewis, A. P.; Mabey, J. E.; Scordis, P.; Selley, J. N.; Wright, W. PRINTS-S: the database formerly known as PRINTS. *Nucleic Acids Res* 2000, 28, 225–227.

Bateman, A.; Birney, E.; Durbin, R.; Eddy, S. R.; Finn, R. D.; Sonnhammer, E. L. Pfam 3.1: 1313 multiple alignments and profile HMMs match the majority of proteins. *Nucleic Acids Res* 1999, 27, 260–262.

Bateman, A.; Birney, E.; Durbin, R.; Eddy, S. R.; Howe, K. L.; Sonnhammer, E.L. The Pfam protein families database. *Nucleic Acids Res* 2000, 28, 263–266.

Best, C.; Scott, D. The preparation of insulin. *J Biol Chem* 1923, 57, 709–723.

Bhogal, N.; Balls, M. Translation of new technologies: from basic research to drug discovery and development. *Curr Drug Discov Technol* 2008, 5, 250–262.

Bleicher, K. H.; Bohm, H. J.; Muller, K.; Alanine, A. I. Hit and lead generation: beyond high-throughput screening. *Nat Rev Drug Discov* 2003, 2, 369–378.

Bowie, J. U.; Luthy, R.; Eisenberg, D. A method to identify protein sequences that fold into a known three-dimensional structure. *Science* 1991, 253, 164–170.

Bugrim, A.; Nikolskaya, T.; Nikolsky, Y. Early prediction of drug metabolism and toxicity: systems biology approach and modeling. *Drug Discov Today* 2004, 9, 127–135.

Capra, J. A.; Laskowski, R. A.; Thornton, J. M.; Singh, M.; Funkhouser T. A. Predicting protein ligand binding sites by combining evolutionary sequence conservation and 3D structure. *PLoS Comput Biol* 2009, 5, e1000585.

Capra, J. A.; Singh, M. Predicting functionally important residues from sequence conservation. *Bioinformatics* 2007, 23, 1875–1882.

Chakraborty, C.; Hsu, C. H.; Wen, Z. H.; Lin, C. S.; Agoramoorthy, G. Zebrafish: a complete animal model for *in vivo* drug discovery and development. *Curr Drug Metab* 2009, 10, 116–124.

Chanumolu, S. K.; Rout, C.; Chauhan, R. S. UniDrug-target: a computational tool to identify unique drug targets in pathogenic bacteria. *PLoS One* 2012, 7, e32833.

Chen, R. J.; Bangsaruntip, S.; Drouvalakis, K. A.; Kam, N. W.; Shim, M.; Li, Y.; Kim, W.; Utz PJ.; Dai, H. Noncovalent functionalization of carbon nanotubes for highly specific electronic biosensors. *Proc Natl Acad Sci USA* 2003, 100, 4984–4989.

Cho, C. R.; Labow, M.; Reinhardt, M.; van Oostrum, J.; Peitsch, M. C. The application of systems biology to drug discovery. *Curr Opin Chem Biol* 2006, 10, 294–302.

Corpet, F.; Servant, F.; Gouzy, J.; Kahn, D. ProDom and ProDom-CG: tools for protein domain analysis and whole genome comparisons. *Nucleic Acids Res* 2000, 28, 267–269.

Domingo-Espin, J.; Unzueta, U.; Saccardo, P.; Rodriguez-Carmona, E.; Corchero, J. L.; Vazquez, E.; Ferrer-Miralles, N. Engineered biological entities for drug delivery and gene therapy protein nanoparticles. *Prog Mol Biol Transl Sci* 2011, 104, 247–298.

Duenas-Gonzalez, A. Paradigms on biogeneric drugs – some views. *Webmed Central Pharmacology* 2010, 1, WMC00801.

Ferrer-Miralles, N.; Corchero, J. L.; Kumar, P.; Cedano, J. A.; Gupta, K. C.; Villaverde, A.; Vazquez, F. Biological activities of histidine-rich peptides; merging biotechnology and nanomedicine. *Microb Cell Fact* 2011, 10, 101.

Fischer, E. Einfluss der konfiguration auf die wirkung der enzyme. *Berichte der Deutschen Chemischen Gesellschaft* 1894, 27, 2985–2993.

Gomase, V. S.; Tagore, S. RNAi – a tool for target finding in new drug development. *Curr Drug Metab* 2008, 9, 241–244.

Grimm, J.; Perez, J. M.; Josephson, L.; Weissleder, R. Novel nanosensors for rapid analysis of telomerase activity. *Cancer Res* 2004, 64, 639–643.

Hall, D. G.; Manku, S.; Wang, F. Solution- and solid-phase strategies for the design, synthesis, and screening of libraries based on natural product templates: a comprehensive survey. *J Comb Chem* 2001, 3, 125–150.

Hambly, K.; Danzer, J.; Muskal, S.; Debe, D. A. Interrogating the druggable genome with structural informatics. *Mol Divers* 2006, 10, 273–281.

Hansson, M.; Nygren, P. A.; Stahl, S. Design and production of recombinant subunit vaccines. *Biotechnol Appl Biochem* 2000, 32, 95–107.

Heinzelmann, E. Biotech in medicine – the topic of the Olten Meeting 2010. *Chimia (Aarau)* 2011, 65, 100–103.

Henikoff, S.; Henikoff, J. G.; Pietrokovski, S. Blocks: a nonredundant database of protein alignment blocks derived from multiple compilations. *Bioinformatics* 1999, 15, 471–479.

Hopkins, A. L.; Groom, C. R. The druggable genome. *Nat Rev Drug Discov* 2002, 1, 727–730.

Houghten, R. A. Parallel array and mixture based synthetic combinatorial chemistry: tools for the next millennium. *Annu Rev Pharmacol Toxicol* 2000, 40, 273–282.

Hu, J.; Chen, C.; Zhang, S.; Zhao, X.; Xu, H.; Lu, JR. Designed antimicrobial and antitumor peptides with high selectivity. *Biomacromolecules* 2011, 12, 3839–3843.

Huang, B.; Schroeder, M. LIGSITEcsc: predicting ligand binding sites using the Connolly surface and degree of conservation. *BMC Struct Biol* 2006, 6, 19.

Jain, K. K. The role of nanobiotechnology in drug discovery. *Adv Exp Med Biol* 2009, 655, 37–43.

Johnson, I. S. Human insulin from recombinant DNA technology. *Science* 1983, 219, 632–637.

Kewal, K. J. Nanotechnology in clinical laboratory diagnostics. *Clin Chim Acta* 2005, 358, 37–54.

Kitano, H. Computational systems biology. *Nature* 2002, 420, 206–210.

Kolaja, K. Drug discovery: computer model predicts side effects. *Nature* 2012, 486, 326–327.

Kumar, N.; Hendriks, B. S.; Janes, K. A.; de Graaf, D.; Lauffenburger, D. A. Applying computational modeling to drug discovery and development. *Drug Discov Today* 2006, 11, 806–811.

Kuroda, D.; Shirai, H.; Jacobson, M. P.; Nakamura, H. Computer-aided antibody design. *Protein Eng Des Sel* 2012, 25(10), 507–521.

Landon, M. R.; Amaro, R. E.; Baron, R.; Ngan, C. H.; Ozonoff, D.; McCammon, J. A.; Vajda, S. Novel druggable hot spots in avian influenza neuraminidase H5N1 revealed by computational solvent mapping of a reduced and representative receptor ensemble. *Chem Biol Drug Des* 2008, 71, 106–116.

Lasagna, L. Clinical testing of products prepared by biotechnology. *Regul Toxicol Pharmacol* 1986, 6, 385–390.

Laurie, A. T.; Jackson, R. M. Q-SiteFinder: an energy-based method for the prediction of protein–ligand binding sites. *Bioinformatics* 2005, 21, 1908–1916.

Le Guilloux, V.; Schmidtke, P.; Tuffery, P. Fpocket: an open source platform for ligand pocket detection. *BMC Bioinformatics* 2009, 10, 168.

Legleiter, J.; Czilli, D. L.; Gitter, B.; De Mattos, R. B.; Holtzman, D.; M, Kowalewski, T. Effect of different antiAbeta antibodies on Abeta fibrillogenesis as assessed by atomic force microscopy. *J Mol Biol* 2004, 335, 997–1006.

Lin, J. H. Pharmacokinetics of biotech drugs: peptides, proteins and monoclonal antibodies. *Curr Drug Metab* 2009, 10, 661–691.

Lin, J. H.; Perryman, A. L.; Schames, J. R.; Mc Cammon, J. A. Computational drug design accommodating receptor flexibility: the relaxed complex scheme. *J Am Chem Soc* 2002, 124, 5632–5633.

Lipinski, C. A.; Lombardo, F.; Dominy, B. W.; Feeney, P. J. Experimental and computational approaches to estimate solubility and permeability in drug discovery and development settings. *Adv Drug Deliv Rev* 2001, 46, 3–26.

Lounkine, E.; Keiser, M. J.; Whitebread, S.; Mikhailov, D.; Hamon, J.; Jenkins, JL.; Lavan, P.; Weber, E.; Doak, A. K.; Cote, S.; Shoichet, B. K.; Urban, L. Large-scale prediction and testing of drug activity on side-effect targets. *Nature* 2012, 486, 361–367.

Luthy, R.; Bowie, J. U.; Eisenberg, D. Assessment of protein models with three-dimensional profiles. *Nature* 1992, 356, 83–85.

Maddox, J. Does folding determine protein configuration. *Nature* 1994, 370, 13.

Mitchell, P. First biotech drug to treat psoriasis. *Nat Biotechnol* 2002, 20, 640–641.

Mukherjee, J.; Tremblay, J. M.; Leysath, C. E.; Ofori, K.; Baldwin, K.; Feng, X.; Bedenice, D.; Webb, R. P.; Wright, P. M.; Smith, L. A.; Tzipori, S.; Shoemaker, C. B. A novel strategy for development of recombinant antitoxin therapeutics tested in a mouse botulism model. *PLoS One* 2012, 7, e29941.

Nayeem, A.; Sitkoff, D.; Krystek, S. Jr. A comparative study of available software for high-accuracy homology modeling: from sequence alignments to structural models. *Protein Sci* 2006, 15, 808–824.

Neumann, H.; Neumann-Staubitz, P. Synthetic biology approaches in drug discovery and pharmaceutical biotechnology. *Appl Microbiol Biotechnol* 2010, 87, 75–86.

Otvos, L. Jr. Peptide-based drug design: here and now. *Methods Mol Biol* 2008, 494, 1–8.

Ozaki, Y.; Ohashi, T.; Kume, S. Potentiation of neutrophil function by recombinant DNA-produced interleukin 1a. *J Leukoc Biol* 1987, 42, 621–627.

Park, J.; Karplus, K.; Barrett, C.; Hughey, R.; Haussler, D.; Hubbard, T.; Chothia, C. Sequence comparisons using multiple sequences detect three times as many remote homologues as pairwise methods. *J Mol Biol* 1998, 284, 1201–1210.

Patschull, A. O.; Gooptu, B.; Ashford, P.; Daviter, T.; Nobeli, I. *In silico* assessment of potential druggable pockets on the surface of alpha1-antitrypsin conformers. *PLoS One* 2012, 7, e36612.

Pei, J.; Grishin, N. V. AL2CO: calculation of positional conservation in a protein sequence alignment. *Bioinformatics* 2001, 17, 700–712.

Rausell, A.; Juan, D.; Pazos, F.; Valencia, A. Protein interactions and ligand binding: from protein subfamilies to functional specificity. *Proc Natl Acad Sci USA* 2010, 107, 1995–2000.

Roy, A.; Kucukural, A.; Zhang, Y. I-TASSER: a unified platform for automated protein structure and function prediction. *Nat Protoc* 2010, 5, 725–738.

Russell, R. B.; Sasieni, P. D.; Sternberg, M. J. Supersites within superfolds – binding site similarity in the absence of homology. *J Mol Biol* 1998, 282, 903–918.

Schreiber, S. L. Target-oriented and diversity-oriented organic synthesis in drug discovery. *Science* 2000, 287, 1964–1969.

Schultz, J.; Copley, R. R.; Doerks, T.; Ponting, C. P.; Bork, P. SMART: a web-based tool for the study of genetically mobile domains. *Nucleic Acids Res* 2000, 28, 231–234.

Sechler, J. M.; Malech, H. L.; White, C. J.; Gallin, J. I. Recombinant human interferon-gamma reconstitutes defective phagocyte function in patients with chronic granulomatous disease of childhood. *Proc Natl Acad Sci USA* 1988, 85, 4874–4878.

Speck-Planche, A.; Kleandrova, V. V.; Luan, F.; Cordeiro, M. N. A ligand-based approach for the *in silico* discovery of multitarget inhibitors for proteins associated with HIV infection. *Mol Biosyst* 2012, 8, 2188–2196.

Steinberg, F. M.; Raso, J. Biotech pharmaceuticals and biotherapy: an overview. *J Pharm Pharm Sci* 1998, 1, 48–59.

Taft, C. A.; Da Silva, V. B.; Da Silva, C. H. Current topics in computer-aided drug design. *J Pharm Sci* 2008, 97, 1089–1098.

Tatusov, R. L.; Galperin, M. Y.; Natale, D. A.; Koonin, E. V. The COG database: a tool for genome-scale analysis of protein functions and evolution. *Nucleic Acids Res* 2000, 28, 33–36.

Tatusov, R. L.; Koonin, E. V.; Lipman, D. J. A genomic perspective on protein families. *Science* 1997, 278, 631–637.

Teli, M. K.; Mutalik, S.; Rajanikant, G. K. Nanotechnology and nanomedicine: going small means aiming big. *Curr Pharm Des* 2010, 16, 1882–1892.

Tseng, Y. Y.; Li, W. H. Evolutionary approach to predicting the binding site residues of a protein from its primary sequence. *Proc Natl Acad Sci USA* 2011, 108, 5313–5318.

Valdar, W. S. Scoring residue conservation. *Proteins* 2002, 48, 227–241.

Venne, K.; Bonneil, E.; Eng, K.; Thibault, P. Improvement in peptide detection for proteomics analyzes using NanoLC-MS and high-field asymmetry waveform ion mobility mass spectrometry. *Anal Chem* 2005, 77, 2176–2186.

Vivona, S.; Gardy, J. L.; Ramachandran, S.; Brinkman, F. S.; Raghava, G. P.; Flower, D. R.; Filippini, F. Computer-aided biotechnology: from immuno-informatics to reverse vaccinology. *Trends Biotechnol* 2008, 26, 190–200.

Wang, K.; Horst, J. A.; Cheng, G.; Nickle, D. C.; Samudrala, R. Protein meta-functional signatures from combining sequence, structure, evolution, and amino acid property information. *PLoS Comput Biol* 2008, 4, e1000181.

Wass, M. N.; Kelley, L. A.; Sternberg, M. J. 3DLigandSite: predicting ligand-binding sites using similar structures. *Nucleic Acids Res* 2010, 38, W469-W473.

Weisel, M.; Proschak, E.; Schneider, G. PocketPicker: analysis of ligand binding-sites with shape descriptors. *Chem Cent J* 2007, 1, 7.

Wilson, S. R. Nanomedicine: fullerene and carbon nanotube biology. In *Perspectives in fullerene nanotechnology,* Osawa, E., Ed.; Kluwer Academic Publishers: Dordrecht, 2002.

Wishart, D. S.; Knox, C.; Guo, A. C.; Cheng, D.; Shrivastava, S.; Tzur, D.; Gautam, B.; Hassanali, M. DrugBank: a knowledgebase for drugs, drug actions and drug targets. *Nucleic Acids Res* 2008, 36, D901-D906.

Wong, S. E.; Lightstone, F. C. Accounting for water molecules in drug design. *Expert Opin Drug Discov* 2011, 6, 65–74.

Xie, L.; Bourne, P. E. Detecting evolutionary relationships across existing fold space, using sequence order-independent profile-profile alignments. *Proc Natl Acad Sci USA* 2008, 105, 5441–5446.

Xu, J.; Hagler, A. *Chemoinformatics and Drug Discovery,* Discovery Partners International, Inc.: San Diego, USA, 2002.

Yona, G.; Linial, N.; Linial, M. ProtoMap: automatic classification of protein sequences and hierarchy of protein families. *Nucleic Acids Res* 2000, 28, 49–55.

Yusibov, V.; Streatfield, S. J.; Kushnir, N. Clinical development of plant-produced recombinant pharmaceuticals: vaccines, antibodies and beyond. *Hum Vaccin* 2011, 7, 313–321.

Zhang, Y. Progress and challenges in protein structure prediction. *Curr Opin Struct Biol* 2008, 18, 342–348.

Zheng, Q.; Kyle, D. J. Accuracy and reliability of the scaling-relaxation method for loop closure: an evaluation based on extensive and multiple copy conformational samplings. *Proteins* 1996, 24, 209–217.

CHAPTER 15

SYSTEMS BIOTECHNOLOGY FOR INDUSTRIAL MICROORGANISMS

ANA M. MARTINS

CONTENTS

15.1   INTRODUCTION

Almost two decades ago, a true scientific revolution began, with the determination of the first complete genome sequence of a living organism, the bacterium *Haemophilus influenzae* (Fleischmann *et al.*, 1995). Shortly after, the scientific community entered the so called "postgenomics era", when the first DNA microarrays were used to simultaneously measure the expression of virtually all the genes of an organism (Chee *et al.*, 1996; De Risi *et al.*, 1997; Lashkari *et al.*, 1997; Lockart *et al.*, 1996; Schena *et al.*, 1995; Shalon *et al.*, 1996). Genomics and transcriptomics studies led to an explosion in the amount of biological data. Soon, proteomics and metabolomics, the study of the whole set of proteins (Proteome) and low molecular weight biomolecules-metabolites (Metabolome) followed. The suffix-ome has since been used in several words to designate the whole, genome-wide study of something (Table 15.1). All 'omics' methods consist of high-throughput, data-driven, holistic approaches with the objective of understanding the cellular processes as a whole. This is the main objective of systems biology, an academic discipline that aims at a systems-level understanding of biological organisms (Kitano, 2002).

TABLE 15.1   Definition of some 'omics' approaches.

'Ome'	Definition	'omics'	Techniques
Genome	Set of all genes of an organism	Genomics	DNA sequencing
Metagenome	Genome of non-cultured microorganisms	Metagenomics	Metagenome isolation, cloning and screening (several technologies)
Transcriptome	Set of all transcripts (RNA) of an organism	Transcriptomics	DNA microarrays
Proteome	Entire set of proteins of an organism	Proteomics	2D-PAGE; combined chromatography (LC) and MS
Metabolome	Whole set of metabolites of an organism	Metabolomics	Combined chromatography (GC, LC) and MS; combined capillary electrophoresis and MS; NMR

TABLE 15.1 *(Continued)*

'Ome'	Definition	'omics'	Techniques
Fluxome	Intracellular metabolic fluxes measured under determined conditions	Fluxomics	^{13}C-based metabolic flux analysis (^{13}C-MFA); flux balance analysis (FBA)
Interactome	Whole set of molecular interactions of a cell	Interactomics	Yeast two-hybrid system, affinity purification, mass spectrometry (for protein-protein interactions); ChIP, ChIP-chip, ChIP-seq (for protein-DNA interactions)

Such a scientific revolution was only possible due to great developments in several scientific areas that had occurred in the previous decades, probably the most important being molecular biology and recombinant DNA technology. The discovery of restriction endonucleases and PCR allowed scientists to manipulate genes and genomes, which was fundamental for what followed. Simultaneously, advances in the scientific technology front, with the development of techniques such as the above mentioned DNA microarrays, mass spectrometry, chromatography, and electrophoresis allowed the progress of 'omics' studies. These methods generate an enormous amount of data, which is not meaningful by itself and cannot be analyzed in the 'traditional' ways, but instead requires powerful computational and statistical tools. The concurrent developments in computer science and mathematical biology led to the emergence of a new discipline, bioinformatics, which made possible to "make sense of" the data by analysis and integration in the form of mathematical models of cellular processes. These models of the cell as a whole allow the prediction of how genetic and/or environmental changes will impact the organism/system (Fig. 15.1).

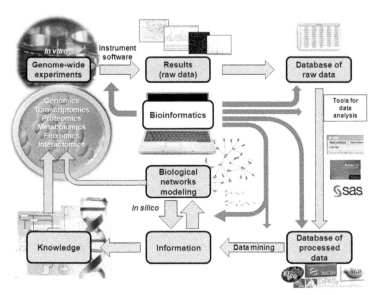

FIGURE 15.1   The systems biology approach. The figure shows the study of a biological problem at a genome-wide level, using 'omics' methods and bioinformatics tools to analyze the raw data generated. Data is deposited in databases and it can be mined by the scientific community to build genome-scale models of cellular processes. Adequate models are used for *in silico* simulations (*dry* experiments), which guide the design of new laboratory (*wet*) experiments. Comparison of *wet* and *dry* data is used to further refine the model, initiating a new cycle in the systems biology process.

The systems biology approach has been applied in basic and applied research, including in biotechnology. The bio-based production of manufacture goods has been increasing steadily, offering a very attractive alternative to chemical processes. In particular, the use of microorganisms for industrial biotechnology applications is very appealing, not only due to the vast knowledge on microbial physiology, but also to the development of new tools and technologies (e.g., genetic engineering by DNA recombinant technology), new approaches for the use of traditional microorganisms (e.g., immobilized cells and enzymes), new habitats where to search for new microbes (e.g., deep sea) with potential biotechnology applications, and also because these bioprocesses are more environmentally friendly. Additionally, microorganisms offer several advantages when compared with plant and animal cell cultures: they have a faster growth rate, they have a much lower space requirement, the control of the growth conditions is much easier, microbial cultures are less prone to contamination, and most microorganisms are amenable to genetic manipulation. Microbial biotechnology applies engineering and biological science concepts to the development of strategies for microbial use to develop new products, and has a wide array of applications, which cover almost the whole spectrum of technology: medical and pharmaceutical industries, food and beverage technology, production of chemicals, proteins, biofuels, etc. Indeed, the vast amount of microbial biochemical reactions is a rich source for the design of processes for the manufacture of desirable products (Otero and Nielsen, 2009). This is the aim of metabolic engineering, defined by Bailey as the improvement of cellular activities by manipulation of enzymatic, transport and regulatory functions of the cell with the help of recombinant DNA technology (Bailey, 1991).

The aim of this chapter is to explore the ways by which systems biology and its bioinformatics tools have been contributing to microbial biotechnology processes, with the establishment of terms such as *industrial systems biology* – approach by which systems biology enhances the development of industrial biotechnology for sustainable chemical production (Papini *et al.*, 2010) – and *systems biotechnology* – application of systems biology to biotechnology, that is, the use of system-level approaches for the modification of metabolic pathways towards an efficient production of targets bioproducts (Park *et al.*, 2005) and the development of bioprocesses to produce new drugs, fine chemicals, fuels, polymers and other biotechnological materials (Lee, 2009). A special focus will be given to the metabolic engineering of efficient microbial cell factories. The chapter begins with an introduction to microbial biotechnology: the use of the *traditional* prokaryotic and eukaryotic cell factories, *Escherichia coli* and *Saccharomyces cerevisiae*, respectively; the more recent use of the nonconventional yeast *Pichia pastoris* for the production of recombinant human proteins; and some examples of other widely used industrial microorganisms. The metabolic engineering for the improvement of industrial strains is discussed from the point of view of systems biology, including the several high-throughput 'omics' techniques, the integration of multi-'omics' data in mathematical models of the cellular metabolism, and how bioinformatics is a fundamental tool by providing databases and software tools for systems biotechnology. A brief reference to the recent field of synthetic biotechnology and how it can be used for the successful engineering of microbial cell factories will also be mentioned. Finally, a series of concrete examples on the use of systems bio(techno)logy in industrial microbio(techno)logy will be presented and discussed.

15.2 MICROBIAL BIOTECHNOLOGY: THE USE OF MICROORGANISMS AS CELL FACTORIES

Microbial biotechnology is the branch of biotechnology that studies the use of microbial cells as factories for the production of both natural products and recombinant proteins, where *cell factory* refers to the relevance of the microorganism's genetics and metabolism to the production process (De Felice *et al.*, 2008). Microorganisms have been widely used as cell factories for the industrial production of proteins, biopharmaceuticals, fine chemicals, food and beverage products, bioinsecticides and biofertilizers and even biofuels. Microbial biotechnology, however, is not a recent field: its origins can be traced back to early human history and probably began with agriculture. The use of microorganism – the first fermentations to produce beer – can be traced back to 7000 BC in Sumeria and Babylon. In Ancient Egypt drying and salting was already used for food conservation, several alcoholic beverages were obtained by fermentation of sugar-containing juices and leavened bread was produced. The history of microbiology in general, and of industrial microbiology in particular, is vast (Demain, 2010).

Nowadays, an immense amount of different microorganisms are used in industrial processes, from the traditional model organisms *Escherichia coli* and *Saccharomyces cerevisiae*, to *Aspergillus niger* for the production of antibiotics and *Pichia pastoris* for recombinant protein production, and several hyperthermophilic Archaea microorganisms (Egorova and Antranikian, 2005) (Table 15.2). The choice of a production host is a key step in the design of the industrial process, and depends on the target product. Industrial enzymes, for example, may be produced in natural producers such as filamentous fungi and *Bacillus subtilis* or in recombinant hosts such as *E. coli* and *Saccharomyces cerevisiae*, yeast being preferred if the product is to be secreted from the cells (Mattanovich *et al.*, 2012).

TABLE 15.2 High-throughput analyses used in microbial biotechnology for metabolic engineering and strain improvement*.

Microorganism	Approach	Description
Aspergillus nidulans	T + F	Functional genomics analyses to understand regulation of metabolism and to determine which factors should be considered when industrial cell factories are designed for polyketide production (Panagiotou *et al.*, 2009).

TABLE 15.2 *(Continued)*

Microorganism	Approach	Description
Aspergillus niger	G + T	Comparison of wild-type citric acid-producing strain and industrial enzyme-producing strains to establish a firm comparative genomics foundation within the same species (Andersen *et al.*, 2011).
	G + M + IS	Identification of a gene deletion pair predicted to increase succinate production (Meijer *et al.*, 2009).
Bacillus subtilis	M + F	Metabolic responses to co-feeding of different carbon substrates in a riboflavin-producing strain (Dauner *et al.*, 2002).
Clostridium acetobu-tylicum	P	Comparative analysis of a wild-type strain and a mutant with enhanced butanol tolerance and yield (Mao *et al.*, 2010).
	F	Genome-wide flux analysis of a mutant strain overproducing butanol under various batch conditions (Lee *et al.*, 2009).
	T + F	Comparative analysis of wild type and related recombinant strains to identify flux-controlling steps and unknown regulatory mechanisms (Tummala et al., 2003b).
Corynebacterium glu-tamicum	G	Use of genome information to generate a L-lysine producing mutant (Ohnishi *et al.*, 2002).
		Comparative analysis of wild-type and recombinant strains for gene target identification for L-methione biosynthesis (Ruckert *et al.*, 2003).
	T	Comparative analysis of wild type and mutant strains to find an optimal temperature for L-lysine production (Ohnishi *et al.*, 2003).
		Comparative analysis of wild type and mutant strains to target genes for increased L-lysine production (Sindelar and Wendisch, 2007)
	F	Overproduction of NADPH and improved L-lysine biosynthesis in a mutant strain overexpressing fructose-1,6-bisphosphatase (Becker et al., 2005).
		Overproduction of glutamate by a knockout mutant for pyruvate kinase (Sawada *et al.*, 2010).
		Overproduction of L-lysine by strains with a modified tricarboxylic acid cycle (Becker *et al.*, 2009).
		Overproduction of NADPH and improved L-lysine biosynthesis in a knockout mutant for 6-phosphogluconate dehydrogenase (Ohnishi *et al.*, 2005).
	T + M + F	Unraveling regulatory mechanisms of a L-lysine producing strain (Kromer *et al.*, 2004).
	M + F	Discovery of a novel biosynthetic pathway for isoleucine in deletion mutants (Kromer *et al.*, 2006).

TABLE 15.2   *(Continued)*

Microorganism	Approach	Description
E. coli	G	Comparative analysis of wild type and recombinant strains to identify and amplify target genes for optimal insulin-like growth factor I production (Ohnishi *et al.*, 2003).
	T	Comparative analysis of xylitol-producing and non-producing conditions for recombinant strains and gene disruption to find the best bioconversion rate (Hibi *et al.*, 2007).
		Engineering of strains to overproduce recombinant insulin-like growth factor 1 by co-expression of down-regulated genes identified by transcriptome profiling (Choi *et al.*, 2003).
	P	Analysis of a recombinant strain lacking phosphoglucoisomerase and determination of optimal glucose concentration for the production of PHB (Kabir and Shimizu, 2003).
		Comparison of the proteome profiles of a wild-type and an engineered strain to unravel the importance of specific enzymes in the production of PHB (Han *et al.*, 2001).
		Comparative analysis of control and recombinant producing strains to unravel target genes to increase the production of soluble recombinant humanized antibody fragment yield (Aldor *et al.*, 2005).
		Comparative analysis of wild type and recombinant strains to identify and amplify key genes for increased leptin production (Han *et al.*, 2003).
	G + T + P	Comparative analysis of a L-threonine overproducing strain and the parent strain to identify target genes (Lee *et al.*, 2003).
	G + F + *IS*	Comparative analysis of *E. coli* and *Mannheimia succiniciproducens* genomes combined with *in silico* metabolic analysis of knockout strains to identify genes for increased succinate production (Lee *et al.*, 2005).
		Identification of gene-knockout strains for maximum lycopene production and validation in controlled culture conditions (Alper *et al.*, 2005).
	T + P	Analysis of the transcriptome and metabolome in high cell density cultures (cell production) (Yoon *et al.*, 2003).
	F + *IS*	*In silico* prediction of target sites for metabolic engineering to increase succinate production (Wang *et al.*, 2006).
Penicillium chrysogenum	P	Comparison of the proteome profile of wild-type and metabolically-engineered strains for antibiotic production (Jami *et al.*, 2010)

TABLE 15.2 *(Continued)*

Microorganism	Approach	Description
Pichia pastoris	T	Analysis of the differential transcriptome of overproducing and non-overproducing strains to identify target genes with potential function in the secretion of recombinant human antibody Fab fragment (Gasser *et al.*, 2007).
Saccharomyces cerevisiae	T	A small increase in the transcription of the gene encoding a phosphoglucomutase (*PGM2*) was linked to an improved galactose utilization (Bro and Nielsen, 2004).
		Comparative transcriptome profiling of a yeast lab strain and an industrial brewing strain to identify genes involved in ethanol stress tolerance (Hirasawa *et al.*, 2007).
		Gene expression during wine fermentation (Pizarro *et al.*, 2007).
		Study of the response of an engineered strain to high levels of lactic acid to identify main regulatory targets (Abbott *et al.*, 2008).
	P	Analysis of strains engineered to use xylose as a carbon source (Pizarro et al., 2007).
		Comparative proteomic profilings of a robust industrial yeast strain in different industrial fermentation processes (Cheng *et al.*, 2008).
	M	Comparative metabolite profiling of a lab strain and an industrial strain, for analysis of the stress response during very high gravity ethanol fermentation (Devantier *et al.*, 2005).
		Study of the metabolite profile of a recombinant xylose-fermenting strain to unravel strategies for increased acetic and formic acid tolerance (Hasunuma *et al.*, 2011).
	F	Comparative study of the flux profiles of two recombinant strains for efficient xylose uptake design (Grotkjaer *et al.*, 2005).
		Metabolic flux analysis of a glycerol-overproducing strain (Kleijn *et al.*, 2007).
		Metabolic flux analysis of a malic acid-overproducing strain (Zelle *et al.*, 2008).
	T + P	Comparison between two commercial wine-producing strains with different fermentative behaviors (Zuzuarregui *et al.*, 2006).
		Comparison between an engineered yeast strain growing on xylose and a strain growing on glucose to understand signaling and regulation processes (Salusjarvi *et al.*, 2008).
	F	Flux analysis of a strain engineered to produce PHB (Carlson *et al.*, 2002)

TABLE 15.2   *(Continued)*

Microorganism	Approach	Description
Streptomyces coeli-color	G	Genome sequence analysis of 5 different species of *Streptomyces* (including *S. coelicolor*) to catalog genome components with a potential functional role in gene regulatory and metabolic networks (Zhou *et al.*, 2012)
	T	Analysis of gene expression under different environmental conditions and in different deletion mutants (Bucca *et al.*, 2003; Huang et al., 2001; Karoonuthaisiri *et al.*, 2005; Mehra *et al.*, 2006).

*Iwatani *et al.* 2008; Park *et al.* 2008, G - genomics, T - transcriptomics, P - proteomics, M - metabolomics, F - fluxomics, *IS* - *in silico* prediction.

15.2.1   THE TRADITIONAL CELL FACTORIES: ESCHERICHIA COLI AND SACCHAROMYCES CEREVISIAE

The yeast *S. cerevisiae* and the bacteria *E. coli* are the most used microorganisms in industrial biotechnology. For comprehensive reviews on the use and engineering of these microorganisms as cell factories, refer to Nielsen and Jewett (2008) for *S. cerevisiae* and Waegeman and Soetaert (2011) for recombinant protein production in *E. coli*. The *E. coli* was the first bacterial system used for the production of recombinant proteins, mainly biopharmaceuticals. Before *E. coli* and recombinant DNA technology, insulin, for example, was obtained from the pancreas of animals in a low yield process. With engineered *E. coli* it became possible to obtain high amounts of insulin in fermentation processes where high cell densities are attained.

The advantages of using *E. coli* cells as factories are well known: (i) there is a vast knowledge about *E. coli* at the physiological, biochemical, and genetic levels; (ii) cultures are easy to maintain in lab conditions; (iii) can be grown in simple, nonexpensive media; (iv) the specific growth rate is high, so high cell density cultures can be obtained in a short period of time; (v) under controlled conditions (e.g., computer-controlled bioreactors) the performance of these cell factories is highly reproducible; (vi) molecular biology tools to quickly and accurately modify the genome are well known by the scientific community; (vii) the promoter control is simple; (viii) the plasmid copy number can be easily changed (Swartz, 1996). One of the main disadvantages of using *E. coli* as cell factories in industrial biotechnology is that the recombinant proteins are usually produced as inactive, aggregated, insoluble inclusion bodies. To obtain the protein in its native form further processing is needed, and this costs time and money. *S. cerevisiae* strains have been used since the beginning of the 1980s for the production of human recombinant proteins and most biopharmaceuticals approved for human use until today were produced in these cell factories (Porro and Branduardi, 2009). *S. cerevisiae* has been also used in large-scale fermentation processes for the production of fuels, chemicals, nutritional compounds and food ingredients (Milne *et al.*, 2009). Examples of compounds produced by industrial biotechnology using *S. cerevisiae* include ethanol, glycerol, malic acid, lactic acid, ascorbic acid, flavonoids, β-carotene, terpenoids and polyketides (Papini *et al.*, 2010).

There are several advantages in using yeasts as cell factories: (i) yeasts are among the most studied living organisms, being extremely well characterized at the genetic, biochemical and physiological levels; (ii) yeast cells are a relevant model system for the much more complex higher eukaryotes, including human cells; (iii) the *S. cerevisiae* genome is fully sequenced and most of it is annotated; (iv) yeast cultures are easy to maintain and grow in lab conditions, using simple, inexpensive media; (v) yeast cells are amenable to genetic manipulation, with a vast array of expression vectors, markers and efficient transformation protocols available and the possibility of stably integrating genes into its genome; (vi) yeasts constitute the largest eukaryotic toolbox, with available collections of deletion mutants (Giaever *et al.*, 2002) and comprehensive sets of tagged open-reading frames for protein studies (Gelperin *et al.*, 2005); (vii) yeasts have predictable fermentation capabilities (Nielsen and Jewett, 2008); and (viii) they are nonpathogenic and considered by the American Food and Drug Administration (FDA) as a organisms "generally regarded as safe" (GRAS). One obvious advantage over *E. coli* is that yeasts are eukaryotic organisms, hence can perform protein posttranslational modifications, which often are essential for the function of the protein being expressed. However, in this aspect, *S. cerevisiae* also poses several problems because the posttranslational glycozylation of proteins performed by this yeast is very different from the one presented by human proteins, which may lead to the production of nonfunctional recombinant proteins, cause immunological problems or both (Cregg *et al.*, 2009; Papini *et al.*, 2010). For this reason, currently *S. cerevisiae* is mainly used as a platform for the production of chemicals, and alternative yeasts have been proposed as cell factories for recombinant protein production. One of the most promising candidates is the methylotrophic yeast *Pichia pastoris*, which has been increasingly used as the cell factory for the production of recombinant human proteins. Other "nonconventional" yeasts used in microbial biotechnology include *Hansenula polymorpha, Arxula adeninivorans, Candida boidinii, Pichia methanolica* and *Kluyveromyces lactis* (Porro and Branduardi, 2009).

15.2.2 PICHIA PASTORIS: CELL FACTORIES FOR THE PRODUCTION OF RECOMBINANT PROTEINS

The methylotrophic yeast *Pichia pastoris* presents all the advantages that an eukaryote offers for the production of human recombinant protein. Additionally, it has several advantages when compared to *S. cerevisiae:* (i) it can be grown to high cell densities in protein-free media; (ii) the protein yield is high due to the very strong and tightly regulated promoter of the gene encoding alcohol oxidase I (*AOX1*), the first enzyme of the methanol utilization pathway (Cregg *et al.*, 2000); (iii) the glycozylation pattern is more similar to that present in human proteins; (iv) multiple copies of the foreign gene can be integrated into the genome; (v) it is amenable to mutagenesis (Swartz, 1996); and (vi) it strongly prefers respiratory growth, which facilitates its culturing at high cell densities, unlike fermentative yeasts (Cregg *et al.*, 2000). Furthermore, it possesses an efficient secretory system and many molecular biology tools are already available for *P. pastoris* (Cregg *et al.*, 2009; Krainer *et al.*, 2012). All of these aspects make *P. pastoris* the microorganism of election for the production of recombinant proteins (Porro and Branduardi, 2009). A report by Cregg and co-workers in 2000 already lists more than 200 recombinant proteins produced in *P. pastoris*, from bacteria, fungi, protists, plants, invertebrates and vertebrates (including humans) and viral proteins (Cregg *et al.*, 2000).

15.2.3 OTHER BIOTECHNOLOGICALLY RELEVANT MICROORGANISMS

There is a vast collection of other microorganisms being used as cell factories in biotechnology, including bacterial species, fungi, and archaea. *Corynebacterium glutamicum* has been the subject of several metabolic engineering projects. These bacteria are key producers of amino acids, one of the main products of biotechnology, with annual productions in the order of 1.5 million tons for L-glutamate and 850,000 tons for L-lysine (Sanchez and Demain, 2008). The use of *C. glutamicum* in industrial processes offers several advantages: (i) its genome is fully sequenced (Kalinowski *et al.*, 2003); (ii) there is a wide knowledge of its metabolic networks and their regulation; (iii) there are several tools for the genetic manipulation of these bacteria, including well-established plasmid vector systems, DNA transfer methods, cloning of heterologous genes, mutations in transport systems and introduction of new protein secretion systems, gene replacement and genome rearrangement methods; (iv) the microbial production of amino acids is stereospecific, that is, *C. glutamicum* only produces the L-forms of the amino acids, which makes downstream processing much simpler (Sanchez and Demain, 2008; Nesvera and Patek, 2011).

Bacillus species have also been widely used as cell factories mainly for the production of extracellular enzymes, but also for the production of vitamins, human recombinant proteins, antibiotics, bioinsecticides and nucleotides (Schallmey *et al.*, 2004). These bacteria have several of the main desirable characteristics of industrial microbes: (i) high growth rates; (ii) wide knowledge at the biochemical, physiological and genetics levels; (iii) complete genome sequence for several species including *Bacillus subtilis* (Kunst *et al.*, 1997); (iv) ability to produce and secrete large amounts of extracellular enzymes of industrial importance, particularly for detergent manufacture; (v) some species, such as *Bacillus subtilis* are classified by the FDA as GRAS.

The first fermentation to be developed in a large-scale, during World War I, was the acetone-butanol-ethanol fermentation (ABE) performed by clostridial species, the most important being *Clostridium acetobutylicum*. Due to the current interest in biofuels, the metabolic engineering of clostridial species for increased butanol production has been pursued (Lutke-Eversloh and Bahl, 2011). Novel bacterial hosts such as cold-adapted marine bacteria *Pseudoalteromonas haloplanktis* have also been described as a potential source of enzymes with optimal activities at low temperature, which allows saving energy in industrial processes (Parrilli *et al.*, 2008). On the other hand, thermostable enzymes for use in several industrial processes that require high temperature may be obtained from extremophilic bacteria such as *Thermotoga maritima* and *Aquifex aeolicus*. Mycelial bacteria such as Streptomycetes are extremely important for industrial microbiology, because they are the main source of small bioactive compounds such as antibiotics, antifungals, anticancer agents and immunosuppressants (Schrempf and Dyson, 2011). Although there are still major technical and computational challenges to overcome, systems biology of *Streptomyces* is emergent due to the availability of genome sequences (Bentley *et al.*, 2002; Zhou *et al.*, 2012), and other high-throughput data.

The filamentous fungus *Aspergilus* has been widely used for the industrial production of metabolites (e.g., citric acid, gluconic acid) and enzymes. The use of *Aspergillus* spp. as a cell factory has several advantages, including: (i) it can degrade a wide variety of natural organic substrates; (ii) it is considered GRAS; (iii) fermentation technologies for *Aspergillus* are well developed; (iv) several species have their genome sequenced and the fungal community is making efforts to develop new tools and technologies for genetics and 'omics' approaches (Andersen and Nielsen, 2009; Papini *et al.*, 2010).

Archaea microorganisms have also been successfully used as cell factories, mainly for the production of extremozymes, enzymes capable of sustaining harsh conditions, which would easily denaturate conventional enzymes (Egorova and Antranikian, 2005). Extremozymes used in industry include: starch-processing enzymes, for starch-processing industries; cellulose-degrading enzymes, used in alcohol production, juice and detergent industries; xylan-degrading enzymes, used in food, feed, pulp and paper industries; alcohol dehydrogenases used in chemical industries; and DNA-processing enzymes, which were fundamental for the development of PCR, such as the thermostable DNA polymerases from *Pyrococcus furiosus* (Pfu DNA polymerase) and *Thermus aquaticus* (Taq DNA polymerase) (Egorova and Antranikian 2005, 2007). Methane production by archaea such as *Methanosarcina barkeri* and *Methanococcus maripalidus* was also described (Wirth *et al.*, 2012).

15.2.4 STRAIN IMPROVEMENT: OBTAINING EFFICIENT CELL FACTORIES

Strain improvement is a process, which consists in the isolation, characterization and manipulation of microbial strains for improved biotechnological applications. The ultimate goal is to obtain cell factories that are highly productive and robust rendering the production process cheaper than chemical synthesis. This requires an efficient design and optimization strategies. Industrial microbial strains can be obtained and improved by a variety of different approaches, from isolation *de novo* (natural screening) to random mutagenesis (treatment with mutagen chemicals, UV light and/or X-rays) followed by selection of the desired phenotypes (Kim *et al.*, 2008). Several industrial-relevant strains have been obtained this way, including the penicillin producer *Penicillium chrysogenum.*

More recently, the implementation of genetic engineering techniques has advanced the metabolic engineering of industrial strains by improving pre-existing metabolic pathways or engineering new ones by introducing heterologous genes (Smid *et al.*, 2005; Liu *et al.*, 2010). Metabolic engineering combines systematic analysis of metabolic and other pathways with molecular biology techniques to improve cellular properties by designing and implementing rational genetic modifications (Koffas *et al.*, 1999). The aim is to increase the metabolic flux towards the desired product and this is accomplished by reducing the bottlenecks in the process by decreasing the formation of by-products, eliminating enzyme bottlenecks due to poor catalytic performance of one or more enzymes in the pathway (Lutke-Eversloh and Bahl, 2011), as well as devizing strategies to deal with stress due to the accumulation of much higher-than-physiological amounts of intermediates and/or of natural or unnatural products (Mukhopadhyay *et al.*, 2008).

The recent review by Sanchez and Demain on strain improvement strategies for primary metabolite production (Sanchez and Demain, 2008) provides a good example of how metabolic engineering can be used to increase the flux of a metabolic pathway for the increased production of a target metabolite. The regulatory mechanisms involved in the biosynthesis of primary metabolites have been well studied and involve the regulation of carbon, nitrogen, sulfur and phosphorus sources, as well as substrate induction and feedback mechanisms (Sanchez and Demain, 2008). Figure 15.2 shows a hypothetical metabolic pathway and provides an example of how metabolism can be engineered with the help of molecular biology tools, to change regulation patterns. If the aim of the process is to produce the compound G, several strategies can be used: (i) amplify the amount of a rate-limiting enzyme E_2; (ii) engineer E_2 to decrease the negative feedback regulation by the final product G; (iii) amplify the first enzyme after a branch point, for example, E_6, so that the flux through $E_6 \rightarrow E_7 \rightarrow E_8$ is higher than the flux through $E_4 \rightarrow E_5$ leading to a higher production of G; (iv) amplify the first enzyme of the pathway E1 to increase the flow of A through this pathway; (v) increase excretion of G by modifying the membrane composition or, if the transport is mediated by a membrane protein, increase the activity of this carrier; in both cases the intracellular concentration of G will be lowered, decreasing the negative feedback regulation over E_2; (vi) delete the enzyme E_4 (by using a strain knocked out for the gene encoding this protein); this results in the accumulation of D which will be metabolized to G via $E_6 \rightarrow E_7 \rightarrow E_8$. Thus, systems biology and synthetic biology play key roles in the metabolic engineering process of microorganisms to be used as cell factories (Nielsen and Jewett, 2008; Prather and Martin, 2008).

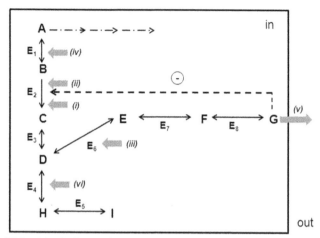

FIGURE 15.2 Metabolic engineering strategies. The figure shows a hypothetical metabolic pathway where the compound A is metabolized to I and G. Metabolites – A, B, C, D, E, F, G, H, I; Enzymes – E_1 to E_8. Arrows represent the reactions; dashed arrows represent a different metabolic pathway where A is also used. Block gray arrows show some of the engineering strategies that can be used to increase the production of the target compound G.

15.3   SYSTEMS BIOLOGY AND THE BIOINFORMATICS TOOLBOX

Systems biology aims at understanding biological processes from a systems point of view, instead of focusing on the parts (Kitano, 2002). Although systems biology has its roots in the 1940s, being based in concepts such as cybernetics and homeostasis, only in the 1990s it became possible for this discipline to show its potential, due to the 'omics' revolution. The release of the first whole-genome sequence of a living organism, *Haemophilus influenzae* (Fleischmann *et al.*, 1995), was followed by several other genomes, including the first eukaryotic genome – *Saccharomyces cerevisiae* – in 1996 (Goffeau *et al.*, 1996). Currently, and according to the NCBI – National Center for Biotechnology Information (http://www.ncbi.nlm.nih.gov/genome), around 2200 microbial genomes have been fully sequenced (Table 15.3). Additionally, there is a vast amount of genome-wide data (gene expression, protein and metabolite profiling, protein-protein interactions, etc.) about these microbes.

TABLE 15.3 Some biotechnologically relevant microorganisms with fully sequenced genome and their corresponding genome-scale metabolic models (GSMMs).

Microorganism	Year	Genome size (Kb)	Reference	GSMMs (reference)	Main biotechnological applications
Synechocystis sp.	1996	3,573	Kaneko *et al.*, 1996	Montagud *et al.*, 2010	Biofuels
Methanococcus jannaschii	1996	1,664	Bult *et al.*, 1996	Tsoka *et al.*, 2004	Cofactors, coenzymes, enzymes
Saccharomyces cerevisiae	1996	12,069	Goffeau *et al.*, 1996	Duarte *et al.*, 2004; Forster *et al.*, 2003; Herrgard *et al.*, 2008; Nookaew *et al.*, 2008	Bioethanol, primary metabolites and recombinant protein production, food and beverage industries
Escherichia coli	1997	4,639	Blattner *et al.*, 1997	Edwards and Palsson, 2000; Feist *et al.*, 2007; Reed *et al.*, 2003	Primary metabolites and recombinant proteins production
Bacillus subtilis	1997	4,214	Kunst *et al.*, 1997	Henry *et al.*, 2009	Extracellular enzymes, recombinant proteins, antibiotics production
Archaeoglobus fulgidus	1997	2,178	Klenk *et al.*, 1997	Biomodels database BMID000000140871*	Heat-stable enzymes, detoxification of metal-contaminated samples
Aquifex aeolicus	1998	1,551	Deckert *et al.*, 1998	Biomodels database BMID000000141549*	Thermostable enzymes
Thermotoga maritime	1999	1,860	Nelson *et al.*, 1999	Zhang *et al.*, 2009	Energy production, thermostable enzymes
Deinococcus radiodurans	1999	3,060	White *et al.*, 1999	Biomodels database BMID000000141241*	Bioremediation of radioactive waste
Clostridium acetobutylicum	2001	3,940	Nolling *et al.*, 2001	Lee *et al.*, 2008; Senger and Papoutsakis, 2008a; Senger and Papoutsakis, 2008b	Acetone, ethanol, hydrogen, butanol (biofuel) production
Halobacterium salinarum	2000	2,571	Ng *et al.*, 2000	Gonzalez *et al.*, 2008	Bacteriorhodopsin
Lactococcus lactis	2001	2,365	Bolotin *et al.*, 2001	Oliveira *et al.*, 2005	Lactic acid, dairy industry (cheese)

TABLE 15.3 *(Continued)*

Pseudomonas putida	2002	6,180	Nelson *et al.*, 2002	Nogales *et al.*, 2008; Puchalka *et al.*, 2008	Fine chemicals and biopolymers production, bioremediation
Streptomyces coelicolor	2002	8,667	Bentley *et al.*, 2002	Borodina *et al.*, 2005	Antibiotics, other bioactive compounds
Corynebacterium glutamicum	2003	3,309	Kalinowski *et al.*, 2003	Kjeldsen and Nielsen, 2009; Mazumdar *et al.*, 2009	Amino acids, organic acids, vitamins, ethanol
Geobacter sulfurreducens	2003	3,814	Methe *et al.*, 2003	Mahadevan *et al.*, 2006	Bioremediation of heavy metals, fuel cell development
Lactobacillus plantarum	2003	3,308	Kleerebezem *et al.*, 2003	Teusink *et al.*, 2005	Lactic acid
Acinetobacter baylyi	2004	3,598	Barbe *et al.*, 2004	Durot *et al.*, 2008	Enzymes and biopolymers production, bioremediation
Desulfovibrio vulgaris	2004	3,570	Heidelberg *et al.*, 2004	Stolyar *et al.*, 2007	Methane, bioremediation
Mannheimia succiniciproducens	2004	2,314	Hong *et al.*, 2004	Hong *et al.*, 2004; Kim *et al.*, 2007	Succinate
Methanococcus maripaludis	2004	1,661	Hendrickson *et al.*, 2004	Stolyar *et al.*, 2007	Methane
Streptococcus thermophiles	2004	1,796	Bolotin *et al.*, 2004	Pastink *et al.*, 2009	Dairy industries (yoghurt, cheese)
Aspergillus oryzae	2005	7,900	Machida *et al.*, 2005	Vongsangnak *et al.*, 2008	Fermented foods, industrial enzymes
Gluconobacter oxydans	2005	2,702	Prust *et al.*, 2005	Biomodels database BMID000000140867*	Vitamin C and other metabolites production, biosensor technology
Zymomonas mobilis	2005	2,056	Seo *et al.*, 2005	Lee *et al.*, 2010	Bioethanol
Lactobacillus delbrueckii bulgaricus	2006	1,865	van de Guchte *et al.*, 2006	Biomodels database BMID000000141606*	Dairy industries (yoghurt)

TABLE 15.3 *(Continued)*

Methanosarcina barkeri	2006	4,837	Maeder *et al.*, 2006	Feist *et al.*, 2006	Methane production, bioremediation
Aspergillus niger	2007	33,900	Pel *et al.*, 2007	Andersen *et al.*, 2008; David *et al.*, 2003	Organic acids, enzymes, pharmaceuticals
Bacillus thuringiensis	2007	5,310	Challacombe *et al.*, 2007	Biomodels database BMID000000141257*	Biopesticides
Chlamydomonas reinhardtii	2007	110,000	Merchant *et al.*, 2007	Chang *et al.*, 2011	Hydrogen (biofuel)
Lactobacillus helveticus	2008	2,080	Callanan *et al.*, 2008	Biomodels database BMID000000140544*	Dairy industries (cheese)
Acetobacter pasteurianus	2009	2,900	Azuma *et al.*, 2009	Biomodels database BMID000000141614*	Vinegar
Geobacter metallireducens	2009	3,997	Aklujkar *et al.*, 2009	Sun *et al.*, 2009	Bioremediation of heavy metals, fuel cell development
Pichia pastoris	2009	9,430	De Schutter *et al.*, 2009	Caspeta *et al.*, 2012; Chung *et al.*, 2010; Sohn *et al.*, 2010	Recombinant proteins

*indicates a non-published (to the best of our knowledge) model deposited in the Biomodels database (http://www.ebi.ac.uk/biomodels-main/); the number after BMID indicates the model ID in the database.

As mentioned previously, these advances in biological knowledge would not have been possible without the fast and astonishing advances in molecular biology and information technology. For instance, the first IBM hard-drive had a storage capacity of only 5 MB while currently much smaller hard drives can store up to 4 TB of information, 1 million times more! The evolution of computer science and technology led to the emergence of bioinformatics, an interdisciplinary research area at the interface between the biological and computational sciences. Its ultimate goal is to uncover the wealth of biological information hidden in the mass of biological data obtained by 'omics' – data mining – and get a clearer insight into the fundamental biology of organisms (Ouzounis and Valencia, 2003).

15.3.1 GENOMICS AND FUNCTIONAL GENOMICS

The genomics of industrial microorganisms has been widely reported in several articles published in the journal *Microbial Biotechnology*. Some examples are the genomics of marine microbes (Heidelberg *et al.*, 2010) including deep sea microbes such as proteobacterial species (Siezen and Wilson, 2009); the genomics of microorganisms involved in dairy fermentations and probiotics, which includes several lactic acid bacteria (Siezen and Bachmann, 2008; Kant *et al.*, 2011; Siezen and Wilson, 2011); wine genomics, which includes grapes, and several enological yeasts and bacteria (Siezen, 2008); the genomics of microalgae, which are promising biofuel producers (Brooijmans and Siezen, 2011). Although the first report of the genome sequence of the yeast *S. cerevisiae* dates from 1996 (Goffeau *et al.*, 1996), a recent paper reports the sequencing of the strain CEN.PK 113–7D which is widely used for metabolic engineering and systems biology research in industry and academia (Nijkamp *et al.*, 2012).

The genome sequence is the starting point for metabolic engineering strategies. Several studies have compared the genome sequences of wild-type and mutant microbial strains with a desired phenotype to identify target genes and control points. For example, Ohnishi and co-workers used comparative genomics of a wild-type *C. glutamicum* strain and a lysine-overproducing mutant to identify target genes for overproduction of L-lysine (Ohnishi *et al.*, 2002). Additionally, whole-genome sequences allow the construction of genome-scale mathematical models of cellular metabolism, which can be used for *in silico* predictions prior to genetically engineering a relevant microbial strain (Lee *et al.*, 2005). The annotation of genomes is the basis for functional genomics, a discipline that focuses on the development of experimental and theoretical tools for the determination of gene function (Bruggeman and Westerhoff, 2007). Functional genomics is extremely important for the development of genome-scale metabolic models (GSMMs).

15.3.2   METAGENOMICS: GENOMICS OF UNCULTURED MICROORGANISMS

Despite the huge diversity of prokaryotic and eukaryotic microorganisms, only a small fraction (<1%) can be cultivated under lab conditions (Langer *et al.*, 2006). Typically, around 99% of the microbial diversity of a certain habitat is not accessible using traditional microbial cultivation techniques, i.e., they cannot be cultured in the lab (Amann *et al.*, 1995). The metagenome is defined as the complete set of genomes of all microorganisms living in a certain habitat. These genomes can be accessed by metagenomics, the application of genomics technology to metagenome DNA (Langer *et al.*, 2006). The methodologies used in metagenomics for the extraction, cloning and analysis of the metagenome of a certain habitat are beyond the scope of this review; they have been previously described in several relevant papers (Ekkers *et al.*, 2012; Schmeisser *et al.*, 2007; Singh *et al.*, 2009; Thomas *et al.*, 2012).

Libraries of metagenomes from several different environments have been created, including from soil, sediments, marine environments, freshwater, guts of animals, the Artic, glacial ice, and acidic and hypersaline environments (Ekkers *et al.*, 2012). Clearly, these libraries provide a massive biological and molecular diversity (Torsvik *et al.*, 2002). Just as an example, more than 10^6 novel ORFs (open-reading frames) have been identified in a metagenomics study of marine prokaryotic plankton collected from the Sargasso Sea (Venter *et al.*, 2004).

Metagenomics is a powerful mining resource for biotechnological industries, opening a new world of possibilities in industrial microbiology (Ekkers *et al.*, 2012). Indeed, several industries have been interested in the exploitation of metagenomes for the search of "commercially useful" enzymes (CUEs) and the discovery of new biomolecules (Lorenz and Eck, 2005). CUEs have a wide array of applications, from biosensors to energy and environmental technologies, food and nutrition, and medicine (Ekkers *et al.*, 2012). CUEs include lipases, polysaccharide-degrading enzymes, enzymes involved in the biosynthesis of vitamins, oxidoreductases and dehydrogenases, proteases, etc. (Schmeisser *et al.*, 2007; Singh *et al.*, 2009). Recently, a database for CUEs discovered by a metagenomics approach – MetaBioME – was described by Sharma and co-workers (Sharma *et al.*, 2009; Singh *et al.*, 2009). Metagenomics also provides a powerful source for the discovery of bioactive compounds, mainly from soil microbes, including novel antibiotics, anticancer agents and immunosupressants (Langer *et al.*, 2006; Singh *et al.*, 2009).

15.3.3 TRANSCRIPTOMICS

Transcriptomics refers to the parallel analysis of the transcriptome, that is, the mRNA levels in a cell under certain conditions, using DNA microarrays (DNA chips). By comparing transcriptome profiles between two different strains, samples obtained in different conditions and/or in time-course experiments, it is possible to unravel the regulatory networks of the cell (Lee *et al.*, 2005). Transcriptomics, by itself or in combination with other 'omics' approaches, has been widely used in microbial biotechnology (Table 15.2), allowing the collection of a vast amount of information concerning the physiology, genetics and regulatory circuits of industrial microorganisms, such as: (i) *S. cerevisiae* (the database Yeast Microarray Global Viewer (Le Crom *et al.*,

2002) provides an extensive list of results published in this area); (ii) *E. coli*, for example the databases, GenExpDB, PortEco (McIntosh *et al.*, 2011); (iii) *Aspergillus* species, the Filamentous Fungal Gene Expression Database contains information about gene expression in *Aspergillus* and other filamentous fungus (Zhang and Townsend, 2010) (Table 15.4).

TABLE 15.4   Relevant databases and other bioinformatics tools for systems biotechnology.

Database	Name	URL
Genomics		
CMR	Comprehensive Microbial Resource – database of all sequenced bacterial genomes of the J. Craig Venter Institute	http://www.jcvi.org/cms/ research/projects/cmr/ overview/
DDBJ	DNA Data Bank of Japan – genome sequences databases of Japan	http://www.ddbj.nig.ac.jp/
E. coli Genome Project	Database for the *E. coli* K-12 genome from the University of Wisconsin – Madison	http://www.genome.wisc. edu/
EMBL-EBI Genome	Genome sequence databases of EMBL-EBI (European Bioinformatics Institute)	http://www.ebi.ac.uk/Da- tabases/genomes.html
GOLD	Genomes Online Database – database of genome and metagenome sequencing projects, and their associated metadata	http://www.genomeson- line.org
IMG	Integrated Microbial Genomes – data-base of microbial and microbial com-munities genome sequences of the US Department of Energy	http://img.jgi.doe.gov/ cgi-bin/w/main.cgi
MBGD	Microbial Genome Database for Com-parative Analysis – comparative analysis of fully sequenced microbial genomes (Japan)	http://mbgd.genome.ad.jp/
Microbial Genome Viewer	A web-based visualization tool that allows the combination of complex genomic data in an interactive way	http://mgv2.cmbi.ru.nl/ genome/index.html
NCBI – Genome	NCBI's databases for information on genomes – includes sequences, maps, chromosomes, assemblies, and annota-tions	http://www.ncbi.nlm.nih. gov/sites/genome
SGD	Saccharomyces Genome Database – comprehensive integrated genome-wide biological information for budding yeast	http://www.yeastgenome. org/

TABLE 15.4 *(Continued)*

Transcriptomics		
ArrayExpress	Database of functional genomics experiments of EMBL	http://www.ebi.ac.uk/arrayexpress/
DAVID	Database for Annotation, Visualization and Integrated Discovery – functional genomics tools	http://david.abcc.ncifcrf.gov/
FFGED	Filamentous Fungal Gene Expression Database	http://bioinfo.townsend.yale.edu/
GenExpDB	*E. coli* Gene Expression Database	http://genexpdb.ou.edu/main/
GEO	Gene Expression Omnibus – public functional genomics data repository	http://www.ncbi.nlm.nih.gov/geo/
PortEco: Expression	Post-publication microarray gene expression data from *E. coli* experiments	http://expression.porteco.org/
SMD	Stanford Microarray Database – microarray research database for researchers at Stanford University	http://smd.stanford.edu/
WebArray	Online microarray data analysis	http://www.webarraydb.org/webarray/index.html
yMGV	Yeast Microarray Global Viewer – database of the summary of all published results in yeast transcriptome	http://transcriptome.ens.fr/ymgv/
Proteomics		
EMBL-EBI PRIDE	Proteomics Identification Database – public data repository for proteomics data	http://www.ebi.ac.uk/pride/
GelBank	Database of 2D-gel electrophoresis images from the Argonne National Lab	http://gelbank.anl.gov/overview/index.asp
NCBI – Protein	NCBI's databases of protein sequence of several organisms	http://www.ncbi.nlm.nih.gov/sites/entrez?db=protein
PIR	Protein Information Resources – resource for protein sequences and functional information	http://pir.georgetown.edu/
SWISS-2DPAGE	Data on proteins identified by 2D-PAGE and SDS-PAGE	http://world-2dpage.expasy.org/swiss-2dpage/
SWISS-PROT	Annotated and non-redundant protein sequence database of the Swiss Institute of Bioinformatics	http://web.expasy.org/docs/swiss-prot_guideline.html
wwPDB	Worldwide Protein Data Bank – database for macromolecular structural data	http://www.wwpdb.org/

TABLE 15.4   *(Continued)*

Metabolomics and fluxomics		
BioCarta	Online maps of metabolic and signaling pathways	http://www.biocarta.com/
BRENDA	Braunschweig Enzyme Database – comprehensive information on enzymes	http://www.brenda-enzymes.org/
Chebi	Chemical Entities of Biological Interest – EMBL-EBI dictionary of metabolites	http://www.ebi.ac.uk/chebi/
ECMDB	*Escherichia coli* Metabolome Database	http://www.ecmdb.ca/
KEGG	Kyoto Encyclopedia of Genes and Genomes	http://www.genome.jp/kegg/
Ligand.Info	Small-Molecule Meta-Database – compilation of various publicly available databases of small molecules	http://www.ligand.info/
MetaCyc	Database of non-redundant, experimentally elucidated metabolic pathways	http://metacyc.org/
PathCase	Pathways Database System	http://nashua.case.edu/pathways/
Reactome	Open-source, manually curated and peer-reviewed pathway database	http://www.reactome.org
UMM-BBD	Biocatalysis/biodegradation database of the University of Minnesota – database of microbial biocatalytic reactions and biodegradation pathways	http://umbbd.ethz.ch/
YMDB	Yeast Metabolome Database	http://www.ymdb.ca/
Interactomics		
BioGrid	Database of protein and genetic interactions	http://thebiogrid.org/
STRING	Database of known and predicted direct and indirect protein interactions	http://string-db.org/
YEASTRACT	Yeast Search for Transcriptional Regulators And Consensus Tracking; curated repository of more than 48333 regulatory associations between transcription factors (TF) and target genes in *Saccharomyces cerevisiae*	http://www.yeastract.com/
Metagenomics		
JGI – Metagenome	DOE Joint Genome Institute – database of Metagenomic Programs	http://genome.jgipsf.org/programs/metagenomes/metagenomic-projects.jsf
MetaBioME	Web resource to find novel homologs for known commercially useful enzymes in metagenomic datasets	http://metasystems.riken.jp/metabiome/

TABLE 15.4 *(Continued)*

Models		
BIGG	Biochemical Genetic and Genomic knowledgebase of large scale metabolic reconstructions.	http://bigg.ucsd.edu/
BioModels DB	A repository for annotated published models from EMBL-EBI	http://www.ebi.ac.uk/biomodels-main/
JWS	JWS Online Model Database – database of curated models of biological systems	http://jjj.biochem.sun.ac.za/database/
PMDB	Protein Model Database – 3D protein models obtained by structure prediction	http://mi.caspur.it/PMDB/
SBML models for KEGG	Model repository of the portal site for systems biology, containing files converted from the KEGG database	http://systems-biology.org/resources/model-repositories/000275.html
Metadatabases – databases of databases		
BioCyc	Collection of databases on pathways and genomes	http://biocyc.org/
MetaBase	A database of more than 2000 biological databases	http://metadatabase.org/wiki/Main_Page
NCI 2DWG	NCI 2DWG Image Meta-Database – compilation of various publicly available 2D gel images	http://www-lecb.ncif-crf.gov/2dwgDB/
ProteomicWorld	Database with links to proteomics databases	http://www.pro-teomicworld.org/DatabasePage.html

Transcriptomics has been used in several studies with *E. coli.* Choi and co-workers, for example, used transcriptome analysis to examine the metabolic changes in recombinant strains producing insulin-like growth factor I (IGF-1) in high-density cultures. Their results led to the engineering of an overproducing strain by coexpression of genes, which were down-regulated in the transcript profiling study (Choi *et al.*, 2003). Global gene expression analysis has also been successfully used to eliminate bottlenecks in biotechnological processes, as exemplified by the study of *E. coli* grown in acetate, which showed that with this relatively poor carbon source, genes involved in cell growth were downregulated (Oh *et al.*, 2002), indicating that metabolic engineering strategies should seek the maintenance of low acetate levels during fermentation to increase the process yield (Park *et al.*, 2005). One example of the application of transcriptomics to the metabolic engineering of *S. cerevisiae* is the work of Bro and Nielsen (2004) which describes a strategy to increase the flow through the galactose utilization pathway in this microorganism. This pathway is an important target in industrial microbiology, because galactose is a component of several industrial media and wild-type strains of *S. cerevisiae* are unable to use this sugar as fast as glucose (Table 15.2).

15.3.4   PROTEOMICS

Although genomics and transcriptomics are both essential in systems biology, the levels of mRNA in a cell usually show a poor correlation with the protein levels (Gygi *et al.*, 1999) and the actual metabolite production (Daran-Lapujade *et al.*, 2004). The expression level of a gene does not indicate the amount of protein produced nor its location, and it may not imply biological activity or functional relationship with the set of cellular metabolites (Zhang *et al.*, 2010). While genes are static, proteins are the real *workhorses* of the cell, putting into action the information encoded in the genes. Proteins participate in virtually all cellular processes from transport across cellular membranes to gene regulation (transcription factors) and biocatalysis (enzymes). Therefore, the determination of a cell proteome (the whole set of proteins) is extremely important and provides a much better picture of a cellular process than the study of the genome/transcriptome.

Proteomics, the study of the cell proteome, consists in the simultaneous analysis of the cellular set of proteins combining technologies such as 2D-PAGE (2-dimensional polyacrylamide gel electrophoresis) for protein separation followed by mass spectrometry (MS) for protein identification. Other proteomics methods include isotope-coded affinity tags, high-affinity epitope tags and yeast two-hybrid systems (May *et al.*, 2011). Unlike transcriptomics, proteomics can provide information about protein synthesis and degradation rates, posttranscriptional and posttranslational modifications, and protein-protein interactions, making it a widely used approach in microbial biotechnology. A recent review (Han *et al.*, 2010) highlights the importance of proteomics studies for biotechnological applications, providing several interesting applications: (i) identification of highly expressed proteins in a 2D-PAGE gel, followed by the analysis of their promoter regions and use of these promoters to develop efficient high-level expression systems for biotechnology applications; (ii) identification of chaperones as proteins, which are highly induced in stress conditions; chaperones fold and stabilize proteins, being essential for the quality control of the proteome and can be used in biotechnology to improve the yield and control the quality and quantity of soluble recombinant proteins; (iii) identification of key enzymes in a metabolic pathway (e.g., enzymes catalyzing rate-controlling steps) and subsequent development of strategies for optimization of the pathway; (iv) identification of fusion partners for enhanced solubility of aggregation-prone, recombinant proteins in *E. coli*; these partners were identified as proteins significantly increased on 2D-gels when *E. coli* cells were exposed to guanidine hydrochloride or heat-shock stress; (vi) identification of bacterial extracellular proteins that can be used as excretion fusion partners for the extracellular production of recombinant proteins; excretion of industrially important intracellular proteins significantly simplifies the downstream process of protein purification and increases the chance of a correct protein folding. Proteome analysis *per se* is limited (just like any other 'omics'), but this approach has been useful for the metabolic engineering of industrial microbes, for example for the overproduction of the biodegradable polymer poly(3-hydroxybutirate) (PHB) (Han *et al.*, 2001) and the human protein leptin (Han *et al.*, 2003) (Table 15.2). For the analysis of proteomics data several bioinformatics tools are available, as reported by Mueller *et al.* (2008).

15.3.5 METABOLOMICS

Metabolomics is defined as the comprehensive (qualitative and quantitative) analysis of the metabolome, i.e., the complete set of all low molecular weight molecules (metabolites) present in and around growing cells at a given time during their growth or production cycle (Mashego *et al.*, 2007). Metabolomics aims at the understanding of how changes in metabolite concentrations in the cells are related to certain phenotypes (Mapelli *et al.*, 2008). Metabolomics appears as one of the last steps of the 'omics' cascade, hence it is closest to the cell phenotype than transcriptomics and proteomics, providing a direct snapshot into the physiological status of the cell at a certain time point in a specific condition (Borodina *et al.*, 2005). Metabolomics mainly relies on mass spectrometry (MS) technology, coupled to liquid (LC) or gas chromatography (GC). Other techniques include nuclear magnetic resonance (NMR) (Wishart, 2008) and capillary electrophoresis-electrospray ionization-mass spectrometry (CE-ESI-MS) (Britz-McKibbin, 2011). Several databases (Table 15.4) and software tools are available for the identification of metabolomics signals. A comprehensive list of resources for metabolomics can be found in a recent review by Werner *et al.* (2008). An example of metabolic pathway engineering based on metabolomics was recently reported by Hasunuma *et al.* (2011). In this study, the effect of acetic acid on xylose-fermenting *S. cerevisiae* strains, was investigated by metabolite profiling. Acetic acid is a toxic growth and metabolism inhibitor compound present in lignolcellulose hydrolysates used as industrial substrates, hence the results obtained in this work are important to unravel target genes that can be manipulated to increase yeast tolerance to acetic acid during bioethanol production.

15.3.6 FLUXOMICS

Fluxomics is closely related to metabolomics. Metabolic flux analysis has been used for the determination of the fluxome – the absolute intracellular fluxes through large networks of central carbon metabolism (Kohlstedt et al., 2010). Intracellular fluxes are the in vivo reaction rates through metabolic pathways in intact living cells and they fundamentally represent the cell phenotype, an integrated regulated network of genes, transcripts, proteins and metabolites (Nielsen, 2003). Fluxomics aims at the cell-wide quantification of intracellular fluxes, and it is based on isotope-based metabolic flux analysis (^{13}C-MFA, ^{13}C-based metabolic flux analysis) and computational methods such as flux balance analysis (FBA) (Feng et al., 2010; Kohlstedt et al., 2010). FBA uses the few experimental data available to place constraints in the flux model, obtaining a more realistic view of the fluxes (Lee et al., 2005). Isotopomer experiments allow the experimental determination of metabolic fluxes by feeding the cells a labeled substrate, usually ^{13}C-glucose. As the labeled substrate is metabolized, its distribution can be measured through NMR and MS (Papini et al., 2010) and used to calculate the intracellular flux ratios.

Fluxomics allows the physiological prediction and enzymatic rate quantification in metabolic networks as well as the identification of metabolic interactions. Hence, this approach has been widely used in metabolic engineering of microbial strains: (i) it can detect the bottlenecks in biotechnological production processes, (ii) it has been employed for the unraveling of pathway function, discovery of new pathways, (iii) it allows the study of the robustness and rigidity of metabolic networks, (iv) it allows the rational modification of industrial microbial strains, and (v) it can be used for the analysis of global physiological changes resulting from genetic engineering strategies (Feng et al., 2010; Kohlstedt et al., 2010). In summary, metabolic flux analysis is a core methodology for metabolic engineering of industrial microorganisms, mainly focusing on flux optimization strategies (Park et al., 2008).

Models based on FBA were used to guide, for example, the genetic engineering of E. coli strains for the production of aromatic (Liao et al., 1996) and other amino acids (Park and Lee, 2008), lactic acid, succinate and 1,3-propanediol (Fong et al., 2005) and lycopene (Alper et al., 2005a, 5005b).

B. subtilis strains have been used in biotechnology for the production of riboflavin. The *B. subtilis* FBA model was used to quantify growth maintenance coefficients, maximum growth yield and specific riboflavin production rate in continuous cultures (Sauer *et al.*, 1996, 1997). In *S. cerevisiae,* FBA models have revealed target genes in the TCA and glyoxylate cycles, which can be engineered for the overexpression of succinate (Otero *et al.*, 2007). Another example is the use of *in silico*-driven metabolic engineering of yeast cells to overproduce sesquiterpenes, compounds that can be used as antiseptics, antibacterials, or antiinflammatories (Asadollahi *et al.*, 2009). Flux analysis has also been widely used for the enhancement of ethanol production by yeast (Feng *et al.*, 2010). In a very interesting study by Papp *et al.* (2004), FBA was used to show that 80% of yeast genes are redundant, i.e., they are not essential for cell viability in laboratory conditions.

15.3.7 INTERACTOMICS

Proteins, being the workforce of a cell, have to interact with other macromolecules, including other proteins. The whole set of protein-protein interactions of a cell is called the protein interactome. Protein-protein interactions form protein interaction networks, the most studied in biology. Techniques used to determine interaction between proteins include the yeast two-hybrid system for the study of binary interactions, among several other methods (Stynen *et al.*, 2012). Genome-wide determination of protein-protein interactions has been carried out for biotechnologically relevant microorganisms, such as *E. coli* (Arifuzzaman *et al.*, 2006) and *S. cerevisiae* (Ito *et al.*, 2000, 2001; Krogan *et al.*, 2006; Tarassov *et al.*, 2008; Uetz *et al.*, 2000).

Interactomics information is extremely important for metabolic engineering to maximize pathway fluxes through engineered spatial organization. Usually enzymes are not isolated but arranged into multienzyme complexes, which increases the local concentrations of intermediates, avoids the loss of reactive intermediates, and prevents the divergence of intermediates to other pathways (Srere, 1987). The genome-wide *in vivo* screen for protein-protein interaction in *S. cerevisiae* reported by Tarassov and co-workers identified 2770 interactions among 1124 endogenously expressed proteins, emphasizing the significance of interactomics in cellular processes (Tarassov *et al.*, 2008).

Proteins also interact with DNA: gene expression is regulated by transcription factors (TF), proteins, which usually can bind to multiple sites in a genome, forming complex gene regulatory networks. The *in vivo* regulation of target genes by transcription factors has been studied by chromatin immunoprecipitation (ChIP), which directly determines if a certain TF is binding to the promoter region of a certain gene under a defined condition (Collas, 2010). The ChIP-chip technique combines ChIP with DNA microarrays (chip) to allow the determination of DNA-protein interactions on a genome-wide scale (Wu *et al.*, 2006). More recently, ChIP-sequencing (ChIP-seq), a technique, which combines ChIP with ultra-high-throughput massively parallel sequencing, has increasingly being used for mapping the *in vivo* protein-DNA interactions on a genome scale (Johnson *et al.*, 2007). The gene regulatory network of the model organism and cell factory *S. cerevisiae*, for example, has been widely studied and it includes more than 48,000 known regulatory associations between TF and target genes, evidencing a complex gene regulatory network for an organism with only 6,000 genes. Information about this network can be obtained in the database YEASTRACT (Teixeira *et al.*, 2006).

15.3.8 THE MULTI-'OMICS' APPROACH: INTEGRATING 'OMICS' DATA IN SYSTEMS BIOLOGY STUDIES

The examples provided previously illustrate how some biotechnologically relevant strains have been successfully developed based on the results of just one 'omics' approach (Table 15.2). However, several reports have been published with the combination of two or more 'omics' approaches, showing that a single approach is not enough to understand the system as a whole. One example, reported by Kromer and co-workers (Kromer *et al.*, 2004) combined transcriptomics, metabolomics, and fluxomics to study the different growth stages of a batch culture of a lysine-producing strain of *Corynebacterim glutamicum*. This study showed that there is a good correlation between gene expression and flux data for certain pathways but not for others, providing good evidence that studying just one layer of the system is not enough to unravel the full complexity of a cellular process.

Systems biology requires a multi-'omics' approach, with each 'omics' study providing a layer of information for the reconstruction of the biological network structure. The aim of systems biology is then to integrate all of these networks-gene network, signal transduction networks, protein interaction networks and metabolic networks-to understand how cells work (Almaas, 2007). Zhang and co-workers recently published a review on the multi-'omics' approach to systems microbiology, which includes several computational methodologies for the integration of 'omics' data (Zhang *et al.*, 2010). Metabolic engineering in microbial biotechnology tends towards a system-level knowledge of metabolism, which integrates datasets of 'omics' (genomics, transcriptomics, proteomics, metabolomics) with flux information-fluxomics (Kohlstedt *et al.*, 2010) and computational analysis (Lee *et al.*, 2005). This combination grants an understanding of the regulation of microbial metabolic networks and is the starting point for metabolic engineering of biotechnology-relevant microbial strains (Table 15.2).

15.3.9   DATABASES

The large-scale study of biological systems using genome-wide technologies results in the generation of large amounts of new data, making it impossible to report this data in the *traditional* way, that is, explicit publication in scientific journals (e.g., genome sequences). Hence, databases became essential as public repositories of data and as a tool to share knowledge within the scientific community. The function of biological databases is to make the data available to researchers in a computer-readable form. Most biological databases are available through browsable websites and usually data can be downloaded in a variety of formats. Databases are available for all types of 'omics' data, as shown in Table 15.4. Additionally, there are databases of biological models and metadatabases-database of databases, such as the recently reported *MetaBase*, the database of biological databases with a format similar to Wikipedia (Bolser *et al.*, 2012). To be fully useful, databases should be accessible by sophisticated computational and visualization tools that allow data analysis and visualization (Zhang *et al.*, 2010).

15.3.10   BUILDING GENOME-SCALE MODELS OF THE CELLULAR METABOLISM

A systems biology approach greatly relies on the modeling of biological systems. Models are used to integrate data, interpret data, and make predictions based on the data (Smid *et al.*, 2005). Mathematical modeling has been very efficient in the simulation of complex biological systems. A genome-scale mathematical model of the metabolism (GSMM), also called genome – scale *in silico* metabolic model – is a stoichiometric representation of the metabolic reactions in a cell. It uses mathematical equations, parameters and certain constraints. The reconstruction of a GSMM is a multistep process (Edwards *et al.*, 2002; Liu *et al.*, 2010; Thiele and Palsson, 2010; Viswanathan *et al.*, 2008) which is primarily based in the annotated genome of the microorganism of interest (Smid *et al.*, 2005).

The first metabolic models of cell metabolism were kinetic models, developed with the aim to understand the dynamic behavior of the pathways (Westerhoff and Palsson, 2004). Unlike GSMMs, these models take into account activities of enzymes and concentrations of metabolites (substrates, products, effectors). However, due to the difficulty in obtaining all the parameters measured under the same (physiological) conditions, these models mostly represent only a small part of metabolism, for example, glycolysis (Teusink *et al.*, 2000). A strategy to obtain wider models would be to merge several small kinetic models, such as the combination of a model of yeast glycolysis and glycerol synthesis for high/medium glucose concentrations, as reported by Martins and co-workers (Martins *et al.*, 2004). Still, this approach has not been widely used by the scientific community (Smid *et al.*, 2005), although a database for these small-scale kinetic models is publicly available-JWS (Java Web Simulation) Online (Snoep and Olivier, 2002).

GSMMs are very important for microbial biotechnology because they allow a comprehensive understanding of microbial physiology (Liu *et al.*, 2010). Besides the elucidation of the properties of the biological networks, these models allow the prediction, *in silico*, of the effect of genetic/environmental changes on the cell phenotype (Milne *et al.*, 2009). *In silico* metabolic simulations allow the calculation of maximum yields, the study of the regulation of key enzymes and the impact of deletion or introduction of new pathways (Smid *et al.*, 2005), guiding the design of *wet* experiments. Genome sequencing and functional genomics data made possible the building of metabolic network models for several biotechnologically relevant microorganisms (Table 15.3). The application of genome-scale *in silico* modeling for industrial and medical biotechnology has been recently reviewed; several interesting examples of the use of GSMMs to improve industrial processes in food production (using GSMMs for *Lactobacillus plantarum*, *Lactococcus lactis* and *Streptococcus thermophiles*), biopolymers (*Pseudomonas putida*), biofuels (e.g., acetone-butanol-ethanol production by *Clostridium acetobutylicum*) and bioremediation (*Acinetobacter baylyi*, *Geobacter metallireducens*, *Geobacter sulfurreducens*), were reported (Milne *et al.*, 2009; Smid *et al.*, 2005). As an example, let us consider the modeling of the model organism and widely used cell factory, the yeast *S. cerevisiae*. The first published model of *S. cerevisiae* metabolism (model *iFF708*) was also the first comprehensive network for an eukaryote organism, and included only 2 cell compartments, the mitochondria and the cytosol, a total of 1,175 reactions and 584 metabolites, which corresponded to only 16% of all yeast ORFs (Forster *et al.*, 2003). The second

GSMM for *S. cerevisiae* (model *iND750*) included eight compartments (extracellular space, cytosol, mitochondrion, peroxisome, nucleus, endoplasmic reticulum, Golgi apparatus, and vacuole) and 1,149 reactions (Duarte *et al.*, 2004). Finally, a consensus model of the previous two was generated, comprising 1,761 reactions and 1,168 metabolites (Herrgard *et al.*, 2008). Nookaew and co-workers used *iFF708* as a template to build *iIN800*, a model that comprised a more rigorous and detailed description of lipid metabolism, and included a total of 1,446 reactions and 1,013 metabolites (Nookaew *et al.*, 2008). These, and several others GSMMs, are publicly available in databases such as the BioModels Database, a repository of curated quantitative models of biochemical and cellular systems (Le Novere *et al.*, 2006; Li *et al.*, 2010).

Bioinformatics plays a fundamental role in the integration of data from genome-wide studies, with the aim to build models of biological systems, from the cellular to the organ level (e.g., the model of the heart) (Noble, 2002), with the ultimate goal to build models of multicellular organisms, including the *virtual human* (Kell, 2007). Several modeling software tools are available for this purpose (Table 15.5). One of the most widely used modeling package software is Gepasi, a biochemical kinetics simulator, for the simulation of the kinetics of metabolic reactions (Mendes, 1993). Closely related is Copasi, a modeling tool for biochemical networks launched in 2006 (Hoops *et al.*, 2006). Recently, Condor-Copasi was presented. This server-based software tool integrates Copasi with Condor, a high-throughput computing environment, and allows the users to run several simulations in parallel. It is especially useful for larger, complex models (Kent *et al.*, 2012). Other available tools for mathematic modeling and simulations include: SCAMP, a simulator of metabolic and chemical networks (Sauro, 1993); BioSpice, a very comprehensive software toolset for the modeling and simulation of spatio-temporal processes in living cells (Kumar and Feidler, 2003); DBSolve, a mathematical simulation workbench for development and analysis of kinetic models of biochemical pathways, which allows the dynamic visualization of the simulation results (Gizzatkulov *et al.*, 2010); MetaFluxNet, a commercial software package that allows quantitative *in silico* simulations of metabolic pathways and design of metabolic engineering strategies (Lee *et al.*, 2003). Recently, Feng and co-workers reported a new web platform MicrobesFlux for the design of drafts of microorganism metabolic models directly from the KEGG database (Feng *et al.*, 2012).

TABLE 15.5   Modeling software tools for systems biology.

et al	Description	Reference
BioMet Toolbox	Web-based resource for stoichiometric analysis and for integration of transcriptome and interactome data	Cvijovic *et al.*, 2010
BioSpice	Open source framework and software toolset for Systems Biology, assists in the modeling and simulation of spatio-temporal processes in living cells	Kumar and Feidler, 2003
CARMEN	Comparative Analysis and *in silico* Reconstruction of organism-specific Metabolic Networks; software for *in silico* reconstruction of metabolic networks to interpret genome data in a functional context	Schneider *et al.*, 2010
Condor-Copasi	Server-based software tool that integrates COPASI, a biological pathway simulation tool, with Condor, a high-throughput computing environment	Kent *et al.*, 2012
Copasi	Software application for simulation and analysis of biochemical networks and their dynamics	Hoops *et al.*, 2006
DBSolve	Software package for kinetic modeling with dynamic visualization of the simulation results	Gizzatkulov *et al.*, 2010
E-CELL	Software environment for whole-cell simulation	Tomita *et al.*, 1999
ERGO	Genome Analysis and Discovery System is an *in silico* systems biology platform for the comprehensive analysis of organisms, allowing the automatic reconstruction of biological models	Overbeek *et al.*, 2003
Gepasi	Software system for modeling chemical and biochemical reaction networks	Mendes, 1993
GEMSiRV	Software platform for Genome-scale Metabolic model Simulation, Reconstruction and Visualization	Liao *et al.*, 2012
KEGGConverter	Web-based application, uses as source the KEGG Pathways Database to construct integrated pathway models fully functional for simulation purposes	Moutselos *et al.*, 2009

TABLE 15.5 *(Continued)*

et al	Description	Reference
METANNOGEN	Software for the reconstruction of metabolic networks based on the KEGG database of biochemical reactions	Gille *et al.*, 2007
MicrobesFlux	Web platform for drafting metabolic models from the KEGG database	Feng *et al.*, 2012
OptFlux	Open-source software platform for *in silico* metabolic engineering	Rocha *et al.*, 2010
SBMLmerge	Software which assists in the combination of models of biological subsystems to larger biochemical networks	Schulz *et al.*, 2006
SCAMP	General-purpose simulator of metabolic and chemical networks	Sauro, 1993
SEBINI	Software Environment for BIological Network Inference; platform for the accurate reconstruction of biological networks	Taylor *et al.*, 2006

In the future, it is expected that data concerning interactions (e.g., protein-protein interactions) and regulation will be routinely integrated in these models. Indeed, it is fundamental to account for the tight regulation of metabolism when doing metabolic engineering, since the manipulation of obvious targets does not always confer the desired phenotype, and may have unpredictable side effects (Papini *et al.*, 2010). The integration of regulatory and signaling levels in an *E. coli* model was recently reported (Covert *et al.*, 2008), as well as an integrative model of regulatory and metabolic networks for *S. cerevisiae* (Herrgard *et al.*, 2006). It is expected that these models will make predictions much more accurate, leading to the creation of more and better cell factories for biotechnology purposes. Additionally, by complementing GSMMs with genome-wide metabolomics, fluxomics and thermodynamics data, it will be possible to develop genome-scale kinetic models (Jamshidi and Palsson, 2008). These models of cellular metabolism will enable the prediction of time-course metabolite concentrations, but to build them it will be necessary to obtain the stoichiometry and kinetics for each cellular reaction under physiological conditions, which constitutes a very challenging task.

15.4 THE PROMISES OF SYNTHETIC BIOLOGY

The aim of metabolic engineering, as discussed previously, is to improve the cellular traits to create microbial phenotypes that are efficient cell factories without compromising the cell viability and avoiding the production of undesired products. Traditionally this has been done by 'playing' with one or a few genes, mainly by deletion or overexpression, using recombinant DNA technology. However, it is well established that the control of the metabolism is not done through a few "rate-limiting" (regulatory) enzymes, but it is distributed, rendering these traditional strategies limited. Engineering cells on a deeper level requires complex modifications of the genetic equipment, which can be done with the help of a recent field of synthetic biology.

Synthetic biology, as defined on the community website (http://syntheticbiology.org/), comprises "the design and fabrication of biological components and systems that do not already exist in the natural world" and/or the "redesign and fabrication of existing biological systems." Originally "synthetic biology" was used as a synonym of bioengineering of bacterial systems using recombinant DNA technology, which includes the concept of metabolic engineering. Currently, synthetic biology goes beyond this and offers a multitude of possibilities for (microbial) biotechnology, with the aim to design entire pathways *in silico* – *de novo* metabolic engineering-and implement them in the target host (Klein-Marcuschamer *et al.*, 2010; Medema *et al.*, 2012).

Synthetic biology leverages advances in computational biology, molecular biology, protein engineering, and systems biology to design, synthesize, and assemble genetic elements for manipulating cell phenotypes (Picataggio, 2009). Besides all the factors that contributed to the development of systems biology (e.g., recombinant DNA technology, 'omics,' *in silico* strategies, etc.), advances in synthetic biology have been possible by the use of synthetic DNA technology, that is, *de novo* oligonucleotide synthesis. What began as an exotic research interest of organic chemists (Brown, 1993) nowadays allows the rapid and cheap synthesis of DNA fragments large enough to encode full-length genes and even entire genomes (Gibson *et al.*, 2008a; Picataggio, 2009). The most used strategy involves an initial synthesis of short oligonucleotide sequences, followed by their assemblage into gene-length sequences by PCR (Stemmer *et al.*, 1995). This strategy was used to synthesize the entire genome of *Mycoplasma genitalium* (Gibson *et al.*, 2008a, 2008b), raizing several comments and concerns from the scientific community and the general public (*And man made life: Artificial life, the stuff of dreams and nightmares, has arrived*, The Economist, May 10th 2010, http://www.economist.com/ node/16163154).

Evidently, synthetic biology requires state-of-the-art computational tools (Medema *et al.*, 2012) for the 'omics' data mining, to find an optimal solution for the choice of the pathway, the enzymes involved and the host organism. These tools are used in the several steps required for *de novo* metabolic engineering (Fig. 15.3): (i) identification of all possible metabolic pathways that lead to the synthesis of a target compound; (ii) list of the pathways in a priority order, based on defined criteria; (iii) use of comparative modeling strategies to predict maximum fluxes in candidate hosts; (iv) pathway selection; (vi) pathway integration-how to integrate all the selected parts in functional transcriptional units optimized for a specific host. The pathway integration requires a fine-tuning at the genetic level of transcription (e.g., use novel synthetic or modified promoters to modulate transcription) and translation (e.g., codon optimization for a highly efficient gene expression, optimization of mRNA sequence and structure and use of RNA silencing for balanced gene expression) (Klein-Marcuschamer *et al.*, 2010).

E. coli and *S. cerevisiae* are the most widely used host organisms for *de novo* biosynthetic pathways. A mevalonate pathway for the production of terpenoids, commercially important compounds that are usually produced by plants in low concentrations, has been engineered in *E. coli* using yeast parts (Martin *et al.*, 2003); terpenoid production in *E. coli* was also achieved by engineering two heterologous pathways that use plant cytochrome P-450 (Chang *et al.*, 2007); the production of D-phenylglycine (a side chain building block for semisynthetic penicillins and cephalosporins) in *E. coli* was achieved by building a synthetic pathway using genes from *Amycolatopsis orientalis*, *Streptomyces coelicolor*, and *Pseudomonas putida* (Muller *et al.*, 2006). *S. cerevisiae* cells were engineered with a pathway to produce vanillin, a very important compound for the global market for flavor and fragrances, using genes from the mold *Podospora pauciseta*, the bacterium *Nocardia* genus and *Homo sapiens* (Hansen *et al.*, 2009). More recently, synthetic biology has been applied to other microorganisms. For example, the application of synthetic biology to the engineering of *Steptomyces* species, the major industrial producers of secondary metabolites, has been widely reported (Medema *et al.*, 2011). Also, the development of a plasmid-based synthetic biology framework for *C. glutamicum* was recently reported by Ravasi and co-workers (Ravasi *et al.*, 2012). All of the plasmid components in this platform are flanked by unique restriction sites to facilitate the evaluation of regulatory sequences and to allow the easy assembly of different genetic parts.

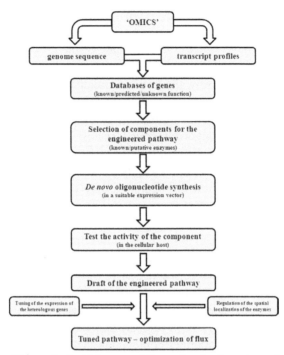

FIGURE 15.3 The synthetic biology workflow, showing how to engineer a synthetic pathway. Steps include: choice of the pathway and host; selection of the *parts* – genes, which encode required activities; *de novo* synthesis of the parts in a suitable expression vector; testing the parts in the candidate host(s); integration of all the tested and selected parts in functional transcriptional units; fine tunning of the system to achieve an optimal flux pathway.

Finally, synthetic biology may soon fulfill the dream of metabolic engineering: the construction of highly efficient cell factories with a minimal genome, that is, strains from which all the nonessential genes are deleted while retaining those genes that are essential for the cell survival, robustness and the production of the desired target (Kolisnychenko *et al.*, 2002). Because all the information for the construction of a cell is encoded in the DNA and since the large-scale DNA synthesis is now possible, the creation of synthetic organisms with a minimal genome may become a reality in a near future. This would overcome problems such as the possible interaction of novel intermediates in synthetic pathways with the host enzymes, which would divert them from the target pathway, possibly leading to the formation of toxic compounds or to the conversion to a metabolic dead-end (Dietz and Panke, 2010). However, engineering such a "minimal cell" is an extremely difficult task, as shown by early studies of genome reduction in *E. coli* that showed that the targeted removal of only small portions of the genome was enough to negatively impact biotechnologically important properties (Hashimoto *et al.*, 2005). Another problem is the possible loss of robustness due to deletion of genes that may not be essential under certain culture conditions, but essential for others. Despite all the difficulties, the possibility is real. It is worth reminding that the chemical synthesis of a complete genome (*Mycoplasma genitalium*) was already attained (Gibson *et al.*, 2008a) and that the *Mycoplasma mycoides* genome was successfully transplanted into a different, although related, bacterial species, *Mycoplasma capricolum* (Lartigue *et al.*, 2007).

15.5   CONCRETE APPLICATIONS IN INDUSTRIAL MICROBIAL BIOTECHNOLOGY

The applications of 'omics' technology to metabolic engineering of microorganisms are numerous and have increased substantially over the years. The main aim of microbial biotechnology is to engineer the metabolism of the microorganisms being used as cell factories, to obtain a desired phenotype (Patil *et al.*, 2005). This is done using molecular biology tools with the help of bioinformatics.

15.5.1   *CORYNEBACTERIUM GLUTAMICUM – THE AMINO ACID CELL FACTORY*

C. glutamicum, a bacteria widely used for the industrial production of amino acids, is the most studied organism with regard to metabolic flux analysis for rational strain optimization (Wittmann, 2010). Wendish and co-workers described how a genome-wide approach allows the creation of models of metabolism of *C. glutamicum*, and how these allow the engineering of the strains to improve the industrial production of amino acids (Wendisch *et al.*, 2006). Several studies using a multi-'omics' approach have revealed key pathways that can be modified to obtain improved production strains. For example, the engineering of lysine-overproducing strains has been achieved by: (i) single gene modifications (e.g., the amplification of the pentose phosphate pathway by mutation of the gene encoding 6-phosphogluconate dehydrogenase, *gnd* (Ohnishi *et al.*, 2005) or the overexpression of the gene encoding fructose 1,6-bisphosphatase (Becker *et al.*, 2005) both lead to an increased supply of NADPH used in lysine biosynthesis); (ii) targeted decrease of competing enzymes such as pyruvate dehydrogenase (Becker *et al.*, 2010), phosphoenolpyruvate carboxylase (Riedel *et al.*, 2001) and the TCA cycle (Becker *et al.*, 2009). Similarly, strains overproducing the amino acid glutamate were engineered based on flux predictions; one example is the increase in the production of glutamate under biotin-limiting conditions in strains deleted for pyruvate kinase (Sawada *et al.*, 2010).

15.5.2 *CLOSTRIDIAL FERMENTATION FOR PRODUCTION OF BIOGAS*

Process development for the production of biofuels (e.g., ethanol, butanol) is becoming more and more important in biotechnology, due to the increased consumption, shortage and increasing prices of fossil fuels. Anaerobic bacteria, such as the solventogenic strains of *Clostridium* spp., are able to naturally produce butanol, and the acetone-butanol-ethanol (ABE) fermentation was the first large-scale fermentation process to be developed (Jones and Woods, 1986). *Clostridium acetobutylicum* is the model organism for the clostridia species and has received renewed attention due to the interest in the production of biofuels. Several 'omics' studies have been conducted to unravel the mechanisms and regulation of several physiological aspects in different growth conditions, for example, batch and chemostat cultures (Lutke-Eversloh and Bahl, 2011; Papoutsakis, 2008). These results have been integrated in GSMMs of *C. acetobutylicum* developed by different groups (Lee *et al.*, 2008; Senger and Papoutsakis, 2008a, 2008b).

Several metabolic engineering strategies have been used to improve butanol production by *C. acetobutylicum* (Bruant *et al.*, 2010; Lutke-Eversloh and Bahl, 2011) (Table 15.2). This is not easily achieved due to the complexity of the fermentative pathways in these bacteria (Fig. 15.4) and their life cycle. The main goal of the rational design and improvement of engineered strains is to increase the metabolic flux towards the production of the desired compound. In the case of ABE, the aim is to reduce the fermentative branches towards acetate, butyrate, acetone and ethanol production, increasing the flux through the alcohol dehydrogenase (Adh)-catalyzed reactions that lead to butyraldehyde and butanol production. Some of the strategies used were: (i) the reduction of flux to acid formation (acetate and butyrate) by using gene-knockout mutant strains for the enzymes catalyzing these reactions (*buk1* and *pta*) and overexpressing the gene encoding alcohol dehydrogenase (Green *et al.*, 1996); (ii) employment of antisense RNA to decrease the expression of enzymes involved in acetone formation (*ctfB*) combined with *Adh* overexpression (Sillers *et al.*, 2009; Tummala *et al.*, 2003a; Tummala *et al.*, 2003c).

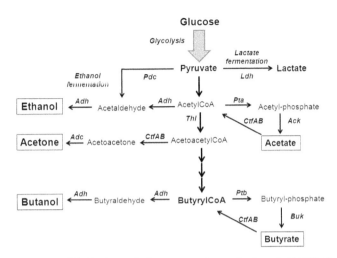

FIGURE 15.4 Simplified view of the acetone-butanol-ethanol (ABE) fermentation pathway. *Adc*, acetoacetate decarboxylase, *Adh*, alcohol/aldehyde dehydrogenase, *Ack*, acetate kinase, *Buk*, butyrate kinase, *CtfAB*, acetoacetyl-CoA:acyl-CoA transferase, *Ldh*, lactate dehydrogenase, *Pdc*, pyruvate decarboxylase, *Pta*, phosphotransacetylase, *Ptb*, phosphotransbutyrylase, *Thl*, thiolase.

Other clostridial species have been searched for the industrial production of biogas. Recently, a new solventogenic clostridial species, *Clostridium carboxidivorans*, was described, which is capable of producing butanol from synthesis gas (syngas, mainly composed of carbon monoxide, CO, and hydrogen, H_2) obtained from the gasification of organic biomass (Bruant *et al.*, 2010). This new species is very promising because it uses CO as a carbon source instead of carbon feedstock, which is expensive and poses several other problems. In this study, the genome sequence of *C. carboxidivorans* was determined and it was shown that the acetone branch is absent in this species, providing useful information for the development of metabolic engineering strategies to improve the species natural capacity to produce potential biofuels from syngas.

15.5.3 METABOLIC ENGINEERING OF ANTIBIOTIC-PRODUCING MICROBES

The *Penicillium chrysogenum* strains currently used in industry are derived from strains that were obtained by classical strain improvement techniques, that is, random mutagenesis followed by intelligent selection. These strains produce around 100,000-fold more penicillin than the original strain discovered by Fleming in 1928 (Rokem *et al.*, 2007). The problem with this traditional approach is that it is time-consuming, carries a risk of accumulation of unwanted mutations and may not lead to the desired phenotype, even after several rounds of mutation. The wide knowledge of physiology, biochemistry and genetics obtained from genome-wide studies for antibiotic-producing strains allowed the engineering of superior production strains in the last two decades (Rokem *et al.*, 2007). Strategies include: (i) increasing the flux through biosynthetic pathways by overexpressing key enzymes, positive regulators of the pathway or both, and/or disrupt negative regulators; (ii) increasing the precursor supply, for example, CoA thioesters, arginine, glyceraldehyde-3-phosphate, aromatic amino acids, etc.; (iii) reducing the formation of by-products, which are usually highly similar to the desired product, hence increasing the costs of downstream processing; (iv) increasing the amount of oxygen in the cultures of the antibiotic producing filamentous microorganisms (e.g., engineering the gene encoding a hemoglobin from *Vitreoscilla* sp. (De Modena *et al.*, 1993); (v) using well known heterologous hosts such as *Streptomyces coelicolor, Streptomyces lividans* and other closely related genera strains, or even the model organisms *E. coli* and *S. cerevisiae* (Rokem *et al.*, 2007). Another interesting example of the application of systems biotechnology to engineer antibiotic-producing strains was reported by Panagiotou *et al.* (2009). A systems biology approach, combining genome-wide expression profiling, fluxomics and physiology knowledge was used to study the production of the polyketide 6-methylsalycilic acid in engineered strains of *Aspergillus nidulans*. Fungal polyketides are compounds with antimicrobial, antifungal, immunosuppressant and cholesterol lowering activities, widely used in human therapeutics.

15.5.4   USING SYNTHETIC BIOLOGY TO ENGINEER UNNATURAL PATHWAYS IN S. CEREVISIAE

As discussed in this chapter, *S. cerevisiae* has been widely used in biotechnological production processes, either for the manufacture of recombinant proteins or chemicals, mainly primary metabolites. Furthermore, introduction of foreign genes by transformation allows yeast cells to produce unnatural yeast metabolites and recombinant proteins, such as lactic acid (Porro *et al.*, 1995) and human insulin (Thim *et al.*, 1987), respectively. The use of synthetic biology tools for the engineering of entire pathways into *S. cerevisiae* has also been described. A recent review discusses the biosynthesis of secondary metabolites in yeast, including flavonoids, isoprenoids, carotenes, alkaloids and polyketides (Siddiqui *et al.*, 2012). The building parts of these synthetic pathways are genes from plants (e.g., poplar, potato, soybean, grape vine, *Arabidopsis thaliana*, amongt others), other yeasts, bacteria and even archaea (*Sulfolobus acidocaldarius*).

Perhaps the most discussed example of the use of synthetic biology for the engineering of a nonnatural compound in yeast, is the production of the antimalaria drug precursor artemisinic acid (Ro *et al.*, 2006). Artemisinin, the active antimalaria drug is synthesized by the plant *Artemisia annua* L., but the extraction process is long (12–18 months from planting to production) leading to shortage of supplies and high costs. Artimisinin can also be chemically synthesized, but the process is difficult and also expensive, while a semisynthetic process involving the production of artemisinic acid by *S. cerevisiae* followed by the chemical synthesis of artemisinin is cost-effective and environmental friendly. The engineering process involved: (i) the over-expression of selected genes of the yeast mavelonate pathway to increase the biosynthesis of the intermediate farnesyl pyrophosphate, (ii) the introduction of the *A. annua* gene encoding amorphadiene synthase for the production of amorphadiene, and (iii) the introduction of an *A. annua* gene encoding a novel cytochrome P450 which catalyzes the 3-step oxidation of amorphadiene to artesiminic acid. The engineered strain produced up to 100 mg/L of artesiminic acid (Ro *et al.*, 2006). A different strategy for the semisynthetic production of artemisinin using engineered *E. coli* was recently proposed (Westfall *et al.*, 2012). This involved the overexpressing of all the genes from the yeast mavelonate pathway and the inclusion of the *A. annua* genes leading to the production of amorphadiene and artemisinic acid. The fact that the engineered strain produced much higher levels of amorphadiene than artemisinic acid, led the authors to pursue the production of this compound and to develop a chemical strategy for the synthesis of the later fiom the former. An impressive amount of 40 g/L of amorphadiene was reached after further metabolic engineering strategies such as the elimination of galactose utilization by disruption of the *GAL1* gene, codon optimization for the heterologous genes, and bioprocess improvement tactics.

S. cerevisiae is the perfect host for the engineering of such complex pathways due to the several robust molecular biology tools available to perform genetic engineering in this yeast (Papini *et al.*, 2010). These well-known tools even allowed the assembly of the first complete synthetic genome (from *Mycoplasma genitalum*) in yeast (Gibson *et al.*, 2008a, 2008b). Additionally, yeast cells were used to assemble a synthetic *Mycoplasma mycoides* genome that was subsequently transplanted into the original cell (Lartigue *et al.*, 2009).

15.6   CONCLUSION

Microbial cell factories are continuously being improved by metabolic engineering strategies. Systems biology applied to microbial biotechnology-*microbial systems biology*-plays a central role in this process by integrating results from multiple-'omics' studies with data from the literature, creating genome-wide models of microbial cells relevant to biotechnology processes. The models integrate data, interpret data and allow *in silico* predictions. The integration of a regulatory level in current GSMMs will allow a much deeper understanding of microbial physiology, metabolism and genetics. This, in turn, will make model-based predictions much more accurate, guiding *wet*-lab experiments to new results, in an iterative process, which leads to the creation of improved industrial microbial strains for more efficient, cheaper and faster biotechnological processes.

The emerging field of synthetic biology opens new perspectives in microbial biotechnology. It is now possible to engineer entire pathways in a host for the production of metabolites that are not naturally produced in that host, such as the antimalarial precursor artemisinic acid in yeast. Synthetic biology tools also allowed the creation of the first synthetic genomes, using well-known yeast molecular biology tools. It is very possible that the future of microbial biotechnology will encompass the use of highly efficient cell factories with a synthetic minimal genome.

KEYWORDS

- **Antibiotic-producing microbes**
- **Cell factories**
- ***Escherichia coli***
- **Industrial microorganisms**
- **Metagenomics**
- **Recombinant proteins**
- ***Saccharomyces cerevisiae***
- **Strain improvement**
- **Synthetic biology**
- **Systems biotechnology**

REFERENCES

Abbott, D. A.; Suir, E.; van Maris A. J.; Pronk, J. T. Physiological and transcriptional responses to high concentrations of lactic acid in anaerobic chemostat cultures of *Saccharomyces cerevisiae*. *Appl Environ Microbiol* 2008, 74, 5759–5768.

Aklujkar, M.; Krushkal, J.; Di Bartolo, G.; Lapidus, A.; Land, M. L.; Lovley, D. R. The genome sequence of *Geobacter metallireducens*: features of metabolism, physiology and regulation common and dissimilar to *Geobacter sulfurreducens*. *BMC Microbiol* 2009, 9, 109.

Aldor, I. S.; Krawitz, D. C.; Forrest, W.; Chen, C.; Nishihara, J. C.; Joly, J. C.; Champion, K. M. Proteomic profiling of recombinant *Escherichia coli* in high-cell-density fermentations for improved production of an antibody fragment biopharmaceutical. *Appl Environ Microbiol* 2005, 71, 1717–1728.

Almaas, E. Biological impacts and context of network theory. *J Exp Biol* 2007, 210, 1548–1558.

Alper, H.; Jin, Y. S.; Moxley, J. F.; Stephanopoulos, G. Identifying gene targets for the metabolic engineering of lycopene biosynthesis in *Escherichia coli*. *Metab Eng* 2005a, 7, 155–164.

Alper, H.; Miyaoku, K.; Stephanopoulos, G. Construction of lycopene-overproducing *E. coli* strains by combining systematic and combinatorial gene knockout targets. *Nat Biotechnol* 2005b, 23, 612–616.

Amann, R. I.; Ludwig, W.; Schleifer, K. H. Phylogenetic identification and *in situ* detection of individual microbial cells without cultivation. *Microbiol Rev* 1995, 59, 143–169.

Andersen, M. R.; Nielsen, J. Current status of systems biology in *Aspergilli*. *Fungal Genet Biol* 2009, 46(S1), 180–190.

Andersen, M. R.; Nielsen, M. L.; Nielsen, J. Metabolic model integration of the bibliome, genome, metabolome and reactome of *Aspergillus niger*. *Mol Syst Biol* 2008, 4, 178.

Andersen, M. R.; Salazar, M. P.; Schaap, P. J.; van de Vondervoort, P. J.; Culley, D.; Thykaer, J.; Frisvad, J. C.; Nielsen, K. F.; Albang, R.; Albermann, K.; Berka, R. M.; Braus, G. H.; Braus-Stromeyer, S. A.; Corrochano, L. M.; Dai, Z.; van Dijck, P. W.; Hofmann, G.; Lasure, L. L.; Magnuson, J. K.; Menke, H.; Meijer, M.; Meijer, S. L.; Nielsen, J. B.; Nielsen, M. L.; van Ooyen, A. J.; Pel, H. J.; Poulsen, L.; Samson, R. A.; Stam, H.; Tsang, A.; van den Brink, J. M.; Atkins, A.; Aerts, A.; Shapiro, H.; Pangilinan, J.; Salamov, A.; Lou, Y.; Lindquist, E.; Lucas, S.; Grimwood, J.; Grigoriev, I. V.; Kubicek, C. P.; Martinez, D.; van Peij, N. N.; Roubos, J. A.; Nielsen, J.; Baker, S. E. Comparative genomics of citric-acid-producing *Aspergillus niger* ATCC 1015 versus enzyme-producing CBS 51388. *Genome Res* 2011, 21, 885–897.

Arifuzzaman, M.; Maeda, M.; Itoh, A.; Nishikata, K.; Takita, C.; Saito, R.; Ara, T.; Nakahigashi, K.; Huang, H. C.; Hirai, A.; Tsuzuki, K.; Nakamura, S.; Altaf-Ul-Amin, M.; Oshima, T.; Baba, T.; Yamamoto, N.; Kawamura, T.; Ioka-Nakamichi, T.; Kitagawa, M.; Tomita, M.; Kanaya, S.; Wada, C.; Mori, H. Large-scale identification of protein-protein interaction of *Escherichia coli* K-12. *Genome Res* 2006, 16, 686–691.

Asadollahi, M. A.; Maury, J.; Patil, K. R.; Schalk, M.; Clark, A.; Nielsen, J. Enhancing sesquiterpene production in *Saccharomyces cerevisiae* through *in silico* driven metabolic engineering. *Metab Eng* 2009, 11, 328–334.

Azuma, Y.; Hosoyama, A.; Matsutani, M.; Furuya, N.; Horikawa, H.; Harada, T.; Hirakawa, H.; Kuhara, S.; Matsushita, K.; Fujita, N.; Shirai, M. Whole-genome analyzes reveal genetic instability of *Acetobacter pasteurianus*. *Nucleic Acids Res* 2009, 37, 5768–5783.

Bailey, J. E. Toward a science of metabolic engineering. *Science* 1991, 252, 1668–1675.

Barbe, V.; Vallenet, D.; Fonknechten, N.; Kreimeyer, A.; Oztas, S.; Labarre, L.; Cruveiller, S.; Robert, C.; Duprat, S.; Wincker, P.; Ornston, L. N.; Weissenbach, J.; Marliere, P.; Cohen, G. N.; Medigue, C. Unique features revealed by the genome sequence of *Acinetobacter* sp. ADP1, a versatile and naturally transformation competent bacterium. *Nucleic Acids Res* 2004, 32, 5766–5779.

Becker, J.; Buschke, N.; Bucker, R.; Wittman, C. Systems level engineering of *Corynebacterium glutamicum* – reprogramming translational efficiency for superior production. *Chem Eng Life Sci* 2010, 10, 430–438.

Becker, J.; Klopprogge, C.; Schroder, H.; Wittmann, C. Metabolic engineering of the tricar-boxylic acid cycle for improved lysine production by *Corynebacterium glutamicum*. *Appl Environ Microbiol* 2009, 75, 7866–7869.

Becker, J.; Klopprogge, C.; Zelder, O.; Heinzle, E.; Wittmann, C. Amplified expression of fruc-tose 1,6-bisphosphatase in *Corynebacterium glutamicum* increases *in vivo* flux through the pentose phosphate pathway and lysine production on different carbon sources. *Appl Environ Microbiol* 2005, 71, 8587–8596.

Bentley, S. D.; Chater, K. F.; Cerdeno-Tarraga, A. M.; Challis, G. L.; Thomson, N. R.; James, K. D.; Harris, D. E.; Quail, M. A.; Kieser, H.; Harper, D.; Bateman, A.; Brown, S.; Chandra, G.; Chen, C. W.; Collins, M.; Cronin, A.; Fraser, A.; Goble, A.; Hidalgo, J.; Hornsby, T.; Howarth, S.; Huang, C. H.; Kieser, T.; Larke, L.; Murphy, L.; Oliver, K.; O'Neil, S.; Rabbin-owitsch, E.; Rajandream, M. A.; Rutherford, K.; Rutter, S.; Seeger, K.; Saunders, D.; Sharp, S.; Squares, R.; Squares, S.; Taylor, K.; Warren, T.; Wietzorrek, A.; Woodward, J.; Barrell, B. G.; Parkhill, J.; Hopwood, D. A. Complete genome sequence of the model actinomycete *Streptomyces coelicolor* A3(2). *Nature* 2002, 417, 141–147.

Blattner, F. R.; Plunkett, G. III.; Bloch, C.; Perna, N. T.; Burland, V.; Riley, M.; Collado-Vides, J.; Glasner, J. D.; Rode, C. K.; Mayhew, G. F.; Gregor, J.; Davis, N. W.; Kirkpatrick, H. A.; Goeden, M. A.; Rose, D. J.; Mau, B.; Shao, Y. The complete genome sequence of *Escherichia coli* K-12. *Science* 1997, 277, 1453–1462.

Bolotin, A.; Quinquis, B.; Renault, P.; Sorokin, A.; Ehrlich, S. D.; Kulakauskas, S.; Lapidus, A.; Goltsman, E.; Mazur, M.; Pusch, G. D.; Fonstein, M.; Overbeek, R.; Kyprides, N.; Purnelle, B.; Prozzi, D.; Ngui, K.; Masuy, D.; Hancy, F.; Burteau, S.; Boutry, M.; Delcour, J.; Goffeau, A.; Hols, P. Complete sequence and comparative genome analysis of the dairy bacterium *Streptococcus thermophilus*. *Nat Biotechnol* 2004, 22, 1554–1558.

Bolotin, A.; Wincker, P.; Mauger, S.; Jaillon, O.; Malarme, K.; Weissenbach, J.; Ehrlich, S. D.; Sorokin, A. The complete genome sequence of the lactic acid bacterium *Lactococcus lactis* ssp. *lactis* IL1403. *Genome Res* 2001, 11, 731–753.

Bolser, D. M.; Chibon, P. Y.; Palopoli, N.; Gong, S.; Jacob, D.; Del Angel, V. D.; Swan, D.; Bassi, S.; Gonzalez, V.; Suravajhala, P.; Hwang, S.; Romano, P.; Edwards, R.; Bishop, B.; Eargle, J.; Shtatland, T.; Provart, N. J.; Clements, D.; Renfro, D. P.; Bhak, D.; Bhak, J. Me-taBase – the wiki-database of biological databases. *Nucleic Acids Res* 2012, 40, D1250–1254.

Borodina, I.; Krabben, P.; Nielsen, J. Genome-scale analysis of *Streptomyces coelicolor* A3(2) metabolism. *Genome Res* 2005, 15, 820–829.

Britz-McKibbin, P. Capillary electrophoresis-electrospray ionization-mass spectrometry (CE-ESI-MS)-based metabolomics. *Methods Mol Biol* 2011, 708, 229–246.

Bro, C.; Nielsen, J. Impact of 'ome' analyses on inverse metabolic engineering. *Metab Eng* 2004, 6, 204–211.

Brooijmans, R. J.; Siezen, R. J.; Genomics of microalgae, fuel for the future. *Microb Biotechnol* 2011, 3, 514–522.

Brown, D. M. A brief history of oligonucleotide synthesis. In *Methods in molecular biology, Protocols for oligonucleotides and analogs*, Vol 20; Humana Press: Totowa, 1993; 1–17.

Bruant, G.; Levesque, M. J.; Peter, C.; Guiot, S. R.; Masson, L. Genomic analysis of carbon monoxide utilization and butanol production by *Clostridium carboxidivorans* strain P7. *PLoS One* 2010, 5, e13033.

Bruggeman, F. J.; Westerhoff, H. V. The nature of systems biology. *Trends Microbiol* 2007, 15, 45–50.

Bucca, G.; Brassington, A. M.; Hotchkiss, G.; Mersinias, V.; Smith, C. P. Negative feedback regulation of dnaK, clpB and lon expression by the DnaK chaperone machine in *Streptomyces coelicolor*, identified by transcriptome and *in vivo* DnaK-depletion analysis. *Mol Microbiol* 2003, 50, 153–166.

Bult, C. J.; White, O.; Olsen, G. J.; Zhou, L.; Fleischmann, R. D.; Sutton, G. G.; Blake, J. A.; FitzGerald, L. M.; Clayton, R. A.; Gocayne, J. D.; Kerlavage, A. R.; Dougherty, B. A.; Tomb, J. F.; Adams, M. D.; Reich, C. I.; Overbeek, R.; Kirkness, E. F.; Weinstock, K. G.; Merrick, J. M.; Glodek, A.; Scott, J. L.; Geoghagen, N. S.; Venter, J. C. Complete genome sequence of the methanogenic archeon, *Methanococcus jannaschii. Science* 1996, 273, 1058–1073.

Callanan, M.; Kaleta, P.; O'Callaghan, J.; O'Sullivan, O.; Jordan, K.; McAuliffe, O.; Sangrador-Vegas, A.; Slattery, L.; Fitzgerald, G. F.; Beresford, T.; Ross, R. P. Genome sequence of *Lactobacillus helveticus*, an organism distinguished by selective gene loss and insertion sequence element expansion. *J Bacteriol* 2008, 190, 727–735.

Carlson, R.; Fell, D. F. Metabolic pathway analysis of a recombinant yeast for rational strain development. *Biotechnol Bioeng* 2002, 79, 121–134.

Caspeta, L.; Shoaie, S.; Agren, R.; Nookaew, I.; Nielsen, J. Genome-scale metabolic reconstructions of *Pichia stipitis* and *Pichia pastoris* and *in silico* evaluation of their potentials. *BMC Syst Biol* 2012, 6, 24.

Challacombe, J. F.; Altherr, M. R.; Xie, G.; Bhotika, S. S.; Brown, N.; Bruce, D.; Campbell, C. S.; Campbell, M. L.; Chen, J.; Chertkov, O.; Cleland, C.; Dimitrijevic, M.; Doggett, N. A.; Fawcett, J. J.; Glavina, T.; Goodwin, L. A.; Green, L. D.; Han, C. S.; Hill, K. K.; Hitchcock, P.; Jackson, P. J.; Keim, P.; Kewalramani, A. R.; Longmire, J.; Lucas, S.; Malfatti, S.; Martinez, D.; McMurry, K.; Meincke, L. J.; Misra, M.; Moseman, B. L.; Mundt, M.; Munk, A. C.; Okinaka, R. T.; Parson-Quintana, B.; Reilly, L. P.; Richardson, P.; Robinson, D. L.; Saunders, E.; Tapia, R.; Tesmer, J. G.; Thayer, N.; Thompson, L. S.; Tice, H.; Ticknor, L. O.; Wills, P. L.; Gilna, P.; Brettin, T. S. The complete genome sequence of *Bacillus thuringiensis* Al Hakam. *J Bacteriol* 2007, 189, 3680–3681.

Chang, M. C.; Eachus, R. A.; Trieu, W.; Ro, D. K.; Keasling, J. D. Engineering *Escherichia coli* for production of functionalized terpenoids using plant P450s. *Nat Chem Biol* 2007, 3, 274–277.

Chang, R. L.; Ghamsari, L.; Manichaikul, A.; Hom, E. F.; Balaji, S.; Fu, W.; Shen, Y.; Hao, T.; Palsson, B. O.; Salehi-Ashtiani, K.; Papin, J. A. Metabolic network reconstruction of *Chlamydomonas* offers insight into light-driven algal metabolism. *Mol Syst Biol* 2011, 7, 518.

Chee, M.; Yang, R.; Hubbell, E.; Berno, A.; Huang, X. C.; Stern, D.; Winkler, J.; Lockhart, D. J.; Morris, M. S.; Fodor, S. P. Accessing genetic information with high-density DNA arrays. *Science* 1996, 274, 610–614.

Cheng, J. S.; Qiao, B.; Yuan, Y. J. Comparative proteome analysis of robust *Saccharomyces cerevisiae* insights into industrial continuous and batch fermentation. *Appl Microbiol Biotechnol* 2008, 81, 327–338.

Choi, J. H.; Lee, S. J.; Lee, S. Y. Enhanced production of insulin-like growth factor I fusion protein in *Escherichia coli* by coexpression of the down-regulated genes identified by transcriptome profiling. *Appl Environ Microbiol* 2003, 69, 4737–4742.

Chung, B. K.; Selvarasu, S.; Andrea, C.; Ryu, J.; Lee, H.; Ahn, J.; Lee, D. Y. Genome-scale metabolic reconstruction and *in silico* analysis of methylotrophic yeast *Pichia pastoris* for strain improvement. *Microb Cell Fact* 2010, 9, 50.

Collas, P. The current state of chromatin immunoprecipitation. *Mol Biotechnol* 2010, 45, 87–100.

Covert, M. W.; Xiao, N.; Chen, T. J.; Karr, J. R. Integrating metabolic, transcriptional regulatory and signal transduction models in *Escherichia coli*. *Bioinformatics* 2008, 24, 2044–2050.

Cregg, J. M.; Cereghino, J. L.; Shi, J.; Higgins, D. R. Recombinant protein expression in *Pichia pastoris*. *Mol Biotechnol* 2000, 16, 23–52.

Cregg, J. M.; Tolstorukov, I.; Kusari, A.; Sunga, J.; Madden, K.; Chappell, T. Expression in the yeast *Pichia pastoris*. *Methods Enzymol* 2009, 463, 169–189.

Cvijovic, M.; Olivares-Hernandez, R.; Agren, R.; Dahr, N.; Vongsangnak, W.; Nookaew, I.; Patil, K. R.; Nielsen, J. BioMet Toolbox: genome-wide analysis of metabolism. *Nucleic Acids Res* 2010, 38, W144–149.

Daran-Lapujade, P.; Jansen, M. L.; Daran, J. M.; van Gulik, W.; de Winde, J. H.; Pronk, J. T. Role of transcriptional regulation in controlling fluxes in central carbon metabolism of *Saccharomyces cerevisiae*. A chemostat culture study. *J Biol Chem* 2004, 279, 9125–9138.

Dauner, M.; Sonderegger, M.; Hochuli, M.; Szyperski, T.; Wuthrich, K.; Hohmann, H. P.; Sauer, U.; Bailey, J. E. Intracellular carbon fluxes in riboflavin-producing *Bacillus subtilis* during growth on two-carbon substrate mixtures. *Appl Environ Microbiol* 2002, 68, 1760–1771.

David, H.; Akesson, M.; Nielsen, J. Reconstruction of the central carbon metabolism of *Aspergillus niger*. *Eur J Biochem* 2003, 270, 4243–4253.

De Felice, M.; Mattanovich, D.; Papagianni, M.; Wegrzyn, G.; Villaverde, A. The scientific impact of microbial cell factories. *Microb Cell Fact* 2008, 7, 33.

De Schutter, K.; Lin, Y. C.; Tiels, P.; Van Hecke, A.; Glinka, S.; Weber-Lehmann, J.; Rouze, P.; Van de Peer, Y.; Callewaert, N. Genome sequence of the recombinant protein production host *Pichia pastoris*. *Nat Biotechnol* 2009, 27, 561–566.

Deckert, G.; Warren, P. V.; Gaasterland, T.; Young, W. G.; Lenox, A. L.; Graham, D. E.; Over-beek, R.; Snead, M. A.; Keller, M.; Aujay, M.; Huber, R.; Feldman, R. A.; Short, J. M.; Olsen, G. J.; Swanson, R. V. The complete genome of the hyperthermophilic bacterium *Aquifex aeolicus*. *Nature* 1998, 392, 353–358.

Demain, A. L. History of industrial biotechnology In *Industrial Biotechnology, Sustainable Growth and Economic Success*, Wim Sotaert, E. J. V., Ed.; Wiley-VCH: Weinheim, 2010; 17–77.

De Modena, J. A.; Gutierrez, S.; Velasco, J.; Fernandez, F. J.; Fachini, R. A.; Galazzo, J. L.; Hughes D. E.; Martin, J. F. The production of cephalosporin C by *Acremonium chrysogenum* is improved by the intracellular expression of a bacterial hemoglobin. *Biotechnology (NY)* 1993, 11, 926–929.

De Risi, JL.; Iyer, V. R.; Brown, P. O. Exploring the metabolic and genetic control of gene expression on a genomic scale. *Science* 1997, 278, 680–686.

Devantier, R.; Scheithauer, B.; Villas-Boas, S. G.; Pedersen, S.; Olsson, L. Metabolite profiling for analysis of yeast stress response during very high gravity ethanol fermentations. *Biotechnol Bioeng* 2005, 90, 703–714.

Dietz, S.; Panke, S. Microbial systems engineering: first successes and the way ahead. *Bioessays* 2010, 32, 356–362.

Duarte, N. C.; Herrgard, M. J.; Palsson, B. O. Reconstruction and validation of *Saccharomyces cerevisiae* iND750, a fully compartmentalized genome-scale metabolic model. *Genome Res* 2004, 14, 1298–1309.

Durot, M.; Le Fevre, F.; de Berardinis, V.; Kreimeyer, A.; Vallenet, D.; Combe, C.; Smidtas, S.; Salanoubat, M.; Weissenbach, J.; Schachter, V. Iterative reconstruction of a global metabolic model of *Acinetobacter baylyi* ADP1 using high-throughput growth phenotype and gene essentiality data. *BMC Syst Biol* 2008, 2, 85.

Edwards, J. S.; Covert, M.; Palsson, B. Metabolic modeling of microbes: the flux-balance approach. *Environ Microbiol* 2002, 4, 133–140.

Edwards, J. S.; Palsson, B. O. The *Escherichia coli* MG1655 *in silico* metabolic genotype: its definition, characteristics, and capabilities. *Proc Natl Acad Sci USA* 2000, 97, 5528–5533.

Egorova, K.; Antranikian, G. Industrial relevance of thermophilic Archaea. *Curr Opin Microbiol* 2005, 8, 649–655.

Egorova, K.; Antranikian, G. Biotechnology. In *Archaea: Evolution, Physiology, and Molecular Biology*, Garrett, R. A., Klenk, H. -P., Eds.; Blackwell Publishing Ltd: Malden, 2007; 295–321.

Ekkers, D. M.; Cretoiu, M. S.; Kielak, A. M.; Elsas, J. D. The great screen anomaly – a new frontier in product discovery through functional metagenomics. *Appl Microbiol Biotechnol* 2012, 93, 1005–1020.

Feist, A. M.; Henry, C. S.; Reed, J. L.; Krummenacker, M.; Joyce, A. R.; Karp, P. D.; Broadbelt, L. J.; Hatzimanikatis, V.; Palsson, B. O. A genome-scale metabolic reconstruction for *Escherichia coli* K-12 MG1655 that accounts for 1260 ORFs and thermodynamic information. *Mol Syst Biol* 2007, 3, 121.

Feist, A. M.; Scholten, J. C.; Palsson, B. O.; Brockman, F. J.; Ideker, T. Modeling methanogenesis with a genome-scale metabolic reconstruction of *Methanosarcina barkeri*. *Mol Syst Biol* 2006, 2, 2006.0004.

Feng, X.; Page, L.; Rubens, J.; Chircus, L.; Colletti, P.; Pakrasi, H. B.; Tang, Y. J. Bridging the gap between fluxomics and industrial biotechnology. *J Biomed Biotechnol* 2010, 460717.

Feng, X.; Xu, Y.; Chen, Y.; Tang, Y. J. MicrobesFlux: a web platform for drafting metabolic models from the KEGG database. *BMC Syst Biol* 2012, 6, 94.

Fleischmann, R. D.; Adams, M. D.; White, O.; Clayton, R. A.; Kirkness, E. F.; Kerlavage, A. R.; Bult, C. J.; Tomb, F.; Dougherty, B. A.; Merrick, J. M. Whole-genome random sequencing and assembly of *Haemophilus influenzae* Rd. *Science* 1995, 269, 496–512.

Fong, S. S.; Burgard, A. P.; Herring, C. D.; Knight, E. M.; Blattner, F. R.; Maranas, C. D.; Palsson, B. O. *In silico* design and adaptive evolution of *Escherichia coli* for production of lactic acid. *Biotechnol Bioeng* 2005, 91, 643–648.

Forster, J.; Famili, I.; Fu, P.; Palsson, B. O.; Nielsen, J. Genome-scale reconstruction of the *Saccharomyces cerevisiae* metabolic network. *Genome Res* 2003, 13, 244–253.

Gasser, B.; Sauer, M.; Maurer, M.; Stadlmayr, G.; Mattanovich, D. Transcriptomics-based identification of novel factors enhancing heterologous protein secretion in yeasts. *Appl Environ Microbiol* 2007, 73, 6499–6507.

Gelperin, D. M.; White, M. A.; Wilkinson, M. L.; Kon, Y.; Kung, L. A.; Wise, K. J.; Lopez-Hoyo, N.; Jiang, L.; Piccirillo, S.; Yu, H.; Gerstein, M.; Dumont, M. E.; Phizicky, E. M.; Snyder, M.; Grayhack, E. J. Biochemical and genetic analysis of the yeast proteome with a movable ORF collection. *Genes Dev* 2005, 19, 2816–2826.

Giaever, G.; Chu, A. M.; Ni, L.; Connelly, C.; Riles, L.; Veronneau, S.; Dow, S.; Lucau-Danila, A.; Anderson, K.; Andre, B.; Arkin, A. P.; Astromoff, A.; El-Bakkoury, M.; Bangham, R.; Benito, R.; Brachat, S.; Campanaro, S.; Curtiss, M.; Davis, K.; Deutschbauer, A.; Entian, K. D.; Flaherty, P.; Foury, F.; Garfinkel, D. J.; Gerstein, M.; Gotte, D.; Guldener, U.; Hegemann, J. H.; Hempel, S.; Herman, Z.; Jaramillo, D. F.; Kelly, D. E.; Kelly, S. L.; Kotter, P.; La Bonte, D.; Lamb, D. C.; Lan, N.; Liang, H.; Liao, H.; Liu, L.; Luo, C.; Lussier, M.; Mao, R.; Menard, P.; Ooi, S. L.; Revuelta, J. L.; Roberts, C. J.; Rose, M.; Ross-Macdonald, P.; Scherens, B.; Schimmack, G.; Shafer, B.; Shoemaker, D. D.; Sookhai-Mahadeo, S.; Storms, R. K.; Strathern, J. N.; Valle, G.; Voet, M.; Volckaert, G.; Wang, C. Y.; Ward, T. R.; Wilhelmy, J.; Winzeler, E. A.; Yang, Y.; Yen, G.; Youngman, E.; Yu, K.; Bussey, H.; Boeke, J. D.; Snyder, M.; Philippsen, P.; Davis, R. W.; Johnston, M. Functional profiling of the *Saccharomyces cerevisiae* genome. *Nature* 2002, 418, 387–391.

Gibson, D. G.; Benders, G. A.; Andrews-Pfannkoch, C.; Denisova, E. A.; Baden-Tillson, H.; Zaveri, J.; Stockwell, T. B.; Brownley, A.; Thomas, D. W.; Algire, M. A.; Merryman, C.; Young, L.; Noskov, V. N.; Glass, J. I.; Venter, J. C.; Hutchison, C. A. III.; Smith, H. O. Complete chemical synthesis, assembly, and cloning of a *Mycoplasma genitalium* genome. *Science* 2008a, 319, 1215–1220.

Gibson, D. G.; Benders, G. A.; Axelrod, K. C.; Zaveri, J.; Algire, M. A.; Moodie, M.; Montague, M. G.; Venter, J. C.; Smith, H. O.; Hutchison, C. A. III. One-step assembly in yeast of 25 overlapping DNA fragments to form a complete synthetic *Mycoplasma genitalium* genome. *Proc Natl Acad Sci USA* 2008b, 105, 20404–20409.

Gille, C.; Hoffmann, S.; Holzhutter, H. G. METANNOGEN: compiling features of biochemical reactions needed for the reconstruction of metabolic networks. *BMC Syst Biol* 2007, 1, 5.

Gizzatkulov, N. M.; Goryanin, I. I.; Metelkin, E. A.; Mogilevskaya, E. A.; Peskov, K. V.; Demin, O. V. DBSolve Optimum: a software package for kinetic modeling, which allows dynamic visualization of simulation results. *BMC Syst Biol* 2010, 4, 109.

Goffeau, A.; Barrell, B. G.; Bussey, H.; Davis, R. W.; Dujon, B.; Feldmann, H.; Galibert, F.; Hoheisel, J. D.; Jacq, C.; Johnston, M.; Louis, E. J.; Mewes, H. W.; Murakami, Y.; Philippsen, P.; Tettelin, H.; Oliver, S. G. Life with 6000 genes. *Science* 1996, 274, 563–547.

Gonzalez, O.; Gronau, S.; Falb, M.; Pfeiffer, F.; Mendoza, E.; Zimmer, R.; Oesterhelt, D. Reconstruction, modeling and analysis of *Halobacterium salinarum* R-1 metabolism. *Mol Biosyst* 2008, 4, 148–159.

Green, E. M.; Boynton, Z. L.; Harris, L. M.; Rudolph, F. B.; Papoutsakis, E, T.; Bennett, G. N. Genetic manipulation of acid formation pathways by gene inactivation in *Clostridium acetobutylicum* ATCC 824. *Microbiology* 1996, 142, 2079–2086.

Grotkjaer, T.; Christakopoulos, P.; Nielsen, J.; Olsson, L. Comparative metabolic network analysis of two xylose fermenting recombinant *Saccharomyces cerevisiae* strains. *Metab Eng* 2005, 7, 437–444.

Gygi, S. P.; Rochon, Y.; Franza, B. R.; Aebersold, R. Correlation between protein and mRNA abundance in yeast. *Mol Cell Biol* 1999, 19, 1720–1730.

Han, M. J.; Jeong, K. J.; Yoo, J. S.; Lee, S. Y. Engineering *Escherichia coli* for increased productivity of serine-rich proteins based on proteome profiling. *Appl Environ Microbiol* 2003, 69, 5772–5781.

Han, M. J.; Lee, S. Y.; Koh, S. T.; Noh, S. G.; Han, W. H. Biotechnological applications of microbial proteomes. *J Biotechnol* 2010, 145, 341–349.

Han, M. J.; Yoon, S. S.; Lee, S. Y. Proteome analysis of metabolically engineered *Escherichia coli* producing Poly(3-hydroxybutyrate). *J Bacteriol* 2001, 183, 301–308.

Hansen, E. H.; Moller, B. L.; Kock, G. R.; Bunner, C. M.; Kristensen, C.; Jensen, O. R.; Okkels, F. T.; Olsen, C. E.; Motawia, M. S.; Hansen, J. *De novo* biosynthesis of vanillin in fission yeast (*Schizosaccharomyces pombe*) and baker's yeast (*Saccharomyces cerevisiae*). *Appl Environ Microbiol* 2009, 75, 2765–2774.

Hashimoto, M.; Ichimura, T.; Mizoguchi, H.; Tanaka, K.; Fujimitsu, K.; Keyamura, K.; Ote, T.; Yamakawa, T.; Yamazaki, Y.; Mori, H.; Katayama, T.; Kato, J. Cell size and nucleoid organization of engineered *Escherichia coli* cells with a reduced genome. *Mol Microbiol* 2005, 55, 137–149.

Hasunuma, T.; Sanda, T.; Yamada, R.; Yoshimura, K.; Ishii, J.; Kondo, A. Metabolic pathway engineering based on metabolomics confers acetic and formic acid tolerance to a recombinant xylose-fermenting strain of *Saccharomyces cerevisiae*. *Microb Cell Fact* 2011, 10, 2.

Heidelberg, J. F.; Seshadri, R.; Haveman, S. A.; Hemme, C. L.; Paulsen, I. T.; Kolonay, J. F.; Eisen, J. A.; Ward, N.; Methe, B.; Brinkac, L. M.; Daugherty, S. C.; Deboy, R. T.; Dodson, R. J.; Durkin, A. S.; Madupu, R.; Nelson, W. C.; Sullivan, S. A.; Fouts, D.; Haft, D. H.; Selengut, J.; Peterson, J. D.; Davidsen, T. M.; Zafar, N.; Zhou, L.; Radune, D.; Dimitrov, G.; Hance, M.; Tran, K.; Khouri, H.; Gill, J.; Utterback, T. R.; Feldblyum, T. V.; Wall, J. D.; Voordouw, G.;

Fraser, C. M. The genome sequence of the anaerobic, sulfate-reducing bacterium *Desulfovibrio vulgaris* Hildenborough. *Nat Biotechnol* 2004, 22, 554–559.

Heidelberg, K. B.; Gilbert, J. A.; Joint, I. Marine genomics: at the interface of marine microbial ecology and biodiscovery. *Microb Biotechnol* 2010, 3, 531–543.

Hendrickson, E. L.; Kaul, R.; Zhou, Y.; Bovee, D.; Chapman, P.; Chung, J.; Conway de Macario, E.; Dodsworth, J. A.; Gillett, W.; Graham, D. E.; Hackett, M.; Haydock, A. K.; Kang, A.; Land, M. L.; Levy, R.; Lie, T. J.; Major, T. A.; Moore, B. C.; Porat, I.; Palmeiri, A.; Rouse, G.; Saenphimmachak, C.; Soll, D.; Van Dien, S.; Wang, T.; Whitman, W. B.; Xia, Q.; Zhang, Y.; Larimer, F. W.; Olson, M. V.; Leigh, J. A. Complete genome sequence of the genetically tractable hydrogenotrophic methanogen *Methanococcus maripaludis*. *J Bacteriol* 2004, 186, 6956–6969.

Henry, C. S.; Zinner, J. F.; Cohoon, M. P.; Stevens, R. L. iBsu1103: a new genome-scale metabolic model of *Bacillus subtilis* based on SEED annotations. *Genome Biol* 2009, 10, R69.

Herrgard, M. J.; Lee, B. S.; Portnoy, V.; Palsson, B. O. Integrated analysis of regulatory and metabolic networks reveals novel regulatory mechanisms in *Saccharomyces cerevisiae*. *Genome Res* 2006, 16, 627–635.

Herrgard, M. J.; Swainston, N.; Dobson, P.; Dunn, W. B.; Arga, K. Y.; Arvas, M.; Bluthgen, N.; Borger, S.; Costenoble, R.; Heinemann, M.; Hucka, M.; Le Novere, N.; Li, P.; Liebermeister, W.; Mo, M. L.; Oliveira, A. P.; Petranovic, D.; Pettifer, S.; Simeonidis, E.; Smallbone, K.; Spasic, I.; Weichart, D.; Brent, R.; Broomhead, D. S.; Westerhoff, H. V.; Kirdar, B.; Penttila, M.; Klipp, E.; Palsson, B. O.; Sauer U.; Oliver, S. G.; Mendes, P.; Nielsen, J.; Kell, D. B. A consensus yeast metabolic network reconstruction obtained from a community approach to systems biology. *Nat Biotechnol* 2008, 26, 1155–1160.

Hibi, M.; Yukitomo, H.; Ito, M.; Mori, H. Improvement of NADPH-dependent bioconversion by transcriptome-based molecular breeding. *Appl Environ Microbiol* 2007, 73, 7657–7663.

Hirasawa, T.; Yoshikawa, K.; Nakakura, Y.; Nagahisa, K.; Furusawa, C.; Katakura, Y.; Shimizu, H.; Shioya, S. Identification of target genes conferring ethanol stress tolerance to *Saccharomyces cerevisiae* based on DNA microarray data analysis. *J Biotechnol* 2007, 131, 34–44.

Hong, S. H.; Kim, J. S.; Lee, S. Y.; In, Y. H.; Choi, S. S.; Rih, J. K.; Kim, C. H.; Jeong, H.; Hur, C. G.; Kim, J. J. The genome sequence of the capnophilic rumen bacterium *Mannheimia succiniciproducens*. *Nat Biotechnol* 2004, 22, 1275–1281.

Hoops, S.; Sahle, S.; Gauges, R.; Lee, C.; Pahle, J.; Simus, N.; Singhal, M.; Xu, L.; Mendes, P.; Kummer, U. COPASI – a COmplex PAthway SImulator. *Bioinformatics* 2006, 22, 3067–3074.

Huang, J.; Lih, C. J.; Pan, K. H.; Cohen, S. N. Global analysis of growth phase responsive gene expression and regulation of antibiotic biosynthetic pathways in *Streptomyces coelicolor* using DNA microarrays. *Genes Dev* 2001, 15, 3183–3192.

Ito, T.; Chiba, T.; Ozawa, R.; Yoshida, M.; Hattori, M.; Sakaki, Y. A comprehensive two-hybrid analysis to explore the yeast protein interactome. *Proc Natl Acad Sci USA* 2001, 98, 4569–4574.

Ito, T.; Tashiro, K.; Muta, S.; Ozawa, R.; Chiba, T.; Nishizawa, M.; Yamamoto, K.; Kuhara, S.; Sakaki, Y. Toward a protein-protein interaction map of the budding yeast: a comprehensive system to examine two-hybrid interactions in all possible combinations between the yeast proteins. *Proc Natl Acad Sci USA* 2000, 97, 1143–1147.

Iwatani, S.; Yamada, Y.; Usuda. Y. Metabolic flux analysis in biotechnology processes. *Biotechnol Lett* 2008, 30, 791–799.

Jami, M. S.; Barreiro, C.; Garcia-Estrada, C.; Martin, J. F. Proteome analysis of the penicillin producer *Penicillium chrysogenum*: characterization of protein changes during the industrial strain improvement. *Mol Cell Proteomics* 2010, 9, 1182–1198.

Jamshidi, N.; Palsson, B. O. Formulating genome-scale kinetic models in the postgenome era. *Mol Syst Biol* 2008, 4, 171.

Johnson, D. S.; Mortazavi, A.; Myers, R. M.; Wold, B. Genome-wide mapping of *in vivo* protein-DNA interactions. *Science* 2007, 316, 1497–1502.

Jones, D. T.; Woods, D. R. Acetone-butanol fermentation revisited. *Microbiol Rev* 1986, 50, 484–524.

Kabir, M. M.; Shimizu, K. Fermentation characteristics and protein expression patterns in a recombinant *Escherichia coli* mutant lacking phosphoglucose isomerase for poly(3-hydroxybutyrate) production. *Appl Microbiol Biotechnol* 1986, 62, 244–255.

Kalinowski, J.; Bathe, B.; Bartels, D.; Bischoff, N.; Bott, M.; Burkovski, A.; Dusch, N.; Eggeling, L.; Eikmanns, B. J.; Gaigalat, L.; Goesmann, A.; Hartmann, M.; Huthmacher, K.; Kramer, R.; Linke, B.; McHardy, A. C.; Meyer, F.; Mockel, B.; Pfefferle, W.; Puhler, A.; Rey, D. A.; Ruckert, C.; Rupp, O.; Sahm, H.; Wendisch, V. F.; Wiegrabe, I.; Tauch, A. The complete *Corynebacterium glutamicum* ATCC 13032 genome sequence and its impact on the production of L-aspartate-derived amino acids and vitamins. *J Biotechnol* 2003, 104, 5–25.

Kaneko, T.; Sato, S.; Kotani, H.; Tanaka, A.; Asamizu, E.; Nakamura, Y.; Miyajima, N.; Hirosawa, M.; Sugiura, M.; Sasamoto, S.; Kimura, T.; Hosouchi, T.; Matsuno, A.; Muraki, A.; Nakazaki, N.; Naruo, K.; Okumura, S.; Shimpo, S.; Takeuchi, C.; Wada, T.; Watanabe, A.; Yamada, M.; Yasuda, M.; Tabata, S. Sequence analysis of the genome of the unicellular cyanobacterium *Synechocystis* sp. strain PCC6803 II – sequence determination of the entire genome and assignment of potential protein-coding regions. *DNA Res* 1996, 3, 109–136.

Kant, R.; Blom, J.; Palva, A.; Siezen, R. J.; de Vos, W. M. Comparative genomics of *Lactobacillus*. *Microb Biotechnol* 2011, 4, 323–332.

Karoonuthaisiri, N.; Weaver, D.; Huang, J.; Cohen, S. N.; Kao, C. M. Regional organization of gene expression in *Streptomyces coelicolor*. *Gene* 2005, 353, 53–66.

Kell, D. B. The virtual human: towards a global systems biology of multiscale, distributed biochemical network models. *IUBMB Life* 2007, 59, 689–695.

Kent, E.; Hoops, S.; Mendes, P. Condor-COPASI: high-throughput computing for biochemical networks. *BMC Syst Biol* 2012, 6, 91.

Kim, T. Y.; Kim, H. U.; Park, J. M.; Song, H.; Kim, J. S.; Lee, S. Y. Genome-scale analysis of *Mannheimia succiniciproducens* metabolism. *Biotechnol Bioeng* 2007, 97, 657–671.

Kim, T. Y.; Sohn, S. B.; Kim, H. U.; Lee, S. Y. Strategies for systems-level metabolic engineering. *Biotechnol J* 2008, 3, 612–623.

Kitano, H. Systems biology: a brief overview. *Science* 2002, 295, 1662–1664.

Kjeldsen, K. R.; Nielsen J. *In silico* genome-scale reconstruction and validation of the *Corynebacterium glutamicum* metabolic network. *Biotechnol Bioeng* 2009, 102, 583–597.

Kleerebezem, M.; Boekhorst, J.; van Kranenburg, R.; Molenaar, D.; Kuipers, O. P.; Leer, R.; Tarchini, R.; Peters, S. A.; Sandbrink, H. M.; Fiers, M. W.; Stiekema, W.; Lankhorst, R. M.;

Bron, P. A.; Hoffer, S. M.; Groot, M. N.; Kerkhoven, R.; de Vries, M.; Ursing, B.; de Vos, W. M.; Siezen, R. J. Complete genome sequence of *Lactobacillus plantarum* WCFS1. *Proc Natl Acad Sci USA* 2003, 100, 1990–1995.

Kleijn, R. J.; Geertman, J. M.; Nfor, B. K.; Ras, C.; Schipper, D.; Pronk, J. T.; Heijnen, J. J.; van Maris, A. J.; van Winden, W. A. Metabolic flux analysis of a glycerol-overproducing *Saccharomyces cerevisiae* strain based on GC-MS, LC-MS and NMR-derived C-labeling data. *FEMS Yeast Res* 2007, 7, 216–231.

Klein-Marcuschamer, D.; Yadav, V. G.; Ghaderi, A.; Stephanopoulos, G. N. *De novo* metabolic engineering and the promise of synthetic DNA. *Adv Biochem Eng Biotechnol* 2010, 120, 101–131.

Klenk, H. P.; Clayton, R. A.; Tomb, J. F.; White, O.; Nelson, K. E.; Ketchum, K. A.; Dodson, R. J.; Gwinn, M.; Hickey, E. K.; Peterson, J. D.; Richardson, D. L.; Kerlavage, A. R.; Graham, D. E.; Kyrpides, N. C.; Fleischmann, R. D.; Quackenbush, J.; Lee, N. H.; Sutton, G. G.; Gill, S.; Kirkness, E. F.; Dougherty, B. A.; McKenney, K.; Adams, M. D.; Loftus, B.; Peterson, S.; Reich, C. I.; McNeil, L. K.; Badger, J. H.; Glodek, A.; Zhou, L.; Overbeek, R.; Gocayne, J. D.; Weidman, J. F.; McDonald, L.; Utterback, T.; Cotton, M. D.; Spriggs, T.; Artiach, P.; Kaine, B. P.; Sykes, S. M.; Sadow, P. W.; D'Andrea, K. P.; Bowman, C.; Fujii, C.; Garland, S. A.; Mason, T. M.; Olsen, G. J.; Fraser, C. M.; Smith, H. O.; Woese, C. R.; Venter, J. C. The complete genome sequence of the hyperthermophilic, sulfate-reducing archeon *Archaeoglobus fulgidus*. *Nature* 1997, 390, 364–370.

Koffas, M.; Roberge, C.; Lee, K.; Stephanopoulos, G. Metabolic engineering. *Annu Rev Biomed Eng* 1999, 1, 535–557.

Kohlstedt M.; Becker, J.; Wittmann, C. Metabolic fluxes and beyond-systems biology understanding and engineering of microbial metabolism. *Appl Microbiol Biotechnol* 2010, 88, 1065–1075.

Kolisnychenko, V.; Plunkett, G. III.; Herring, C. D.; Feher, T.; Posfai, J.; Blattner, F. R.; Posfai, G. Engineering a reduced *Escherichia coli* genome. *Genome Res* 2002, 12, 640–647.

Krainer, F. W.; Dietzsch, C.; Hajek, T.; Herwig, C.; Spadiut, O.; Glieder, A. Recombinant protein expression in *Pichia pastoris* strains with an engineered methanol utilization pathway. *Microb Cell Fact* 2012, 11, 22.

Krogan, N. J.; Cagney, G.; Yu, H.; Zhong, G.; Guo, X.; Ignatchenko, A.; Li, J.; Pu, S.; Datta, N.; Tikuisis, A. P.; Punna, T.; Peregrin-Alvarez, J. M.; Shales, M.; Zhang, X.; Davey, M.; Robinson, M. D.; Paccanaro, A.; Bray, J. E.; Sheung, A.; Beattie, B.; Richards, D. P.; Canadien, V.; Lalev, A.; Mena, F.; Wong, P.; Starostine, A.; Canete, M. M.; Vlasblom, J.; Wu, S.; Orsi, C.; Collins, S. R.; Chandran, S.; Haw, R.; Rilstone, J. J.; Gandi, K.; Thompson, N. J.; Musso, G.; St Onge, P.; Ghanny, S.; Lam, M. H.; Butland, G.; Altaf-Ul, A. M.; Kanaya, S.; Shilatifard, A.; O'Shea, E.; Weissman, J. S.; Ingles, C. J.; Hughes, T. R.; Parkinson, J.; Gerstein, M.; Wodak, S. J.; Emili, A.; Greenblatt, J. F. Global landscape of protein complexes in the yeast *Saccharomyces cerevisiae*. *Nature* 2006, 440, 637–643.

Kromer, J. O.; Heinzle, E.; Schroder, H.; Wittmann, C. Accumulation of homolanthionine and activation of a novel pathway for isoleucine biosynthesis in *Corynebacterium glutamicum* McbR deletion strains. *J Bacteriol* 2006, 188, 609–618.

Kromer, J. O.; Sorgenfrei, O.; Klopprogge, K.; Heinzle, E.; Wittmann, C. In-depth profiling of lysine-producing *Corynebacterium glutamicum* by combined analysis of the transcriptome, metabolome, and fluxome. *J Bacteriol* 2004, 186, 1769–1784.

Kumar, S. P.; Feidler, J. C. BioSPICE: a computational infrastructure for integrative biology. *OMICS* 2003, 7, 225.

Kunst, F.; Ogasawara, N.; Moszer, I.; Albertini, A. M.; Alloni, G.; Azevedo, V.; Bertero, M. G.; Bessieres, P.; Bolotin, A.; Borchert, S.; Borriss, R.; Boursier, L.; Brans, A.; Braun, M.; Brignell, S. C.; Bron, S.; Brouillet, S.; Bruschi, C. V.; Caldwell, B.; Capuano, V.; Carter, N. M.; Choi, S. K.; Codani, J. J.; Connerton, I. F.; Cummings, N. J.; Daniel, R. A.; Denizot, F.; Devine, K..M.; Düsterhöft, A.; Ehrlich, S. D.; Emmerson, P. T.; Entian, K. D.; Errington, J.; Fabret, C.; Ferrari, E.; Foulger, D.; Fritz, C.; Fujita, M.; Fujita, Y.; Fuma, S.; Galizzi, A.; Galleron, N.; Ghim, S. –Y.; Glaser, P.; Goffeau, A.; Golightly, E. J.; Grandi, G.; Guiseppi, G.; Guy, B. J.; Haga, K.; Haiech, J.; Harwood, C. R.; Hénaut, A.; Hilbert, H.; Holsappel, S.; Hosono, S.; Hullo, M. -S.; Itaya, M.; Jones, L.; Joris, B.; Karamata, D.; Kasahara, Y.; Klaerr-Blanchard, M.; Klein, C.; Kobayashi, Y.; Koetter, P.; Koningstein, G.; Krogh, S.; Kumano, M.; Kurita, K.; Lapidus, A.; Lardinois, S.; Lauber, J.; Lazarevic, V.; Lee, S. -M.; Levine, A.; Liu, H.; Masuda, S.; Mauël, C.; Médigue, C.; Medina, N.; Mellado, R. P.; Mizuno, M.; Moestl, D.; Nakai, S.; Noback, M.; Noone, D.; O'Reilly, M.; Ogawa, K.; Ogiwara, A.; Oudega, B.; Park, S. -H.; Parro, V.; Pohl, T. M.; Portetelle, D.; Porwollik, S.; Prescott, A. M.; Presecan, E.; Pujic, P.; Purnelle, B.; Rapoport, G.; Rey, M.; Reynolds, S.; Rieger, M.; Rivolta, C.; Rocha, E.; Roche, B.; Rose, M.; Sadaie, Y.; Sato, T.; Scanlan, E.; Schleich, S.; Schroeter, R.; Scoffone, F.; Sekiguchi, J.; Sekowska, A.; Seror, S. J.; Serror, P.; Shin, B. -S.; Soldo, B.; Sorokin, A.; Tacconi, E.; Takagi, T.; Takahashi, H.; Takemaru, K.; Takeuchi, M.; Tamakoshi, A.; Tanaka, T.; Terpstra, P.; Tognoni, A.; Tosato, V.; Uchiyama, S.; Vandenbol, M.; Vannier, F.; Vassarotti, A.; Viari, A.; Wambutt, R.; Wedler, E.; Wedler, H.; Weitzenegger, T.; Winters, P.; Wipat, A.; Yamamoto, H.; Yamane, K.; Yasumoto, K.; Yata, K.; Yoshida, K.; Yoshikawa, H. -F.; Zumstein, E.; Yoshikawa, H.; Danchin, A. The complete genome sequence of the gram-positive bacterium *Bacillus subtilis*. *Nature* 1997, 390, 249–256.

Langer, M.; Gabor, E. M.; Liebeton, K.; Meurer, G.; Niehaus, F.; Schulze, R.; Eck, J.; Lorenz, P. Metagenomics: an inexhaustible access to nature's diversity. *Biotechnol J* 2006, 1, 815–821.

Lartigue, C.; Glass, J. I.; Alperovich, N.; Pieper, R.; Parmar, P. P.; Hutchison, C. A. III.; Smith, H. O.; Venter, J. C. Genome transplantation in bacteria: changing one species to another. *Science* 2007, 317, 632–638.

Lartigue, C.; Vashee, S.; Algire, M. A.; Chuang, R. Y.; Benders, G. A.; Ma, L.; Noskov, V. N.; Denisova, E. A.; Gibson, D. G.; Assad-Garcia, N.; Alperovich, N.; Thomas, D. W.; Merryman, C.; Hutchison, C. A. III.; Smith, H. O.; Venter, J. C.; Glass, J. I. Creating bacterial strains from genomes that have been cloned and engineered in yeast. *Science* 2009, 325, 1693–1696.

Lashkari, D. A.; De Risi, J. L.; Mc Cusker, J. H.; Namath, A. F.; Gentile, C.; Hwang, S. Y.; Brown, P. O.; Davis, R. W. Yeast microarrays for genome wide parallel genetic and gene expression analysis. *Proc Natl Acad Sci USA* 1997, 94, 13057–13062.

Le Crom, S.; Devaux, F.; Jacq, C.; Marc, P. yMGV: helping biologists with yeast microarray data mining. *Nucleic Acids Res* 2002, 30, 76–79.

Le Novere, N.; Bornstein, B.; Broicher, A.; Courtot, M.; Donizelli, M.; Dharuri, H.; Li, L.; Sauro, H.; Schilstra, M.; Shapiro, B.; Snoep, J. L.; Hucka, M. BioModels Database: a free, centralized database of curated, published, quantitative kinetic models of biochemical and cellular systems. *Nucleic Acids Res* 2006, 34, D689–691.

Lee, D. Y.; Yun, H.; Park, S.; Lee, S. Y. MetaFluxNet: the management of metabolic reaction information and quantitative metabolic flux analysis. *Bioinformatics* 2003, 19, 2144–2146.

Lee, J.; Yun, H.; Feist, A. M.; Palsson, B. O.; Lee, S. Y. Genome-scale reconstruction and *in silico* analysis of the *Clostridium acetobutylicum* ATCC 824 metabolic network. *Appl Microbiol Biotechnol* 2008, 80, 849–862.

Lee, J. H.; Lee, D. E.; Lee, B. U.; Kim, H. S. Global analyzes of transcriptomes and proteomes of a parent strain and an L-threonine-overproducing mutant strain. *J Bacteriol* 2003, 185, 5442–5451.

Lee, J. Y.; Jang, Y. S.; Lee, J.; Papoutsakis, E. T.; Lee, S. Y. Metabolic engineering of *Clostridium acetobutylicum* M5 for highly selective butanol production. *Biotechnol J* 2009, 4, 1432–1440.

Lee, K. Y.; Park, J. M.; Kim, T. Y.; Yun, H.; Lee, S. Y. The genome-scale metabolic network analysis of *Zymomonas mobilis* ZM4 explains physiological features and suggests ethanol and succinic acid production strategies. *Microb Cell Fact* 2010, 9, 94.

Lee, S. J.; Lee, D. Y.; Kim, T. Y.; Kim, B. H.; Lee, J.; Lee, S. Y. Metabolic engineering of *Escherichia coli* for enhanced production of succinic acid, based on genome comparison and *in silico* gene knockout simulation. *Appl Environ Microbiol* 2005, 71, 7880–7887.

Lee, S. Y. Systems biotechnology. *Genome Inform* 2009, 23, 214–216.

Lee, S, Y.; Lee, D, Y.; Kim, T. Y. Systems biotechnology for strain improvement. *Trends Biotechnol* 2005, 23, 349–358.

Li, C.; Donizelli, M.; Rodriguez, N.; Dharuri, H.; Endler, L.; Chelliah, V.; Li, L.; He, E.; Henry, A.; Stefan, M. I.; Snoep, J. L.; Hucka, M.; Le Novere, N.; Laibe, C. BioModels Database: an enhanced, curated and annotated resource for published quantitative kinetic models. *BMC Syst Biol* 2010, 4, 9.

Liao, J. C.; Hou, S. Y.; Chao, Y. P. Pathway analysis, engineering, and physiological considerations for redirecting central metabolism. *Biotechnol Bioeng* 1996, 52, 129–140.

Liao, Y. C.; Tsai, M. H.; Chen, F. C.; Hsiung, C. A. GEMSiRV: a software platform for GEnome-scale metabolic model simulation, reconstruction and visualization. *Bioinformatics* 2012, 28, 1752–1758.

Liu, L.; Agren, R.; Bordel, S.; Nielsen, J. Use of genome-scale metabolic models for understanding microbial physiology. *FEBS Lett* 2010, 584, 2556–2564.

Lockhart, D. J.; Dong, H.; Byrne, M. C.; Follettie, M. T.; Gallo, M. V.; Chee, M. S.; Mittmann, M.; Wang, C.; Kobayashi, M.; Horton, H.; Brown, E. L. Expression monitoring by hybridization to high-density oligonucleotide arrays. *Nat Biotechnol* 1996, 14, 1675–1680.

Lorenz, P.; Eck, J. Metagenomics and industrial applications. *Nat Rev Microbiol* 2005, 3, 510–516.

Lutke-Eversloh, T.; Bahl, H. Metabolic engineering of *Clostridium acetobutylicum*: recent advances to improve butanol production. *Curr Opin Biotechnol* 2011, 22, 634–647.

Machida, M.; Asai, K.; Sano, M.; Tanaka, T.; Kumagai, T.; Terai, G.; Kusumoto, K.; Arima, T.; Akita O.; Kashiwagi, Y.; Abe, K.; Gomi, K.; Horiuchi, H.; Kitamoto, K.; Kobayashi, T.; Takeuchi, M.; Denning, D. W.; Galagan, J. E.; Nierman, W. C.; Yu, J.; Archer, D. B.; Bennett,

J. W.; Bhatnagar, D.; Cleveland, T. E.; Fedorova, N. D.; Gotoh, O.; Horikawa, H.; Hosoyama, A.; Ichinomiya, M.; Igarashi, R.; Iwashita, K.; Juvvadi, P. R.; Kato, M.; Kato, Y.; Kin, T.; Kokubun, A.; Maeda, H.; Maeyama, N.; Maruyama, J.; Nagasaki, H.; Nakajima, T.; Oda, K.; Okada, K.; Paulsen, I.; Sakamoto, K.; Sawano, T.; Takahashi, M.; Takase, K.; Terabayashi, Y.; Wortman, J. R.; Yamada, O.; Yamagata, Y.; Anazawa, H.; Hata, Y.; Koide, Y.; Komori, T.; Koyama, Y.; Minetoki, T.; Suharnan, S.; Tanaka, A.; Isono, K.; Kuhara, S., Ogasawara, N.; Kikuchi, H. Genome sequencing and analysis of *Aspergillus oryzae*. *Nature* 2005, 438, 1157–1161.

Maeder, D. L.; Anderson, I.; Brettin, T. S.; Bruce, D. C.; Gilna, P.; Han, C. S.; Lapidus, A.; Metcalf, W. W.; Saunders, E.; Tapia, R.; Sowers, K. R. The *Methanosarcina barkeri* genome: comparative analysis with *Methanosarcina acetivorans* and *Methanosarcina mazei* reveals extensive rearrangement within methanosarcinal genomes. *J Bacteriol* 2006, 188, 7922–7931.

Mahadevan, R.; Bond, D. R.; Butler, J. E.; Esteve-Nunez, A.; Coppi, M. V.; Palsson, B. O.; Schiling, C. H.; Lovley, D. R. Characterization of metabolism in the Fe(III)-reducing organism *Geobacter sulfurreducens* by constraint-based modeling. *Appl Environ Microbiol* 2006, 72, 1558–1568.

Mao, S.; Luo, Y.; Zhang, T.; Li, J.; Bao, G.; Zhu, Y.; Chen, Z.; Zhang, Y.; Li, Y.; Ma, Y. Proteome reference map and comparative proteomic analysis between a wild type *Clostridium acetobutylicum* DSM 1731 and its mutant with enhanced butanol tolerance and butanol yield. *J Proteome Res* 2010, 9, 3046–3061.

Mapelli, V.; Olsson, L.; Nielsen, J. Metabolic footprinting in microbiology: methods and applications in functional genomics and biotechnology. *Trends Biotechnol* 2008, 26, 490–497.

Martin, V. J.; Pitera, D. J.; Withers, S. T.; Newman, J. D.; Keasling, J. D. Engineering a mevalonate pathway in *Escherichia coli* for production of terpenoids. *Nat Biotechnol* 2003, 21, 796–802.

Martins, A. M.; Camacho, D.; Shuman, J.; Sha, W.; Mendes, P.; Shulaev, V. A systems biology study of two distinct growth phases of *Saccharomyces cerevisiae* cultures. *Current Genomics* 2004, 5, 649–663.

Mashego, M. R.; Rumbold, K.; De Mey, M.; Vandamme, E.; Soetaert, W.; Heijnen, J. J. Microbial metabolomics: past, present and future methodologies. *Biotechnol Lett* 2007, 29, 1–16.

Mattanovich, D.; Branduardi, P.; Dato, L.; Gasser, B.; Sauer, M.; Porro, D. Recombinant protein production in yeasts. *Methods Mol Biol* 2012, 824, 329–358.

May, C.; Brosseron, F.; Chartowski, P.; Schumbrutzki, C.; Schoenebeck, B.; Marcus, K. Instruments and methods in proteomics. *Methods Mol Biol* 2011, 696, 3–26.

Mazumdar, V.; Snitkin, E. S.; Amar, S.; Segre, D. Metabolic network model of a human oral pathogen. *J Bacteriol* 2009, 191, 74–90.

McIntosh, B. K.; Renfro, D. P.; Knapp, G. S.; Lairikyengbam, C. R.; Liles, N. M.; Niu, L.; Supak, A. M.; Venkatraman, A.; Zweifel, A. E.; Siegele, D. A.; Hu, J. C. EcoliWiki: a wiki-based community resource for *Escherichia coli*. *Nucleic Acids Res* 2011, 40, D1270–1277.

Medema, M. H.; Breitling, R.; Takano, E. Synthetic biology in Streptomyces bacteria. *Methods Enzymol* 2011, 497, 485–502.

Medema, M. H.; van Raaphorst, R.; Takano, E.; Breitling, R. Computational tools for the synthetic design of biochemical pathways. *Nat Rev Microbiol* 2012, 10, 191–202.

Mehra, S.; Lian, W.; Jayapal, K. P.; Charaniya, S. P.; Sherman, D. H.; Hu, W. S. A framework to analyze multiple time series data: a case study with *Streptomyces coelicolor*. *J Ind Microbiol Biotechnol* 2006, 33, 159–172.

Meijer, S.; Nielsen, M. L.; Olsson, L.; Nielsen, J. Gene deletion of cytosolic ATP: citrate lyase leads to altered organic acid production in *Aspergillus niger*. *J Ind Microbiol Biotechnol* 2009, 36, 1275–1280.

Mendes, P. GEPASI: a software package for modeling the dynamics, steady states and control of biochemical and other systems. *Comput Appl Biosci* 1993, 9, 563–571.

Merchant, S. S.; Prochnik, S. E.; Vallon, O.; Harris, E. H.; Karpowicz, S. J.; Witman, G. B.; Terry, A.; Salamov, A.; Fritz-Laylin, L. K.; Marechal-Drouard, L.; Marshall, W. F.; Qu, L. H.; Nelson, D. R.; Sanderfoot, A. A.; Spalding, M. H.; Kapitonov, V. V.; Ren, Q.; Ferris, P.; Lindquist, E.; Shapiro, H.; Lucas, S. M.; Grimwood, J.; Schmutz, J.; Cardol, P.; Cerutti, H.; Chanfreau, G.; Chen, C. L.; Cognat, V.; Croft, M. T.; Dent, R.; Dutcher, S.; Fernandez, E.; Fukuzawa, H.; Gonzalez-Ballester, D.; Gonzalez-Halphen, D.; Hallmann, A.; Hanikenne, M.; Hippler, M.; Inwood, W.; Jabbari, K.; Kalanon, M.; Kuras, R.; Lefebvre, P. A.; Lemaire, S. D.; Lobanov, A. V.; Lohr, M.; Manuell, A.; Meier, I.; Mets, L.; Mittag, M.; Mittelmeier, T.; Moroney, J. V.; Moseley, J.; Napoli, C.; Nedelcu, A. M.; Niyogi, K.; Novoselov, S. V.; Paulsen, I. T.; Pazour, G.; Purton, S.; Ral, J. P.; Riano-Pachon, D. M.; Riekhof, W.; Rymarquis, L.; Schroda, M.; Stern, D.; Umen, J.; Willows, R.; Wilson, N.; Zimmer, S. L.; Allmer, J.; Balk, J.; Bisova, K.; Chen, C. J.; Elias, M.; Gendler, K.; Hauser, C.; Lamb, M. R.; Ledford, H.; Long, J. C.; Minagawa, J.; Page, M. D.; Pan, J.; Pootakham, W.; Roje, S.; Rose, A.; Stahlberg, E.; Terauchi, A. M.; Yang, P.; Ball, S.; Bowler, C.; Dieckmann, C. L.; Gladyshev, V. N.; Green, P.; Jorgensen, R.; Mayfield, S.; Mueller-Roeber, B.; Rajamani, S.; Sayre, R. T.; Brokstein, P.; Dubchak, I.; Goodstein, D.; Hornick, L.; Huang, Y. W.; Jhaveri, J.; Luo, Y.; Martinez, D.; Ngau, W. C.; Otillar, B.; Poliakov, A.; Porter, A.; Szajkowski, L.; Werner, G.; Zhou, K.; Grigoriev, IV.; Rokhsar, D. S.; Grossman, A. R. The *Chlamydomonas* genome reveals the evolution of key animal and plant functions. *Science* 2007, 318, 245–250.

Methe, B. A.; Nelson, K. E.; Eisen, J. A.; Paulsen, I. T.; Nelson, W.; Heidelberg, J. F.; Wu, D.; Wu, M.; Ward, N.; Beanan, M. J.; Dodson, R. J.; Madupu, R.; Brinkac, L. M.; Daugherty, S. C.; DeBoy, R. T.; Durkin, A. S.; Gwinn, M.; Kolonay, J. F.; Sullivan, S. A.; Haft, D. H.; Selengut, J.; Davidsen, T. M.; Zafar, N.; White, O.; Tran, B.; Romero, C.; Forberger, H. A.; Weidman, J.; Khouri, H.; Feldblyum, T. V.; Utterback, T. R.; Van Aken, S. E.; Lovley, D. R.; Fraser, C. M. Genome of *Geobacter sulfurreducens*: metal reduction in subsurface environments. *Science* 2003, 302, 1967–1969.

Milne, C. B.; Kim, P. J.; Eddy, J. A.; Price, N. D. Accomplishments in genome-scale *in silico* modeling for industrial and medical biotechnology. *Biotechnol J* 2009, 4, 1653–1670.

Montagud, A.; Navarro, E.; Fernandez de Cordoba, P.; Urchueguia, J. F.; Patil, K. R. Reconstruction and analysis of genome-scale metabolic model of a photosynthetic bacterium. *BMC Syst Biol* 2010, 4, 156.

Moutselos, K.; Kanaris, I.; Chatziioannou, A.; Maglogiannis, I.; Kolisis, F. N. KEGGconverter: a tool for the *in silico* modeling of metabolic networks of the KEGG Pathways database. *BMC Bioinformatics* 2009, 10, 324.

Mueller, L. N.; Brusniak, M. Y.; Mani, D. R.; Aebersold, R. An assessment of software solutions for the analysis of mass spectrometry based quantitative proteomics data. *J Proteome Res* 2008, 7, 51–61.

Mukhopadhyay, A.; Redding, A. M.; Rutherford, B. J.; Keasling J. D. Importance of systems biology in engineering microbes for biofuel production. *Curr Opin Biotechnol* 2008, 9, 228–234.

Muller, U.; van Assema, F.; Gunsior, M.; Orf, S.; Kremer, S.; Schipper, D.; Wagemans, A.; Townsend, C. A.; Sonke, T.; Bovenberg, R.; Wubbolts, M. Metabolic engineering of the *E coli* L-phenylalanine pathway for the production of D-phenylglycine (D-Phg). *Metab Eng* 2006, 8, 196–208.

Nelson, K. E.; Clayton, R. A.; Gill, S. R.; Gwinn, M. L.; Dodson, R. J.; Haft, D. H.; Hickey, E. K.; Peterson, J. D.; Nelson, W. C.; Ketchum, K. A.; McDonald, L.; Utterback, T. R.; Malek, J. A.; Linher, K. D.; Garrett, M. M.; Stewart, A. M.; Cotton, M. D.; Pratt, M. S.; Phillips, C. A.; Richardson, D.; Heidelberg, J.; Sutton, GG.; Fleischmann, R. D.; Eisen, J. A.; White, O.; Salzberg, S. L.; Smith, H. O.; Venter, J. C.; Fraser, C. M. Evidence for lateral gene transfer between Archaea and bacteria from genome sequence of *Thermotoga maritima*. *Nature* 1999, 399, 323–329.

Nelson, K. E.; Weinel, C.; Paulsen, I. T.; Dodson, R. J.; Hilbert, H.; Martins dos Santos, V. A.; Fouts, D. E.; Gill, S. R.; Pop, M.; Holmes, M.; Brinkac, L.; Beanan, M.; De Boy, R. T.; Daugherty, S.; Kolonay, J.; Madupu, R.; Nelson, W.; White, O.; Peterson, J.; Khouri, H.; Hance, I.; Chris Lee, P.; Holtzapple, E.; Scanlan, D.; Tran, K.; Moazzez, A.; Utterback, T.; Rizzo, M.; Lee, K.; Kosack, D.; Moestl, D.; Wedler, H.; Lauber, J.; Stjepandic, D.; Hoheisel, J.; Straetz, M.; Heim, S.; Kiewitz, C.; Eisen, J. A.; Timmis, K. N.; Dusterhoft, A.; Tummler B.; Fraser, C. M. Complete genome sequence and comparative analysis of the metabolically versatile *Pseudomonas putida* KT2440. *Environ Microbiol* 2002, 4, 799–808.

Nesvera, J.; Patek, M. Tools for genetic manipulations in *Corynebacterium glutamicum* and their applications. *Appl Microbiol Biotechnol* 2011, 90, 1641–1654.

Ng, W. V.; Kennedy, S. P.; Mahairas, G. G.; Berquist, B.; Pan, M.; Shukla, H. D.; Lasky, S. R.; Baliga, N. S.; Thorsson, V.; Sbrogna, J.; Swartzell, S.; Weir, D.; Hall, J.; Dahl, T. A.; Welti, R.; Goo, Y. A.; Leithauser, B.; Keller, K.; Cruz, R.; Danson, M. J.; Hough, D. W.; Maddocks, D. G.; Jablonski, P. E.; Krebs, M. P.; Angevine, C. M.; Dale, H.; Isenbarger, T. A.; Peck, R. F.; Pohlschroder, M.; Spudich, J. L.; Jung, K. W.; Alam, M.; Freitas, T.; Hou, S.; Daniels, C. J.; Dennis, P. P.; Omer, A. D.; Ebhardt, H.; Lowe, T. M.; Liang, P.; Riley, M.; Hood, L.; Das Sarma, S. Genome sequence of *Halobacterium* species NRC-1. *Proc. Natl Acad Sci USA* 2000, 97, 12176–12181.

Nielsen, J. It is all about metabolic fluxes. *J Bacteriol* 2003, 185, 7031–7035.

Nielsen, J.; Jewett, M. C. Impact of systems biology on metabolic engineering of *Saccharomyces cerevisiae*. *FEMS Yeast Res* 2008, 8, 122–131.

Nijkamp, J. F.; van den Broek, M.; Datema, E.; de Kok, S.; Bosman, L.; Luttik, M. A.; Daran-Lapujade, P.; Vongsangnak, W.; Nielsen, J.; Heijne, W. H.; Klaassen, P.; Paddon, C. J.; Platt, D.; Kotter, P.; van Ham, R. C.; Reinders, M. J.; Pronk, J. T.; de Ridder, D.; Daran, J. M. *De novo* sequencing, assembly and analysis of the genome of the laboratory strain *Saccharomyces cerevisiae* CENPK113–7D, a model for modern industrial biotechnology. *Microb Cell Fact* 2012, 11, 36.

Noble, D. Modeling the heart – from genes to cells to the whole organ. *Science* 2002, 295, 1678–1682.

Nogales, J.; Palsson, B. O.; Thiele, I. A genome-scale metabolic reconstruction of *Pseudomonas putida* KT2440: iJN746 as a cell factory. *BMC Syst Biol* 2008, 2, 79.

Nolling, J.; Breton, G.; Omelchenko, M. V.; Makarova, K. S.; Zeng, Q.; Gibson, R.; Lee, H. M.; Dubois, J.; Qiu, D.; Hitti, J.; Wolf, Y. I.; Tatusov, R. L.; Sabathe, F.; Doucette-Stamm, L.; Soucaille, P.; Daly, M. J.; Bennett, G. N.; Koonin, E. V.; Smith, D. R. Genome sequence and comparative analysis of the solvent-producing bacterium *Clostridium acetobutylicum*. *J Bacteriol* 2001, 183, 4823–4838.

Nookaew, I.; Jewett, M. C.; Meechai, A.; Thammarongtham, C.; Laoteng, K.; Cheevadhanarak, S.; Nielsen, J.; Bhumiratana, S. The genome-scale metabolic model iIN800 of *Saccharomyces cerevisiae* and its validation: a scaffold to query lipid metabolism. *BMC Syst Biol* 2008, 2, 71.

Oh, M. K.; Rohlin, L.; Kao, K. C.; Liao, J. C. Global expression profiling of acetate-grown *Escherichia coli*. *J Biol Chem* 2002, 277, 13175–13183.

Ohnishi, J.; Hayashi, M.; Mitsuhashi, S.; Ikeda, M. Efficient 40 degrees C fermentation of L-lysine by a new *Corynebacterium glutamicum* mutant developed by genome breeding. *Appl Microbiol Biotechnol* 2003, 62, 69–75.

Ohnishi, J.; Katahira, R.; Mitsuhashi, S.; Kakita, S.; Ikeda, M. A novel gnd mutation leading to increased L-lysine production in *Corynebacterium glutamicum*. *FEMS Microbiol Lett* 2005, 242, 265–274.

Ohnishi, J.; Mitsuhashi, S.; Hayashi, M.; Ando, S.; Yokoi, H.; Ochiai, K.; Ikeda, M. A novel methodology employing *Corynebacterium glutamicum* genome information to generate a new L-lysine-producing mutant. *Appl Microbiol Biotechnol* 2002, 58, 217–223.

Oliveira, A. P.; Nielsen, J.; Forster, J. Modeling *Lactococcus lactis* using a genome-scale flux model. *BMC Microbiol* 2005, 5, 39.

Otero, J. M.; Nielsen, J. Industrial systems biology. *Biotechnol Bioeng* 2009, 105, 439–460.

Otero, J. M.; Olssona, L.; Nielsen, J. Metabolic engineering of *Saccharomyces cerevisiae* microbial cell factories for succinic acid production. *J Biotechnol* 2007, 131, 205.

Ouzounis, C. A.; Valencia, A. Early bioinformatics: the birth of a discipline – a personal view. *Bioinformatics* 2003, 19, 2176–2190.

Overbeek, R.; Larsen, N.; Walunas, T.; D'Souza, M.; Pusch, G.; Selkov, E. Jr.; Liolios, K.; Joukov, V.; Kaznadzey, D.; Anderson, I.; Bhattacharyya, A.; Burd, H.; Gardner, W.; Hanke, P.; Kapatral, V.; Mikhailova, N.; Vasieva, O.; Osterman, A.; Vonstein, V.; Fonstein, M.; Ivanova, N.; Kyrpides, N. The ERGO genome analysis and discovery system. *Nucleic Acids Res* 2003, 31, 164–171.

Panagiotou, G.; Andersen, M. R.; Grotkjaer, T.; Regueira, T. B.; Nielsen, J.; Olsson, L. Studies of the production of fungal polyketides in *Aspergillus nidulans* by using systems biology tools. *Appl Environ Microbiol* 2009, 75, 2212–2220.

Papini, M.; Salazar, M.; Nielsen, J. Systems biology of industrial microorganisms. *Adv Biochem Eng Biotechnol* 2010, 120, 51–99.

Papoutsakis, E. T. Engineering solventogenic clostridia. *Curr Opin Biotechnol* 2008, 19, 420–429.

Papp, B.; Pal, C.; Hurst, L. D. Metabolic network analysis of the causes and evolution of enzyme dispensability in yeast. *Nature* 2004, 429, 661–664.

Park, J. H.; Lee, S. Y. Towards systems metabolic engineering of microorganisms for amino acid production. *Curr Opin Biotechnol* 2008, 19, 454–460.

Park, J. H.; Lee, S. Y.; Kim, T. Y.; Kim, H. U. Application of systems biology for bioprocess development. *Trends Biotechnol* 2008, 26, 404–412.

Park, S. J.; Lee, S. Y.; Cho, J.; Kim, T. Y.; Lee, J. W.; Park, J. H.; Han, M. J. Global physiological understanding and metabolic engineering of microorganisms based on omics studies. *Appl Microbiol Biotechnol* 2005, 68, 567–579.

Parrilli, E.; De Vizio, D.; Cirulli, C.; Tutino, M. L. Development of an improved *Pseudoalteromonas haloplanktis* TAC125 strain for recombinant protein secretion at low temperature. *Microb Cell Fact* 2008, 7, 2.

Pastink, M. I.; Teusink, B.; Hols, P.; Visser, S.; de Vos, W. M.; Hugenholtz, J. Genome-scale model of *Streptococcus thermophilus* LMG18311 for metabolic comparison of lactic acid bacteria. *Appl Environ Microbiol* 2009, 75, 3627–3633.

Patil. K. R.; Rocha, I.; Forster, J.; Nielsen, J. Evolutionary programming as a platform for *in silico* metabolic engineering. *BMC Bioinformatics* 2005, 6, 308.

Pel, H. J.; de Winde, J. H.; Archer, D. B.; Dyer, P. S.; Hofmann, G.; Schaap, P. J.; Turner, G.; de Vries, R. P.; Albang, R.; Albermann, K.; Andersen, M. R.; Bendtsen, J. D.; Benen, J. A.; van den Berg, M.; Breestraat, S.; Caddick, M. X.; Contreras, R.; Cornell, M.; Coutinho, P. M.; Danchin, E. G.; Debets, A. J.; Dekker, P.; van Dijck, P. W.; van Dijk, A.; Dijkhuizen, L.; Driessen, A. J.; de Enfert, C.; Geyzens, S.; Goosen, C.; Groot, G. S.; de Groot, P. W.; Guillemette, T.; Henrissat, B.; Herweijer, M.; van den Hombergh, J. P.; van den Hondel, C. A.; van der Heijden, R. T.; van der Kaaij, R. M.; Klis, F. M.; Kools, H. J.; Kubicek, C. P.; van Kuyk, P. A.; Lauber, J.; Lu, X.; van der Maarel, M. J.; Meulenberg, R.; Menke, H.; Mortimer, M. A.; Nielsen, J.; Oliver, S. G.; Olsthoorn, M.; Pal, K.; van Peij, N. N.; Ram, A. F.; Rinas, U.; Roubos, J. A.; Sagt, C. M.; Schmoll, M.; Sun, J.; Ussery, D.; Varga, J.; Vervecken, W.; van de Vondervoort, P. J.; Wedler, H.; Wosten, H. A.; Zeng, A. P.; van Ooyen, A. J.; Visser, J.; Stam, H. Genome sequencing and analysis of the versatile cell factory *Aspergillus niger* CBS 51388. *Nat Biotechnol* 2007, 25, 221–231.

Picataggio, S. Potential impact of synthetic biology on the development of microbial systems for the production of renewable fuels and chemicals. *Curr Opin Biotechnol* 2009, 20, 325–329.

Pizarro, F.; Vargas, F. A.; Agosin, E. A systems biology perspective of wine fermentations. *Yeast* 2007, 24, 977–991.

Porro, D.; Brambilla, L.; Ranzi, B. M.; Martegani, E.; Alberghina, L. Development of metabolically engineered *Saccharomyces cerevisiae* cells for the production of lactic acid. *Biotechnol Prog* 1995, 11, 294–298.

Porro, D.; Branduardi, P. Yeast cell factory: fishing for the best one or engineering it? *Microb Cell Fact* 2009, 8, 51.

Prather, K. L.; Martin, C. H. *De novo* biosynthetic pathways: rational design of microbial chemical factories. *Curr Opin Biotechnol* 2008, 19, 468–474.

Prust, C.; Hoffmeister, M.; Liesegang, H.; Wiezer, A.; Fricke, W. F.; Ehrenreich, A.; Gottschalk, G.; Deppenmeier, U. Complete genome sequence of the acetic acid bacterium *Gluconobacter oxydans. Nat Biotechnol* 2005, 23, 195–200.

Puchalka, J.; Oberhardt, M. A.; Godinho, M.; Bielecka, A.; Regenhardt, D.; Timmis, K. N.; Papin, J. A.; Martins dos Santos, V. A. Genome-scale reconstruction and analysis of the *Pseudomonas putida* KT2440 metabolic network facilitates applications in biotechnology. *PLoS Comput Biol* 2008, 4, e1000210.

Ravasi, P.; Peiru, S.; Gramajo, H.; Menzella, H. G. Design and testing of a synthetic biology framework for genetic engineering of *Corynebacterium glutamicum. Microb Cell Fact* 2012, 11, 147.

Reed, J. L.; Vo, T. D.; Schiling, C. H.; Palsson, B. O. An expanded genome-scale model of *Escherichia coli* K-12 (iJR904 GSM/GPR). *Genome Biol* 2003, 4, R54.

Riedel, C.; Rittmann, D.; Dangel, P.; Mockel, B.; Petersen, S.; Sahm, H.; Eikmanns, B. J. Characterization of the phosphoenolpyruvate carboxykinase gene from *Corynebacterium glutamicum* and significance of the enzyme for growth and amino acid production. *J Mol Microbiol Biotechnol* 2001, 3, 573–583.

Ro, D. K.; Paradise, E. M.; Ouellet, M.; Fisher, K. J.; Newman, K. L.; Ndungu, J. M.; Ho, K. A.; Eachus, R. A.; Ham, T. S.; Kirby, J.; Chang, M. C.; Withers, S. T.; Shiba, Y.; Sarpong, R.; Keasling, J. D. Production of the antimalarial drug precursor artemisinic acid in engineered yeast. *Nature* 2006, 440, 940–943.

Rocha, I.; Maia, P.; Evangelista, P.; Vilaca, P.; Soares, S.; Pinto, J. P.; Nielsen, J.; Patil, K. R.; Ferreira E. C.; Rocha, M. OptFlux: an open-source software platform for *in silico* metabolic engineering. *BMC Syst Biol* 2010, 4, 45.

Rokem, J. S.; Lantz, A. E.; Nielsen, J. Systems biology of antibiotic production by microorganisms. *Nat Prod Rep* 2007, 24, 1262–1287.

Ruckert, C.; Puhler, A.; Kalinowski, J. Genome-wide analysis of the L-methionine biosynthetic pathway in *Corynebacterium glutamicum* by targeted gene deletion and homologous complementation. *J Biotechnol* 2003, 104, 213–228.

Salusjarvi, L.; Kankainen, M.; Soliymani, R.; Pitkanen, J. P.; Penttila, M.; Ruohonen, L. Regulation of xylose metabolism in recombinant *Saccharomyces cerevisiae. Microb Cell Fact* 2008, 7, 18.

Sanchez, S.; Demain, A. L. Metabolic regulation and overproduction of primary metabolites. *Microb Biotechnol* 2008, 1, 283–319.

Sauer, U.; Hatzimanikatis, V.; Bailey, J, E.; Hochuli, M.; Szyperski, T.; Wuthrich, K. Metabolic fluxes in riboflavin-producing *Bacillus subtilis. Nat Biotechnol* 1997, 15, 448–452.

Sauer, U.; Hatzimanikatis, V.; Hohmann, H. P.; Manneberg, M.; van Loon, A. P.; Bailey, J. E. Physiology and metabolic fluxes of wild-type and riboflavin-producing *Bacillus subtilis. Appl Environ Microbiol* 1996, 62, 3687–3696.

Sauro, H. M. SCAMP: a general-purpose simulator and metabolic control analysis program. *Comput Appl Biosci* 1993, 9, 441–450.

Sawada, K.; Zen, I. S.; Wada, M.; Yokota, A. Metabolic changes in a pyruvate kinase gene deletion mutant of *Corynebacterium glutamicum* ATCC 13032. *Metab Eng* 2010, 12, 401–407.

Schallmey, M.; Singh, A.; Ward, O. P. Developments in the use of *Bacillus* species for industrial production. *Canadian Journal of Microbiology* 2004, 50, 1–17.

Schena, M.; Shalon, D.; Davis, R. W.; Brown, P. O. Quantitative monitoring of gene expression patterns with a complementary DNA microarray. *Science* 1995, 270, 467–470.

Schmeisser, C.; Steele, H.; Streit, W. R. Metagenomics, biotechnology with nonculturable microbes. *Appl Microbiol Biotechnol* 2007, 75, 955–962.

Schneider, J.; Vorholter, F. J.; Trost, E.; Blom, J.; Musa, Y. R.; Neuweger, H.; Niehaus, K.; Schatschneider, S.; Tauch, A.; Goesmann, A. CARMEN – Comparative Analysis and *in silico* Reconstruction of organism – specific MEtabolic Networks. *Genet Mol Res* 2010, 9, 1660–1672.

Schrempf, H.; Dyson, P. Streptomycetes. *Microb Biotechnol* 2011, 4, 138–140.

Schulz, M.; Uhlendorf, J.; Klipp, E.; Liebermeister, W. SBMLmerge, a system for combining biochemical network models. *Genome Inform* 2006, 17, 62–71.

Senger, R. S.; Papoutsakis, E. T. Genome-scale model for *Clostridium acetobutylicum*: Part I. Metabolic network resolution and analysis. *Biotechnol Bioeng* 2008a, 101, 1036–1052.

Senger, R. S.; Papoutsakis, E. T. Genome-scale model for *Clostridium acetobutylicum*: Part II. Development of specific proton flux states and numerically determined subsystems. *Biotechnol Bioeng* 2008b, 101, 1053–1071.

Seo, J. S.; Chong, H.; Park, H. S.; Yoon, K. O.; Jung, C.; Kim, J. J.; Hong, J. H.; Kim, H.; Kim, J. H.; Kil, J. I.; Park, C. J.; Oh, H. M.; Lee, J. S.; Jin, S. J.; Um, H. W.; Lee, H. J.; Oh, S. J.; Kim, J. Y.; Kang, H. L.; Lee, S. Y.; Lee, K. J.; Kang, H. S. The genome sequence of the ethanologenic bacterium *Zymomonas mobilis* ZM4. *Nat Biotechnol* 2005, 23, 63–68.

Shalon, D.; Smith, S. J.; Brown, P. O. A DNA microarray system for analyzing complex DNA samples using two-color fluorescent probe hybridization. *Genome Res* 1996, 6, 639–645.

Sharma, V. K.; Kumar, N.; Prakash, T.; Taylor, T. D. MetaBioME: a database to explore commercially useful enzymes in metagenomic datasets. *Nucleic Acids Res* 2009, 38, D468–472.

Siddiqui, M. S.; Thodey, K.; Trenchard, I.; Smolke, C. D. Advancing secondary metabolite biosynthesis in yeast with synthetic biology tools. *FEMS Yeast Res* 2012, 12, 144–170.

Siezen, R. J. Wine genomics. *Microb Biotechnol* 2008, 1, 97–103.

Siezen, R. J.; Bachmann, H. Genomics of dairy fermentations. *Microb Biotechnol* 2008, 1, 435–442.

Siezen, R. J.; Wilson, G. Genomics of deep-sea and subseafloor microbes. *Microb Biotechnol* 2009, 2, 157–163.

Siezen, R. J.; Wilson, G. Probiotics genomics. *Microb Biotechnol* 2011, 3, 1–9.

Sillers, R.; Al-Hinai, M. A.; Papoutsakis, E. T. Aldehyde-alcohol dehydrogenase and/or thiolase overexpression coupled with CoA transferase downregulation lead to higher alcohol titers and selectivity in *Clostridium acetobutylicum* fermentations. *Biotechnol Bioeng* 2009, 102, 38–49.

Sindelar, G.; Wendisch, V. F. Improving lysine production by *Corynebacterium glutamicum* through DNA microarray-based identification of novel target genes. *Appl Microbiol Biotechnol* 2007, 76, 677–689.

Singh, J.; Behal, A.; Singla, N.; Joshi, A.; Birbian, N.; Singh, S.; Bali, V.; Batra, N. Metagenomics: concept, methodology, ecological inference and recent advances. *Biotechnol J* 2009, 4, 480–494.

Smid, E.J.; Molenaar, D.; Hugenholtz, J.; de Vos, W. M.; Teusink, B. Functional ingredient production: application of global metabolic models. *Curr Opin Biotechnol* 2005, 16, 190–197.

Snoep, J. L.; Olivier, B. G. Java Web Simulation (JWS); a web based database of kinetic models. *Mol Biol Rep* 2002, 29, 259–263.

Sohn, S. B.; Graf, A. B.; Kim, T. Y.; Gasser, B.; Maurer, M.; Ferrer, P.; Mattanovich, D.; Lee, S. Y. Genome-scale metabolic model of methylotrophic yeast *Pichia pastoris* and its use for *in silico* analysis of heterologous protein production. *Biotechnol J* 2010, 5, 705–715.

Srere, P. A. Complexes of sequential metabolic enzymes. *Annu Rev Biochem* 1987, 56, 89–124.

Stemmer, W. P.; Crameri, A.; Ha, K. D.; Brennan, T. M.; Heyneker, H. L. Single-step assembly of a gene and entire plasmid from large numbers of oligodeoxyribonucleotides. *Gene* 1995, 164, 49–53.

Stolyar, S.; Van Dien, S.; Hillesland, K. L.; Pinel, N.; Lie, T. J.; Leigh, J. A.; Stahl, D. A. Metabolic modeling of a mutualistic microbial community. *Mol Syst Biol* 2007, 3, 92.

Stynen, B.; Tournu, H.; Tavernier, J.; Van Dijck, P. Diversity in genetic *in vivo* methods for protein-protein interaction studies: from the yeast two-hybrid system to the mammalian split-luciferase system. *Microbiol Mol Biol Rev* 2012, 76, 331–382.

Sun, J.; Sayyar, B.; Butler, J. E.; Pharkya, P.; Fahland, T. R.; Famili, I.; Schiling, C. H.; Lovley, D. R.; Mahadevan, R. Genome-scale constraint-based modeling of *Geobacter metallireducens*. *BMC Syst Biol* 2009, 3, 15.

Swartz, J. *Escherichia coli* recombinant DNA technology. In *Escherichia coli and Salmonella: Cellular and Molecular Biology*, Neidhardt, F. C., Ed.; American Society of Microbiology: Washington DC, 1996; 1693–1771.

Tarassov, K.; Messier, V.; Landry, C. R.; Radinovic, S.; Serna Molina, M. M.; Shames, I.; Malitskaya, Y.; Vogel, J.; Bussey, H.; Michnick, S. W. An *in vivo* map of the yeast protein Interactome. *Science* 2008, 320, 1465–1470.

Taylor, R. C.; Shah, A.; Treatman, C.; Blevins, M. SEBINI: Software Environment for BIological Network Inference. *Bioinformatics* 2006, 22, 2706–2708.

Teixeira, M. C.; Monteiro, P.; Jain, P.; Tenreiro, S.; Fernandes, A. R.; Mira, N. P.; Alenquer, M.; Freitas, A. T.; Oliveira, A. L.; Sa-Correia, I. The YEASTRACT database: a tool for the analysis of transcription regulatory associations in *Saccharomyces cerevisiae*. *Nucleic Acids Res* 2006, 34, D446–451.

Teusink, B.; Passarge, J.; Reijenga, C. A.; Esgalhado, E.; van der Weijden. C. C.; Schepper, M.; Walsh, M. C.; Bakker, B. M.; van Dam, K.; Westerhoff, H. V.; Snoep, J. L. Can yeast glycolysis be understood in terms of *in vitro* kinetics of the constituent enzymes? Testing biochemistry. *Eur J Biochem* 2000, 267, 5313–5329.

Teusink, B.; van Enckevort, F. H.; Francke, C.; Wiersma, A.; Wegkamp, A.; Smid, E. J.; Siezen, R. J. *In silico* reconstruction of the metabolic pathways of *Lactobacillus plantarum*: comparing predictions of nutrient requirements with those from growth experiments. *Appl Environ Microbiol* 2005, 71, 7253–7262.

Thiele, I.; Palsson, B. O. A protocol for generating a high-quality genome-scale metabolic reconstruction. *Nat Protoc* 2010, 5, 93–121.

Thim, L.; Hansen, M. T.; Sorensen, A. R. Secretion of human insulin by a transformed yeast cell. *FEBS Lett* 1987, 212, 307–312.

Thomas, T.; Gilbert, J.; Meyer, F. Metagenomics – a guide from sampling to data analysis. *Microb Inform Exp* 2012, 2, 3.

Tomita, M.; Hashimoto, K.; Takahashim, K.; Shimizu, T. S.; Matsuzaki, Y.; Miyoshi, F.; Saito, K.; Tanida, S.; Yugi, K.; Venter, J. C.; Hutchison, C. A. III. E-CELL: software environment for whole-cell simulation. *Bioinformatics* 1999, 15, 72–84.

Torsvik, V.; Ovreas, L.; Thingstad, T. F. Prokaryotic diversity – magnitude, dynamics, and controlling factors. *Science* 2002, 296, 1064–1066.

Tsoka, S.; Simon, D.; Ouzounis, C. A. Automated metabolic reconstruction for *Methanococcus jannaschii. Archaea* 2004, 1, 223–229.

Tummala, S. B.; Junne, S. G.; Papoutsakis, E. T. Antisense RNA downregulation of coenzyme A transferase combined with alcohol-aldehyde dehydrogenase overexpression leads to predominantly alcohologenic *Clostridium acetobutylicum* fermentations. *J Bacteriol* 2003a, 185, 3644–3653.

Tummala, S. B.; Junne, S. G.; Paredes, C. J.; Papoutsakis, E. T. Transcriptional analysis of product-concentration driven changes in cellular programs of recombinant *Clostridium acetobutylicum* strains. *Biotechnol Bioeng* 2003b, 84, 842–854.

Tummala, S. B.; Welker, N. E.; Papoutsakis, E. T. Design of antisense RNA constructs for downregulation of the acetone formation pathway of *Clostridium acetobutylicum. J Bacteriol* 2003c, 185, 1923–1934.

Uetz, P.; Giot, L.; Cagney, G.; Mansfield, T. A.; Judson, R. S.; Knight, J. R.; Lockshon, D.; Narayan, V.; Srinivasan, M.; Pochart, P.; Qureshi-Emili, A.; Li, Y.; Godwin, B.; Conover, D.; Kalbfleisch, T.; Vijayadamodar, G.; Yang, M.; Johnston, M.; Fields, S.; Rothberg, J, M. A comprehensive analysis of protein-protein interactions in *Saccharomyces cerevisiae. Nature* 2000, 403, 623–627.

van de Guchte, M.; Penaud, S.; Grimaldi, C.; Barbe, V.; Bryson, K.; Nicolas, P.; Robert, C.; Oztas S.; Mangenot, S.; Couloux, A.; Loux, V.; Dervyn, R.; Bossy, R.; Bolotin, A.; Batto, J. M.; Walunas, T.; Gibrat, J. F.; Bessieres, P.; Weissenbach, J.; Ehrlich, S. D.; Maguin, E. The complete genome sequence of *Lactobacillus bulgaricus* reveals extensive and ongoing reductive evolution. *Proc Natl Acad Sci USA* 2006, 103, 9274–9279.

Venter, J. C.; Remington, K.; Heidelberg, J. F.; Halpern, A. L.; Rusch, D.; Eisen, J.A.; Wu D.; Paulsen I.; Nelson, K. E.; Nelson, W.; Fouts, D. E.; Levy, S.; Knap, A. H.; Lomas, M. W.; Nealson, K.; White, O.; Peterson, J.; Hoffman, J.; Parsons, R.; Baden-Tillson, H.; Pfannkoch, C.; Rogers, Y. H.; Smith, H. O. Environmental genome shotgun sequencing of the Sargasso Sea. *Science* 2004, 304, 66–74.

Viswanathan, G. A.; Seto, J.; Patil, S.; Nudelman, G.; Sealfon, S. C. Getting started in biological pathway construction and analysis. *PLoS Comput Biol* 2008, 4, e16.

Vongsangnak, W.; Olsen, P.; Hansen, K.; Krogsgaard, S.; Nielsen, J. Improved annotation through genome-scale metabolic modeling of *Aspergillus oryzae. BMC Genomics* 2008, 9, 245.

Waegeman, H.; Soetaert, W. Increasing recombinant protein production in *Escherichia coli* through metabolic and genetic engineering. *J Ind Microbiol Biotechnol* 2011, 38, 1891–1910.

Wang, Q.; Chen, X.; Yang, Y.; Zhao, X. Genome-scale *in silico* aided metabolic analysis and flux comparisons of *Escherichia coli* to improve succinate production. *Appl Microbiol Biotechnol* 2006, 73, 887–894.

Wendisch, V. F.; Bott, M.; Kalinowski, J.; Oldiges, M.; Wiechert, W. Emerging *Corynebacterium glutamicum* systems biology. *J Biotechnol* 2006, 124, 74–92.

Werner, E.; Heilier, J. F.; Ducruix, C.; Ezan, E.; Junot, C.; Tabet, J. C. Mass spectrometry for the identification of the discriminating signals from metabolomics: current status and future trends. *J Chromatogr B Analyt Technol Biomed Life Sci* 2008, 871, 143–163.

Westerhoff, H. V.; Palsson, B. O. The evolution of molecular biology into systems biology. *Nat Biotechnol* 2004, 22, 1249–1252.

Westfall, P. J.; Pitera, D. J.; Lenihan, J. R.; Eng, D.; Woolard, F. X.; Regentin, R.; Horning, T.; Tsuruta, H.; Melis, D. J.; Owens, A.; Fickes, S.; Diola, D.; Benjamin, K. R.; Keasling, J. D.; Leavell, M. D.; McPhee, D. J.; Renninger, N. S.; Newman, J. D.; Paddon, C. J. Production of amorphadiene in yeast, and its conversion to dihydroartemisinic acid, precursor to the antimalarial agent artemisinin. *Proc Natl Acad Sci USA* 2012, 109, E111–118.

White, O.; Eisen, J. A.; Heidelberg, J. F.; Hickey, E. K.; Peterson, J. D.; Dodson, R. J.; Haft, D. H.; Gwinn, M. L.; Nelson, W. C.; Richardson, D. L.; Moffat, K. S.; Qin, H.; Jiang, L.; Pamphile, W.; Crosby, M.; Shen, M.; Vamathevan, J. J.; Lam, P.; McDonald, L.; Utterback, T.; Zalewski, C.; Makarova, K. S.; Aravind, L.; Daly, M, J.; Minton, K. W.; Fleischmann, R. D.; Ketchum, K. A.; Nelson, K. E.; Salzberg, S.; Smith, HO.; Venter, J. C.; Fraser, C. M. Genome sequence of the radioresistant bacterium *Deinococcus radiodurans* R1. *Science* 1999, 286, 1571–1577.

Wirth, R.; Kovacs, E.; Maroti, G.; Bagi, Z.; Rakhely, G.; Kovacs, K. L. Characterization of a biogas-producing microbial community by short-read next generation DNA sequencing. *Biotechnol Biofuels* 2012, 5, 41.

Wishart, D. S. Quantitative metabolomics using NMR. *Trends Anal Chem* 2008, 27, 228–237.

Wittmann, C. Analysis and engineering of metabolic pathway fluxes in *Corynebacterium glutamicum*. *Adv Biochem Eng Biotechnol* 2010, 120, 21–49.

Wu, J.; Smith, L. T.; Plass, C.; Huang, T. H. ChIP-chip comes of age for genome-wide functional analysis. *Cancer Res* 2006, 66, 6899–6902.

Yoon, S. H.; Han, M. J.; Lee, S. Y.; Jeong, K. J.; Yoo, J. S. Combined transcriptome and proteome analysis of *Escherichia coli* during high cell density culture. *Biotechnol Bioeng* 2003, 81, 753–767.

Zelle, R. M.; de Hulster, E.; van Winden, W. A.; de Waard, P.; Dijkema, C.; Winkler, A. A.; Geertman, J. M.; van Dijken, J. P.; Pronk, J. T.; van Maris, A. J. Malic acid production by *Saccharomyces cerevisiae*: engineering of pyruvate carboxylation, oxaloacetate reduction, and malate export. *Appl Environ Microbiol* 2008, 74, 2766–2777.

Zhang, W.; Li, F.; Nie, L. Integrating multiple 'omics' analysis for microbial biology: application and methodologies. *Microbiology* 2010, 156, 287–301.

Zhang, Y.; Thiele, I.; Weekes, D.; Li, Z.; Jaroszewski, L.; Ginalski, K.; Deacon, A. M.; Wooley, J.; Lesley, S. A.; Wilson, I. A.; Palsson, B.; Osterman, A.; Godzik, A. Three-dimensional structural view of the central metabolic network of *Thermotoga maritima*. *Science* 2009, 325, 1544–1549.

Zhang, Z.; Townsend, J. P. The filamentous fungal gene expression database (FFGED). *Fungal Genet Biol* 2010, 47, 199–204.

Zhou, Z.; Gu, J.; Li, Y. Q.; Wang, Y. Genome plasticity and systems evolution in *Streptomyces*. *BMC Bioinformatics* 2012, 13, S8.

Zuzuarregui, A.; Monteoliva, L.; Gil, C.; del Olmo, M. Transcriptomic and proteomic approach for understanding the molecular basis of adaptation of *Saccharomyces cerevisiae* to wine fermentation. *Appl Environ Microbiol* 2006, 72, 836–847.

CHAPTER 16

BIOINFORMATICS AND BIOTECHNOLOGY IN HUMAN GENETIC RESEARCH: A CURRENT SCENARIO

SATHIYA MARAN, THIRUMULU PONNURAJ KANNAN, and
TEGUH HARYO SASONGKO

CONTENTS

16.1   INTRODUCTION

The hereditary information of all living organisms is carried by deoxyribo-
nucleic acid (DNA) molecules, which are held by complementary chains of
nucleotides into a double-helical structure. The ability of molecular biologists
to rapidly sequence these nucleotides has created many new areas of biomedical researches and initiated the sequencing techniques and technologies. In the
current era, DNA sequencing has become a standard laboratory technique and
has resulted in the recent sequencing of the complete genomes for numerous
organisms. Most notable was the Human Genome Project that mapped 30000
functional human genes and sequenced approximately 3 billion DNA (van
Ommen, 2002). These breakthrough-sequencing results have paved way for
the availability of tremendous amounts of data. Moreover, the increasingly
expanding data's necessitated an exhaustive, reliable and reproducible application of skills for the storage and interpretation of datasets. These skills when
applied to the domain of functional genomics, is described as *bioinformatics*.
The Human Genome Project has made available the genetic data such as DNA
sequences, physical maps, genetic maps, gene polymorphisms, protein structures, gene expression profiles and protein interaction effects (Kohane and
Butte, 2002). The collections of these huge datasets had further produced an
urgent need for systematic quantitative analysis.

16.2   BIOINFORMATICS AND STATISTICAL GENETICS

Bioinformatics being a cross-disciplinary field was first initiated by Margaret
O. Dayhoff, Walter M. Fitch and Russell F. Doolittle in the early 1960. The
National Center for Biotechnology Information (NCBI) defined Bioinformatics as the field of science in which biology, computer science and information technologies are merged to form a 'single discipline' (NCBI 2004). This
'single discipline' encompasses biochemistry, molecular biology, molecular
evolution, thermodynamics, biophysics, molecular engineering and statistical
mechanics. Being a thriving field that is currently in the forefront of science
and technology, bioinformatics are widely being used in the maintenance of
databases that stores biological information such as nucleotide sequences,
amino acid sequences, protein domains and protein structures. Bioinformatics
are also extensively used in development of complex interfaces for data access and data analysis (Lindblom and Robinson, 2011).

Genetics is the branch of biology that deals with heredity and attempts
to explain the similarities and differences that exist between parents and

offspring. Meanwhile, statistical genetics, which is also known as genetic epidemiology is the study of role of genetic factors in addressing issues related to health and disease in the aspects of family and populations (Mortan, 1978). In recent years, the scope of genetic epidemiology has expanded to include common diseases for, which many genes are involved in a gene and each of the genes contributes to the disease (polygenic, multifactorial or multigenic disorders). This hypothesis had developed rapidly in the first decade of the twenty-first century after the completion of the Human Genome Project.

16.3   GENETICS IN BIOTECHNOLOGY

The Guide to Biotechnology (2007) reported that biotechnology has flourished since the prehistoric times. According to this guide, humans learnt to use biotechnology when they could plant their own crops and breed their own animals. The discovery of using fruit juice to ferment into wine, converting milk into cheese or yogurt and fermenting malts and hops into beer had initiated the study of biotechnology (Peters, 1993). Biotechnology was described by the United Nation Convention (UNC) as any technological application that uses biological systems, living organisms, or derivatives thereof, to make or modify products or processes for specific use. Currently, biotechnology is widely being applied in agriculture, food science, and medicine. Pharmaceutical companies have incorporated biotechnology into the industry in many ways. The traditional pharmaceutical companies had been focussing on small scale drugs in treating diseases and illness (Strickland, 2007). The pioneering pharmaceutical companies were then developed into large scale Biopharmaceutical companies targeting large biological molecules such as proteins. This advancement had further led the scientist in understanding the underlying disease mechanisms and pathways. Modern biotechnology is often associated with the use of genetically altered microorganisms such as *E. coli* or yeast for production of insulin and antibiotics. Genetically altered mammalian cells, such as Chinese Hamster Ovary (CHO) cells, are also widely used to manufacture pharmaceuticals. Another promising new biotechnology application is the development of plant-made pharmaceuticals (Strickland, 2007). Besides the pharmaceutical filed, biotechnology is also making trademarks in medical therapies in order to treat diseases like diabetes, hepatitis B, hepatitis C, cancers, arthritis, hemophilia, bone fractures, multiple sclerosis, cardiovascular as well as molecular diagnostic devices (Strickland, 2007).

16.4 BIOTECHNOLOGY TOOLS IN GENETIC RESEARCH

In the research field, biotechnology is being used to gain insight into the precise details of cell processes, the mechanics of cell division, specialization of undifferentiated cells, and coordinating cell activities and cell response to the environmental changes. Tools of biotechnology have become important research gear in many branches of science other than cell and molecular biology, such as chemistry, engineering, materials science, ecology, evolution and computer science. The biotech-driven discoveries in these fields help the biotech industry and others discover and develop products, as well as help industries improve their performance in areas such as environmental stewardship and workplace safety.

16.4.1 STEM CELL TECHNOLOGY

Stem cells are biological cells found in all multicellular organisms that divide to differentiate into diverse specialized cell types (Fig. 16.1). In mammals, there are two broad types of stem cells: embryonic stem cells, which are isolated from the inner cell mass of blastocysts, and adult stem cells, which are found in various tissues. Embryonic stem cells (ESCs) are pluripotent, meaning they are able to differentiate into all derivatives of the three primary germ layers: ectoderm, endoderm and mesoderm. Meanwhile, adult stem cells, which are also known as somatic stem cells are undifferentiated cells, found throughout the body after development, that multiply by cell division to replenish dying cells and regenerate damaged tissues.

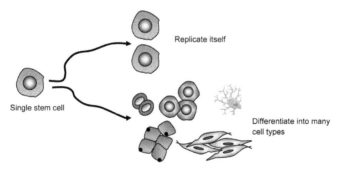

FIGURE 16.1   Differentiation of stem cell into diverse specialized cell types.

Cell culture was defined as the growth and maintenance of cells in a controlled environment outside of an organism and culturing of stem cells is the first step in establishing a stem cell line (The National Academy Press, 2001). Once a stable stem cell line has been established, the process of causing the stem cells to differentiate into specialized cell types is then initiated. Stem cells offer opportunities for scientific advances in addressing many of the biology's most fundamental questions. The stem cell research aids in clarifying the role of genes in human development and how genetic mutations affects the normal processes (The National Academy Press, 2001). Stems cells also can be used to study how infectious agents invade and attack human cells, to investigate the genetic and environmental factors that are involved in cancer and other diseases and to interpret the process of aging (The National Academy Press, 2001). An added advantage of embryonic stem cell is its capability to divide for long periods of time and produce a variety of cell types, whereby this could provide a valuable source of human cells for testing of drugs and measuring the effects of toxins on normal tissues without human volunteers. Nuclear transfer technology is used to produce stem cells that can be applied in testing drugs for disorders that are of genetic origin. For example, in studying the progression of Alzheimer's and Parkinson's diseases, the cells of an Alzheimer's patient is used to create stem cell lines with nuclear transfer, and the development of the disease in a culture dish and test drugs that regenerate lost nerve cells with no danger to the patient can be traced.

16.4.2 MOLECULAR CLONING

The National Academy Press (2001) reported molecular cloning to be the primary "driving force" of the biotechnology revolution and has made remarkable discoveries routine. Molecular cloning (Fig. 16.2) involves insertion of a new piece of DNA into a cell in such a way that it can be maintained, replicated and studied. Molecular cloning refers to the plasmid containing the new DNA and the cells or organisms, such as bacteria, containing the new piece of DNA. Cell division "amplifies" the amount of available DNA. Molecular cloning provides researchers with an unlimited amount of a specific piece of genetic material to manipulate and study. Up to date, molecular cloning has enabled the identifying, localization and characterization of genes in order to create genetic maps and sequences of complex diseases (Mullis, 1990). One of the primary applications of molecular cloning is to identify the protein product of a particular gene and to associate that protein with the appearance of a certain trait. While this is useful for answering certain questions, genes do not act in isolation of one another. To truly understand gene function, we need to monitor the activity of many genes simultaneously, which is possible by the use of microarray technology.

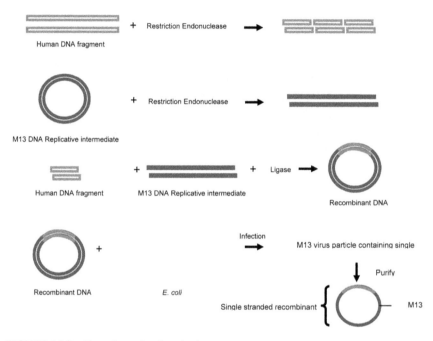

FIGURE 16.2   Steps in molecular cloning.

16.4.3   MICROARRAY TECHNOLOGY

DNA microarray is a relatively new multiplex technology used in the field of molecular biology and medicine (Ganguly *et al.*, 2009). It consists of an arrayed series of thousands of microscopic spots of DNA oligonucleotides. This can be a short section of a gene or DNA elements that are used as probes to hybridize a cDNA or cRNA sample (called target) under high-stringency conditions. The principle behind microarray technique involves matching unknown DNA and known DNA based on base pairing (A–T and G–C) or hybridization to identify the unknown (Ganguly *et al.*, 2009). Microarray is widely being applied in many research fields especially in order to detect mutations in disease-related genes, monitor gene expression, diagnose infectious diseases and identify the best antibiotic treatment, identify genes important to crop productivity and improves screening for microbes used in environmental clean-up (Ganguly *et al.*, 2009).

DNA microarray technology helps in understanding the molecular basis and the underlying mechanisms of gene functions and expression levels. Microarray based techniques are making landmarks in disease diagnosing, drug discovery and toxicology research. In terms of drug discovery, DNA microarray technology plays important role in elucidating the complex mechanisms of diseases such as heart diseases, mental illness, and infectious disease and cancer studies. Recently, the advancements in DNA microarray technology have enabled the classification of types of cancer to the organs in which the tumors develop. With the rapid evolution of microarray technology, it will be made possible for the researchers to classify the types of cancer on the basis of the patterns of gene activity in the tumor cells. This will largely benefit the pharmaceutical community to develop cancer targeted drugs and treatments.

Microarray technology has extensive application in pharmacogenomics. It is the study of correlations between therapeutic responses to drugs and the genetic profiles of the patients. Comparative analysis of the genes from a diseased and a normal cell will help the identification of the biochemical constitution of the proteins synthesized by the diseased genes. The researchers can use this information to synthesize drugs that combat the proteins and reduce their effect. Toxicogenomics establishes the correlation between responses to toxicants and the changes in the genetic profiles of the cells exposed. Microarray technology provides a robust platform in determining the impact of toxins on the cells.

In the recent years, microarray technology has been extensively used by the scientific community and consequently led to generation of data related to gene expression, SNPs genotyping and mutation analysis. For easing the accessibility to this data, the NCBI has formulated the Gene Expression Omnibus (GEO) (www.ncbi.nlm.nih.gov/geo/). The microarray experiments can even be narrowly categorized into microarray expression analysis, microarray for SNP analysis, and comparative genomic hybridization. Gene expression profiling or microarray analysis has enabled the measurement of thousands of genes in a single RNA sample. In this method, the cDNA derived from the mRNA of known genes is immobilized and the sample has genes from both the normal and the diseased tissues. Fluorescent spots with more intensity will be observed for diseased tissue's gene if the gene is over expressed in the diseased condition. This expression pattern is then compared to the expression pattern of a gene responsible for a disease. A single base difference between two sequences is known as Single Nucleotide Polymorphism (SNP). Microarray based approach has been used in SNP genotyping in identifying ones susceptibility towards diseases. Comparative genomic hybridization is used for the identification in the increase or decrease of the important chromosomal fragments harboring genes involved in a disease.

16.4.4   QUANTITATIVE REAL-TIME POLYMERASE CHAIN REACTION

Quantitative Real-Time Polymerase Chain Reaction (qPCR) is an extension of conventional Polymerase Chain Reaction (PCR). While in PCR, results are displayed as present or absent with very limited accuracy of quantification based on amplification intensity, qPCR allows real-time monitoring of amplification that occurs throughout PCR reaction and finally provide quantification of the amplicons produced, be it absolute or relative quantification. In real-time PCR, the *exponential phase of PCR* is monitored as it occurs, using fluorescently labeled molecules. During the exponential phase, the amount of *PCR amplicon* present in the reaction tube is directly proportional to the amount of starting material specific to the PCR primer pair (or target sequence). Thus, the amount of emitted fluorescence is directly proportional to the amount of amplicon, which, in turn, is proportional to the starting amount of target sequence. In such a way, quantitative real-time PCR is beneficial in, for example, measuring copy number of a target sequence and viral load (Coleman, 2004; Serre, 2006). Quantitative applications of real-time PCR include gene copy number analysis or dosage analysis of any locus, gene expression analysis and quantitative DNA methylation analysis. Melting analysis of this application could, however, be useful to also detect the presence of nucleotide variations and allelic discrimination, though with low resolution.

16.4.5   HIGH RESOLUTION MELTING ANALYSIS

High Resolution Melting Analysis (HRM or HRMA) is a relatively recent technology for fast, high-throughput postPCR analysis of changes in nucleic acid sequences. It enables researchers to rapidly screen and categorize nucleic acid changes, such as single nucleotide polymorphisms (SNPs), disease-causing genetic mutations and genetic variation in a population (e.g., viral diversity) prior to sequencing. HRMA involves amplification of the region of interest, using standard PCR techniques, in the presence of a specialized double-stranded DNA (dsDNA) binding dye. This specialized dye is highly fluorescent when bound to dsDNA and poorly fluorescent in the unbound state. This change allows the user to monitor the DNA amplification during PCR (as in quantitative PCR). After completion of the PCR step, the amplified target is gradually denatured by increasing the temperature in small increments, in order to produce a characteristic melting profile; this is termed melting analysis. The amplified target denatures gradually, releasing the dye, which results in a drop in fluorescence. When set up correctly, HRM is sensitive enough to allow the detection of a single base change between otherwise identical nucleotide sequences.

In High Resolution Melting concept the denaturing temperature increments are typically set to be extremely tight, between 0.008–0.2°C. This is to allow detailed analysis of the melting behavior in between temperature, so that slight differences in melting temperature could be discriminated among different sequence species. The extremely tight temperature increments is in contrast to the currently known concept as Low Resolution Melting commonly applied in quantitative PCR technology where the increments are set only at 0.5°C (Kapa Biosystems; www.kapabiosystems.com). In principle, HRMA allows identification of variations in nucleic acid sequences either those predetermined or unknown. This extends to wide application of HRMA which includes SNP genotyping, mutation discovery, heterozygosity screening, DNA fingerprinting, haplotype blocks characterization, DNA methylation analysis, DNA mapping, species identification, viral/bacterial population diversity investigation, HLA compatibility typing and association study (case/control). Limitations, however, of this technology is that it mainly centers on the identification of heterozygotes. It is true that heterozygotes could be clearly discriminated from homozygotes. However, different heterozygotes may produce melting curves so similar to each other. Besides, it has been shown that some homozygotes require mixing for identification (Wittwer, 2009).

16.4.6   MULTIPLE LIGATION-DEPENDENT PROBE AMPLIFICATION

Multiple Ligation-dependent Probe Amplification (MLPA) is basically a hybrid technique that combines an *oligonucleotide ligation assay (OLA)* and *quantitative-fluorescent PCR (QF-PCR)*. It is ideally suited to the measurement of exon copy number in multiexon genes, but it could also be used for the quantitative assessment of any locus. Genomic DNA is denatured and a mixture of oligonucleotide probes is hybridized to the DNA. Each MLPA probe consists of two oligonucleotides that take part in a one-cycle OLA reaction. The two oligonucleotides have generic sequences attached to the 5′ end of the upstream primer and to the 3′ end of the downstream primer. Thus, after the initial OLA step, a PCR is then carried out using primers complementary to the generic sequences on the ends of OLA primers. In this way, multiple different target sequences can be investigated, which are all amplified using the same PCR primers. As one of the generic primers is fluorescently labeled, then the fluorescence resulting from a particular amplicon (MLPA probe) is proportional to the amount of starting target DNA, and the copy number can be determined. The lengths of the various probes are engineered in such a way that individual probes differ by a few base pairs and can be easily separated on a fluorescent DNA analyzer (sequencing instrument) (Frayling, 2004).

What has differed MLPA from conventional multiplex PCR reaction is that amplification occur on the probes that are hybridized to the target sequence, not the target sequences themselves, thus allowing equal efficiency across multiple amplification reaction. Typical amplification product of MLPA reactions, especially those resulted from the MLPA SALSA kit as provided by MRC-Holland, which is currently the sole supplier of this technology, range between 130 and 480 nt in length and can be analyzed by capillary electrophoresis. Comparing the peak pattern obtained to that of reference samples indicates, which sequences show aberrant copy numbers. Before data analysis, there are four major steps involved in MLPA reactions: 1) DNA denaturation and hybridization of MLPA probes where DNA strands separate and probes attached to their target sequences; 2) ligation reaction that connect the adjacent ends of both probes; 3) PCR reaction that amplifies the hybridized probes using uniform primers; and 4) separation of amplification products by electrophoresis whereby amplification products are separated according to their sizes.

During data analysis, normalization of MLPA data is essential because variations in experimental conditions may lead to quantitative differences. MLPA data can be normalized only among samples, which were extracted using the same method, run within the same experiment and run within the same probemix lot. There are two steps involved in data normalization; 1) intrasample normalization, in, which the peak area generated by each probe is expressed as part of the whole, and 2) intersample comparison, whereby results of the first step of each peak area were compared with reference samples, though in some cases reference samples can be omitted. However, blank control containing no DNA is always necessary (MRC-Holland; www.mlpa.com).

16.4.7   MINI-SEQUENCING FOR MULTIPLEX GENOTYPING

Most of the known human genetic diseases, especially those of single-gene nature, are caused by point mutations, that is, mutations implicated a single nucleotide change. Moreover, a growing number of single nucleotide poly-morphisms (SNPs) have also been shown to predispose to common, mul-tifactorial diseases. Most conventional identification of single nucleotide variations, be it disease-causing mutations or disease-predisposing SNPs, are single-plex. This hampers analysis of many numbers of samples in a reason-able amount of time, as are the cases for molecular diagnosis of certain ge-netic diseases or SNPs association studies. Mini-sequencing, which is based on fluorescent labeling of single-nucleotide primer extension, allows multi-plex analysis of nucleotide variations. In the mini-sequencing primer exten-sion assays, the DNA synthesis reaction catalyzed by the DNA polymerases is used to distinguish between sequence variants. The concept of the assays is to anneal a detection primer to the nucleic acid sequence immediately 3' of the nucleotide position to be analyzed and to extend this primer with a single labeled nucleoside triphosphate that is complementary to the nucleotide to be detected using a DNA polymerase.

The reaction depends on a primer extension starting from the purified am-plified target. A specific mini-sequencing primer, which is exactly one base short of the polymorphic site, is used for each nucleotide to be studied. Primer extension reactions are carried out using fluorescent-labeled dideoxynucleo-tides (F ddNTP). Each of the four different ddNTPs are labeled differently, thus different ddNTP shall show different signal. In this way, the incorporated nucleotide exactly one base position after the primer could be identified. The use of ddNTP is to ensure that the reaction discontinued exactly at one nucleo-tide extension. Since purification steps are not necessary, the product obtained is immediately separated electrophoretically; the color of the peaks obtained makes it possible to identify the SNP or mutation. A major advantage of the mini-sequencing reaction principle over hybridization with allele-specific oli-gonucleotide (ASO) probes is that the distinction between the sequence vari-ants is based on the high accuracy of the nucleotide incorporation reaction catalyzed by a DNA polymerase, instead of on differences in thermal stability between mismatched and perfectly matched hybrids formed with the ASO probes. Consequently, the mini-sequencing assays allow excellent discrimi-nation between the homozygous and heterozygous genotypes and the assays are robust and insensitive to small variations in the reaction conditions. More-over, the same reaction conditions can be employed for detecting any variable nucleotide irrespectively of the nucleotide sequence flanking the variable site. These features are of central importance when designing multiplex assays for the simultaneous detection of many SNPs per sample. Because of the high specificity of the primer extension reaction, it is a useful tool for accurate quantitative PCR-based analysis and for detection of sequences present as a minority of a sample (Syvanen, 1999).

Although most applications of this methodology requires predetermined known single nucleotide variations for identification, a study on families with multiple endocrine neoplasia type 2 shows that unknown variations in a targeted codon could be identified. This could be done by separately interrogating the 3 nucelotides within the target codon by different primers (Bugalho *et al.*, 2002).

16.5 CONCEPTION AND INCORPORATION OF BIOINFORMATICS INTO HUMAN GENETICS

The basic concept of Bioinformatics involves discovery, development and implementation of computational algorithm and software tools that aids in the understanding of biological process. Developing countries with agriculture as backbone such as India, the advancements in bioinformatics play a key role in increasing the nutrient levels, increasing gross capital and establishing disease resistant. Pharmaceutical sectors can make use of bioinformatics in reducing the time and cost involved in drug discovery process, custom designing of drugs and also personalized medicines (Guide to Biotechnology, 2007).

16.5.1 SEQUENCE DATABASES

The exponential growth in molecular sequence data due to availability of DNA sequencing, made ways to databases such as GenBank, EMBL (European Molecular Biology Laboratory nucleotide sequence database), DDBJ (DNA Data Bank of Japan), PIR (Protein Information Resource) and SWISS-PROT. Computational methods and algorithms were then constructed for data retrieval and analysis, similarity searches, structural predictions and functional predictions. GenBank (Genetic Sequence Databank) is a comprehensive public database of nucleotide sequences and supporting bibliographic and biological annotation. NCBI builds GenBank primarily from the submission of sequence data from authors and from the bulk submission of expressed sequence tag (EST), genome survey sequence (GSS) and other high-throughput data from sequencing centers. GenBank consist of accession numbers and gene names, phylogenetic classification and references to published literature (http://www.ncbi.nlm.nih.gov/genbank/). EMBL Nucleotide Sequence Database is a comprehensive database of DNA and RNA sequences collected from the scientific literature and patent applications and directly submitted from researchers and sequencing groups. Data collection is done in collaboration with GenBank (USA) and the DDBJ. The database currently contains nearly more than 2 million bases (http://www.ebi.ac.uk/embl/).

UniProt is a protein sequence database that provides the scientific community with a comprehensive, high-quality and freely accessible resource of protein sequence and functional information (http://www.uniprot.org/). PROSITE dictionary of sites and patterns in proteins was prepared by Amos Bairoch, University of Geneva. It consists of documentation entries describing protein domains, families and functional sites as well as associated patterns and protein profiles (http://prosite.expasy.org/). ExPASy-ENZYME is a repository of information relative to the nomenclature of enzymes. The 'ENZYME' data bank containing EC number, recommended name, alternative names, catalytic activity, cofactors, pointers to the SWISS-PROT entry(s) that correspond to the enzyme, pointers to disease(s) associated with a deficiency of the enzyme (http://enzyme.expasy.org/). Protein Data Bank (PDB) archive contains information's on experimentally determined structures of proteins, nucleic acids, and complex assemblies (http://www.rcsb.org/pdb/home/home.do). The Genome Database (GDB) is a public repository of data on human genes, clones, STSs, polymorphisms and maps. GDB entries are highly cross-linked to each other, to literature citations and to entries in other databases, including the sequence databases, OMIM, and the Mouse Genome Database (http://www.gdb.org).

Online Mendelian Inheritance in Man (OMIM) is a comprehensive, authoritative, and timely compendium of human genes and genetic phenotypes. OMIM contains information on all known mendelian disorders and over 12,000 genes. OMIM focuses on the relationship between phenotype and genotype. It is updated daily, and the entries contain copious links to other genetics resources (http://www.ncbi.nlm.nih.gov/omim/). The Protein Information Resource (PIR) is an integrated public resource of protein informatics that supports genomic and proteomic research and scientific discovery. PIR contains over 283,000 sequences covering the entire taxonomic and family classifications are used for sensitive identification, consistent annotation and detection of annotation errors. Protein sequence databases are classified as primary, secondary and composite depending upon the content protein sequences as 'raw' data. Secondary databases (like Prosite) contain the information derived from protein sequences (http://pir.georgetown.edu/). UCSC Genome Browser is an interactive website offering access to genome sequence data from a variety of vertebrate and invertebrate species and major model organisms, integrated with a large collection of aligned annotations. The browser is a graphical viewer optimized to support fast interactive performance and is an open-source, web-based tool suite built on top of a MySQL database for rapid visualization, examination, and querying of the data at many levels (http://genome.ucsc.edu/). The Mouse Genome Database (MGD) is an integrated data resource for mouse genetic, genomic, and biological information. MGD includes a variety of data, ranging from gene characterization and genomic structures, to orthologous relationships between mouse genes and those of other mammalian species, to maps (genetic, cytogenetic, physical), to descriptions of mutant phenotypes, to characteristics of inbred strains, to information about biological reagents such as clones and primers (http://www.nih.gov/science/models/mouse/resources/mgd.html). A *Caenorhabditis elegans* Database (ACeDB) containing data from the Caenorhabditis Genetics Center (funded by the NIH National Center for Research Resources), the *C. elegans* genome project (funded by the MRC and NIH), and the worm community, was originally developed for the *C. elegans* genome project and later on this tool was generalized to be much flexible and currently is used for different genomic databases ranging from bacteria, fungi, plants and human (http://www.acedb.org/introduction.shtml). MEDLINE is a NLM's premier bibliographic database covering the fields of medicine, nursing, dentistry, veterinary medicine, and the preclinical sciences. Journal articles are indexed for MEDLINE, and their citations are searchable, using NLM's controlled vocabulary, MeSH (Medical Subject Headings). MEDLINE contains all citations published in Index Medicus, and corresponds in part to the International Nursing Index and the Index to Dental Literature. Citations include the English abstract when published with the article (approximately 70% of the current file) (http://www.nlm.nih.gov/bsd/pmresources.html).

16.5.2   PATHWAY DATABASES

Kyoto Encyclopaedia of Genes and Genomes (KEGG) (http://www.kegg.jp/) was developed by the Bioinformatics Center of Kyoto University and the Human Genome Center, University of Tokyo. This database has been extended to give a detailed understanding on biology of genome sequences (Nagasaki *et al.*, 2009). KEGG also focuses and covers yeast, mouse and human metabolic pathways. BioCyc is a high-quality database that focuses on metabolic pathways and EcoCyc, MetaCyc and HumanCyc has been associated with this database (http://www.biocyc.org/). Ingenuity Pathways Knowledge Base (IPKB) was created by Ingenuity Systems Inc. The database consists of gene regulatory and signaling pathways (Nagasaki *et al.*, 2009). The IPKB allows access to a wide variety of information's including on integrated biology and chemical knowledge in a single platform. IPKB uses Ingenuity Pathway Analysis (IPA) for viewing and analyzing pathways (http://www.ingenuity.com/). Ingenuity Pathway Analysis (IPA) is software used for displaying the pathway data's from IPKB. IPA automatically generates pathways that are related for a set of gene when given a query (http://www.ingenuity.com/products/pathways_analysis.html). TRANSPATH is a gene regulatory and signaling pathway database created by BIOBASE (http://www.biobase-international.com/) for the transcription and proteomic analysis. This database consists of a total of more than 9800 molecules, 1800 genes and 11400 human reactions (http://www.biobase.de/pages/products/databases.html). Pathway Studio is the viewer for Adrianne Genomics' ResNet. The uniqueness of this database is that, it could add new molecules of interest and user's information into the pathway (http://www.ariadnegenomics.com/products/pathway-studio/).

16.5.3   DATA MINING

Data mining is a process of discovering the meaningful patterns and relationship within a large database through automated analysis (Roy, 2009). Data mining provides two basic information, which are descriptive information and predictive information. The descriptive data mining calculates values that represent the overall characteristics of the datasets. Meanwhile, data mining using known variables to predict the outcome based on pattern suggested by the attributes is known as 'predictive' (Roy, 2009). Data mining being a process of discovering new patterns and building models from a given datasets, involves many steps. These steps are such as data selections, data cleaning, data transformation, data mining and finally postprocessing and interpretation of data (Zaki and Sequeira, 2006).

16.5.3.1   DATA SELECTIONS, DATA CLEANING AND DATA TRANSFORMATION

Collection of a dataset that captures the possible situation that are highly associated with the problem being analyzed is important. The clean-up session, which involves removal of inconsistent data's, noise, outliers and missing values should also be carried out. Another important task of data mining is the normalization step that deals with effluence of datasets.

16.5.3.2   FEATURE SELECTIONS AND DIMENSION REDUCTION

Even after the data clean-up task as mentioned earlier undertaken, noise irrelevant to problem being analyzed might still be present. This noise might interfere with the data mining steps thus producing irrelevant associations of the problem being analyzed. Therefore, it is wise to perform dimension reduction in order to noise from the dataset.

16.5.3.3   MINING ALGORITHMS AND RESULTS INTERPRETATIONS

Data mining algorithms are mathematical and statistical algorithms that transform the 'cases' in the original data source into data mining methodology. An algorithm refers to a logical sequence of steps through, which a task can be performed (Roy, 2009). The different types of algorithms widely used in data mining are (i) decision tree algorithms used for creating repeating series of branches, and (ii) cluster algorithms for grouping data into clusters, association analysis and regression analysis. The challenges in data mining are to integrate data from various sources, define a robust algorithm with visualization front end, result interpretations and fine tuning the algorithms. The scope for improving the existing methods and incorporating a new method is still at great (Buehler and Rashidi, 2005).

16.5.4   GENOME DATABASE MINING

Genome Database mining is a process of identification of sequence elements such as genes, introns, exons, promoters, upstream regulatory elements, enhancers, silencers, repetitive DNA sequence such as telomeres, centromeres, retrotransposons, minisatellites, and microsatellites (Roy, 2009). The popular genome database mining tools are such as BLAST and FASTA.

Basic Local Alignment Search Tool (BLAST) is supported by the NCBI and is capable of searching major sequence databases such as SWISS-PROT and PDB (Buehler and Rashidi, 2005). BLAST programs uses substitution scoring matrix for the alignment scanning and extension phase (Buehler and Rashidi, 2005). Substitution scoring matrix is a scoring method used in alignment of one residue or nucleotide against another. These matrices are relatively known as Dayhoff, MDM and PAM (Buehler and Rashidi, 2005). BLAST programs are designed to enhance speed and maximize the sensitivity of distance sequence relationship, which allows the identification of closest sequence homology (Roy, 2009). BLASTp allows searching of protein query sequence against a protein database. This enables the identification of all possible sequence homologs for a given protein query sequence (http://blast.ncbi. nlm.nih.gov/Blast). BLASTx is to search a translated nucleotide sequence against a protein database. BLASTx is useful in identification of nucleotide sequencing errors by comparing the translated nucleotide query sequence to its potential protein homologs in a protein sequence database. BLASTn is used to search translated nucleotide query sequence against a nucleotide database. tBLASTn can be used to search translated nucleotide sequence in a given nucleotide database against a protein query sequence. This program is very useful in finding protein sequencing errors by comparing the protein query sequence to the potential translated nucleotide homologs in a given nucleotide database. tBLASTx allows the user to search the six frame translation of a nucleotide query sequence against the six frame translation of a nucleotide sequence entries in a given nucleotide database.

Fast-all Sequence Alignment Algorithm (FASTA) was designed to search for protein, oligonucleotide and oligopeptide sequence library for homologous sequences (Roy, 2009). FASTA focuses on a specific region in a pair of sequence that shares the high-density identity using a robust scoring method (Roy, 2009). There are many different programs of FASTA and all these programs calculate the 'local similarity score' (http://www.genome.jp/tools/fasta/). TFASTA which compares a protein sequence to a DNA sequence and the DNA sequence is translated into all six frames and the protein query sequence is compared to each of the six derived protein sequence. LFASTA compares two sequences to identify region of sequence similarity. This program can simultaneously report several sequence alignments and similarity scores. PLFASTA is much identical to LFASTA but this program is present in a dot-matrix like plot of similar regions rather than actual alignments. RDF2 uses Monte Carlo analysis to evaluate the statistical significance of a similarity score.

16.5.5 GENE MINING

Gene mining refers to a procedure of searching for a novel gene in a target genome by screening the databases of genome wide collections of ESTs from other genomes for homologies with EST or cDNA sequence retrieved from a target cell, tissue, organ or organism (Roy, 2009).

16.6   BIOINFORMATICS ALGORITHMS AND SCRIPTING LANGUAGES

In the current years of genomic revolution and the concomitant development of bioinformatics methodologies, the downside, which is lack of quality and irreproducibility has been questioned (Butte *et al.*, 2001). This is particularly true in microarray-based researches, which postulates broad and unsubstantial claims about gene functions and its underlying mechanisms. In order to surpass these drawbacks, the following statistical algorithms have been programmed.

16.6.1   EXPECTATION-MAXIMIZATION ALGORITHMS

Expectation-Maximization (EM) Algorithm is a general computational method for calculating maximum likelihood of an incomplete data. The basic approach is based on the intuitive notion that (i) the values of missing data can be estimated without iteration and (ii) 'filling-in' the missing data with the known parameters by setting it to be equal to the expected values (Laird, 2002). EM algorithms are being applied extensively within the field of genetics. The gene counting algorithm to estimate the allele frequency is one of the earliest uses of EM algorithms (Ceppellini *et al.*, 1955). The wide application of EM algorithm in genetic field is largely in segregation and linkage analysis. Segregation analysis is a method of determining whether transmission pattern of a genetic trait is consistent with Mendelian expectations of dominant, recessive or codominant. The nature of segregation analysis is complicated due to involvement of numerous parameters (Laird, 2002). Linkage analysis is a study that seeks to locate the position of a diseased gene by correlating inheritance of the disease with inheritance of a 'marker' with the known genomic position (Laird, 2002). Recombinant event is the separation of parental chromosome during the formation of gametes. Recombinant occurs so that the chromosomes inherited by the child consist of a combination of disease and marker alleles that did not exist within the parent. If the disease and the marker allele inherited by the child are identical to those located on one of the parental chromosome, then no recombinant event has occurred. The recombinant fraction is calculated by determining the distance between two locations on the chromosome statistically, which then gives the probability of the recombinant event between the marker and the disease gene during gamete formation (Laird, 2002). Larger recombination fraction indicates that the marker and the disease gene are located far (Laird, 2002). In linkage analysis, EM algorithm is applied in 'filling-in' the missing data, which would make the estimation of the recombination fraction trivial (Laird, 2002). EM algorithm is also used in haplotype frequency estimation. In certain cases when the person is double heterozygote (A/a and B/b), the haplotype cannot be reconstructed thus EM algorithm is used in this estimation (Laird, 2002).

16.6.2 *PROGRAMMING LANGUAGES IN BIOINFORMATICS*

Scripting languages are widely being used by the bioinformaticians for scientific researches in the field of biological sciences. Open Bioinformatics Foundation (O|B|F), a working coalition of several of the bioinformatics programming language projects focuses on development and promotion of open source software development in the life science field. The primary activity of the O|B|F is to support the Bioinformatics Open Source Conference (BOSC) (Open Bioinformatics Foundation, 2006). In bioinformatics, it is crucial to find the definitive patterns in data files, and these data files are constructed in multi format. In order to read these definitive patterns, a regular expression is crucial. These expressions are such as Python, Perl, Ruby, Java and the C programming.

16.6.2.1 PYTHON

Python is an object-oriented scripting language, which is used for processing text files and computerizing common tasks. The flexibility of Python makes it portable to be used in multiple platforms including UNIX variants, Windows, and Macintosh (Chapman and Chang, 2000). Besides Python's usage in scripting languages, Python also has added advantage in advanced numerical capabilities through the Scientific Tools for Python (SciPy) project (de Hoon *et al.*, 2004). The Biopython project was initiated at 1999 and is modeled after the Bioperl project. Biopython project largely focuses on creating parsers for biological data and designing a useful interface to represent sequences (Chapman and Chang, 2000). One of the unique features of the Biopython project is the use of a standard event-oriented parser design (http://biopython.org/).

16.6.2.2 PERL

Perl was first developed by Larry Wall in the late 1980s for NASA system administrating (Sheppard, 2000). Since then, it has moved into several other areas: automating system administration, web programming, bioinformatics, data mugging and software development. Bioperl code is an extensive library of core modules written in Perl to support the processing, manipulating, and managing of biological information in the form of sequences (Stajich and Birney, 2000). The Bioperl project attempts to emulate the object oriented programming paradigm through the use of Perl modules and by adhering to three design principles. The first principle is to separate the interface from the implementation. The second principle is to provide a base framework for the respective operation by generalizing common routines into a single module. The third and final principle is to use the Factory and Strategy patterns as defined by Erich Gamma (Stajich et al., 2002). More information can be obtained from http://www.bioperl.org/.

16.6.2.3   PHP

Hypertext Preprocessor (PHP) is a "Scripting language" scripting language originally designed for web developing in order to produce "Dynamic Web page" dynamic Web pages. It is one of the first developed server-side scripting languages embedded into an HTML source document. Initially PHP was known as GenePHP, the BioPHP project seeks to extend the PHP language so that it can be used to develop bioinformatics applications. The main purpose of the BioPHP project in bioinformatics field was to act as an adhesive to bind web-based bioinformatics applications and databases (Rasmus, 2007). The advantage of BioPHP is its ability to read biological data's in the GenBank, Swissprot, FASTA and Clustal ALN formats and perform simple sequence analysis tasks (http://genephp.sourceforge.net/).

16.6.2.4   RUBY

Ruby is a programming language, which was designed and developed in the mid-1990s by Yukihiro "Matz" Matsumoto in Japan. Ruby has a dynamic type system and automatic memory management very similar to other programming languages such as Smalltalk, Python, Perl, Lisp, Dylan, Pike, and CLU. BioRuby being a part of the O|B|F, it is primarily supported by the Human Genome Center at the University of Tokyo and the Bioinformatics Center at Kyoto University (Goto et al., 2010).

16.6.3   GENERAL PURPOSE LANGUAGES

The use of programming languages based on its general use and usage has been increasing especially among the researches in the biological science field. Listed below are a number of high-level general purposes programming languages.

16.6.3.1 C/C++

C is a general-purpose computer programming language developed by Dennis Ritchie for use with the UNIX operating system (Giannini *et al.*, 2004). Initially C was designed for implementing system software and then gradually C was applied in developing portable application software (Lawlis, 1997). In 2004, the Laboratory of DNA Information Analysis at the University of Tokyo has produced an open source C library of the most commonly used clustering algorithms (de Hoon *et al.*, 2004). These clustering routines are used to analyze the gene expression data. C language was then extended to be further used in Perl and Python programs. C++ in an intermediate-level language program as it comprises a combination of both high-level and low-level language features (Schildt, 1998). The main goal of C++ is to implement algorithms that will be used in developing applications in the field of bioinformatics (Della Vedova and Dondi, 2003).

16.6.3.2 JAVA

Java platform encompasses an object oriented programming language in co-operating sets of libraries for network communications and graphical user interface designs. Example of Java based web interface is the protein interaction network visualization module by the MapView Tool at the Human Genome Database (Letovsky *et al.*, 1998). The benefit of Java programs compared to C/C++ is the 'write once run anywhere' paradigm.

16.6.3.3 HASKELL

Haskell is a general purpose language, which is purely used for functional programming. Robert Giegerich's research group employed Haskell to implement dynamic programming algorithms; including programming that involves RNA folding grammars (www.biowiki.org/BioHaskell). The main implementation of Haskell is as an interpreter and native-code compiler that runs on most platforms.

16.6.3.4   LISP

The name *Lisp* was derived from "LISt Processing." LISP is a web-based programming environment that enables biologists to analyze biological systems by combining knowledge and data through direct end-user programming (Massar *et al.*, 2005). Lisp was originally created as a practical mathematical notation for computer programs and later on it became the programming language for artificial intelligence (AI) research. As one of the earliest programming languages, Lisp pioneered many ideas in computer science, including tree data structures, automatic storage management, dynamic typing, and the self-hosting compiler.

16.6.3.5   PROLOG

Prolog is a general-purpose logic programming language, which is highly integrated with artificial intelligence and computational linguistics (Clocksin and Mellish, 2003). Initially, Prolog was aimed at natural language processing and then the language was further developed into theorem proving, expert systems, games, automated answering systems, ontologies and sophisticated control systems (Stickel, 1988; Merritt, 1989). Modern Prolog environments support creating graphical user interfaces, as well as administrative and networked applications (http://bioprolog.org/).

16.6.3.6   XML

The *eXtensible Markup Language* (XML) was designed to describe data using hierarchical structure. An XML document uses document type definition (DTD) or an XML scheme to describe data and it is designed to be self-descriptive. Many programming languages has incorporated XML parsers and it is also being widely used in biological databanks such as GenBank. The key advantage of XML is that, the user can create their own tags to represent the meaning and structure of the datasets (Wu and Baker, 2005). Common XML definitions being developed are such as Genome Annotation Markup Elements (GAME), Bioinformatic Sequence Markup Language (BSML), BIOpolymer Markup Language (BIOML) and BioPerl XML (Wu and Baker, 2005).

16.6.3.7　LITTLE B

Little b is a domain-specific modeling programming language, which was designed based on LISP to build modular mathematical models of biological systems. It was designed and authored by Aneil Mallavarapu at the Virtual Cell Program at Harvard Medical School.The little b allows biologists to build models quickly and easily from shared parts, and to allow theorists to program new ways of describing complex systems. The language draws on techniques from artificial intelligence and symbolic mathematics, and provides syntactic conveniences derived from object-oriented languages (www.littleb.org/).

16.7　FUNCTIONAL GENOMICS

The vast amount of data generated by the Human Genome Project and the Single Nucleotide Polymorphisms (SNPs) genotyping in human complex diseases had made it clear for concomitant statistical advances in mapping of these complex traits. The mapping methodologies are substantially being used are such as linkage analysis, linkage disequilibrium, haplotype analysis and Markov Chain Monte Carlo.

16.7.1　LINKAGE ANALYSIS

Linkage analysis is defined as a mathematical procedure that analyzes the meiotic recombination frequencies between pairs of genes to determine whether how closely the two loci are linked. Identifying a disease-causing gene in the human genome is very tricky and tedious due to the nature of human genome that contains large number of genes. Traditionally, linkage analysis was used to search for a disease gene by determining the rough location of the gene relative to another DNA sequence called a genetic marker. The location of the diseased gene can be determined by looking for the incidence of the disease in a family tree or pedigree (Fig. 16.3).

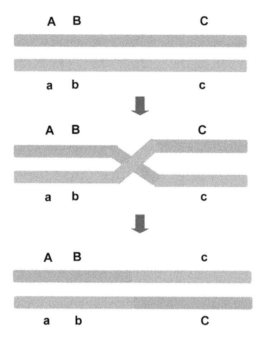

FIGURE 16.3   Diagram representing linkage analysis. If A represents the diseased gene and B and C are the genetic markers, recombination is likely to occur more frequent between A and C compared to between A and B.

16.7.2   LINKAGE DISEQUILIBRIUM

The nonrandom association between the alleles at two or more genetic loci in a natural breeding population is known as Linkage Disequilibrium (LD) (Chakravarti, 2005). An association of genetic markers with certain disease is largely attributed by both the genetic and nongenetic causes. Genetic markers play a functional role in the disease or the marker and the disease suscepti-bility loci are in close proximity and statistically associated is known a LD (Schaid, 2005). LD occurs when an allele transpire due to mutation and thus giving a new characterization to the chromosome. These characters are then transmitted *en bloc* from one generation to another (Schaid, 2005). LD may diminish over the generation due to recombination but the loci that is located close to the mutant allele are less likely to be affected due to this event thus maintains a strong association (Schaid, 2005). The presence of these LD's enables obtaining of information regarding the location of the disease susceptibility genes. The concept of LD was initially postulated by population genetics in understanding the consequences of random mating on the distribution of alleles. Later on, with the availability of high-resolution genetic maps in human, empirical studies of LD have been initiated. These studies not only are shedding lights on the distribution of alleles but also on population genetic mechanisms that lead to LD. These studies had further suggested that LD can be highly used in inferring the location of disease causing gene in a population (Chakravarti, 2005). There is currently a major interest in estimating the nature and extend of LD in isolated populations with different founding and demographic values.

16.7.3 HAPLOTYPE ANALYSIS

Haplotype analysis examines the genetic informatics through a pedigree and provides a useful visualization of the gene flow. Haplotypes are a set of alleles that are inherited from the same parents. Thus, each individual carries two haplotypes with one of maternal origin and one of paternal origin, respectively. Figure 16.4 illustrates the haplotype pedigrees. Haplotype analysis gives a more precise localization of a putative trait locus compared to other linkage analysis. The rapid change in genetic advances and computer technology had lead to the evolvement of haplotype analysis. This evolution shifted the *ad hoc* qualitative methodology to quantitative estimations based on maximum likelihood considerations.

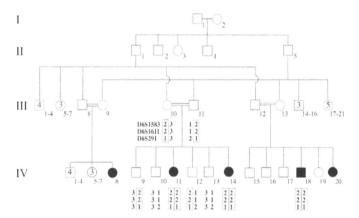

FIGURE 16.4 The pedigree shows a consanguineous Israeli Muslim Arab family segregating early onset autosomal recessive retinitis pigmentosa (family TB13) (Abassi *et al.*, 2008). The mutation-bearing haplotype is marked by a box. Males are represented by squares; females are represented by circles. Affected individuals are marked by black shadings.

16.7.4 MARKOV CHAIN MONTE CARLO

Markov Chain Monte Carlo (MCMC) is a powerful technique for performing integration by simulation and has been revolutionized by Bayesian statistics. Initially the detailed introductory material of MCMC and the biostatistical application was briefed by Gelman and Rubin (1992) and Gilks *et al.* (1996). The Monte Carlo techniques in biostatistical analysis are used in situations in which the analytic solutions of the problems are either intractable or time consuming (O'Neil *et al.*, 2005). In this situation, instead of calculating the exact quantity, simulations are used to produce stochastic approximations. Common uses of Monte Carlo techniques are in hypothesis testing, bootstrap distribution and Bayesian calculations (O'Neil *et al.*, 2005). In segregation analysis, MCMC methods are applied in parameter applications. In this approach, the latent genotypes and polygenotypes are imputed against the phenotypic information. The MCMC approach therefore is much feasible for realistic segregation analysis. This method also allows performing of Bayesian estimations and inference (Thomas and Gauderman, 2005). The MCMC technique has been vastly applied in statistics and biostatistics. Gilks *et al.* (1996) explained the application of MCMC in vaccine efficacy, pharmacokinetics and genetics. Meanwhile Berry and Stangl (2002) reported its application in clinical trial designs, meta-analysis and change point analysis.

16.8 PROGNOSTICS AND ENCUMBRANCE OF BIOINFORMATICS IN HUMAN GENETICS

The field of biomedical informatics is relatively young and keeps growing, thus making ways to huge and complex datasets generation. These huge datasets cause challenges mainly in data capturing, management, analysis and mining. As the need for these data management increases, the knowledge and understanding to develop formats of representing these datasets also increases. The need is especially pressing in the field of human genetics with its focus on DNA sequences. Recent advancements, especially in next generation sequencing (NGS), bio-ontologies, Semantic Web and hospital information technology (IT) systems focussing on electronic health records, lead to generation of a mass of data in human genetics scientists and clinicians (Lindblom and Robinsons, 2011).

Francis Ouellette, the director of the UBC Bioinformatics Centre reported that the future of bioinformatics is in its data integration, especially in integration of clinical and genomic data in predicting genetic mutations. In terms of agriculture, integration of GIS data such as maps and weather monitoring systems allows in prediction of successful agriculture outcomes and observation. Large-scale comparative genomics of modeling and visualization of networks of complex systems could be used in the future to predict interactions of the cells and drugs. This can provide guidelines to choose the right drug and avoid adverse drug effects such as patient allergy or drug interaction, or by receiving pharmacogenetic and pharmacogenomic information important for drug dosing, efficacy or toxicity (Lindblom and Robinsons, 2011). The advancement in techniques related to genome analysis has advantaged the pharmacogenomic field, particularly in identification of new pharmacogenomic biomarkers that can be used as predictive tools for improved drug response thus reducing adverse drug reactions. Such biomarkers mainly originate from genes encoding drug-metabolizing enzymes, drug transporters, drug targets and human leukocyte antigens (Sim *et al.*, 2011).

A lot of attention is being given to the importance of patients' genetic data especially in clinical decision making, in disease risk assessment, diagnosis and drug therapy. Thus priories the need for electronic health records, which stores phenotypic data, clinical data, previous and current diagnoses, previous and current medication and accessibility to genetic test results (Taschner and den Dunnen, 2011). Besides, the above-mentioned opportunities and advantages of the bioinformatics field, this field also possess a number of worth mentioning hurdles. The main challenges are the need for faster computers, technological advances in disk storage space and increased bandwidth. As this scenario is advancing through time, lag in expertise has lead to real gaps in the knowledge of bioinformatics in the research community. Looking into the human genetics aspects, the ideal research question will be on how to computationally compare complex biological observations of gene expression patterns, SNP profiles and transcriptomic networks.

16.9   CONCLUSION

This chapter had identified a number of important areas in which bioinformatics is providing an indispensable contribution to research and clinical care in human genetics. Reports have stated that bioinformatics field will grow in parallel with advances in next-generation sequencing and other high-throughput technologies in biomedicine. In order to achieve this, the geneticists and bioinformaticians should work hand in hand and are able to cope with the challenges in data analysis. Primary approach in training the bioinformaticians, geneticist and clinician should be put forward especially in statistics, genomics and bioinformatics (Lindblom and Robinsons, 2011).

KEYWORDS

- **Bioinformatics**
- **Biotechnology**
- **Gene mining**
- **Human genetics**
- **Microarray**
- **Molecular cloning**
- **Statistical genetics**
- **Stem cell**

REFERENCES

Bugalho, M. J.; Domingues, R.; Sobrinho, L. The minisequencing method: a simple strategy for genetic screening of MEN 2 families. *BMC Genet* 2002, 3, 8.

Ceppellini, R.; Siniscalco, M.; Smith, C. A. The estimation of gene frequencies in a random-mating population. *Ann Hum Genet* 1955, 20(2), 97–115.

Chakravarti, A. Disease marker association. In: *Database annotation in molecular biology, principles and practice*, Arthur, M.L., Ed.; John Wiley and Sons Ltd: England, 2005; 206–217.

Clocksin, W. F., Mellish, C. F. *Programming in Prolog: using the ISO Standard*, 5th Ed.; Spring-er-Verlag: Berlin, 2003.

Coleman, W. B.; Tsongalis, G. J. *The polymerase chain reaction in molecular diagnostics for the clinical laboratorian*, 2nd Ed.; Humana Press Inc.: New Jersey, 2004.

De Hoon, M. J.; Imoto, S.; Nolan, J.; Miyano, S. Open source clustering software. *Bioinformatics* 2004, 20(9), 1453–1454.

Della Vedova, G.; Dondi, R. A library of efficient bioinformatics algorithms. *Appl Bioinformatics* 2003, 2(2), 117–121.

Frayling, I. M.; Monk, E.; Butler, R. *PCR based methods for mutation detection in molecular diagnostics for the clinical laboratories*, 2nd Ed.; Humana Press Inc: New Jersey, 2004.

Ganguly, S.; Paul, I.; Mukhopadhayay, S. K. DNA microarray technique for diagnosis of various animal infections – a brief introductory preview. *Indian Pet Journal* 2009, 10, 5.

Giannini, M. C/C++. In *The internet encyclopedia*, Bidgoli, H., Ed.; John Wiley and Sons Ltd: England, 2004; p 164.

Goto, N.; Prins, P.; Nakao, M.; Bonnal, R.; Aerts, J.; Katayama, T. BioRuby: bioinformatics software for the Ruby programming language. *Bioinformatics* 2010, 26(20), 2617–2619.

Hooman, H. R.; Buehler, L. K. *Bioinformatics basics: applications in biological science and medicine*, CRC Press: Boca Raton, 2005.

Kohane, I. S.; Butte, A. Bioinformatics. In *Biostatistical genetics and genetic epidemiology*, Elston, R., Olson, J., Palmer, L., Eds.; Wiley: England, 2002; 61–69.

Laird, N. M.; Agorithm, E. M. In *Biostatistical genetics and genetic epidemiology*, Elston, R., Olson, J., Palmer, L., Eds.; Wiley: England, 2002; 232–245.

Letovsky, S. I.; Cottingham, R. W.; Porter, C. J.; Li, P. W. GDB: the human genome database. *Nucleic Acids Res* 1998, 26(1), 94–99.

Lindblom, A.; Robinson, P. N. Bioinformatics for human genetics: promises and challenges. *Hum Mutat* 2011, 32(5), 495–500.

Massar, J. P.; Travers, M.; Elhai, J.; Shrager, J. BioLingua: a programmable knowledge environment for biologists. *Bioinformatics* 2005, 21(2), 199–207.

Merritt, D. *Building expert systems in Prolog*. Springer-Verlag: Berlin, 1989.

Morton, N. E. Genetic epidemiology. *Ann Hum Genet* 1997, 61(1), 1–13.

Mullis, K. B. Recombinant DNA technology and molecular cloning. *Sci Am* 1990, 262, 36.

Nagasaki, M.; Saito, A.; Doi, A.; Matsuno, H.; Miyano, S. *Foundations of system biology – using cell illustrator and pathway databases*, Springer-Verlag: London, 2009.

Peters, P. A. *Guide to genetic engineering*, Wm. C. Brown Publishers, Inc.: Dallas, 1993.

Roy D. *Bioinformatics*, Alpha Science International Limited: Oxford, 2009.

Schaid, D. S. Disease marker association. In: *Database annotation in molecular biology, principles and practice*, Arthur, M. L., Ed.; John Wiley and Sons Ltd, England, 2005; 206–217.

Schildt, H. *C++, the complete reference*, 3rd Ed.; Osborne McGraw-Hill: New York, 1998.

Serre, J. L. Techniques and tools in molecular biology used in genetic diagnoses. In *Diagnostic Techniques in Genetics*, Serre, J. L., Ed.; John Wiley and Sons Ltd, England, 2006, 1–59.

Sim, S. C.; Altman, R. B.; Ingelman-Sundberg, M. Databases in the area of pharmacogenetics. *Hum Mutat* 2011, 32(5), 526–531.

Stajich, J. E.; Block, D.; Boulez, K.; Brenner, S. E.; Chervitz, S. A.; Dagdigian, C.; Fuellen, G.; Gilbert, J. G.; Korf, I.; Lapp, H.; Lehvaslaiho, H.; Matsalla Mungall, C. J.; Osborne, B. I.; Pocock, M. R.; Schattner, P.; Senger, M.; Stein, L. D.; Stupka, E.; Wilkinson, M. D.; Birney, E. The Bioperl toolkit: Perl modules for the life sciences. *Genome Res* 2002, 12(10), 1611–1618.

Stickel, M. E. A prolog technology theorem prover: implementation by an extended prolog compiler. *Automated Reasoning* 1998, 4(4), 353–380.

Strickland, D. *Guide to Biotechnology*. Biotechnology Industry Organization: Washington DC, 2007.

Syvanen, A. C. From gels to chips: "minisequencing" primer extension for analysis of point mutations and single nucleotide polymorphisms. *Hum Mutat* 1999, 13(1), 1–10.

Taschner, P. E.; den Dunnen, J. T. Describing structural changes by extending HGVS sequence variation nomenclature. *Hum Mutat* 2011, 32(5), 507–511.

Thomas, D. C.; Gauderman, W. J. Gibbs sampling methods in genetics. In *Markov Chain Monte Carlo in Practice*, Vol. 2; Gilks, W. R., Richardson, S., Spiegelhalter, D. J., Eds.; Chapman and Hall: London, 2005, 419–440.

van Ommen, G. J. The Human Genome Project and the future of diagnostics, treatment and prevention. *J Inherit Metab Dis* 2002, 25(3), 183–188.

Wittwer, C. T.High-resolution DNA melting analysis: advancements and limitations. *Hum Mutat* 2009, 30(6), 857–859.

Wu, C. H.; Baker, W. C. Annotation and databases status and prospects. In *Database annotation in molecular biology, principles and practice,* Arthur, M. L., Ed.; John Wiley and Sons: England, 2005.

Zaki, M. J.; Sequeira, K. *Data mining in computational biology,* CRC Press: Boca Raton, 2006.

INDEX